EVAPOTRANSPIRATION – REMOTE SENSING AND MODELING

Edited by **Ayse Irmak**

INTECHOPEN.COM

Evapotranspiration – Remote Sensing and Modeling
Edited by Ayse Irmak

CBS, Edition **2017**

Published by InTech

Janeza Trdine 9, 51000 Rijeka, Croatia

Publishing Process Manager Dragana Manestar

Technical Editor Teodora Smiljanic

Cover Designer InTech Design Team

Image Copyright Selyutina Olga, 2011. Used under license from Shutterstock.com

Additional hard copies can be obtained from orders@intechopen.com

Evapotranspiration – Remote Sensing and Modeling, Edited by Ayse Irmak
p. cm.
ISBN 978-953-307-808-3

INTECH

open science | open minds

Contents

Contents

Contents

Preface

The transfer of liquid water from soil to vapor in the atmosphere (Evapotranspiration) is one of the most profound and consequential processes on Earth. Evapotranspiration (ET), along with evaporation from open water, supplies vapor to the atmosphere to replace that condensed as precipitation. The flux of water through plants via transpiration transports minerals and nutrients required for plant growth and creates a beneficial cooling process to plant canopies in many climates. At the global scale, ET measures nearly one hundred trillion cubic meters per year and is the largest component of the hydrologic cycle, following precipitation. The large spatial variability in water consumption from land surfaces is strongly related to vegetation type, vegetation amount, soil water holding characteristics, and of course, precipitation or irrigation amount. There are very strong feedbacks from all of these factors and consequent ET rates. In this book, Evapotranspiration is defined as the aggregate sum of evaporation (E) direct from the soil surface and the surfaces of plant canopies and transpiration (T), where T is the evaporation of water from the plant system via the plant leaf, stem and root-soil system.

In addition to consuming enormous amounts of water, ET substantially modifies the Earth's energy balance through its consumption of enormous amounts of energy during conversion of liquid water to vapor. Each cubic meter of water evaporated requires 2.45 billion Joules of energy. That consumption of energy cools the evaporating surface and reduces the heating of air by the surface. On a global basis, the cooling effect to the land surface is measured in trillions of GigaJoules per day. Much of that 'latent' energy absorbed by ET later reenters the surface energy balance when the vapor recondenses as precipitation.

Even though the magnitude of ET is enormous over the Earth's surface, and even though ET has such high bearing on vegetation growth and health, its spatial distribution and magnitudes are poorly understood and poorly quantified. Although man has been able to estimate general magnitudes of ET via its strong correlation with precipitation for centuries, it has only been during the past thirty years, with the advent of satellites and remote sensing technologies, along with sophisticated modeling approaches, that we have been able to view and quantify the complex and variable geospatial structure of ET. The combination of thermally-equipped satellites, such as Landsat, AVHRR, MODIS and ASTER, and the improved ability to simulate the energy balance at the Earth's surface has enabled a substantial revolution in 'mapping' of ET over large, variable landscapes.

This edition of *Evapotranspiration* contains 23 chapters, covering a broad range of topics related to the modeling and simulation of ET, as well as to the remote sensing of ET. Both of these areas are at the forefront of technologies required to quantify the highly spatial ET from the Earth's surface. The chapters cover mechanics of ET simulation, including ET from partially vegetated surfaces and the modeling of stomatal conductance for natural and agricultural ecosystems, ET estimation using soil water balance, weather data and vegetation cover, ET estimation based on the Complementary Relationship, and adaptability of woody plants in conditions of soil aridity. Modeling descriptions include chapters focusing on distributed benchmarking frameworks for ET models, Hargreaves and other temperature-radiation based methods, Fuzzy-Probabilistic calculations, a hybrid-method for modeling evaporation and ET, and estimation of ET using water balance modeling. One chapter provides a critical review of methods for estimation of actual ET in hydrological models. In addition to that, six chapters describe modeling applications for determining ET patterns in alpine catchments, ET assessment and water resource management planning under shortage conditions, estimation of the annual and interannual variation of potential ET, impacts of irrigation on hydrologic change in a highly cultivated basin, ET of grasslands and pastures in north-eastern part of Poland, and climatological aspects of water balance components for Croatia.

Remote sensing based approaches are described in five chapters that include deriving crop ET from satellite data, integration with other information sources and an assessment of ET using MODIS products with energy balance algorithms. Importantly, the book includes two chapters describing an overview of recommended guidelines for operational remote sensing of ET, and a review of operational remote sensing-based energy balance models including SEBAL and METRIC, and specific challenges and insights for their application.

These 23 chapters represent the current state of the art in ET modeling and remote sensing applications, and provide valuable insights and experiences of developers and appliers of the technologies that have been gained over decades of development work, experimentation and modeling. This text provides valuable background information and theory for university students and courses on ET, as well as guidance and ideas for those that apply these modern methods. I wish to express my thanks to the authors of all chapters for making these timely and very useful contributions available, and to all anonymous reviewers of chapters. I also wish to thank Mr Baburao Kamble, University of Nebraska, for assistance in the handling of chapter manuscripts during reviews and for providing technical assistance.

Dr. Ayse Irmak
School of Natural Resources and Civil Engineering, Center for Advanced Land Management Information Technologies (CALMIT), University of Nebraska-Lincoln, USA

1

Assessment of Evapotranspiration in North Fluminense Region, Brazil, Using Modis Products and Sebal Algorithm

José Carlos Mendonça[1], Elias Fernandes de Sousa[2], Romísio Geraldo Bouhid André[3], Bernardo Barbosa da Silva[4] and Nelson de Jesus Ferreira[5]

[1]*Laboratório de Meteorologia (LAMET/UENF). Rod. Amaral Peixoto, Av. Brennand s/n Imboassica, Macaé, RJ*

[2]*Laboratorio de Engenharia Agrícola (LEAG/UENF); Avenida Alberto Lamego, CCTA, sl 209, Parque Califórnia, Campos dos Goytacazes, RJ*

[3]*Instituto Nacional de Meteorologia (INMET/MAPA); Eixo Monumental, Via S1 - Sudoeste, Brasília, DF*

[4]*Departamento de Ciências Atmosféricas (DCA/UFCG); Avenida Aprígio Veloso, Bodocongó, Campina Grande, PB*

[5]*Centro de Previsão de Tempo e Estudos Climáticos (CPTEC/INPE); Av. dos Astronautas, Jardim da Granja, São José dos Campos, SP*

Brazil

1. Introduction

North Fluminense Region, Rio de Janeiro State, Brazil (Fig. 1) is known as a sugar cane producer. The production during harvest season 2007/08 were 4 million tons of sugar cane, that were transformed into 4.8 million sacks of sugar, 36,786 liters anhydrous alcohol (ethanol) and 91,008 liters of hydrated alcohol. Economically generated 250 million U. S. dollars (Morgado, 2009). However, this activity is declining in the region due to different factors, including hidric deficit and the use of irrigation techniques may reverse this situation(Azevedo et al., 2002). Some authors (Ide e Oliveira, 1986; Magalhães, 1987) define temperature as a factor of greater importance for sugar cane physiology maturation (ripening) because more the affecting nutrients and water absorption through transpiration flux is a non-controllable condition. Soil humidity is another preponderant factor to sugar cane physiology and varies in function of the cultivation cycle, development stage, climactic conditions and others factors, such as spare water in the soil. The soil moisture content varies during the growth that corresponds to the main cause of production variation. However, the precipitation distribution along the year and spare soil water for the plant disposition are more important in the vegetative cycle of the sugar cane that total precipitation. (Magalhães, 1987).

The physical properties of energy exchange between the plant community and environment such as momentum, latent heat, sensible heat and others are evidenced by the influence they exert on physiological processes of plants and the occurrence of pests and diseases, which affect the productive potential of plants species exploited economically (Frota, 1978). The radiation components measurements of energy balance in field conditions have direct applicability in agricultural practices, especially in irrigation rational planning, appropriate use of land in regional agricultural zoning, weather variations impact on agricultural crops, protecting plants, among others. The knowledge advance in micro-scale weather, as well as the instrumental monitoring technology evolution has allowed a research increase in this area. Energy balance studies on a natural surface based on energy conservation principle. By accounting means for components that make up this balance, can be evaluate the net radiation plots used for the flow of sensible and latent heat.

The analysis of data collected by artificial satellites orbiting planet earth, allows the determination of various physical properties of planet, consequently, spatial and temporal modifications of different ecosystems are able to be identified.

According Moran et al. (1989), estimative of evepotranspiration - ET, based in data collected in meteorological stations have the limitation of representing punctual values that are capable of satisfactory representing local conditions but, if the objective is to obtain analysis of a regional variation of ET using a method with interpolation and extrapolation from micro-meteorological parameters of an specific area, these punctual data may increase the uncertainty of the analysis.

Trying to reduce such uncertainty degree, different algorithms were developed during the last decades to estimate surface energy flux based in the use of remote sensing techniques.

Bastiaanssen (1995) developed the 'Surface Energy Balance Algorithm for Land - SEBAL', with its validation performed in experimental campaigns in Spain and Egypt (arid climate) using Landsat 5 –TM images. This model involves the spatial variability of the most agro-meteorological variables and can be applied to various ecosystems and requires spatial distributed visible, near-infrared and thermal infrared data together with routine weather data. The algorithm computes net radiation flux - Rn, sensible heat flux - H and soil heat flux - G for every pixel of a satellite image and latent heat flux - LE is acquired as a residual in energy balance equation (Equation 01). This is accomplished by firt computing the surface radiation balance, flowed by the surface energy balance. Althoygh SEBAL has been designed to calculate the energy partition at the regional scale with minimum ground data (Teixeira, 2008).

Roerink et al. (1997) also used Landsat 5 –TM images to evaluate irrigation's performance in Argentina and AVHRR/NOAA sensor images in Pakistan. Combination of Landsat 5 - TM and NOAA/AVHRR images were used by Timmermans and Meijerink (1999) in Africa. Latter, Hafeez et al. (2002) used the SEBAL algorithm with the ASTER sensor installed onboard 'Terra' satellite while studying Pumpanga river region in Philippines. These authors concluded that the combination of the high spatial resolution of ETM+ and ASTER sensors, together with the high temporal resolution from AVHRR and MODIS, provided high precision results of water balance and water use studies on regional scale.

In Brazil, several research center are conducting research using the SEBAL algorithm specially 'Federal University of Campina Grande, PB - UFCG', 'National Institute of Space Research - INPE' and others.

Sebal was developed and validated in arid locations and one of its peculiarities is the use of two anchors pixels (hot pixel - LE = 0 and cold pixel - H =0) with the determination or selection of hot pixel easier in dry climates. In humid and sub-humid climates is not easy determine a hot pixel, where the latent heat flux is zero or null.

The objectives of the research described in this work are (i) to evaluate two propositions to estimate the sensible heat flux (H) and (ii) to evaluate two methods for conversion of ETinst values to ET24h on the daily evepotranspiration to estimate evepotranspiration in regional scale using SEBAL algorithm, MODIS images, the two propositions to estimate H and meteorological data of the four surface meteorological stations.

2. Materials and methods

2.1 Study area

The Norte Fluminense region in Rio de Janeiro State, Brazil, has an area of 9.755,1 km^2, corresponding to 22% of the state's total area. Among its agricultural production, sugar cane plantations are predominant as well as cattle production. In the last years irrigation technologies for fruit production are being promoted and implemented by the government. Nowadays, passion fruit, guava, coconut and pineapple plantations extend for more than 4.000 ha (SEAAPI, 2006).

According Koppen, this region's clime is classified as Aw, that is, tropical humid with rainy summers, dry winters and temperatures average above 18 °C during the coolest months. The annual mean temperatures are of 24°C, with a little thermal amplitude and mean rain precipitation values of 1.023 mm (Gomes, 1999).

The area under study is showed in Figure 1, comparing the area of the Norte Fluminense region within the Rio de Janeiro state and the RJ state within Brazil.

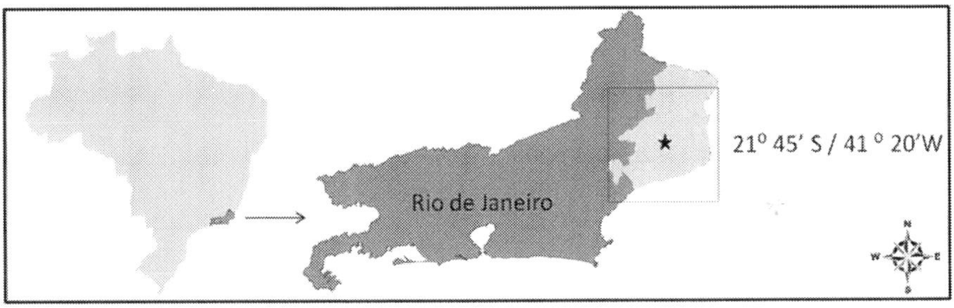

Fig. 1. Study area localization.

2.2 Digital orbital images – MODIS images

Daily MOD09 and MYD09 data (Surface Reflectance - GHK / 500 m and GQK / 250 m) and MOD11A1 and MYD11A1 data (Surface Temperature - LST) were used in this research, totalizing 24 scenes over the 'tile' h14/v11 corresponding to Julian Day 218th, 227th, 230th, 241st, 255th, 285th, 320th and 339th in 2005 and 15th, 36th, 63rd , 102nd, 116th, 139th, 166th, 186th, 189th, 190th, 191st, 200th, 201st, 205th, 208th and 221st in 2006. These days were selected because no cloud covering was registered over the study area during the satellite's course over the area were obtained from the Land Processes Distributed Active Archive Center (LP-DAAC), of the National Aeronautics and Space Administration (NASA), at http://edcimswww.cr.usgs.gov/pub/imswelcome/.

The GHK - 500 m (Blue, Green, Red, Nir, Mir, Fir, Xir) reflectance band were resampled fron 500 m to 250 m. The Red and Nir bands were excluded and GQK (250 m) bands included. This operation aimed to input the value of the red and nir bands in the algorithm. The LST bands were also resampled from 1000 m to 250 m.

The software Erdas Image - Pro, version 8.7 was used for the piles, compositions, clippings and algebra. The Model Maker tool was used to application of the algorithm and the thematic maps were produced using the software ArcGis 9.0.

2.3 Meteorological data

Surface data were collected in two micro-meteorological stations from the Universidade Estadual do Norte Fluminense - UENF, installed over agricultural areas cultivated with sugar cane (geographical coordinates: 21° 43′ 21,8″ S and 41° 24′ 26,1″ W), and 'dwarf green' coconut irrigated (geographical coordinates: 21° 48′ 31,2″ S and 41° 10′ 46,2″ W).

The micrometeorological stations installed in both areas (sugar cane and coconut) were equipped with the following sensor: 1 Net radiometer NR Lite (Kipp and Zonen), 2 Piranometer LI 200 (Li-Cor), 2 Probe HMP45C-L (Vaissala), 2 Met One Anemometer (RN Yong) and 3 HFP01SC_L Soil Healt Flux Plat (Hukseflux). All data from were collected every minute and average values extracted and stores every 15 min in a datalogger CR21X (Sugar cane) and CR 1000 (coconut). Both dataloggers are Campbell Scientific's (USA). The horizontal bars were placed 0.50 m above crop canopy (first level) and 2.0 m between the first and second bars. This standard was maintained all crop cycle and bars relocated where necessary (sugar cane station). In coconut station the relocated was not necessary.

These stations were installed in the center of an area of 5,000 hectare (sugar cane - Santa Cruz Agroindustry) 256 hectare (coconut - Agriculture Taí).

Fig. 2. Localization of the surface micro-meteorological and meteorological stations installed in the study area.

The meteorological stations, both installed on grass (*Paspalum Notatum L.*) are property of research center. The Thies Clima model (Germany) installed at the UENF's Evapotranspiration Station - Pesagro Research Center, (geographical coordinates: 21° 24′ 48″ S and 41° 44′ 48″ W) is an automatic station. Is equipped with 1 Anemometer, 1 Barometer, 1 Termohygrometer, 1 Piranometer and 1 Pluviometer. All sensor are connected to a datalogger model DL 12 - V. 2.00 - Thies Clima, recording values every minute and stored an average every 10 minutes.

The Agrosystem model install at the Meteorological Station of the Experimental Campus 'Dr. Leonel Miranda' - UFRRJ, (geographical coordinates: 21° 17′ 36″ S and 41° 48′ 09″ W) contains 1 Anemometer, 1 Barometer, 1 Termohygrometer, 1 Piranometer and 1 Pluviometer and recording values every minute and stored an average every 10 minutes.

All geographical coordinates are related to Datum WGS 84 - zone 24, with average altitude of 11 m. The localization of the surface stations, where meteorological data used in this study were collected are showed in Figure 2.

2.4 Real evapotranspiration estimation with SEBAL

To calculate surface radiation balance was used the Model Maker tool from the software Erdas Image 8.6. The estimations of the incident solar radiation and the long wave radiation emitted by the atmosphere to the surface were performed in electronic sheet.

To better understand the different phases of the Sebal algorithm using Modis products, a general diagram of the computational routines are shown in Figure 3.

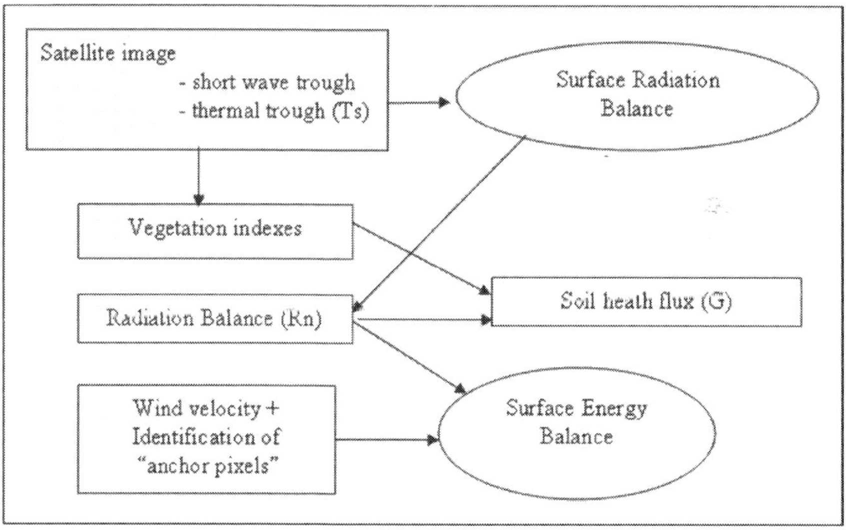

Fig. 3. Diagram of the computational routines for determination of the Surface Energy Balance using SEBAL, form MODIS products. (Modified from Trezza (2002).

A schematic diagram for the estimation of the surface radiation balance (Rn), adapted to MODIS images is showed in Figure 4.

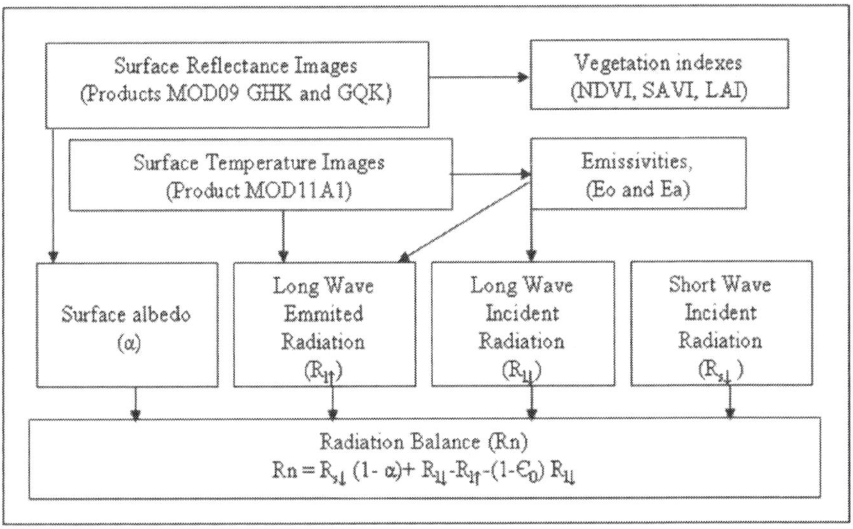

Fig. 4. Diagram showing the process steps of the surface radiation balance adapted for MODIS images.

Detailed processes, as well as the equations for the SEBAL algorithm development, may be obtained in Bastiaanssen et al. (1998). In the present work two propositions were assumed to select the anchor pixels, the first was similar to the one used by Bastiaanssen (1995), with the selection of two pixels with external temperatures (hot pixel/LE = 0 and cool pixel/H = 0). The hot pixel always comprising an area of exposed soil with little vegetation and the cool pixel localized in the interior of a great extension water body. The first proposition was called as 'H_Classic'.

With the hypothesis that the linear relation dT = a + d.Ts would be better represented with the selection of a hot pixel with its energy balance components previously known, specially the sensible heat flux (H) and in regions of humid and sub-humid climate be difficult identifying de hot pixels, which can hardly meet the condition of being dry, or have LE = 0, the second hypothesis was formulated. The criterion used for the selection of the cool pixel was the same as in the first hypothesis, that is, to be localized inside a water body of a great extension, but the selection of the hot pixel, where determination of the H values estimate as residue of the Penman-Monteih FAO56 equation using meteorological data from installed at the UENF's Evapotranspiration Station - Pesagro Research Center. This second hypothesis was called 'H_Pesagro'.

2.5 Latent heat flux (LE)

Latent heat flux (vapor transference to the atmosphere trough the process of vegetal transpiration and soil water evaporation) was computed by the simple difference between the radiation balance cards, soil heat flux and sensible heat flux:

$$LE = Rn - G - H \tag{1}$$

where: LE represents the latent heat flux, Rn is the radiation balance and G is the soil heat flux, all expressed in W m^{-2} and obtained during the course of the satellite over the study area.
The value of the instantaneously latent heat flux (LE_{inst}), integrated at the time (hour) of the satellites passage (mm h^{-1}) is:

$$LE_{inst} = 3600\frac{LE}{\lambda} \tag{2}$$

where: LE_{inst} is the value of instantaneously ET, expressed in mm h^{-1}; LE is the latent heat flux at the moment of the sensor's course and λ is the water vaporization latent heat, expressed by the equation:

$$\lambda = 2,501 - 0,00236\,(Ts - 273,16) * 10^{6} \tag{3}$$

where: Ts is the surface temperature chart (°C) obtained by the product MOD11A1 (K).
With the radiation balance, soil heat flux and latent heat flux charts, the evaporative fraction was obtained and expressed by the equation:

$$\Lambda = \frac{LET}{Rn - G} \tag{4}$$

The evaporative fraction has an important characteristic, it regularity and constancy in clear sky days. In this sense, we can admit that its instantaneously character represents its diurnal mean value satisfactorily, enabling the estimation of daily evapotranspiration by the equation:

$$ET_{24h} = \frac{86400\,\Lambda Rn_{24h}}{\lambda} \tag{5}$$

where: Rn_{24h}, is the mean radiation balance occurred during a period of 24 h, expressed in W.m^{-2}, obtained by the equation:

$$Rn_{24h} = (1 - \alpha)\,Rs24h - 110\,\tau_{sw}24h \tag{6}$$

where: α, is the surface albedo; Rs_{24h}, is the daily mean radiation of short incident wave expressed in W m^{-2} and $\tau_{sw}24h$, is the mean daily atmospheric transmissivity.
To determine Rs_{24h} values, an approximation similar to the method proposed by Lagouarde and Brunet (1983) for the estimation of diurnal cycles of Rn and Rs↓ in clear sky days, was used. With the values of Rn_{24h}, Rs_{24h} and the surface albedo, extracted from the PESAGRO pixel, a linear regression between these values was performed to obtain a regression equation, its coefficients a_1 and b_1 and then to calculate the Rn_{24h} chart as a function of the short wave balance. To determine the linear regression the following equation was used:

$$Rn24h = a_1(1 - \alpha) * Rs24h + b_1 \tag{7}$$

Allen et al. (2002) defined the evaporative fraction of reference (ETrF) as the relation between the ET_{inst} chart and the ETo integrated at the same moment and computed with data obtained from a meteorological station, that is:

$$ETrF = \frac{ET_{inst}}{ET_{FAO56}} \quad (8)$$

This procedure generates a type of hourly-cultive coefficient (kc_h), admitting that this relation represents the daily relation expressed by the equation:

$$Kc\_h = \frac{ETinst}{ETo h} = \frac{ET24}{ETo24} \quad (9)$$

Admitting the relation represented in equation 09 it is possible to obtain the ET_{24h} expressed in mm day^{-1} from the equation:

$$ET_{24h} = ETrF * ETo_{24} \quad (10)$$

In the present work, four values of $ET24h_{SEBAL}$ were estimated for the same day, applying equations 5 and 10 to the 'H_Classic' and H_Pesagro' propositions.

3. Results and discusion

3.1 Daily evapotranspiration (ET_{24h})

3.1.1 Determination of Rn24h values

To determine Rn24h charts, an adaptation proposed by Ataide (2006) for the sinusoidal model estimator of the cycle of radiation balance for clear sky days, based in an approximation similar to the Lagourade and Brunet (1983) method, was adopted.

Looking forward for reliability and applicability in the generation of the Rn24h charts form values of Rs↓24h, a linear regression between the short wave balance and the daily radiation balance was performed, where the regression equation coefficients were determined as a = 0,9111 and b = -23,918.

The coefficients obtained (a and b) are next to the values found by Alados et al. (2003), whit values of a = 0,709 and b = -25,4 where values of global solar radiation (Rg) and not short wave balance (BOC) were used in the linear regression, thus excluding the effect of the surface albedo in the calculation. Considering that values of Rg were determined in a standard meteorological station, installed on a grass field, with values of albedo varying between 20 and 25 %, the coefficients determined by the linear regression between values of BOC and Rn24h tent to be in agreement with the values mentioned by Alados et al. (2003).

Thus, the radiation balance for the daily period (Rn24h) was ultimately determined for each pixel of the study scene by the equation:

$$Rn_{24h} = 0{,}9111 * (1 - \text{chart of albedo}) * Rs{\downarrow}24h - 23{,}918 \quad (11)$$

3.1.2 Determination of the ET24h values

Based on charts of Rn, G, H, LE, Ts and α and values of ETo_{24h} and ETo_{inst}, estimated from data observed at Pesagro's meteorological station, four values of ET24h were estimated for each scene studied: ET24h_'Classic' w/ETrF; ET24h_'Classic' w/Rn24h; ET24h_'H_Pesagro' w/ETrf and ET24h_'H_Pesagro' w/Rn24h.

Mean, maximum and minimum values obtained in charts of daily evapotranspiration (ET24h) estimated with the 'H_Classic' proposition and expressed in mm day^{-1}, are showed in Table 1.

	Mean		Maximum		Minimum	
DJ	Rn 24h	ETr_F	Rn 24h	ETr_F	Rn 24h	ETr_F
218	3,42	4,25	6,51	12,69	0,0	0,0
227	2,88	2,89	6,89	7,64	0,0	0,0
230	3,13	3,19	6,99	7,76	0,0	0,0
241	3,25	2,98	7,39	7,48	0,0	-1,04
255	4,07	3,64	8,25	8,27	0,0	-0,10
285	4,82	4,17	9,63	9,82	0,0	-0,60
320	4,50	3,70	10,65	10,15	0,0	-1,10
339	5,25	4,52	10,75	10,37	0,0	-0,81
15	4,77	4,06	10,91	10,97	0,0	-2,17
36	4,65	4,16	10,12	10,34	0,0	-1,64
63	5,10	5,67	9,40	11,82	0,0	-0,37
102	4,06	3,67	7,75	8,23	0,0	-1,10
116	3,27	3,20	6,93	7,65	0,0	-2,22
139	2,78	2,71	5,94	6,62	0,0	-0,67
166	2,73	2,88	5,45	6,73	0,0	-0,43
186	2,16	2,47	5,48	6,93	0,0	-0,83
189	2,75	2,99	5,61	6,90	0,0	-0,13
190	3,09	2,71	7,23	7,08	0,0	-0,20
191	2,27	2,56	5,68	7,23	0,0	-1,23
200	2,02	2,28	5,51	7,11	0,0	-0,52
201	2,87	3,46	5,84	8,31	0,0	-0,02
205	3,36	4,03	5,84	8,31	1,05	1,07
208	2,54	3,09	6,09	8,26	0,0	-0,98
221	2,88	3,08	6,59	7,82	0,0	-1,25

Table 1. Statistical data of daily evapotranspiration charts (ET24h) of the study area using the 'H_Classic' proposition w/ Rn24h and w/ ETr_F, in mm day^{-1}.

Average mean data showed in Table 1 are similar, with a slight superiority for the values estimated by the method using Rn24h for the ET estimative. Minimum values for ETr_F have negative values. Tasumi et al. (2003), using SEBAL in Idaho, U.S.A., also observed negative values for ET and attributed such results to systematic errors caused by diverse parameterizations used during the process of energy balance estimation.

Average mean, maximum and minimum values obtained in charts of daily evapotranspiration (ET24h) estimated with the "H_Pesagro' proposition, expressed in mm day^{-1}, are showed in Table 2.

	Mean		Maximum		Minimum	
DJ	Rn 24h	ETr_F	Rn 24h	ETr_F	Rn 24h	ETr_F
218	4,45	5,34	6,51	13,38	2,39	2,18
227	4,65	4,61	6,89	7,71	1,87	1,53
230	4,83	4,86	6,99	7,75	0,78	0,60
241	5,84	5,26	7,44	7,53	4,44	3,45
255	6,00	5,30	8,26	8,28	3,78	2,79
285	7,29	6,21	9,75	9,94	5,17	3,82
320	7,31	5,84	10,69	10,39	4,64	3,14
339	7,01	5,99	10,81	10,70	2,17	1,40
15	7,92	6,51	10,96	11,02	2,60	1,48
36	8,12	7,01	10,23	10,84	5,79	4,24
63	6,69	7,35	9,46	12,07	3,90	3,60
102	5,41	4,85	7,75	8,23	0,80	0,50
116	4,62	4,56	6,95	7,65	0,0	-0,46
139	4,27	4,13	5,95	6,62	2,60	2,21
166	4,27	4,11	5,95	6,62	2,61	2,21
186	3,37	3,84	5,48	7,12	1,73	1,75
189	3,86	4,16	5,62	6,94	2,28	2,14
190	5,49	4,74	7,25	7,10	4,41	3,39
191	3,74	4,16	5,70	7,25	0,36	0,27
200	3,29	3,70	5,58	7,20	1,69	1,74
201	3,36	4,03	5,83	8,31	1,05	1,08
205	4,50	5,00	5,95	7,91	3,40	3,26
208	4,33	5,11	6,09	8,35	2,93	3,06
221	4,68	4,93	6,62	7,86	2,88	2,65

Table 2. Statistical data of daily evapotranspiration charts (ET 24h) of the study area using the 'H_Pesagro' proposition w/ Rn 24hs and w/ ETr_F, in mm day^{-1}.

Average mean values of the same magnitude order and with a slight superiority to values estimated using Rn24h are obse4rved in Table 2. In a general way, by the use of the 'Classic' proposal as well as by 'Pesagro' proposal, a higher amplitude of the estimated values is observed when using the method of ETr_F.

Values of ET $24h_{SEBAL}$, observed in pixels where the micro-meteorological and meteorological stations were located (pixels from Pesagro, UFFRJ, Sugar-cane and Coconut), were correlated with values of ETo estimated by the equation of Penman-Monteith_FAO (ETo PM_FAO56) with data observed in Pesagro Station. Figures 5, 6, 7 and 8 show graphical representations of the regression analysis, the adjustment equation and the correlation coefficient (R^2), obtained among the values estimated by SEBAL for all four methods used.

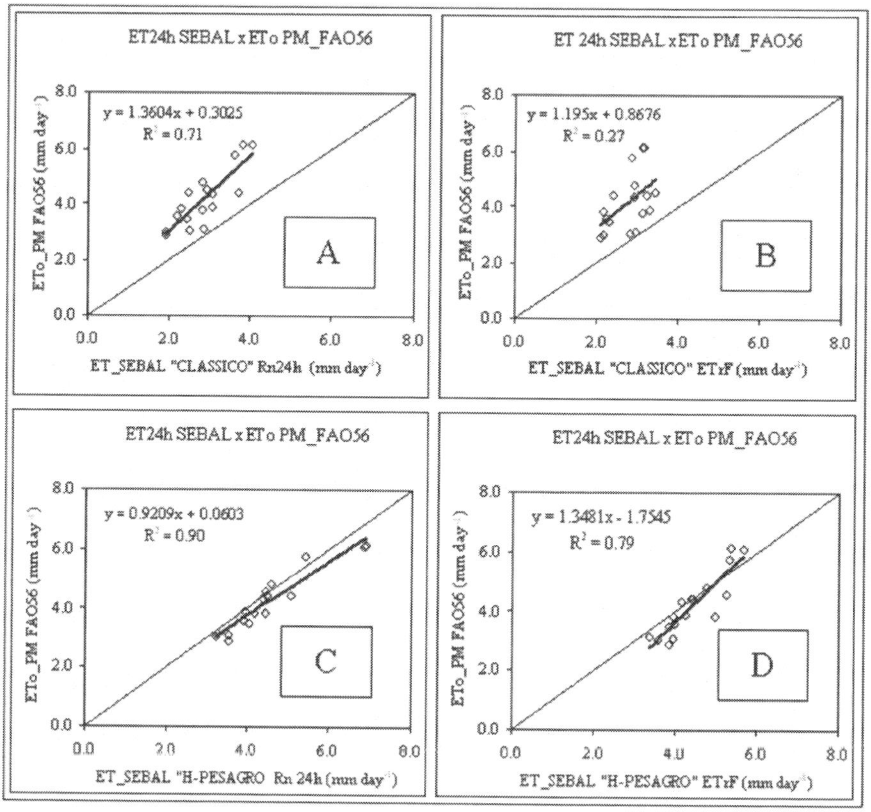

Fig. 5. Correlation between values of ET24h estimated with the method FAO (PM_FAO56) with data collected at PESAGRO station and values of ET24h estimated by SEBAL with propositions "H_Classic" w/Rn24h (A), "H_Classic" w/ETr_F (B), "H_Pesagro" w/Rn24h (C) and "H_Pesagro" w/ETr_F (D) observed in pixel from Pesagro, expressed in mm day^{-1}.

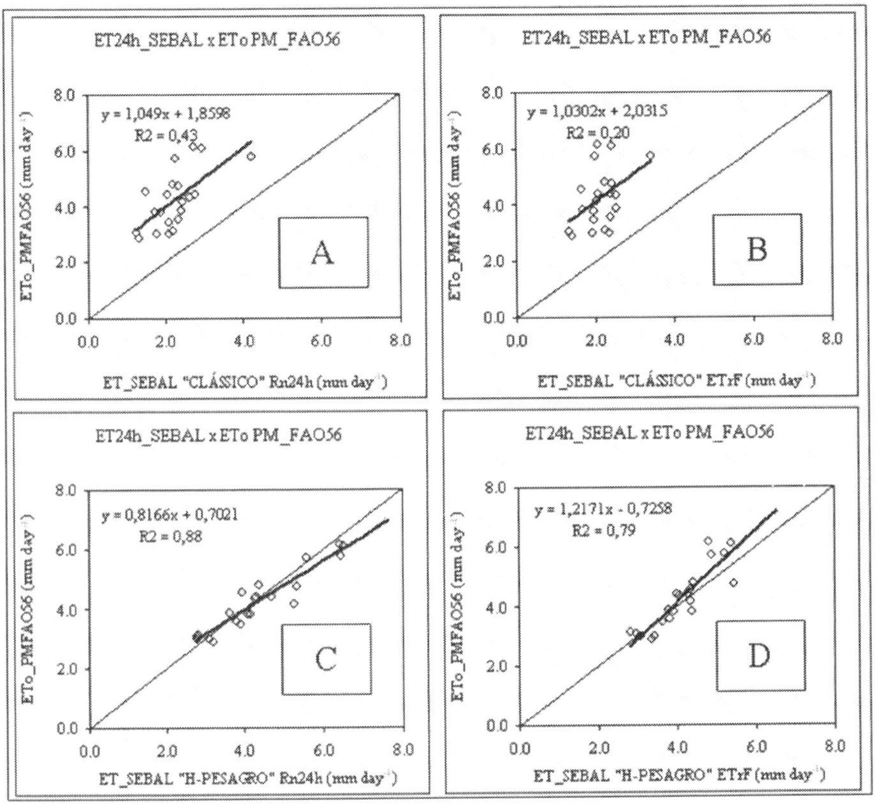

Fig. 6. Correlation between values of ET24h estimated with the method FAO (PM_FAO56) with data collected in PESAGRO station and values of ET24h estimated by SEBAL with propositions "H_Classic" w/Rn24h (A), "H_Classic" w/ETr_F (B), "H_Pesagro" w/Rn24h (C) and "H_Pesagro" w/ETr_F (D) observed in pixel pixel from UFRRJ, expressed in mm day^{-1}.

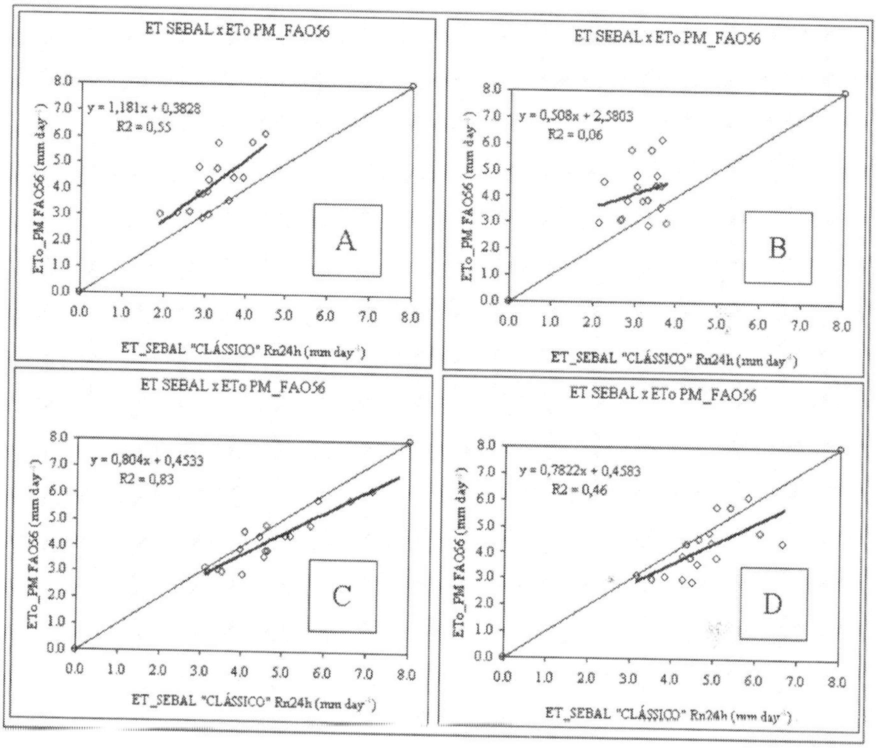

Fig. 7. Correlation between values of ET24h estimated by the method FAO (PM_FAO56) with data collected from PESAGRO station and values of ET24h estimated by SEBAL with propositions "H_Classic" w/Rn24h (A), "H_Classic" w/ETr_F (B), "H_Pesagro" w/Rn24h (C) and "H_Pesagro" w/ETr_F (D) observed in pixel from Sugar-cane (SANTA CRUZ **AGROINDUSTRY**), expressed in mm day^{-1}.

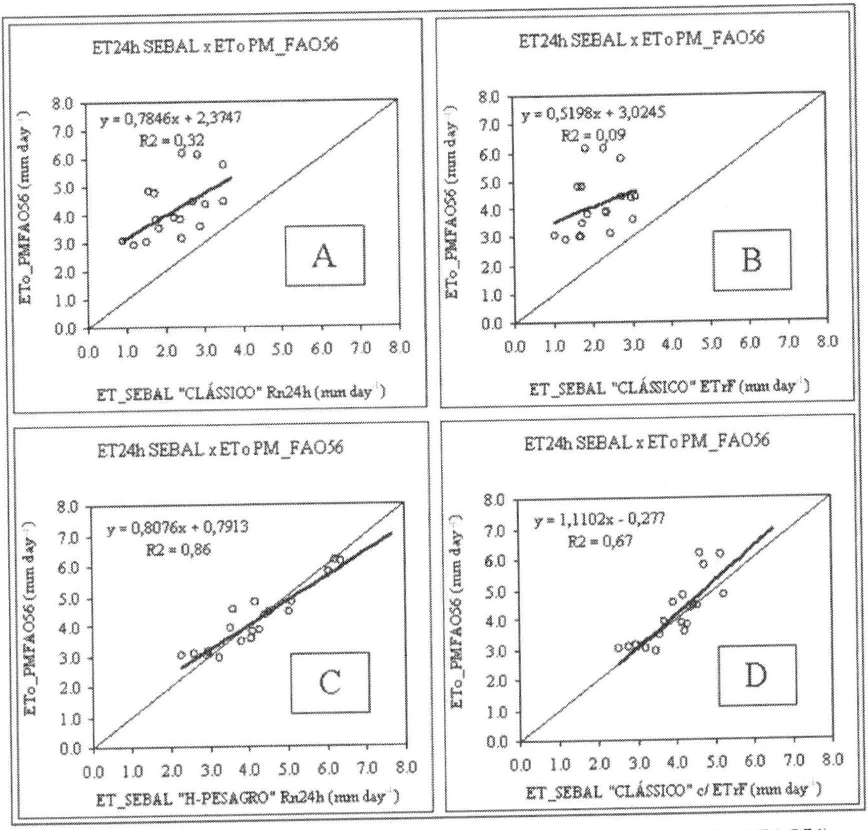

Fig. 8. Correlation between values of ET24h estimated by the method FAO (PM_FAO56) with data collected from PESAGRO station and values of ET24h estimated by SEBAL with propositions "H_Classic" w/Rn24h (A), "H_Classic" w/ETr_F (B), "H_Pesagro" w/Rn24h (C) and "H_Pesagro" w/ETr_F (D) observed in pixel from Coconut (**AGRICULTURE TAÍ**) expressed in mm day^{-1}.

Observing Figures 5, 6, 7, and 8, it is possible to conclude that the proposition 'H_Classic' under estimated values projected by PM_FAO56 method, showing better results for values estimated using Rn24h.

Proposition 'H_Pesagro', although in a slight way, super estimated values of the ETo estimated with data from the meteorological station Pesagro, in all four control points, showing higher correction coefficients than the others with emphasis for the method using Rn24h.

Hafeez et al. (2002) applied SEBAL using MODIS images in Philippines and observed that the ET_SEBAL super estimated in 13,5 % the values of ETo estimated by PM_FAO56, justifying such behavior due to the spatial resolution of 1.000 m of the surface temperature chart (MOD11A1).

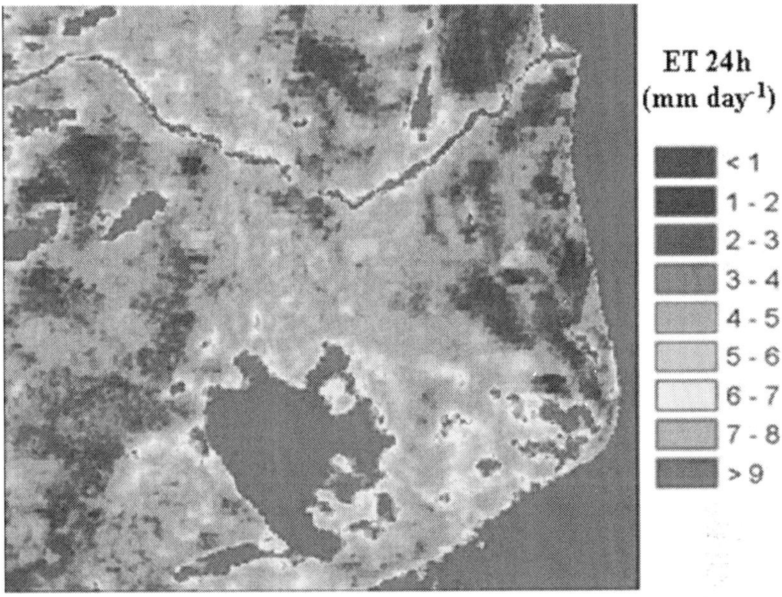

Fig. 9. Images of the daily evapotranspiration for the dry period in the Fluminense North Region, Rio de Janeiro State. DJ 2005218.

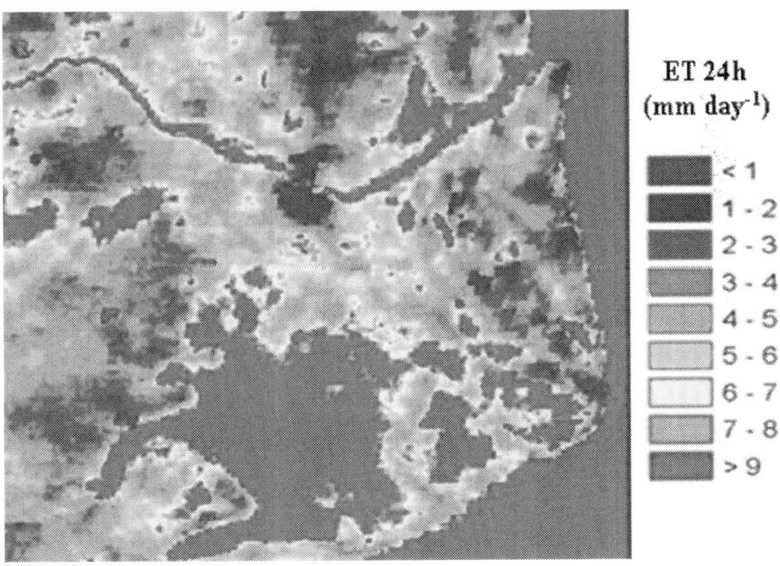

Fig. 10. Images of the daily evapotranspiration for the humid period in the Fluminense North Region, Rio de Janeiro State. DJ 2006015.

Allen et al. (2001), using images of LANDSAT in the basin of river Bear, North-East region of the U.S.A., observed that SEBAL showed a good precision for the estimation of ET, compared with weighing lysimeters, super estimating monthly mean values in 16% and 4 % for seasonal values.
Images of the daily evapotranspiration for the dry and humid periods in the Fluminense North Region, Rio de Janeiro State is showed in Figures 9 (DJ 2005218) and 10. (DJ 2006015).

4. Conclusion

In accordance with the proposed objectives in this work, it is possible to conclude that in conditions de sub-humid climate: For the estimative of sensible heath flux, the use of proposition 'H_Pesagro' resulted more efficient than 'H_Classic'; The method that uses values of mean radiation balance integrated in 24 hours (Rn24h) is more consistent than the method that uses the reference evaporative fraction (ETr_F) for the conversion of instantaneous evapotranspiration values (ET_{inst}) in daily values (ET_{24h}).

5. Acknowledgements

The authors are grateful for the National Counsel for Scientific and Technological Development - CNPq and the Coordenação *de Aperfeiçoamento de Pessoal de Nível Superior - CAPES*, for the financial support and logistics that made this study possible.

6. References

Alados, C.L.; Pueyo, Y.; Giner, M.L.; Navarro, T.; Escos, J.; Barroso, F.; Cabezudo, B.; Emlen, G.M., 2003. Quantitative characterization of the regressive ecological succession by fractal analysis of plant spatial patterns. Ecological Modell. v.163, p.1-17.

Allen, R. G.; Pereira, L. S.; Raes, D.; Smith, M., 1998. Crop evapotranspiration - Guidelines for computing crop water requeriments. FAO Irrigation and Drainage Paper 56, Rome, Italy, 318 p.

Allen, R.G.; Tasumi, M.; Trezza, R.; Bastiaanssen, W.G.M., 2002. SEBAL - Surface Energy Balance Algorithms for Land. Advanced training and users manual, Version *1.0*. University of Idaho, EUA. 97 p.

Ataíde, K.R.P., 2006. Determinação do saldo de radiação e radiação solar global com produtos do sensor MODIS Terra e Aqua. Tese (Mestrado em Meteorologia) - Campina Grande, PB - Universidade Federal de Campina Grande - UFCG, 88p.

Azevedo, H.J.; Silva Neto, R.; Carvalho, A. M.; Viana, J.L.; Mansur, A.F.U., 2002. Uma análise da cadeia produtiva da cana-de-açúcar na Região Norte Fluminense. Observatório sócio-econômico da Região Norte Fluminense - Boletim Técnico n° 6, 51p.

Bastiaanssen, W.G.M., 1995. Regionalization of surface flux densities and moisture indicators in composite terrain. Ph,D Thesis, Wageningen Agricultural University, Wageningen, The Netherlands. 273p.

Bastiaanssen, W.G.M.; Pelgrum, H.; Wang, J.; Ma, Y.; Moreno, J.; Roerink, G. J.; van der Val, T., 1998. A remote sensing surface energy balance algorithm for land (SEBAL):Part 2 validation, Journal of Hidrology, v, 212-213: 213-229.

Frota, P.C.E., 1978. Estudo do calor sensível e latente no interior de uma cultura de milho (Zea mays L.), Dissertação (Mestrado em Agrometeorologia), Piracicaba, SP, Universidade Luiz de Queiroz - ESALQ/USP, 75p.

Gomes, M.C.R., 1999. Efeito da irrigação suplementar na produtividade da cana-de-açúcar em Campos dos Goytacazes, RJ. Dissertação (Mestrado em Produção Vegetal) - Campos dos Goytacazes - RJ, Universidade Estadual do Norte Fluminense - UENF, 51p.

Hafeez, M.M.; Chemin, Y,; Van de Giesen, N.; Bouman, B.A.M., 2002. Field evapotranspiration estimation in Central Luzón, Philippines, using different sensors: Landsat 7 ETM+, Terra Modis and Aster. Symposium on Geospatial Theory, Processing and Applications. Ottawa-Canadian. 7 p.

Ide, B. Y.; Oliveira, M.A. de, 1986. Efeito do clima na produção da cana-de-açúcar. In: Seminário de Tecnologia Agronômica, 3. Piracicaba, SP. Anais.... São Paulo: COPERSUCAR, p.573-583.

Lagouarde, J.P.; Brunet, Y., 1983. A simple model for estimating the daily upward long wave surface radiation flux from NOOA/AVHRR data. International Journal of Remote Sensing. 14(5):907-925.

Land Processes Distributed Active Archive Center - LP-DACC Available online: http://edcimswww.cr.usgs.gov/pub/imswelcome/ (Accessed on 15 April/2005).

Magalhães, A. C. N. 1987. Ecofisiologia da cana-de-açúcar: aspectos do metabolismo do carbono na planta. In: Castro, P.R.C.; Ferreira, S.O.; Yamada, T. (Ed.). Ecofisiologia da Produção Agrícola. Piracicaba, SP: Potafós, p. 113-118.

Moran, M. S.; Jackson, R. D.; Raymond L.; Gay, L.; Slater, P., 1999. Mapping surface energy balance components by combining Landsat Thematic Mapper and groudbase meteorological data. Remote Sensing of Environment. n.30:77-87.

Morgado, I.F., 2009. Agroindústria Sucroalcooleira do Estado do Rio de Janeiro. Universidade Cândido Mendes - UCAM. Available online: http://www.infoagro.ucam-campos.br/agro_in_rio.htm. (Accessed on 25 June/2010).

Paiva, C.M.; Liu, W.T.H.; Franca, G.B.; Filho, O.C. R., 2004. Estimativa das componentes do balanço de energia via satélite através do modelo SEBAL. XIII Congresso Brasileiro de Meteorologia, Fortaleza, CE. Anais.

Roerink, G.J.; Bastiaanssen, W.G.M.; Chambouleyron, J.; Menenti, M., 1997. Relating crop water consumption to irrigation water supply to remote sensing. Water Resources Management. 11: 445-465.

Secretaria Estadual de Agricultura, Pesca e Desenvolvimento do Interior - SEAAPI, RJ (2006). Available online: www.seaapi.rj.gov.br/frutificar. (Accessed on 15 November/ 2006).

Silva, B.B.; Lopes, G.M.; Azevedo, P.V., 2005. Balanço de radiação em áreas irrigadas utilizando imagens Landsat 5 -TM. Revista Brasileira de Meteorologia. V.20 (2): 243-252.

Tasumi, M., 2003. Progess in operational estimation of regional evapotranspiration using satellite imagery. PhD Dissertation. Idaho State University. Idaho. USA. 379 p.

Teixeira, A. H. C., 2008. Measurements and modelling of evepotranspiration to assess agricultural water productivity in basins with changing land use patterns - a case study in the São Francisco River basin, Brazil. PhD Dissertation. Wageningen University.. Nederland. 239 p.

Timmermans, W.J.; Meijerink, A.M.J., 1999. Remotely sensed actual evapotranspiration: implications for groundwater management in Botzwana. Journal of Applied Geohydrology. 1:222-233.

Trezza, R., 2002. Evapotranspiration using a satellite-based energy balance with sandarized ground control. PhD Dissertation. Utah State University. Logan. USA. 247p.

2

Evapotranspiration Estimation Based on the Complementary Relationships

Virginia Venturini[1], Carlos Krepper[1,2] and Leticia Rodriguez[1]
[1]*Centro de Estudios Hidro-Ambientales-Facultad de Ingeniería y Ciencias Hídricas Universidad Nacional del Litoral*
[2]*Consejo Nacional de Investigaciones Científicas y Técnicas*
Argentina

1. Introduction

Many hydrologic modeling and agricultural management applications require accurate estimates of the actual evapotranspiration (ET), the relative evaporation (F) and the evaporative fraction (EF). In this chapter, we define ET as the actual amount of water that is removed from a surface due to the processes of evaporation-transpiration whilst the potential evapotranspiration (Epot) is any other evaporation concept. There are as many potential concepts as developed mathematical formulations. In this chapter, F represents the ratio between ET and Epot, as it was introduced by Granger & Gray (1989). Meanwhile, EF is the ratio of latent flux over available energy.

It is worthy to note that, in general, the available evapotranspiration concepts and models involve three sets of variables, i.e. available net radiation (Rn), atmospheric water vapor content or temperature and the surface humidity. Hence, different Epot formulations were derived with one or two of those sets of variables. For instance, Penman (1948) established an equation by using the Rn and the air water vapor pressure. Priestley & Taylor (1972) derived their formulations with only the available Rn.

In the last three decades, several models have been developed to estimate ET for a wide range of spatial and temporal scales provided by remote sensing data. The methods could be categorized as proposed by Courault et al. (2005).

Empirical and semi-empirical methods: These methods use site specific or semi-empirical relationships between two o more variables. The models proposed by Priestley & Taylor (1972), hereafter referred to as P-T, Jackson et al. (1977); Seguin et al. (1989); Granger & Gray (1989); Holwill & Stewart (1992); Carlson et al. (1995); Jiang & Islam (2001) and Rivas & Caselles (2004), lie within this category.

Residual methods: This type of models commonly calculates the energy budged, then ET is estimated as the residual of the energy balance. The following models are examples of residual methods: The Surface Energy Balance Algorithm for Land (SEBAL) (Bastiaanssen et al., 1998; Bastiaanssen, 2000), the Surface Energy Balance System (SEBS) model (Su, 2002) and the two-source model proposed by Norman et al. (1995), among others.

Indirect methods: These physically based methods involve Soil-Vegetation-Atmosphere Transfer (SVAT) models, presenting different levels of complexity often reflected in the number of parameters. For example, the ISBA (Interactions between Soil, Biosphere, and

Atmosphere) model by Noilhan & Planton (1989), developed to be included within large scale meteorological models, parameterizes the land surface processes. The ISBA Ags model (Calvet et al., 1998) improved the canopy stomatal conductance and CO_2 concentration with respect to the ISBA original model.

Among the first category (Empirical and semi-empirical methods), only few methodologies to calculate ET have taken advantage of the complementary relationship (CR).

It is worth mentioning that there are only two CR approaches known so far, one attributed to Bouchet (1963) and the other to Granger & Gray (1989). Even though various ET models derived from these two fundamental approaches are referenced to throughout the chapter, it is not the intention of the authors to review them in detail.

Bouchet (1963) proposed the first complementary model based on an experimental design. He postulated that, for a large homogeneous surface and in absence of advection of heat and moisture, regional ET could be estimated as a complementary function of Epot and the wet environment evapotranspiration (Ew) for a wide range of available energy. Ew is the ET of a surface with unlimited moisture. Thus, if Epot is defined as the evaporation that would occur over a saturated surface, while the energy and atmospheric conditions remain unchanged, it seems reasonable to anticipate that Epot would decrease as ET increases. The underlying argument is that ET incorporates humidity to the surface sub-layer reducing the possibility for the atmosphere to transport that humidity away from the surface. Bouchet´s idea that Epot and ET have this complementary relationship has been the subject of many studies and discussions, mainly due to its empirical background (Brutsaert & Parlange, 1998; Ramírez et al., 2005). Examples of successful models based on Bouchet's heuristic relationship include those developed by Brutsaert & Stricker (1979); Morton (1983) and Hobbins et al. (2001). These models have been widely applied to a broad range of surface and atmospheric conditions (Brutsaert & Parlange, 1998; Sugita et al., 2001; Kahler & Brutsaert, 2006; Ozdogan et al., 2006; Lhomme & Guilioni, 2006; Szilagyi, 2007; Szilagyi & Jozsa, 2008).

Granger (1989a) developed a physically based complementary relationship after a meticulous analysis of potential evaporation concepts. He remarked that "*Bouchet corrected the misconception that a larger potential evaporation necessarily signified a larger actual evaporation*". The author used the term "potential evaporation" for the Epot and Ew concepts, and clearly presented the complementary behavior of common potential evaporation theories. This author suggested that Ew is the value of the potential evaporation when the actual evaporation rate is equal to the potential rate. The use of two potential parameters, i.e. Epot and Ew, seems to generate a universal relationship, and therefore, universal ET models. Conversely, attempting to estimate ET from only one potential formulation may need site-specific calibration or auxiliary relationships (Granger, 1989b). In addition, the relative evaporation coefficient introduced by Granger & Gray (1989) enhances the complementary relationship with a dimensionless coefficient that yields a simpler complementary model.

The foundation of the complementary relationship is the basis for operational estimates of areal ET by Morton (1983), who formulated the Complementary Relationship Areal Evapotranspiration (CRAE) model. The reliability of the independent operational estimates of areal evapotranspiration was tested with comparable, long-term water budget estimates for 143 river basins in North America, Africa, Ireland, Australia and New Zealand.

A procedure to calculate ET requiring only common meteorological data was presented by Brutsaert & Stricker (1979). Their Advection-Aridity approach (AA) is based on a conceptual model involving the effect of the regional advection on potential evaporation and Bouchet's complementary model. Thus, the aridity of the region is deduced from the regional advection of the drying power of the air. The authors validated their model in a rural watershed finding a good agreement between estimated daily ET and ET obtained with the energy budget method.

Morton's CRAE model was tested by Granger & Gray (1990) for field-size land units under a specific land use, for short intervals of time such as 1 to 10 days. They examined the CRAE model with respect to the algorithms used to describe different terms and its applicability to reduced spatial and temporal scales. The assumption in CRAE that the vapor transfer coefficient is independent of wind speed may lead to appreciable errors in computing ET. Comparisons of ET estimates and measurements demonstrated that the assumptions that the soil heat flux and storage terms are negligible, lead to large overestimation by the model during periods of soil thaw.

Hobbins et al. (2001) and Hobbins & Ramírez (2001) evaluated the implementations of the complementary relationship hypothesis for regional evapotranspiration using CRAE and AA models. Both models were assessed against independent estimates of regional evapotranspiration derived from long-term, large-scale water balances for 120 minimally impacted basins in the conterminous United States. The results suggested that CRAE model overestimates annual evapotranspiration by 2.5% of mean annual precipitation, whereas the AA model underestimates annual evapotranspiration by 10.6% of mean annual precipitation. Generally, increasing humidity leads to decreasing absolute errors for both models. On the contrary, increasing aridity leads to increasing overestimation by the CRAE model and underestimation by the AA model, except at high aridity basins, where the AA model overestimates evapotranspiration.

Three evapotranspiration models using the complementary relationship approach for estimating areal ET were evaluated by Xu & Singh (2005). The tested models were the CRAE model, the AA model, and the model proposed by Granger & Gray (1989) (GG), using the concept of relative evaporation. The ET estimates were compared in three study regions representing a wide geographic and climatic diversity: the NOPEX region in Central Sweden (typifying a cool temperate humid region), the Baixi catchment in Eastern China (typifying a subtropical, humid region), and the Potamos tou Pyrgou River catchment in Northwestern Cyprus (typifying a semiarid to arid region). The calculation was made on a daily basis whilst comparisons were made on monthly and annual bases. The results showed that using the original parameter values, all three complementary relationship models worked reasonably well for the temperate humid region, while their predictive power decreased as soil moisture exerts increasing control over the region, i.e. increased aridity. In such regions, the parameters need to be calibrated.

Ramírez et al. (2005) provided direct observational evidence of the complementary relationship in regional evapotranspiration hypothesized by Bouchet in 1963. They used independent observations of ET and Epot at a wide range of spatial scales. This work is the first to assemble a data set of direct observations demonstrating the complementary relationship between regional ET and Epot. These results provided strong evidence for the complementary relationship hypothesis, raising its status above that of a mere conjecture.

A drawback among the aforementioned complementary ET models is the use of Penman or Penman-Monteith equation (Monteith & Unsworth, 1990) to estimate Epot. Specifically, the Morton's CRAE model (Morton, 1983) uses Penman equation to calculate Epot, and a modified P-T equation to approximate Ew. Brutsaert & Stricker (1979) developed their AA model using Penman for Epot and the P-T equilibrium evaporation to model Ew. At the time those models were developed, networks of meteorological stations constituted the main source of atmospheric data, while the surface temperature (Ts) or the soil temperature were available only at some locations around the World. The advent of satellite technology provided routinely observations of the surface temperature, but the source of atmospheric data was still ancillary. Thus, many of the current remote sensing approaches were developed to estimate ET with little amount of atmospheric data (Price, 1990; Jiang & Islam, 2001).

The recent introduction of the Atmospheric Profiles Product derived from Moderate Resolution Imaging Spectroradiometer (MODIS) sensors onboard of EOS-Terra and EOS-Aqua satellites meant a significant advance for the scientific community. The MODIS Atmospheric profile product provides atmospheric and dew point temperature profiles on a daily basis at 20 vertical atmospheric pressure levels and at 5x5km of spatial resolution (Menzel et al., 2002). When combined with readily available Ts maps obtained from different sensors, this new remote source of atmospheric data provides a new opportunity to revise the complementary relationship concepts that relate ET and Epot (Crago & Crowley, 2005; Ramírez et al., 2005).

A new method to derive spatially distributed EF and ET maps from remotely sensed data without using auxiliary relationships such as those relating a vegetation index (VI) with the land surface temperature (Ts) or site-specific relationships, was proposed by Venturini et al. (2008). Their method for computing ET is based on Granger's complementary relationship, the P-T equation and a new parameter introduced to calculate the relative evaporation (F=ET/Epot). The ratio F can be expressed in terms of Tu, which is the temperature of the surface if it is brought to saturation without changing the actual surface vapor pressure. The concept of Tu proposed by these authors is analogous to the dew point temperature (Td) definition.

Szilagyi & Jozsa (2008) presented a long term ET calculation using the AA model. In their work the authors presented a novel method to calculate the equilibrium temperature of Ew and P-T equation that yields better long-term ET estimates. The relationship between ET and Epot was studied at daily and monthly scales with data from 210 stations distributed all across the USA. They reported that only the original Rome wind function of Penman yields a truly symmetric CR between ET and Epot which makes Epot estimates true potential evaporation values. In this case, the long-term mean value of evaporation from the modified AA model becomes similar to CRAE model, especially in arid environments with possible strong convection. An R^2 of approximately 0.95 was obtained for the 210 stations and all wind functions used. Likewise, Szilagyi & Jozsa in (2009) investigated the environmental conditions required for the complementary ET and Epot relationship to occur. In their work, the coupled turbulent diffusion equations of heat and vapor transport were solved under specific atmospheric, energy and surface conditions. Their results showed that, under near-neutral atmospheric conditions and a constant energy term at the evaporating surface, the analytical solution across a moisture discontinuity of the surface yields a symmetrical complementary relationship assuming a smooth wet area.

Recently, Crago et al. (2010) presented a modified AA model in which the specific humidity at the minimum daily temperature is assumed equal to the daily average specific humidity. The authors also modified the drying power calculation in Penman equation using Monin-Obukhov theory (Monin & Obukhov, 1954). They found promising results with these modifications. Han et al. (2011) proposed and verified a new evaporation model based on the AA model and the Granger's CR model (Granger, 1989b). This newly proposed model transformed Granger´s and AA models into similar, dimensionless forms by normalizing the equations with Penman potential model. The evaporation ratio (i.e. the ratio of ET to Penman potential evaporation) was expressed as a function of dimensionless variables based on radiation and atmospheric conditions. From the validation with ground observations, the authors concluded that the new model is an enhanced Granger`s model, with better evaporation predictions. In addition, the model somewhat approximates the AA model under neither too-wet nor too-dry conditions. As the reader can conclude, the complementary approach is nowadays the subject of many ongoing researches.

2. A review of Bouchet's and Granger's models

Bouchet (1963) set an experiment over a large homogeneous surface without advective effects. Initially, the surface was saturated and evaporated at potential rate. With time, the region dried, but a small parcel was kept saturated (see Figure 1), evaporating at potential rate. The region and the parcel scales were such that the atmosphere could be considered stable. Bouchet described his experiment, dimension and scales as follows[1],

- The energy balance requires the prior definition of the limits of the system. To avoid taking into account the phenomena of accumulation and restoration of heat during the day and night phases, the assessment will cover a period of 24 hours.
- The system includes an ensemble of vegetation, soil, and a portion of the lower atmosphere. The sizes of these layers are such that the daily temperature variations are not significant.
- If this system is located in an area which, for any reason, does not have the same climatic characteristics, there will be exchanges of energy throughout the side "walls" of the system, that need to be analyzed (advection free area).
- Lateral exchanges by conduction in the soil are negligible. The lateral exchanges in the atmosphere due to the homogenization of the air masses will be named as "oasis effect". Given the heterogeneity from one point to another, the lateral exchanges of energies, or the "oasis effect", rule the natural conditions.
- The oasis effect phenomenon can be schematically represented as shown in Figure 1. If in a flat, homogeneous area (brown line in Figure 1), a discontinuity appears, i.e. a change in soil specific heat, moisture or natural vegetation cover, etc. (green line in Figure 1), then a disturbed area is developed in the direction of airflow (gray filled area in Figure 1) where environmental factors are modified from the general climate because of the discontinuity.
- The perturbation raises less in height than in width. It always presents a "flat lens" shape in which the thickness is small compared to the horizontal dimensions.

As mentioned, initially the surface was saturated and evaporated at its potential rate, i.e. at the so-called reference evapotranspiration (or Ew). In this initial condition, Epot = Ew = ET. When ET is lower than Ew due to limited water availability, a certain excess of energy would become available. This remaining energy not used for evaporation may, in tern, warm the lower layer of the atmosphere. The resulting increase in air temperature due to the heating, and the decrease in humidity caused by the reduction of ET, would lead to a new value of Epot larger than Ew by the amount of energy left over.

[1] The following text was translated by the authors of this chapter from Bouchet's original paper (in French).

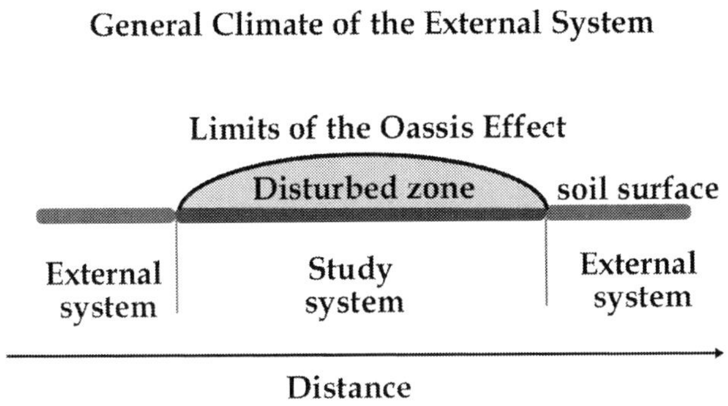

Fig. 1. Reproduction of Bouchet´s schematic representation of the Oasis Effect experiment.

Thus, Bouchet's complementary relationship was obtained from the balance of these evaporation rates,

$$ET + Epot = 2Ew \tag{1}$$

Bouchet postulated that in such a system, under a constant energy input and away from sharp discontinuities, there exists a complementary feedback mechanism between ET and Epot, that causes changes in each to be complementary, that is, a positive change in ET causes a negative change in Epot (Ozdogan et al., 2006), as sketched in Figure 2. Later, Morton (1969) utilized Bouchet's experiment to derive the potential evaporation as a manifestation of regional evapotranspiration, i.e. the evapotranspiration of an area so large that the heat and water vapor transfer from the surface controls the evaporative capacity of the lower atmosphere.

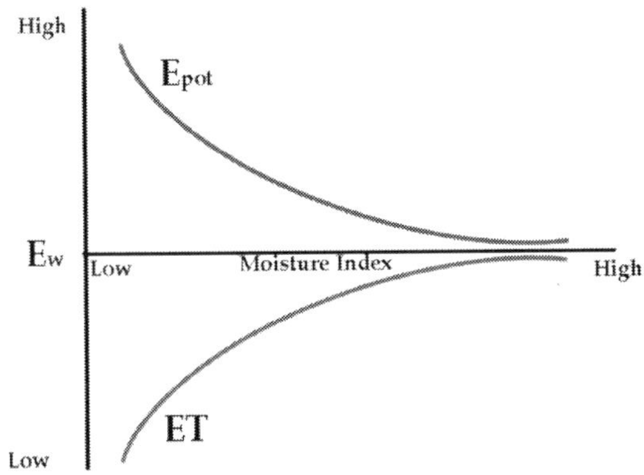

Fig. 2. Sketch of Bouchet´s complementary ET and Epot relationship

The hypothesis asserts that when ET falls below Ew as a result of limited moisture availability, a large quantity of energy becomes available for sensible heat flux that warms and dries the atmospheric boundary layer thereby causing Epot to increase, and *vise versa.*

Equation (1) holds true if the energy budget remains unchanged and all the excess energy goes into sensible heat (Ramírez et al., 2005). It should be noted that Bouchet´s experimental system is the so-called advection-free-surface in P-T formulation.
This relationship assumes that as ET increases, Epot decreases by the same amount, i.e. δET = -δEpot, where the symbol δ means small variations. Bouchet's equation has been widely used in conjunction with Penman (1948) and Priestley-Taylor (1972) (Brutsaert & Stricker, 1979; Morton, 1983; Hobbins el al., 2001).
Granger (1989b) argued that the above relationship lacked a theoretical background, mainly due to Bouchet's symmetry assumption (δET=-δEpot). Nonetheless, the author recognized that Bouchet´s CR set the basis for the complementary behavior between two potential concepts of evaporation and ET. One of the benefits of using two potential evaporation concepts rather than a single one is that the resulting CR would be universal, without the need of tuning parameters from local data.
Granger (1989a) revised the diversity of potential evaporation concepts available at that moment and expertly established an inequity among them. The resulting comparison yielded that Penman (1948) and Priestley & Taylor (1972) concepts are Ew concepts, and that the true potential evaporation would be that proposed by van Bavel (1966). Thus, these parameterizations would result in the following inequity, Epot $\geq$ Ew $\geq$ ET, where Epot would be van Bavel´s concept, Ew could be obtained with either Penman or P-T, knowing that ET-Penman is larger than ET-Priestley-Taylor (Granger, 1989a). Hence, the author postulated that the above inequity comprises Bouchet´s equity (δET = -δEpot) but it is based on a new CR. Granger (1989b) then proposed the following CR formulation,

$$ET + Epot\frac{\gamma}{\Delta} = Ew\left(\frac{\Delta+\gamma}{\Delta}\right) \tag{2}$$

where γ is the psychrometric constant and Δ is the slope of the saturation vapor pressure (SVP) curve.
Equation (2) shows that for constant available energy and atmospheric conditions, $-\gamma/\Delta$ is equal to the ratio δET/δEpot. In addition, this CR is not symmetric with respect to Ew. It can be easily verified that equation (2) is equivalent to equation (1) when $\gamma=\Delta$. The condition that the slope of the SVP curve equals the psychrometric constant is only true when the temperature is near 6 °C (Granger, 1989b). This has been widely tested (Granger & Gray, 1989; Crago & Crowley, 2005; Crago et al., 2005; Xu & Singh, 2005; Venturini et al., 2008; Venturini et al., 2011).

3. Bouchet`s versus Granger`s complementary models

A review of the two complementary models widely used for ET calculations was presented. Both methods are not only conceptually different, but also differ in their derivations. Mathematically speaking, Bouchet's complementary relationship (equation 1) results a simplification of Granger's complementary equation (equation 2) for the case $\Delta=\gamma$. Equations (1) and (2) can also be written, respectively, as follows,

$$\frac{1}{2}ET + \frac{1}{2}Epot = Ew \tag{3}$$

$$\left(\frac{\Delta}{\Delta+\gamma}\right)ET+\left(\frac{\gamma}{\Delta+\gamma}\right)Epot=Ew \tag{4}$$

The re-written Bouchet´s complementary model, equation (3), clearly expresses Ew as the middle point between the ET and the Epot processes. In contrast, the re-written Granger's complementary relationship, equation (4), shows how both, ET and Epot contribute to Ew with different coefficients, the coefficients varying with the slope of the SVP curve at the air temperature Ta, since γ is commonly assumed constant. For clarity, Table 1 summarizes all symbols and definitions used in this Chapter.

Recently, Ramírez et al., (2005) discussed Bouchet's coefficient "2" with monthly average ground measurements. In their application, Epot was calculated with the Penman-Monteith equation and Ew with the P-T model. They concluded that the appropriate coefficient should be slightly lower than 2.

Venturini et al. (2008) and Venturini et al. (2011) introduced the concept of the relative evaporation, F= ET/Epot, proposed earlier by Granger & Gray (1989), along with P-T equation in both CR models. Thus, Epot is replaced by ET/F and Ew is equated to P-T equation. Hence, replacing Epot in equation (3),

$$ET+\frac{ET}{F}=k\,Ew \tag{5}$$

where k is Bouchet´s coefficient, originally assumed k=2

Then, when Ew is replaced in (5) by the P-T equation, results

$$ET\left(1+\frac{1}{F}\right)=ka\ \ (\mathrm{Rn}-\mathrm{G})\frac{\Delta}{\Delta+\gamma} \tag{6}$$

where α is the P-T's coefficient, and the rest of the variables are defined in Table 1. Finally, Bouchet's CR is obtained by rearranging the terms in equation (6),

$$ET=ka\left(\frac{F}{F+1}\right)\left(\frac{\Delta}{\Delta+\gamma}\right)\ (\mathrm{Rn}-\mathrm{G}) \tag{7}$$

Following the same procedure with equation (4), the equivalent equation for Granger´s CR model is,

$$ET=\alpha\left(\frac{F\Delta}{F\Delta+\gamma}\right)\ (\mathrm{Rn}-\mathrm{G}) \tag{8}$$

It should be noted that the underlying assumptions of equation (7) are the same as those behind equation (8), plus the condition that Δ is approximately equal to γ.

Both, equations (7) and (8), require calculating the F parameter, otherwise the equations would have only theoretical advantages and would not be operative models. Venturini et al. (2008) developed an equation for F that can be estimated using MODIS products. Their F method is briefly presented here.

Consider the relative evaporation expression proposed by Granger & Gray (1989),

$$\frac{\mathrm{ET}}{\mathrm{Epot}}=\frac{f_u\,(\mathrm{e_s}-\mathrm{e_a})}{f_u(\mathrm{e_s^*}-\mathrm{e_a})} \tag{9}$$

where f_u is a function of the wind speed and vegetation height, e_s is the surface actual water vapor pressure, e_a is the air actual water vapor pressure, e^*_s is the surface saturation water vapor pressure.

Symbol	Definition
α	Priestley & Taylor's coefficient. α = 1.26
Δ [hPa/°C]	Slope of the saturation water vapor pressure curve
γ [hPa/°C]	Psychrometric constant
λE [W m-2]	Latent heat flux density
e_a [hPa]	Air actual water vapor pressure at Td
e^*_a [hPa]	Air saturation water vapor pressure at Ta
e_s [hPa]	Surface actual water vapor pressure at Tu
e^*_s [hPa]	Surface saturation water vapor pressure at Ts
Ew [W m-2]	Evapotranspiration of wet environment
Epot [W m-2]	Potential evapotranspiration
f_u	Wind function
F	Relative evaporation coefficient of Venturini et al. (2008)
G [W m-2]	Soil heat flux
H [W m-2]	Sensible heat flux
Q [W m-2]	Available energy, (Rn -G)
Rn [W m-2]	Net radiation at the surface
Ta [°K] or [°C]	Air temperature
Td [°K] or [°C]	Dew point temperature
Ts [°K] or [°C]	Surface temperature
Tu [°K] or [°C]	Surface temperature if the surface is brought to saturation without changing e_s

Table 1. Symbols and units

This form of the relative evaporation equation needs readily available meteorological data. A key difficulty in applying equation (9) lies on the estimation of (e_s-e_a), since there is no simple way to relate e_s to any readily available surface temperature. Thus, a new temperature should be defined. Many studies have used temperature as a surrogate for vapor pressure (Monteith & Unsworth, 1990, Nishida et al., 2003). Although the relationship between vapor pressure and temperature is not linear, it is commonly linearized for small temperature differences. Hence, e_s and e_s^* should be related to soil+vegetation at a temperature that would account for water vapor pressure. Figure 3 shows the relationship between e_s, e^*_s and e_a and their corresponding temperatures; where e_u^* is the SVP at an unknown surface temperature Tu.

An analogy to the dew point temperature concept (Td) suggests that Tu would be the temperature of the surface if the surface is brought to saturation without changing the surface actual water vapor pressure. Accordingly, Tu must be lower than Ts if the surface is not saturated and close to Ts if the surface is saturated. Consequently, e_s could be derived from the temperature Tu. Although Tu may not possibly be observed in the same way as Td, it can be derived, for instance, from the slope of the exponential SVP curve as a function of Ts and Td. This calculation is further discussed later in this chapter.

Assuming that the surface saturation vapor pressure at Tu would be the actual soil vapor pressure and that the SVP can be linearized, (e_s -e_a) can be approximated by Δ_1(Tu-Td) and (e^*_s -e_a) by Δ_2(Ts-Td), respectively. Figure 3 shows a schematic of these concepts.

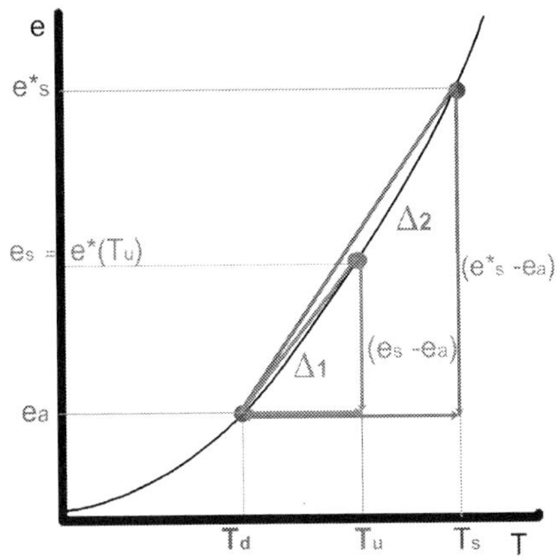

Fig. 3. Schematic of the linearized saturation vapor pressure curve and the relationship between (e_s -e_a) and Δ_1(Tu-Td), and (e^*_s -e_a) and Δ_2(Ts-Td).

Therefore, ET/Epot (see equation 9) can be rewritten as follows,

$$F = \frac{ET}{Epot} = \frac{(Tu - Td)}{(Ts - Td)}\left(\frac{\Delta_1}{\Delta_2}\right) \tag{10}$$

The wind function, f_u, depends on the vegetation height and the wind speed, but it is independent of surface moisture. In other words, it is reasonable to expect that the wind function will affect ET and Epot in a similar fashion (Granger, 1989b), so its effect on ET and Epot cancels out. The slopes of the SVP curve, Δ_1 and Δ_2, can be computed from the SVP first derivative at Td and Ts without adding further complexity to this method. However, Δ_1 and Δ_2 will be assumed approximately equal from now on, as they will be estimated as the first derivative of the SVP at Ta.

The relationship between Ts and Tu can be examined throughout the definition of Tu, which represents the saturation temperature of the surface. For a saturated surface, Tu is expected to be very close or equal to Ts. In contrast, for a dry surface, Ts would be much larger than Tu. Since Epot is larger than or equal to ET, F ranges from 0 to 1. For a dry surface, with Ts >> Tu, (Ts-Td) would be larger than (Tu-Td) and ET/Epot would tend to 0. In the case of a saturated surface with e_s close to e_s^* and Ts close to Tu, (Ts-Td) would be similar to (Tu-Td) and ET/Epot would tend to 1.

The calculation of Tu proposed by Venturini et al. (2008) is presented in the next section, where results from MODIS data are shown. However, it is emphasized that the definition of Tu is not linked to any data source; therefore it can be estimated with different approaches.

4. Complementary models application using remotely sensed data

In order to show the potential of the complementary relationships, equations (7) and (8) were applied to the Southern Great Plains of the USA region and the results compared and analyzed.

4.1 Study area

The Southern Great Plains (SGP) region in the United States of America extends over the State of Oklahoma and southern parts of Kansas. The area broadens in longitude from 95.3° W to 99.5° W and in latitude from 34.5° N to 38.5° N (Figure 4). This region was the first field measurement site established by the Atmospheric Radiation Measurement (ARM) Program. At present, the ARM program has three experimental sites. Scientists from all over the World are using the information obtained from these sites to improve the performance of atmospheric general circulation models used for climate change research. The SGP was chosen as the first ARM field measurement site for several reasons, among them, its relatively homogeneous geography, easy accessibility, wide variability of climate cloud types, surface flux properties, and large seasonal variations in temperature and specific humidity (http://www.arm.gov/sites/sgp).

Most of this region is characterized by irregular plains. Altitudes range from approximately 500 m to 90 m, increasing gradually from East to West. In southwestern Oklahoma, the highest Wichita Mountains rise as much as 800 m above the surrounding landscape (Heilman & Brittin, 1989; Venturini et al., 2008). The climate is semiarid-subtropical. Although the maximum rainfall occurs in summer, high temperatures make summer relatively dry. Average annual temperatures range from 14°C to 18°C. Winters are cold and dry, and summers are warm to hot. The frost-free season stretches from 185 to 230 days. Precipitation ranges from 490 to 740 mm, with most of it falling as rain.

Grass is the dominant prairie vegetation. Most of it is moderately tall and usually grows in bunches. The most prevalent type of grassland is the bluestem prairie (*Andropogon gerardii and Andropogon hallii)*, along with many species of wildflowers and legumes. In many places where grazing and fire are controlled, deciduous forest is encroaching on the prairies.

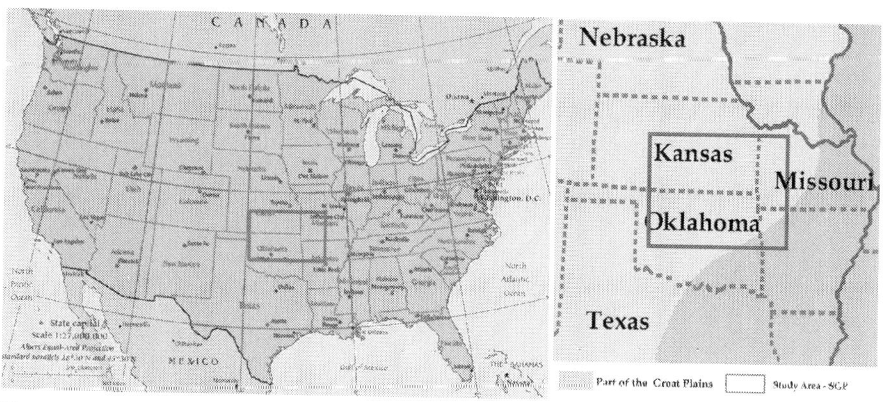

Fig. 4. Study area map

Due to generally favorable conditions of climate and soil, most of the area is cultivated, and little of the original vegetation remains intact. Oak savanna occurs along the eastern border of the region and along some of the major river valleys.

4.2 Ground data availability

The latent heat data was obtained from the ARM program Web site (http://www.arm.gov). The ARM instruments and measurement applications are well established and have been used for validation purposes in many studies (Halldin & Lindroth, 1992; Fritschen & Simpson, 1989). The site and name, elevation, geographic coordinates (latitude and longitude) and surface cover of the stations used in this work are shown in Table 2.

Site	Elevation (m a.m.s.l.)	Lat/Lon	Vegetation Type
Ashton, Kansas E-9	386	37.133 N/97.266 W	Pasture
Coldwater, Kansas E-8	664	37.333 N/99.309 W	Rangeland (grazed)
Cordell, Oklahoma: E-22	465	35.354 N/98.977 W	Rangeland (grazed)
Cyril, Oklahoma: E-24	409	34.883 N/98.205 W	Wheat (gypsum hill)
Earlsboro, Oklahoma: E-27	300	35.269 N/96.740 W	Pasture
Elk Falls, Kansas E-7	283	37.383 N/96.180 W	Pasture
El Reno, Oklahoma: E-19	421	35.557 N/98.017 W	Pasture (ungrazed)
Hillsboro, Kansas E-2	447	38.305 N/97.301 W	Grass
Lamont, Oklahoma: E-13	318	36.605 N/97.485 W	Pasture and wheat
Meeker, Oklahoma: E-20	309	35.564 N/96.988 W	Pasture
Morris, Oklahoma: E-18	217	35.687 N/95.856 W	Pasture (ungrazed)
Pawhuska, Oklahoma: E-12	331	36.841 N/96.427 W	Native prairie
Plevna, Kansas E-4	513	37.953 N/98.329 W	Rangeland (ungrazed)
Ringwood, Oklahoma: E-15	418	36.431 N/98.284 W	Pasture

Table 2. Site name and station name, elevation, latitude, longitude and surface type

The first instrumentation installation to the SGP site took place in 1992, with data processing capabilities incrementally added in the succeeding years. This region has relatively extensive and well-distributed coverage of surface fluxes and meteorological observation stations. In this study, Energy Balance Bowen Ratio stations (EBBR), maintained by the ARM program were used for the validation of surface fluxes. The EBBR system produces 30 minute estimates of the vertical fluxes of sensible and latent heat at the local points. The EBBR fluxes estimates are calculated from observations of net radiation, soil surface heat flux, the vertical gradients of temperature and relative humidity.

4.3 MODIS products

The method proposed here was physically derived from universal relationships. Moreover, data sources do not represent a limitation for the applicability of equations (6) and (8), nonetheless remotely sensed data such as that provided by MODIS scientific team would empower the potential applications of the methods. Hence, the equations applicability using MODIS products was explored. The sensor's bands specifications can be obtained from http://modis.gsfc.nasa.gov/about/specifications.php.

Daytime images for seven days in year 2003 with at least 80% of the study area free of clouds were selected. Table 3 summarizes the images information including date, day of the year, satellite overpass time and image quality.
Geolocation is the process by which scientists specify where a specific radiance signal was detected on the Earth's surface. The MODIS geolocation dataset, called MOD03, includes eight Earth location data fields, e.g. geodetic latitude and longitude, height above the Earth ellipsoid, satellite zenith angle, satellite azimuth, range to the satellite, solar zenith angle, and solar azimuth. Similarly Earth location algorithms are widely used in modeling and geometrically correct image data from the Land Remote Sensing Satellite (Landsat) Multispectral Scanner (MSS), Landsat Thematic Mapper (TM), System pour l'Observation de la Terre (SPOT), and Advanced Very High Resolution Radiometer (AVHRR) missions.

Date in 2003	**Day of the Year (DOY)**	**Overpass time (UTC)**	**Image Quality (% clouds)**
March 23rd	*82*	*17:05*	*18*
March 31st	*90*	*17:55*	*15*
April 1st	*91*	*17:00*	*18*
September 6th	*249*	*17:10*	*6*
September 19th	*262*	*16:40*	*23*
October 12th	*285*	*16:45*	*9*
October 19th	*292*	*16:50*	*6*

Table 3. Date, Day of the Year, overpass time and image quality of the seven study days.

MOD11 is the Land Surface Temperature (LST), and emissivity product, providing per-pixel temperature and emissivity values. Average temperatures are extracted in Kelvin with a day/night LST algorithm applied to a pair of MODIS daytime and nighttime observations. This method yields 1 K accuracy for materials with known emissivities, and the view angle information is included in each LST product. The LST algorithms use other MODIS data as input, including geolocation, radiance, cloud masking, atmospheric temperature, water vapor, snow, and land cover. These products are validated, meaning that product uncertainties are well defined over a range of representative conditions. The theories behind this product can be found in Wan (1999), available at http://modis.gsfc.nasa.gov/data/atbd/atbd_mod11.pdf.
In particular, MODIS Atmospheric Profile product consists on several parameters: total ozone burden, atmospheric stability, temperature and moisture profiles, and atmospheric water vapor. All of these parameters are produced day and night at 5×5 km pixel resolution. There are two MODIS Atmosphere Profile data product files: MOD07_L2, containing data collected from the Terra platform and MYD07_L2 collecting data from Aqua platform. The MODIS temperature and moisture profiles are defined at 20 vertical levels. A simultaneous direct physical solution to the infrared radiative-transfer equation in a cloudless sky is used. The profiles are also utilized to correct for atmospheric effects for some of the MODIS products (e.g., sea-surface temperature and LST, ocean aerosol properties, etc) as well as to characterize the atmosphere for global greenhouse studies. Temperature and moisture profile retrieval algorithms are adapted from the International TIROS Operational Vertical Sounder (TOVS) Processing Package (ITPP), taking into account MODIS' lack of stratospheric channels and far higher horizontal resolution. The profile retrieval algorithm

requires calibrated, navigated, and co-registered 1-km field of the view (FOV) radiances from MODIS channels 20, 22-25, 27-29, and 30-36. The atmospheric water vapor is most directly obtained by integrating the moisture profile through the atmospheric column. Data validation was conducted by comparing results from the Aqua platform with *in situ* data (Menzel et al., 2002). In the present study, air temperature and dew point temperature at 1000 hPa level are used to calculate the vapor pressure deficit. Also the temperatures are assumed to be homogenous over the 5x5 km grid.

5. Results

In this section, the results are divided in two parts. The results of variables and parameters needed to apply the CR models are presented in first place, followed by a comparison of results between equations (7) and (8).

5.1 Variables calculation

In order to apply Bouchet´s and Granger´s CR, Rn, G and F for each pixel of every image of the study area must be computed. The other parameters, Δ and γ, can be assumed constant for the entire region. Alternatively, they can be estimated with spatially distributed information of Ta over the region. The constants α and k are assumed equal to 1.26 and 2, respectively.

The Rn maps were estimated with the methodology published by Bisht et al. (2005), which provides a spatially consistent and distributed Rn map over a large domain for clear sky days. With this method, Rn can be evaluated in terms of its components of downward and upward short wave radiation fluxes, and downward and upward long wave radiation fluxes. Several MODIS data products are utilized to estimate every component. Details of these calculations for the study days presented in this work can be found in Bisht et al. (2005), from where we took the Rn maps.

Soil heat fluxes G were calculated according to Moran et al. (1989) with the daily Normalized Difference Vegetation Index (NDVI) maps (Kogan et al., 2003), calculated with MOD021KM products. The equations used are

$$G = 0.583 \text{ Rn } e^{(2.13 * NDVI)} \qquad \text{for NDVI} > 0 \tag{11}$$

$$G = 0.583 \text{ Rn} \qquad \text{for NDVI} \leq 0 \tag{12}$$

The slope of the SVP curve, Δ, was calculated at Ta using Buck's equation *(Buck, 1981*) and the MODIS Ta product.

In order to determine F, a methodology to estimate Tu is needed. By definition, different types of soils and water content would render different Tu values. Here, it is proposed to estimate the variable Tu from the SVP curve. It can be assumed that e_s is larger or equal to e_a and lower or equal to e^*_s, thus Tu must lie between Ts and Td.

The first derivative of the SVP curve at Ts and at Td represents the slope of the curve between those points. It can also be computed from the linearized SVP curve between the intervals [Tu,Ts] and [Td,Tu], which are symbolized as Δ_1 and Δ_2, respectively. Thus, an expression for Tu is derived from a simple system of two equations with two unknowns, as follows,

$$T_u - \frac{\left(e_s^* - e_a\right) - \Delta_1 Ts + \Delta_2 Td}{\Delta_2 - \Delta_1} \tag{13}$$

There are many published SVP equations that can be used to obtain the derivative of e as function of the temperature. Here, Buck's formulation (Buck, 1981) was chosen for its simple form (equation 14),

$$e = 6.1121 \exp\left(\frac{17.502\ T}{240.97 + T}\right) \tag{14}$$

where "e" is water vapor pressure [hPa] and T is temperature [°C]. Thus, the first derivative of equation 14 is computed at Td and Ts to estimate Δ_1 and Δ_2 in equation (13).

$$\frac{de}{dT} = \left[\frac{4217.45694}{(240.97 + \mathrm{T})^2}\right] * 6.1121 \exp\left(\frac{17.502\ \mathrm{T}}{240.97 + T}\right) \tag{15}$$

The estimation of Tu could be improved by introducing another surface variable, such as soil moisture or any other surface variable that accounts for the surface wetness. However, in order to demonstrate the strength of the CR models, the Tu calculation is kept simple, with minimum data requirements. It is recognized, however, that this calculation simplifies the physical process and may introduce errors and uncertainties to the F ratio.
Figure 5 shows Rn maps obtained for April 1st, 2003 as an example of what can be expected in terms of spatial resolution with Bisht et al. methodology. Figure 6 displays Tu map for the same date obtained with the MOD07 spatial resolution (5x5 km).

5.2 Comparison of the CR models

The results obtained from equations (7) and (8) are compared to demonstrate the strength of the complementary relationship. The contrasted results were computed assuming k=2, α=1.26, γ=0.67 hPa/C, Δ was obtained with Ta maps, estimating F as proposed in Venturini et al. (2008). The resulting ET estimates are shown in Table 4, where average root mean square errors (RMSEs) and biases are about 25 Wm^{-2}, indicating that equation (7), obtained with Bouchet`s complementary model, would lead to larger ET estimates. However, only the "ground truth" would tell which equation is more precise. In this case, the ground truth is considered to be the ground measurements of ET described in section 4.2. Then, observed ET values were compared with the results obtained using equations (7) and (8), (see Figure 7). The overall RMSE is about 52.29 and the bias (Observed-Bouchet) is -37.90 Wm^{-2}. For Granger`s CR, the overall RMSE and bias (Observed-Granger) are 33.89 and -10.96 Wm^{-2} respectively, with an R^2 of about 0.79.

	RMSE	BIAS (Bouchet-Granger)	R^2
DOY82	5.42	0.91	0.990
DOY90	7.38	0.86	0.993
DOY91	13.70	13.01	0.983
DOY 249	31.74	31.56	0.995
DOY 262	25.51	25.33	0.991
DOY 285	26.79	26.40	0.990
DOY 292	28.24	28.11	0.999

Table 4. ET(Wm^{-2}) comparison between Bouchet´s and Granger´s CR.

From Table 4 it can be concluded that Bouchet's simplification results in larger ET estimates, with biases up to approximately 32 Wm^{-2}, than those obtained with Granger`s CR. From Figure 7 it can be seen that Bouchet`s CR overestimates ground observations as well.
Ramírez et al. (2005) derived the value of Bouchet´s k parameter from ground data. The authors presented evidences of the complementary relationship from independent measurements of ET and Epot. Then, k values were calculated for different hypothesis. These authors reported a mean k of about 2.21 and a k variance equal to 0.07 using uncorrected pan evaporation data as a surrogate of Epot.
In this chapter, equations (7) and (8) are equated and k calculated for instantaneous ET values. Thus,

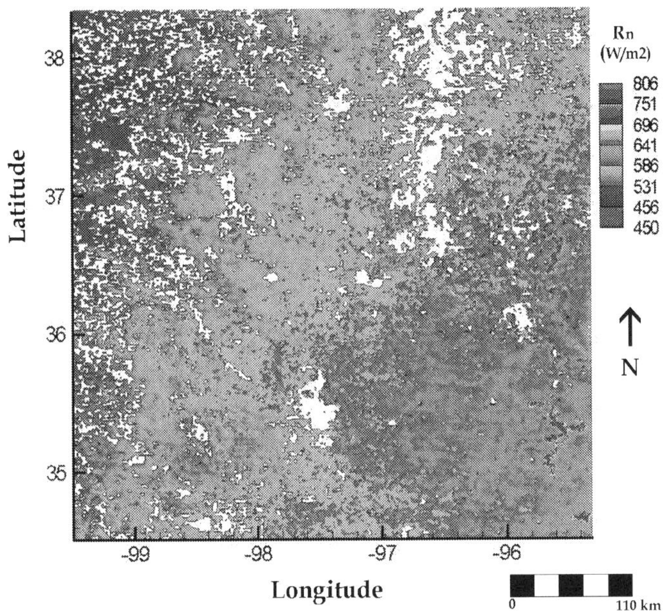

Fig. 5. Net radiation map of the SGP for April 1st, 2003

$$\frac{F\Delta}{F\Delta+\gamma}=\frac{kF\Delta}{(F+1)(\Delta+\gamma)} \tag{16}$$

$$k=\frac{(F+1)(\Delta+\gamma)}{F\Delta+\gamma} \tag{17}$$

Bouchet's coefficient k was calculated for each pixel in every day. The overall mean k value is 2.341, with an overall minimum of 1.784 and a maximum of 2.710, standard deviations varying from 0.025 to 0.078. These results are close to those reported by Ramírez et al. (2005).

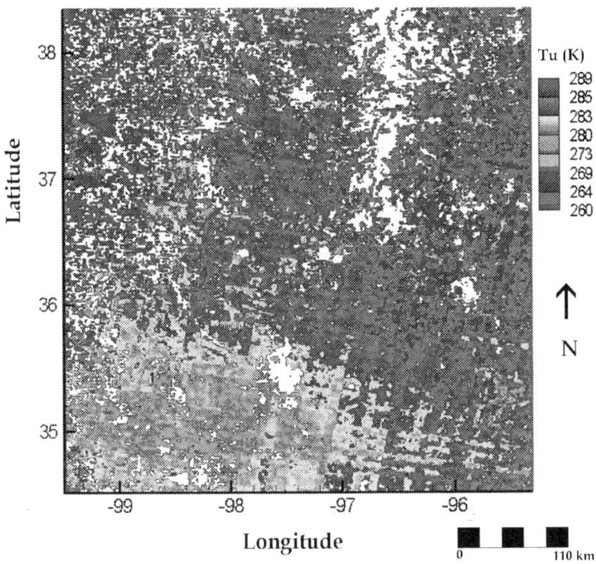

Fig. 6. Tu map of the SGP for April 1st, 2003

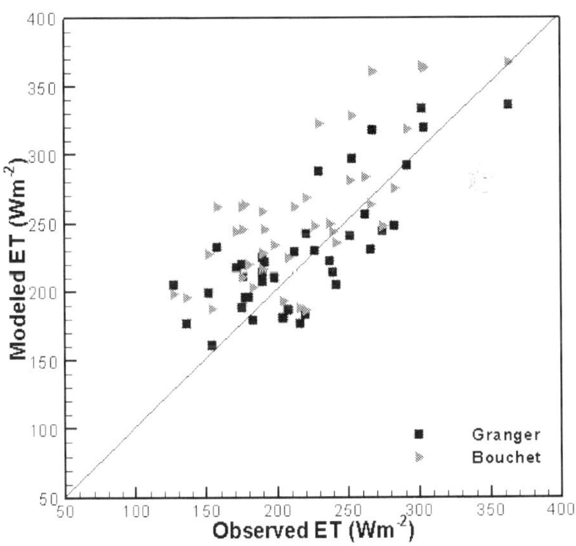

Fig. 7. Comparison between Bouchet's and Granger`s complementary models against ground measurements

Both complementary models yield similar ET estimates, however Granger´s model lead to more accurate results than Bouchet's method. The slope of the SVP curve at the air temperature sets a k value slightly different from 2.

6. Spatial and temporal scales considerations

The complementary theory assumes a surface without advection influences and so does the regional evapotranspiration concept (Penman, 1948; Priestly & Taylor, 1972; Brutsaert & Stricker 1979). In fact, in his original work, Bouchet (1963) described five scales implicated in the oasis effect (see Table 5). Therefore for each scale of heterogeneity (s), we can define the oasis effects that give the lateral energy exchange of Q_1, Q_2, Q_3, Q_4, Q_5. In the development of his theory he assumed that only Q_3 is variable with ET while Q_4 and Q_5 are not affected by changes of ET and Epot associated with water availability. For the other two scales, s_1 and s_2, Q_1 and Q_2 are not involved in the complementary relationship. Bouchet´s experiment established an energy balance over 24 hours, avoiding taking into account the phenomena of accumulation and restoration of heat during the day and night phases. These particular assumptions left smaller time and space scales out of the CR, therefore a review of the scales of applicability of the CR might be interesting.
The "evaporation paradox" mentioned by Brutsaert & Parlange (1998) refers to the seemingly opposing trends observed between pan evaporation and actual evaporation. The authors suggested that the paradox is solved in the CR framework.
The usefulness of the CR for understanding global scale in climate studies have been analized by Brutsaert & Parlange (1998), Szilagyi (2001) and Hobbins et al. (2001), among others. Szilagy & Josza, (2009), coupled Bouchet´s CR with a long-term water-energy balance based on considerations of the precipitation time series and the soil water balance. The authors show that important ecosystem characteristics, such as the maximum soil water storage, can be derived from this "long-term" application of the CR. The scales shown in Table 5 seem to be compatible with those used in the aforementioned works. Nonetheless, the applicability of the CR at small scales is not evident from Bouchet´s publication.
Crago & Crowley (2005) evaluated the complementary relationship at relatively small temporal scales (10 to 30 min) using data from meteorological stations in different grassland sites. The authors demonstrated that the CR holds true also at small scales. Kahler and Brutsaert (2006) used properly scaled data of daily ET and daily pan evaporation observed at two experimental sites to demonstrate the validity of the CR. The CR at daily scales was confirmed by this research. The authors argue that for unscaled daily data of pan evaporation the CR may not be noticeable.

Scale (symbol in the text)	Timescales	Spatial Scales	Effects of oasis corresponding
Molecular - s_1	10^{-9} second	few hundred meters	Q_1
Turbulent - s_2	1 second to some minutes	few hundred meters	Q_2
Convection and related movements - s_3	10 minutes to a few hours	few kilometers	Q_3
Cyclonic - s_4	3 to 4 hours	1000 a 2000 kilometers	Q_4
Global - s_5	10 to 30 hours	5000 to 10000 km	Q_5

Table 5. Translation of Table 1 published by Bouchet in 1963

In a more practical way, the method proposed by Venturini et al. (2008) corrects the ET from a saturated surface with the local surface-atmosphere conditions at the pixel scale. The absence of regional assumptions makes the method applicable to a wide range of spatial scales even though the background of their method is Granger´s CR. Venturini´s method has been applied with instantaneous data, i.e. remotely sensed data with MODIS. The comparison between observed and estimated ET values yields errors of about 15% of observed instantaneous ET(Venturini et al., 2011).

7. References

Bastiaanssen, W.G.M., Menenti, M.A, Feddes, R.A. & Hollslag, A.A.M. (1998). A remote sensing surface energy balance algorithm for land (SEBAL) 1. Formulation. *Journal of Hydrology,* 212, 13, pp. 198-212, ISSN 0022-1694.

Bastiaanssen, W.G.M. (2000). SEBAL-based sensible and latent heat fluxes in the irrigated Gediz Basin, Turkey. *Journal of Hydrology,* 229, pp. 87-100, ISSN 0022-1694.

Bisht, G., Venturini, V., Jiang, L. & Islam, S. (2005). Estimation of Net Radiation using MODIS (Moderate Resolution Imaging Spectroradiometer) Terra Data for clear sky days. *Remote Sensing of Environment,* 97, pp. 52-67, ISSN 0034-4257.

Bouchet, R.J. (1963). Evapotranspiration rèelle et potentielle, signification climatique. *International Association of Scientific Hydrology,* 62, pp. 134-142, ISSN 0262-6667.

Brutsaert, W., & Stricker, H. (1979). An advection-aridity approach to estimate actual regional evapotranspiration. *Water Resources Research,* 15,2, pp. 443–450, ISSN 0043-1397.

Brutsaert W., & Parlange M.B. (1998) Hydrologic cycle explains the evaporation paradox. *Nature,* 396, pp. 30, ISSN 0028-0836.

Buck, A.L. (1981). New equations for computing vapor pressure and enhancement factor. *Journal of Applied Meteorology,* 20, pp. 1527-1532, ISSN 0894-8763.

Calvet, J.C., Noilhan , J. & Besseoulin, P. (1998). Retrieving the root zone soil moisture from surface soil moisture or temperature estimates: A feasibility study on field measurements. *Journal of Applied Meteorology,* 37, pp. 371-386, ISSN 0894-8763.

Carlson, T.N., Gillies, R.R., & Schmugge, T. J. (1995). An interpretation of methodologies for indirect measurement of soil water content. *Agricultural and Forest Meteorology,* 77, pp. 191-205, ISSN 0168-1923.

Courault, D., Seguin, B. & Olioso, A. (2005). Review to estimate Evapotranspiration from remote sensing data: Some examples from the simplified relationship to the use of mesoscale atmospheric models. *Irrigation and Drainage Systems,* 19, pp. 223-249, ISSN 0168-6291.

Crago, R., & Crowley, R. (2005). Complementary relationship for near-instantaneous evaporation. *Journal of Hydrology,* 300, pp. 199-211, ISSN 0022-1694.

Crago, R., Hervol, N., & Crowley, R. (2005). A complementary evaporation approach to the scalar roughness length. *Water Resources Research,* 41, W06017, ISSN 0043-1397.

Crago, R.D., Qualls R.J., & Feller M. (2010) A calibrated advection-aridity evaporation model requiring no humidity data. *Water Resources Research,* 46, W09519, doi:10.1029/2009WR008497, (September, 2010), ISSN. 0043-1397.

Fritschen, L., & Simpson, J. R. (1989). Surface energy and radiation balance systems: General description and improvements. *Journal of Applied Meteorology 28,* 680-689. ISSN 0894-8763.

Granger, R.J. (1989a). An examination of the concept of potential evaporation. *Journal of Hydrology*, 111, pp. 9-19, ISSN 0022-1694.

Granger, R.J., & Gray, D.M. (1989). Evaporation from natural nonsaturated surfaces. *Journal of Hydrology*, 111, pp. 21-29, ISSN 0022-1694.

Granger, R.J. (1989b). A complementary relationship approach for evaporation from nonsaturated surfaces. *Journal of Hydrology*, 111, pp. 31-38, ISSN 0022-1694

Granger, R.J., & Gray, D.M. (1990). Examination of Morton's CRAE model for estimating daily evaporation from field-sized areas. *Journal of Hydrology*, 120, pp. 309-325, ISSN 0022-1694.

Halldin, S & Lindroth, A. (1992). Errors in net radiometry: Comparison and evaluation of six radiometer designs. *Journal of Atmospheric Oceanic Technology*, 9, 762-783, ISSN 0739-0572.

Han, S., Hu, H., Yang, D., & Tian, F. (2011). A complementary relationship evaporation model referring to the Granger model and the advection–aridity model. *Hydrological Processes*, 25, 8, doi:10.1002/hyp.7960, ISSN 0885-6087.

Heilman, J.L. & Brittin, C. L. (1989). Fetch requirements for Bowen ratio measurements of latent and sensible heat fluxes. *Agricultural and Forest Meteorology*, 44, 261-273, ISSN 0168-1923.

Hobbins, M.T., & Ramírez, J.A. (2001). The complementary relationship in estimation of regional evapotranspiration: An enhanced advection-aridity model, *Water Resources Research*, 37,5, pp. 1389-1403, ISSN 0043-1397.

Hobbins, M.T., Ramírez, J.A., Brown T.C. & Classens L.H.J.M. (2001). The complementary relationship in estimation of regional evapotranspiration: The complementary relationship areal evapotranspiration and advection-aridity models, *Water Resources Research*, 37,5, pp. 1367-1487, ISSN 0043-1397.

Holwill, C.J., & Stewart, J.B. (1992). Spatial variability of evaporation derived from Aircraft and ground-based data. *Journal of Geophysical Research*, 97, D17, pp. 19061-19089, ISSN 0148-0227

Jackson, R.D., Reginato, R.J., & Idso, S.B. (1977). Wheat canopy temperature: A practical tool for evaluating water requirements. *Water Resources Research*, 13, pp. 651-656, ISSN 0043-1397

Jiang, L. & Islam, S. (2001). Estimation of surface evaporation map over southern Great Plains using remote sensing data. *Water Resources Research*, *37*(2), 329-340. ISSN 0043-1397.

Kahler, D. M. & Brutsaert, W. (2006). Complementary relatinship between daily evaporation in the environment and pan evaporation. *Water Resources Research*, 42, W05413, doi:10.1029/2005WR004541, ISSN 0043-1397.

Kogan, F., Gitelson, A., Zakarin, E., Spivak, L. & Lebed, L., (2003). AVHRR-Based Spectral Vegetation Index for Quantitative Assessment of Vegetation State and Productivity: Calibration and Validation. *Photogrammetric Engineering & Remote Sensing*, 69, 8, (August 2003) pp. 899-906, ISSN 0099-1112.

Lhomme J.P., & Guilione L. (2006). Comments on some articles about the complementary relationship, Discussion. *Journal of Hydrology*, 323, pp. 1-3, ISSN 0022-1694.

Menzel, W.P., Seemann, S.W., Li, J., & Gumley, L.E. (2002). *MODIS Atmospheric Profile Retrieval Algorithm Theoretical Basis Document, Version 6. Reference Number: ATBD-MOD-07*. NASA. http://modis.gsfc.nasa.gov/data/atbd/atbd_mod07.pdf.

Monin, A.S. & Obukhov, A.M., (1954). Osnovnye zakonomernosti turbulentnogo peremesivanija v prizemnom sloe atmosfery. *Trudy Geofizicheskogo Instituta Akademiya Nauk SSSR*, 24, 151, pp. 163-187.

Monteith, J.L. & Unsworth, M. (1990). *Principles of Environmental Physics* (2nd edition), Butterworth-Heinemann, ISBN: 071312931X, Burlington-MA- USA.

Moran, M.S., Jackson, R.D., Raymond, L.H, Gay, L.W. & Slater, P.N. (1989). Mapping surface energy balance components by combining LandSat thematic mapper and ground-based meteorological data, *Remote Sensing of Environment*, 30, pp.77-87, ISSN 0034-4257.

Morton, F. I. (1969). Potential evaporation as manifestation of regional evaporation. *Water Resources Research*, 5, pp. 1244-1255, ISSN 0043-1397.

Morton, F.I. (1983). Operational estimates of areal evapotranspiration and their significance to the science and practice of hydrology. *Journal of Hydrology*, 66, 1-76, ISSN 0022-1694.

Nishida, K., Nemani, R.R., Running, S.W. & Glassy, J.M. (2003). An operational remote sensing algorithm of land evaporation. *Journal of Geophysical Research*, 108, D9, 4270, doi:10.1029/2002JD002062, ISSN 0148-0227.

Noilhan, J. & Planton, S. (1989). GCM gridscale evaporation from mesoscale modelling. *Journal of Climate*, 8, pp. 206-223, ISSN 0894-8755

Norman, J.M., Kustas, W.P. & Humes, K.S. (1995). Sources approach for Estimating soil and vegetation energy fluxes in observations of directional radiometric surface temperature. *Agricultural Forest and Meteorology*, 77, pp. 263-293, ISSN 0168-1923.

Ozdogan M., Salvucci G.D., & Anderson B.T. (2006). Examination of the Bouchet-Morton complementary relationship using a mesoscale climate model and observations under a progressive irrigation scenario, *Journal of Hydrometeorology*, 7, pp. 235-251, ISSN 1525-755X

Penman, H.L. (1948). Natural evaporation from open water, bare soil and grass. *Proceedings of the Royal Society of London, Series A*, 193,1032,(April, 1948), pp. 120-145, ISSN 1471-2946.

Price, J.C. (1990). Using spatial context in satellite data to infer regional scale evapotranspiration. *IEEE Transactions on Geoscience and Remote Sensing*, 28, 5, pp. 940-948, ISSN 0196-2892

Priestley, C.H.B. & Taylor, R.J. (1972). On the Assessment of Surface Heat Flux and Evaporation Using Large-Scale Parameters. *Monthly Weather Review*, 100, pp. 81–92, ISSN 0027-0644.

Ramírez, J.A., Hobbins, M.T. & Brown T. (2005). Observational evidence of the complementary relationship in regional evaporation lends strong support for Bouchet's hypothesis. *Geophysical Research Letters*, 32, L15401, doi:10.1029/2005GL023549, ISSN 0094-8276.

Rivas, R. & Caselles, V. (2004). A simplified equation to estimate spatial reference evaporation from remote sensing-based surface temperature and local meteorological data. *Remote Sensing of Environment*, 83, pp. 68-76, ISSN 0034-4257.

Seguin, B., Assad, E., Fretaud, J.P., Imbernom, J.P., Kerr, Y., & Lagouarde, J.P. (1989). Use of meteorological satellite for rainfall and evaporation monitoring. *International Journal of Remote Sensing*, 10, pp. 1001-1017, ISSN 0143-1161.

Su, B. (2002). The surface energy balance system (SEBS) for estimation of turbulent heat fluxes. *Hydrology and Earth System Sciences*, 6, pp. 85-99, ISSN 1027-5606.

Sugita, M., Usui, J., Tamagawa, I. & Kaihotsu, I. (2001).Complementary relationship with a convective boundary layer to estimate regional evaporation. *Water Resources Research*, 37,2, pp. 353-365, ISSN 0043-1397.

Szilagyi J., (2001). On Bouchet's complementary hypothesis. *Journal of Hydrology*, 246, pp. 155-158, ISSN 0022-1694.

Szilagyi J. (2007). On the inherent asymmetric nature of the complementary relationship of evaporation, *Geophysical Research Letters*, 34, L02405, ISSN 0094-8276.

Szilagyi J., & Jozsa J. (2008). New findings about the complementary relationship based evaporation estimation methods. *Journal of Hydrology*, 354, pp. 171- 186, ISSN 0022-1694.

Szilagyi J., & Jozsa J. (2009). Analytical solution of the coupled 2-D turbulent heat and vapor transport equations and the complementary relationship of evaporation . *Journal of Hydrology*, 372, pp. 61–67, ISSN 0022-1694.

van Bavel, C.H.M. (1966). Potential evaporation: The combination concept and its experimental verification. *Water Resources Research* , 2, pp. 455-467, ISSN 0043-1397.

Venturini, V., Islam, S., & Rodríguez, L., (2008). Estimation of evaporative fraction and evapotranspiration from MODIS products using a complementary based model. *Remote Sensing of Environment*. 112, pp. 132-141, ISSN 0034-4257.

Venturini, V., Rodriguez L., & Bisht G. (2011). A comparison among different modified Priestley and Taylor´s equation to calculate actual evapotranspiration". *International Journal of Remote Sensing*, In Press, ISSN 0143-1161

Xu, C.Y., & Singh, V.P. (2005). Evaluation of three complementary relationship evapotranspiration models by water balance approach to estimate actual regional evapotranspiration in different climatic regions. *Journal of Hydrology*, 308, pp. 105-121, ISSN 0022-1694.

Wan, Z. (1999). *MODIS Land-Surface Temperature Algorithm Basis Document (LST ATBD)*, version 3.3, NASA, www.icess.ucsb.edu/modis/atbd-mod-11.pdf.

3

Evapotranspiration Estimation Using Soil Water Balance, Weather and Crop Data

Ketema Tilahun Zeleke and Leonard John Wade
School of Agricultural and Wine Sciences, EH Graham Centre for Agricultural Innovation, Charles Sturt University Australia

1. Introduction

The rise in water demand for agriculture, industry, domestic, and environmental needs requires sagacious use of this limited resource. Since agriculture (mainly irrigation) is the major user of water, improving agricultural water management is essential. Efficient agricultural water management requires reliable estimation of crop water requirement (evapotranspiration). Evapotranspiration (ET) is the transfer of water from the soil surface (evaporation) and plants (transpiration) to the atmosphere. ET is a critical component of water balance at plot, field, farm, catchment, basin or global level. From an agricultural point of view, ET determines the amount of water to be applied through artificial means (irrigation). Reliable estimation of ET is important in that it determines the size of canals, pumps, and dams. The use of the terms 'reference evapotranspiration', 'potential evapotranspiration', 'crop evapotranspiration', 'actual evapotranspiration' in this chapter is based on FAO-56 (FAO Irrigation and Drainage publication No 56) (Allen et al., 1998).

There are different methods of determining evapotranspiration: direct measurement, indirect methods from weather data and soil water balance. These methods can be generally classified as empirical methods (eg. Thornthwaite, 1948; Blancy and Criddle, 1950) and physical based methods (eg. Penman, 1948; Montheith, 1981 and FAO Penman Montheith (Allen et al., (1998)). They vary in terms of data requirement and accuracy. At present, the FAO Penman Montheith approach is considered as a standard method for ET estimation in agriculture (Allen et al., 1998). A case study from a semiarid region of Australia will be used to demonstrate ET estimation for a canola (*Brassica napus* L.) crop using soil water balance and crop coefficient approaches. Daily rainfall data, soil moisture measurement data using neutron probe, and AquaCrop (Steduto et al., 2009) -estimated deep percolation below the crop root zone will be used to determine actual evapotranspiration of the crop using soil water balance. Reference evapotranspiration ET_o will be determined using FAO *ET_o calculator* (Raes, 2009). Crop canopy cover measured using a handheld *GreenSeeker™* and expressed as normalized difference vegetation index (NDVI) will be used to interpret evolution of evapotranspiration during the growing season (life cycle) of the canola crop.

2. Field experiment

2.1 Description of study area and field experiment

The study area is in Wagga Wagga, New South Wales (Australia). Wagga Wagga, referred to as 'the capital of Riverina', is located in the Riverina region of NSW. The Riverina extends from the foot hills of the Great Dividing Range in the east to the flat and dry inland plains in the west. Agriculture in the Riverina is significantly diversified with dry land farming of winter cereals and irrigation in Murrumbidgee and Colleambally irrigation areas. It has a Mediterranean type climate with a mixed farming system of winter cereal crops, summer crops, and pastures grazing lands. In addition to the major grain crops of rice, canola, wheat, and maize, the area also produces a quarter of NSW fruit and vegetable production (RDA, 2011). The Riverina region is characterized by the semiarid climate, with hot summers and cool winters (Stern et al., 2000). Seasonal temperature varies little across the region. More consistent rainfall occurs in winter months. Mean annual temperature is 15-18°C. January is the hottest month of the year while July is the coolest. Mean annual rainfall varies from 238 mm in the west to 617 mm in the east. Long term and 2010 mean monthly rainfall, reference evapotranspiration, and temperature are presented in Fig. 1. Rainfall in 2010 was much higher than the long term average while evapotranspiration in 2010 was lower than the long term average.

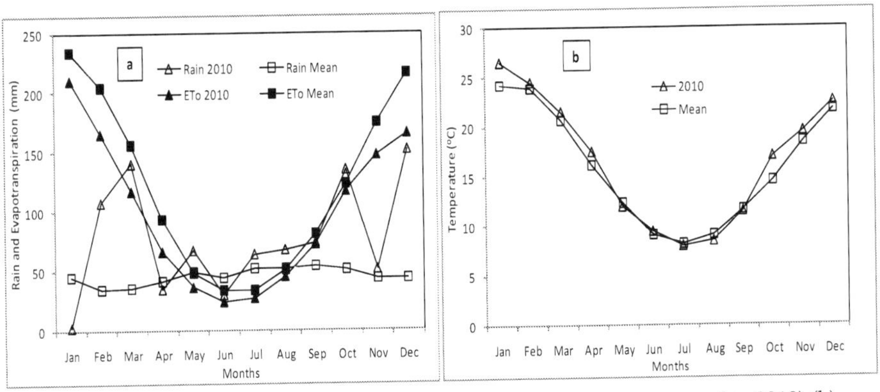

Fig. 1. (a) Rain and reference evapotranspiration ET_o (long term average and in 2010) (b) Monthly average temperature (long term average and in 2010) at Wagga Wagga, NSW (Australia).

A field experiment was carried out during the growing season of 2010 at canola field experimental site of Wagga Wagga Agricultural Research Institute located at Wagga Wagga (35°03'N; 147°21'E; 235 m asl), NSW (Australia). There was enough rainfall (930 mm) in contrast to long term average of 522 mm in 2010 to provode ideal growing conditions. A popular variety of canola (Hyola50) was sown on 30 April 2010. The experiment was conducted on a 24 m x 24 m area. There were 24 plots, 12 experimental plots and 12 buffer plots. The plots were 6 m long with 1 meter buffer on either end. Plot width was 1.8 m with a 0.5 m walking strip between plots for data collection.

About a month before the experimental season, neutron probe access tubes were installed to a depth of 1.5 m for soil moisture measurement. Two access tubes were installed at 2 m from either end of the plot and 2 m from each other. Soil moisture content was measured at 15, 30, 45, 60, 90, and 120 cm depths every two weeks. The probe was calibrated using gravimetric soil moisture measurements done when access tubes were installed on site.

2.2 Weather data

Daily weather data (rainfall, minimum and maximum temperature, solar radiation, relative humidity, and wind speed) were collected from the meteorological station of the Wagga Wagga Agricultural Institute located adjacent to the experimental site. Out of the total annual rainfall of 930 mm, the amount or proportion (in percentage) during the canola growing season (May to November) was 514 mm (53%) while the long term average was 333 mm (64% of the long term average of 522 mm). Monthly average maximum and minimum temperature was 26°C and 3°C respectively. Reference evapotranspiration ET_o was calculated using the procedure described in the FAO Irrigation and Drainage Paper 56 (Allen et al., 1998) with the help of the program FAO *ET_o Calculator* (Raes, 2009).

2.3 Soil hydraulic characteristics

A 1.5m x 1.5m x 1.5m soil trench was dug for soil texture, field capacity (θ_{FC}), and wilting point (θ_{WP}) determination. Soil samples were retrieved from 0-30, 30-60, 60-90, and 90-120 cm depths for soil texture, θ_{FC}, and θ_{WP} determination using standard laboratory procedures hydrometer and pressure plate apparatus apparatus.

2.4 Crop parameters

The following crop phenological stages were recorded during the growing season: planting date, 90% emergence, beginning and end of flowering, senescence and maturity. The canopy cover was measured using *GreenSeeker™*, an Optical Sensor Unit (NTech Industries, Inc., USA). *GreenSeeker™*, is a handheld tool that determines Normalized Difference Vegetative Index (NDVI), is an integrated optical sensing and application system that measures green crop canopy cover.

3. Soil water balance method

Rain or irrigation reaching a unit area of soil surface, may infiltrate into the soil, or leave the area as surface runoff. The infiltrated water may (a) evaporate directly from the soil surface, (b) taken up by plants for growth or transpiration, (c) drain downward beyond the root zone as deep percolation, or (d) accumulate within the root zone. The water balance method is based on the conservation of mass which states that change in soil water content ΔS of a root zone of a crop is equal to the difference between the amount of water added to the root zone, Q_i, and the amount of water withdrawn from it, Q_o (Hillel, 1998) in a given time interval expressed as in Eq. (1).

$$\Delta S = Q_i - Q_o \qquad (1)$$

Eq. (1) can be used to determine evapotranspiration of a given crop as follows

$$ET = P + I + U - R - D - \Delta S \qquad (2)$$

where ΔS = change in root zone soil moisture storage, P = Precipitation, I = Irrigation, U = upward capillary rise into the root zone, R = Runoff, D = Deep percolation beyond the root

zone, ET = evapotranspiration. All quantities are expressed as volume of water per unit land area (depth units).
In order to use Eq. (2) to determine evapotranspiration (ET), other parameters must be measured or estimated. It is relatively easy to measure the amount of water added to the field by rain and irrigation. In agricultural fields, the amount of runoff is generally small so is often considered negligible. When the groundwater table is deep, capillary rise U is negligible. The most difficult parameter to measure is deep percolation D. If soil water potential and moisture content are monitored, D can be estimated using Darcy's Principle. In this study, deep percolation estimated using AquaCrop (Raes et al., 2009), was adopted. Runoff R was also estimated using AquaCrop following USDA curve number approach (Hawkins et al., 1985). The change in soil water storage ΔS is measured using specialized instruments such as neutron probe and time-domain reflectrometer.

4. Crop coefficient method

4.1 Introduction

The crop coefficient approach relates evapotranspiration from a reference crop surface (ET_o) to evapotranspiration from a given crop (ET_c) through a coefficient. Estimation of crop water requirement from weather and crop data is a simpler and cost effective method compared to other methods such as soil water balance method. In this method, potential evapotranspiration of a crop is presumed to be determined by the evaporative demand of the atmosphere and crop characteristics. Evaporative demand of the air is determined as the evapotranspiration from a reference crop. The reference crop is a hypothetical crop (grass or alfalfa) with specific characteristics such as crop height of 0.12 m and albedo of 0.23 (Allen et al., 1998). Penman (1956) defined reference evapotranspiration as "the amount of water transpired in unit time by a shorter green crop, completely shading the ground, of uniform height and never short of water." It is a useful standard of reference for the comparison of different regions and of different measured evapotranspiration values within a given region. As such, ET_o is a climatic parameter expressing the evaporation power of the atmosphere independent of crop type, crop development and management practices (Allen et al., 1998). FAO Penman Montheith approach is considered as the standard method. In this method, reference evapotranspiration ET_o is estimated from weather data as given in Eq. (3).

$$ET_o = \frac{0.408\Delta(R_n - G) + \gamma \frac{900}{T+273} u_2 (e_s - e_a)}{\Delta + \gamma(1 + 0.34u_2)} \quad (3)$$

where ET_o = reference evapotranspiration (mm/day); Rn = net radiation at the crop surface (MJ/m^2 day); G = soil heat flux density (MG/m^2 day); T = air temperature at 2 m height (°C); u_2 = wind speed at 2 m height (m/s); e_s= saturation vapor pressure (kPa); e_a = actual vapor pressure (kPa); e_s-e_a = saturation vapor pressure deficit (kPa); Δ = slope vapor pressure curve (kPa/°C); γ = psychrometric constant (kPa/°C).
Reference evapotranspiration ET_o can be calculated using a spreadsheet or computer programs which are designed for various level of data availability eg. *CROPWAT* (Smith, 1992) and *ET_o Calculator* (Raes, 2009). In this study, the latter program was used. It is important to make clear distinction between reference evapotranspiration ET_o and potential crop evapotranspiration ET_c. The latter is also called maximum crop evapotranspiration.

Evapotranspiration from a given crop grown and managed under standard conditions is called potential crop evapotranspiration ET_c. Standard condition is a disease-free, well-fertilized crops, grown in large fields, under optimum soil water conditions, and achieving full production under the given climatic conditions. ET_o depends evapotranspiration (ET_c) represents the climatic "demand" for water by a given crop. Potential crop depends primarily on the evaporative demand of the air.

4.2 Single crop coefficient method

The single crop coefficient (K_c) method is used to determine soil evaporation and transpiration lumped over a number of days or weeks. The single "time-averaged" K_c curve incorporates averaged transpiration and soil wetting effects into a single K_c factor. The FAO-56 publication divides the crop growth stages into four phenological stages. Initial stage is from planting to 10% ground cover. Development stage is from 10% groundcover to maximum cover. Midseason stage is from the beginning of full cover to the start of senescence. The late season stage is from the start of senescence to full senescence or harvest. The evolution of crop coefficients during these stages is tabulated in FAO-56 for a number of crops including canola. Three coefficients are given for the initial, midseason, and end of season stages as $K_{c\ ini}$, $K_{c\ mid}$, and $K_{c\ end}$ respectively. $K_{c\ ini}$ is assumed to be constant and relatively small (<0.4). The K_c begins to increase during the crop development stage and reaches a maximum value $K_{c\ mid}$ which is relatively constant for most growing and cultural conditions. During the late season period, as leaves begin to age and senesce, the K_c begins to decrease until it reaches a lower value at the end of the growing period equal to $K_{c\ end}$. The K_c during the development is estimated using linear interpolation between $K_{c\ ini}$ and $K_{c\ mid}$. Similarly, K_c during the late season stage is determined using linear interpolation between $K_{c\ mid}$ and $K_{c\ end}$. The value of $K_{c\ ini}$ and $K_{c\ end}$ can vary considerably on a daily basis, depending on the frequency of wetting by irrigation and rainfall. The single crop coefficient method can be used for irrigation planning and design. It is accurate enough for systems with large interval such as surface and set sprinkler irrigation. It is also used for catchment level hydrologic water balance studies (Allen et al., 1998).

In the single crop coefficient method, potential crop evapotranspiration ET_c is estimated from a single crop coefficient (K_c) and reference evapotranspirations ET_o as in Eq. (4).

$$ET_c = ET_0 K_c \tag{4}$$

Eq. (4) gives the potential (maximum) evapotranspiration of the crop when the soil moisture is not limiting. Since localized K_c values are not always available in many parts of the world, the values of K_c as suggested by FAO (Allen et al., 1998) are being widely used to estimate evapotranspiration.

When rainfall amount and irrigation are not sufficient to keep the soil moisture high enough, the soil moisture content in the root zone is reduced to levels too low to sustain the potential crop evapotranspiration ET_c. This results in an evapotranspiration less than the potential, and the plants are said to be under water stress. This evapotranspiration is called actual evapotranspiration (ET_a). In general, the actual evapotranspiration ET_a from various crops will not be equal to the potential value ET_c. Actual evapotranspiration ET_a is generally a fraction of ET_c depending on soil moisture availability. Actual evapotranspiration ET_a from a well-watered crop might generally approach ET_c during the active growing stage, but may fall below during the early growth stage, prior to full canopy coverage, and again toward the end of the growing season as the matured plant starts to dry out (Hillel, 1997). The actual evapotranspiration ET_a is calculated by combining the effects of K_c and soil water stress coefficient (K_s) as shown in Eq. (5).

$$ET_a = ET_o K_c K_s \tag{5}$$

The stress reduction coefficient K_s [0-1] reduces K_c when the average soil water content of the root zone is not high enough to sustain full crop transpiration. The stress coefficient K_s is determined by the amount of moisture the crop depleted from the rootzone of a crop. The amount of water depleted from the rootzone is expressed by root zone depletion D_r, i.e. water storage relative to field capacity. Stress is presumed to initiate when D_r exceeds the readily available water (RAW), Fig. 2. When more than RAW is extracted from the rootzone (D_r >RAW), K_s is expressed (Allen et al., 1998) as

$$K_s = \frac{TAW - D_r}{TAW - RAW} = \frac{TAW - D_r}{(1-p)TAW} \tag{6}$$

Where TAW = total plant available soil water in the root zone (mm), and p = fraction of TAW that a crop can extract from the root zone without suffering water stress. When $D_r \leq$ RAW, K_s =1 indicating no water stress. The total available water in the root zone (TAW, mm) is estimated as the difference between the water content at the field capacity and wilting point

$$TAW = 1000(\theta_{FC} - \theta_{WP})Z_r \tag{7}$$

Where Z_r = effective rooting depth (m); θ_{FC} is soil moisture content at field capacity ($m^3\ m^{-3}$); θ_{WP} is soil moisture content at permanent wilting point ($m^3\ m^{-3}$).

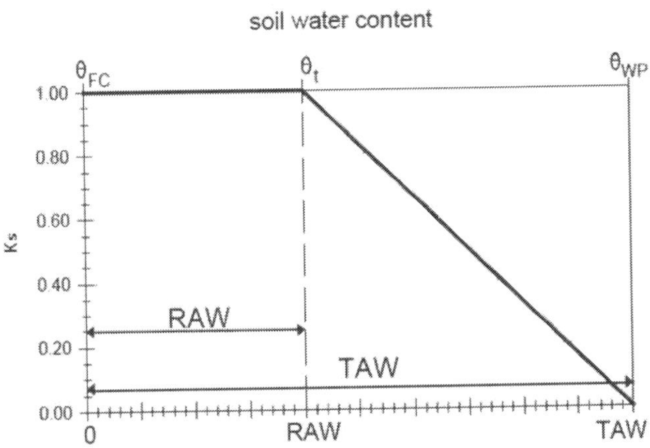

Fig. 2. Schematic of moisture stress coefficient (adapted from Allen et al., 1998).

Readily available water (RAW) is the amount of water which the crop can extract without experiencing stress. It is expressed as

$$RAW = pTAW \tag{8}$$

Soil moisture depletion fraction (p) is the fraction of soil water in the root zone that can be depleted before stress occurred. It varies from crop to crop and also varies at different growth stages of a given crop. Shallow rooted and sensitive crops such as vegetables have low p value while deep rooted and stress tolerant crops have a higher p value.

Canola crop coefficient values given in FAO 56 (Allen et al., 1998) are $K_{c\ ini}$ = 0.35, $K_{c\ mid}$ = 1.0-1.15, $K_{c\ end}$ = 0.35. These values represent K_c for a sub humid climate with RHmin = 45% and wind speed of 2 m/s. To take account for impacts of differences in aerodynamic roughness between crops and the grass reference with changing climate, the $K_{c\ mid}$ and $K_{c\ end}$ values larger than 0.45 must be adjusted using the following equation:

$$K_c = K_{c\ (tab)} + \left[0.04(u_2 - 2) - 0.004(RH_{min} - 45)\right]\left(\frac{h}{3}\right)^{0.3} \tag{9}$$

Where K_c (tab) is the value of K_c taken from Table 12 of Allen et al. (1998); h is the mean plant height during the mid or late season stage (m); RHmin the mean value for daily minimum relative humidity during the mid or late season growth stages (%) for 20%≤RHmin≤ 80%; u_2 is the mean value for daily wind speed at 2 m during the mid season or late season stages (m/s) for 1m/s ≤ u_2 ≤ 6 m/s. In this study, $K_{c\ ini}$ = 0.35, $K_{c\ mid}$ = 1.10, and $K_{c\ end}$ = 0.35 were used. Accordingly, $K_{c\ mid}$ value was adjusted to 1.08 for RHmin = 48%, u_2 = 1.91 m/s, and plant height of 1.0 m during this stage. Since $K_{c\ end}$ was less than 0.4, it was not necessary to adjust it. Once the K_{cb} values for the initial stage, mid season stage, and end-of-season stage were determined, K_{cb} values for development and late season stages were determined using linear interpolation.

4.3 Dual crop coefficient method

The single coefficient method does not separate evaporation and transpiration components of evapotranspiration. The dual crop coefficient approach calculates the actual increase in K_c for each day as a function of plant development and the wetness of the soil surface. It is best for high frequency irrigation such as microirrigation, centre pivots, and linear move systems (Suleiman et al., 2007). The effects of crop transpiration and soil evaporation are determined separately using two coefficients: the basal crop coefficient (K_{cb}) to describe plant transpiration and the soil water evaporation coefficient (K_e) to describe evaporation from the soil surface, Eq (10). AquaCrop determines crop transpiration (T_r) and soil evaporation (E) by multiplying ETo with their specific coefficients K_{cb} and K_e (Eq. 11) (Steduto et al., 2009).

$$K_c = K_{cb} + K_e, \text{ and} \tag{10}$$

$$ET_c = (K_{cb} + K_e)\, ET_o \tag{11}$$

The range of K_{cb} and K_e is [0-1.4]. When soil moisture is limiting, K_{cb} is multiplied by a coefficient K_s which is equal to 1 when D_r≤RAW and declines linearly to zero when all the available water in the rooting zone has been used. Evapotranspiration under such a condition is calculated using Eq. (12).

$$ET_a = (K_s K_{cb} + K_e)\, ET_o \tag{12}$$

Because the water stress coefficient impacts only crop transpiration, rather than evaporation from the soil, the application using Eq. (12) is generally more valid than is application using

Eq. (5) in the single crop coefficient approach. Allen et al. (1998) reported that in situations where evaporation from soil is not a large component of ET_c, use of Eq. (5) will provide reasonable results. The dual coefficient approach can be summarized into the following three steps: Calculate reference evapotranspiration (ET_o) from climatic data using Eq. (3), calculate individual crops potential evapotranspiration ET_c using Eq. (11), and when the soil moisture content is limited, K_{cb} coefficient is multiplied by stress factors K_s to calculate actual evapotranspiration ET_a using Eq. (12).

4.3.1 Basal crop coefficient

The basal crop coefficient K_{cb} is defined as the ratio of ET_c to ET_o when the soil surface layer is dry but where the average soil water content of the rootzone is adequate to sustain full plant transpiration (Bonder et al., 2007). The dual crop coefficient approach uses daily time step and is readily adapted to spreadsheet program. Some models such as AquaCrop (Steduto et al., 2009) determine crop water productivity from the "productive" component of evapotranspiration i.e. transpiration. AquaCrop requires regression of daily values of biomass and crop transpiration to determine crop water productivity. Therefore, transpiration should be measured or estimated.

FAO-56 has tabulated K_{cb} values for a number of crops, including canola, at the initial, mid season, and end of season stages. Since localized K_{cb} values were not available for the study area, the values of K_{cb} suggested by FAO-56 (Allen et al., 1998) were used. For canola these value were $K_{cb\ ini}$ = 0.15, $K_{cb\ mid}$ = 0.95-1.10, and $K_{cb\ end}$ = 0.25. In this study, K_{cb} of 0.15, 1, and 0.25, respectively, for the initial, mid-season, and end of season stages were selected. The growing season of canola vary from 5 months to 7 months in Australia i.e. 150 -210 days depending on the planting date and the weather conditions (rainfall and temperature) during the season. Initial, development, mid-season, and late season stage lengths for canola grown during the 2010 winter season in Wagga Wagga (Australia) were 10, 64, 84, 48 days respectively.

The values for K_{cb} in the FAO-56 table represent values for a sub humid climate with RH_{min} = 45% and wind speed of 2 m/s. To take account for impacts of differences in aerodynamic roughness between crops and the grass reference, the $K_{cb\ mid}$ and $K_{cb\ end}$ values larger than 0.45 must be adjusted using the following equation:

$$K_{cb} = K_{cb\ (\text{tab})} + \left[0.04(u_2 - 2) - 0.004(RH_{\min} - 45)\right]\left(\frac{h}{3}\right)^3 \tag{13}$$

Where K_{cb} (tab) is the value of $K_{cb\ mid}$ taken from Table 17 of Allen et al. (1998). The other parameters are as defined in Eq. (9). The K_{cb} values for the mid-season stage was adjusted using Eq. (13) to 0.98 for for RH_{min} = 48%, u_2 = 1.91 m/s, and plant height of 1.0 m. Once the K_{cb} values for the initial stage, mid season stage, and end-of-season stage were determined, K_{cb} values for development and late season stages were determined using linear interpolation.

The K_{cb} coefficient for any period (day) of the growing season can be derived by considering that during the initial and mid-season stages K_{cb} is constant and equal to the K_{cb} value of the growth stage under consideration. During the crop development and late season stage, K_{cb} varies linearly between the K_{cb} at the end of the initial stage ($K_{c\ ini}$) and the K_{cb} at the beginning of the midseason stage ($K_{cb\ mid}$). During the mid season stage K_{cb} is constant as $K_{cb\ mid}$. During late season stage, K_{cb} varies linearly between K_{cb} mid and K_{cb}

end. In the case of canola the end of season K_{cb} does not need adjustment since it is 0.25 which is less than 0.45.

4.3.2 Soil evaporation coefficient

Similar to K_{cb}, soil evaporation coefficient K_e needs to be calculated on a daily basis. K_e is a function of soil water characteristics, exposed and wetted soil fraction, and top layer soil water balance (Allen et al., 2005). In the initial stage of crop growth, the fraction of soil surface covered by the crop is small, and thus, soil evaporation losses are considerable. Following rain or irrigation, K_e can be as high as 1. When the soil surface is dry, K_e is small and even zero. K_e is determined using Eq. (14).

$$K_e = \min\{[K_r(Kc\ \max - K_{cb})],[f_{ew}\ Kc\ \max]\} \tag{14}$$

Where K_c max = maximum value of crop coefficient K_c following rain or irrigation; K_r = evaporation reduction coefficient which depends on the cumulative depth of water depleted; and f_{ew} = fraction of the soil that is both wetted and exposed to solar radiation. K_c max represents an upper limit on evaporation and transpiration from the cropped surface. K_c max ranges [1.05-1.30] (Allen et al., 2005). Its value is calculated for initial, development, mid-season, or late season using Eq. 15.

$$K_{c\ \max} = \max\left(\left\{1.2+\left[0.04(u_2-2)-0.004(RH_{\min}-45)\right]\left(\frac{h}{3}\right)^{0.3}\right\},\{K_{cb}+0.05\}\right) \tag{15}$$

Evaporation occurs predominantly from the exposed soil fraction. Hence, evaporation is restricted at any moment by the energy available at the exposed soil fraction, i.e. K_e cannot exceed f_{ew} x K_c max. The calculation of K_e consists in determining K_c max, K_r, and f_{ew}. K_c max for initial, development, midseason, and late season stages were calculated to be 1.196, 1.181, 1.187, and 1.195 respectively.

4.3.3 Evaporation reduction coefficient

The estimation of evaporation reduction coefficient K_r requires a daily water balance computation for the surface soil layer. Evaporation from exposed soil takes place in two stages: an energy limiting stage (Stage 1) and a falling rate stage (Stage 2) (Ritchie 1972) as indicated in Fig. 3. During stage 1, evaporation occurs at the maximum rate limited only by energy availability at the soil surface and therefore, K_r = 1. As the soil surface dries, the evaporation rate decreases below the potential evaporation rate (K_c max - K_{cb}). K_r becomes zero when no water is left for evaporation in the evaporation layer. Stage 1 holds until the cumulative depth of evaporation D_e is depleted which depends on the hydraulic properties of the upper soil. At the end of Stage 1 drying, D_e is equal to readily evaporable water (REW). REW ranges from 5 to 12 mm and highest for medium and fine textured soils (Table 1 of Allen et al., 2005). The evolution of K_r is presented in Fig. 3.

The second stage begins when D_e exceeds REW. Evaporation from the soil decreases in proportion to the amount of water remaining at the surface layer. Therefore reduction in evaporation during stage 2 is proportional to the cumulative evaporation from the surface soil layer as expressed in Eq. (16).

$$K_r = \frac{TEW - D_{e,j-1}}{TEW - REW} \text{ for } D_{e,j-1} > REW \tag{16}$$

where De, j-1 = cumulative depletion from the soil surface layer at the end of previous day (mm); The TEW and REW are in mm. The amount of water that can be removed by evaporation during a complete drying cycle is estimated as in Eq. (17).

$$TEW = 1000(\theta_{FC} - 0.5\theta_{WP})Z_e \tag{17}$$

Where TEW =maximum depth of water that can be evaporated from the surface soil layer when the layer has been initially completely wetted (mm). θ_{FC} and θwp are in (m^3 m^{-3}) and Ze (m) = depth of the surface soil subject to evaporation. FAO-56 recommended values for Ze of 0.10-0.15m, with 0.10 m for coarse soils and 0.15 m for fine textured soils.

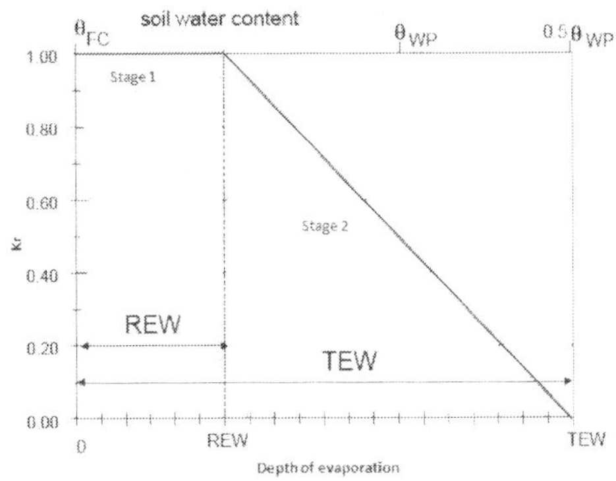

Fig. 3. Soil evaporation reduction coefficient K_r (adapted from Allen et al., 2005). REW stands for readily extractable water and TEW stands for total extractable water.

Calculation of K_e requires a daily water balance for the wetted and exposed fraction of the surface soil layer (f_{ew}). Eq. (18) is used to determine cumulative evaporation from the top soil layer (Allen et al., 2005).

$$D_{e,j} = D_{e,j-1} - (P_j - R_j) - \frac{I_j}{f_w} + \frac{E_j}{f_{ew}} + T_{ei,j} + D_{ei,j} \tag{18}$$

where $D_{e,j-1}$ and $D_{e,j}$ = cumulative depletion at the ends of days j-1 and j (mm); P_j and R_j = precipitation and runoff from the soil surface on day j (mm); I_j = irrigation on day j (mm); E_j = evaporation on day j (i.e., $E_j = K_e \times ET_o$) (mm); $T_{ei,j}$ = depth of transpiration from exposed and wetted fraction of the soil surface layer (f_{ew}) on day j (mm); and $D_{ei,j}$ = deep percolation from the soil surface layer on day j (mm) if soil water content exceeds field capacity (mm). Assuming that the surface layer is at field capacity following heavy rain or irrigation, the minimum value of $D_{e,j}$ is zero and limits imposed are $0 \leq D_{e,j} \leq TEW$. T_{ei} can be ignored except for shallow rooted crops (0.5-0.6m).

Evaporation is greater between plants exposed to sunlight and with air ventilation. The fraction of the soil surface from which most evaporation occurs is $f_{ew} = 1-f_c$.

$$f_{ew} = \min(1-f_c, f_w) \tag{19}$$

Where $1-f_c$ = 1-CC; f_w is fraction of soil surface wetted by irrigation or rainfall; f_w is 1 for rainfall (Table 20 of Allen et al., 1998); f_c is fraction of soil surface covered by vegetation. In this study f_c is the canopy cover measured using *GreenSeeker™*. Values of parameters used in the dual coefficient approach are presented in Table 1.

Parameter	Value
Field capacity, θ_{FC} (m^3 m^{-3})	30.1
Permanent wilting point, θ_{WP} (m^3 m^{-3})	15.0
Effective rooting depth, Z_r (m)	1.00
Depth of the surface soil layer, Z_e (m)	0.15
Total evaporable water, TEW (mm)	33.7
Readily evaporable water, REW (mm)	9
Total available water, TAW (mm)	160
Readily available water, RAW (mm)	96
The ratio of RAW to TAW, p (fraction)	0.6
Wetting fraction, f_w (fraction)	1

Table 1. The parameters of the soil used in the determination of K_s, K_e, and K_r in the FAO dual coefficient method.

The top soil layer (0-0.15 m) of the soil in this study is sandy clay loam. Readily extractable water (REW) is 9 mm for this soil texture (Table 1 of Allen et al., 2005). Field capacity and wilting point of this soil were determined as part of soil hydraulic properties characterization. Canola effective rooting depth was determined as part of National Brasicca Germaplasm Improvement Program (David Luckett, personal communication). Soil moisture content was monitored using on-site calibrated neutron probe. Soil moisture depletion fraction (p) of 0.6 m was taken from FAO-56 publication (Allen et al., 1998). Since the only source of water was rainfall, wetting fraction f_w of 1 was used.

4.4 AquaCrop approach of determining dual evapotranspiration coefficients

Eq. (11) gives evapotranspiration when the soil water is not limiting. When the soil evaporation and transpiration drops below their respective maximum rates, AquaCrop simulates ET_a by multiplying the crop transpiration coefficient with the water stress coefficient for stomatal closure (Ks_{sto}), and the soil water evaporation coefficient with a reduction K_r [0-1] (Steduto et al., 2009) as

$$ET_a = (Ks_{sto}K_{cb} + K_rK_e)\,ET_o \tag{20}$$

AquaCrop calculates basal crop coefficient at any stage as a product of basal crop coefficient at mid-season stage $K_{cb(x)}$ and green canopy cover (CC). For canola $K_{cb(x)}$ = 0.95 was used.

$$K_{cb} = K_{cb(x)} \times CC \tag{21}$$

$$K_e = K_{e(x)} \times (1-CC) \tag{22}$$

Evaporation from a fully wet soil surface is inversely proportional to the effective canopy cover. The proportional factor is the soil evaporation coefficient for fully wet and unshaded soil surface ($K_{e(x)}$) which is a program parameter with a default value of $K_{e(x)}$ = 1.1 (Raes et al., 2009).

During the energy limiting (non-water limiting) stage of evaporation, maximum evaporation (E_x) is given by

$$E_x = K_e\,ET_o = [(1\text{-}CC)K_{ex}]ET_o \quad (23)$$

Where CC is green canopy cover; K_{ex} is soil evaporation coefficient for fully wet and non shaded soil surface (Steduto et al., 2009). In AquaCrop, K_{ex} is a program parameter with a default value of 1.10 (Allen et al., 1998). When the soil water is limiting, actual evaporation rate is given by

$$E_a = K_r E_x \quad (24)$$

Maximum crop transpiration (T_{rx}) for a well-watered crop is calculated as

$$T_{rx} = K_{cb}\,ET_o = [CC\;K_{cbx}]ET_o \quad (25)$$

K_{cbx} is the basal crop coefficient for well-watered soil and complete canopy cover.

5. Results and discussion

5.1 Soil water balance

The actual evapotranspiration determined using soil water balance method is presented in Table 2. Evapotranspiration was determined using Eq. (2) from measurement of 12 neutron probes several times during the season. Deep percolation and runoff were not measured. Therefore, values estimated by AquaCrop (Steduto et al., 2009; Raes et al., 2009) during the canola water productivity simulation were adopted.

DAP*	Rainfall (mm)	Deep percolation (mm)	Runoff (mm)	Change in storage (mm)	Evapotranspiration ET_a using water balance (mm)
0-13	6.5	0	0	-2.1	8.6
14-21	0	0	0	-1.8	1.8
22-28	36.9	4.6	0.5	13.4	18.4
29-35	23.4	24.6	1.4	-10	7.4
36-42	1.8	1.8	0	-3.1	3.1
43-49	6	2.2	0	-1.1	4.9
50-63	21.8	6.7	0	4.6	10.5
64-77	60	20.2	4.1	17.7	18
78-94	3.2	18.9	0	-25.6	9.9
95-118	58.7	21.2	1.6	6.7	29.2
119-143	81	34.3	3.8	-20.8	63.7
144-159	0	1.5	0	-39.6	38.1
160-173	103.9	8.6	14	30.3	51
174-196	31.6	3.8	0	-20.7	48.5
*DAP stands for days after planting				Seasonal	313

Table 2. Evapotranspiration determined using soil water balance method for canola planted on 30 April 2010 at Wagga Wagga (Australia).

The runoff estimated using AquaCrop was low, supporting the consensus that runoff from agricultural land is low. However, deep percolation past the 1.2 m was significant. The actual annual crop evapotranspiration estimated using this method was 313 mm. It can be observed that evapotranspiration was higher during the mid season and highly evaporative months.

5.2 Evapotranspiration coefficient

Single and dual evapotranspiration coefficients and crop canopy cover data are presented in Fig. 4. The K_c and K_{cb} values adopted from FAO-56 publication and adjusted for the local condition are shown in the Figure. The K_c and K_{cb} curves follow similar trend as the measured canopy cover curve. The canopy cover values were higher than the K_c and K_{cb} curves towards the end of the season. This is due to the fact that as an indeterminate crop, canola still had green canopy due to the ample rainfall during this late season stage of the crop. The soil evaporation coefficient K_e was correctly simulated using the top-layer soil water balance model. It can be seen that K_e is high during the initial and late season stages. It remained low and steady during the midseason stage. The higher number of K_e spikes are

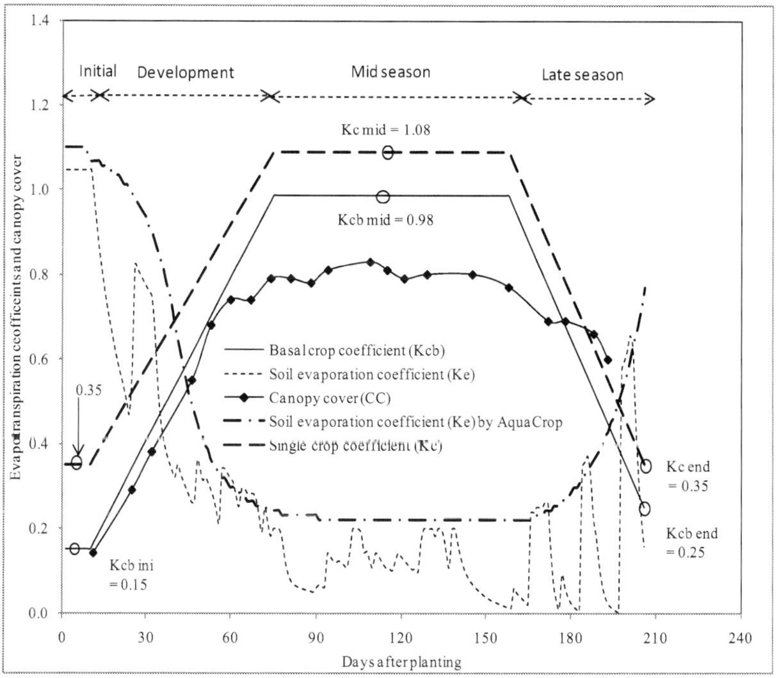

Fig. 4. Single crop coefficient (K_c), basal coefficient (K_{cb}), soil evaporation coefficient (K_e), crop canopy cover (CC) curves for canola having growth stage lengths of 10, 64, 84, and 48 days during initial, development, midseason, and late season stages. Indicated on curve are also single and basal crop coefficient (K_c and K_{cb}) at initial, midseason, and end of season stages. Day of planting is 30 April 2010.

due to frequent rainfall during the season. The K_e value estimated using AquaCrop followed similar trend to the manually calculated using Eq. (14). However, AquaCrop did not simulate response to individual rainfall events.

In the development stage, the soil surface covered by the crop gradually increases and the K_e value decreases. In the midseason stage, the soil surface covered by the crop reaches maximum and water loss is mainly by crop transpiration and K_e is as low as 0.05. In the late season stage, the K_e values are greater than that in the mid-season stage because of the senescence.

Evaporation and transpiration estimated using the dual coefficient approach (Fig. 5) are correctly simulated, with high evaporation during the initial and late stages, and low during the developmental and mid season stages. The fluctuation in the evaporation component is high at these stages and low and steady during the mid season stage except minor spikes after rainfall events. Evaporation during the late stage (late spring months) was high compared with the initial stage which is a winter period. The transpiration component was steady increasing during the crop development stage before reaching a maximum in late mid season stage and declined during the late season stage due to senescence. The trends in evaporation and transpiration were in perfect phase with the weather and crop phenology.

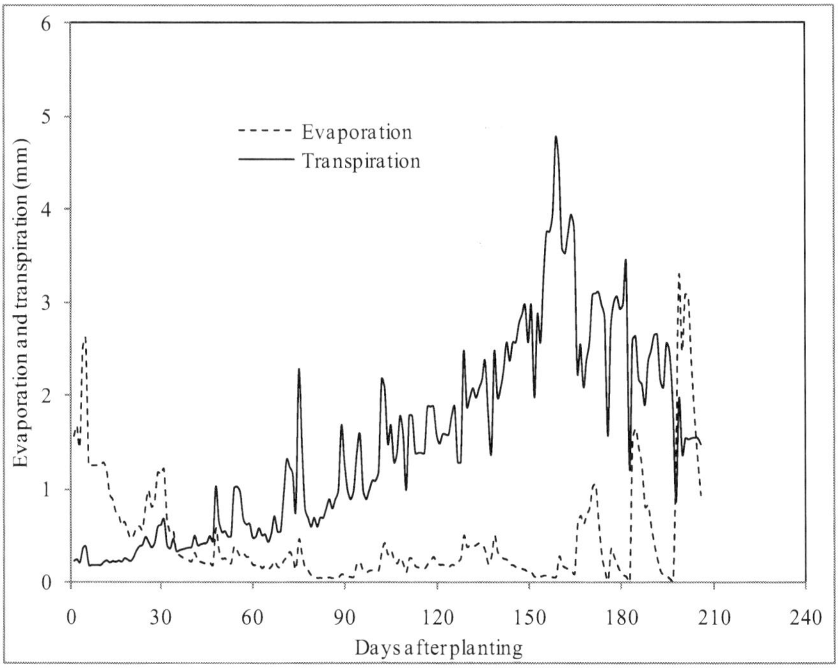

Fig. 5. Daily soil evaporation and transpiration estimated using dual coefficient method for canola planted on 30 April 2010 at Wagga Wagga, NSW (Australia).

Evapotranspiration varies during the growing period of a crop due to variation in crop canopy and climatic conditions (Allen et al., 1998). Variation in crop canopy changes the proportion of evaporation and transpiration components of evapotranspiration. The spikes in basal crop coefficient were high during the initial and crop development phases and decreases as the soil dries (Fig. 4). The spikes decrease as the canopy closes and much of ET is by transpiration. During the late season stage, there were fewer spikes because soil evaporation was low and almost constant. The largest difference between K_c and K_{cb} is found in the initial growth stage where evapotranspiration is predominantly in the form of soil evaporation and crop transpiration. Because crop canopies are near or at full ground cover during the mid-season stage, soil evaporation beneath the canopy has less effect on crop transpiration and the value of K_{cb} in the mid season stage is very close to K_c. Depending on the ground cover, the basal crop coefficient during the mid season stage may be only 0.05-0.10 lower than the K_c value. In this study $K_{cb\ mid}$ is 0.10 lower than $K_{c\ mid}$.

Some studies, carried out in different regions of the world, have compared the results obtained using the approach described by Allen et al. (1998) with those resulting from other methodologies. From this comparison, some limitations should be expected in the application of the dual crop coefficient FAO-56 approach. Dragoni et al. (2004), which measured actual transpiration in an apple orchard in cool, humid climate (New York, USA), showed a significant overestimation (over 15%) of basal crop coefficients by the FAO 56 method compared to measurements (sap flow). This suggests that dual crop coefficient method is more appropriate if there is substantial evaporation during the season and for incomplete cover and drip irrigation.

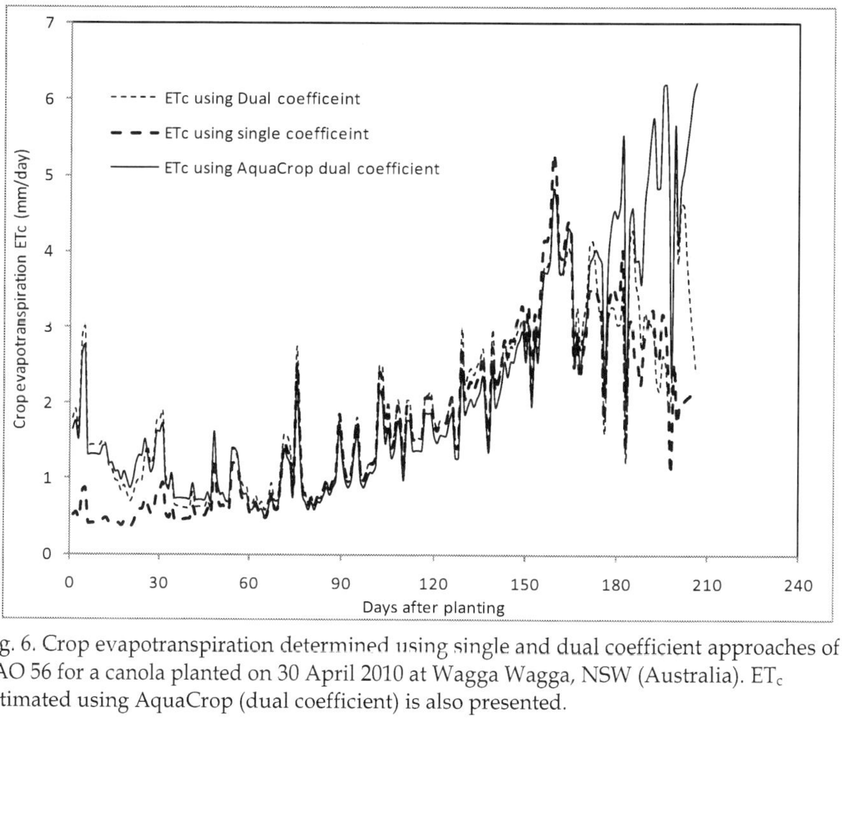

Fig. 6. Crop evapotranspiration determined using single and dual coefficient approaches of FAO 56 for a canola planted on 30 April 2010 at Wagga Wagga, NSW (Australia). ET_c estimated using AquaCrop (dual coefficient) is also presented.

Crop evapotranspiration estimated using single and double coefficients is presented in Fig. 6. ET_c estimated using AquaCrop is also presented in the Figure. It can be observed that ET_c estimated using the three approaches is similar except in the initial and late season stages. During the initial stage, the ET_c estimated using Eq. (14) and AquaCrop (Eqs. 21 and 22) are very close. However, the single coefficient method underestimated ET_c at this stage. During the initial stage when most of the soil is bare, evaporation is high especially if the soil is wet due to irrigation or rainfall. The single crop coefficient approach does not sufficiently take this into account. A similar pattern was observed during the late season stage. However, AquaCrop overestimated ET_c during this stage compared to the other two methods. The annual evapotranspiration estimated using different approaches was as follows: soil water balance (ET_a = 313 mm), single crop coefficient (ET_c = 332 mm), dual coefficient approach (ET_c = 366 mm with E of 79 mm and T of 288 mm), AquaCrop (ET_c = 382 mm with E of 139 mm and T of 243 mm). The evapotranspiration determined using soil water balance method is the "actual" evapotranspiration while the other methods measure potential evapotranspiration ET_c. Soil water depletion (Dr) in Eq. (6) was determined using soil moisture content measured during the season and it was found that Dr<RAW throughout the season indicating that there was no soil moisture stress (K_s = 1). That might be why the ET_c estimated using single coefficient method is close to the ET_c determined using soil water balance method. Approaches using dual coefficient (Eq. 14) and Eqs. (21 and 22) resulted in higher ET_c values. This might be due to the fact that in these approaches, the evaporation during the initial and late season stages was well simulated.

6. Conclusion

Two approaches of estimating crop evapotranspiration were demonstrated using a field crop grown in a semiarid environment of Australia. These approaches were the rootzone soil water balance and the crop coefficient methods. The components of rootzone water balance, except evapotranspiration, were measured/estimated. Evapotranspiration was calculated as an independent parameter in the soil water balance equation. Single crop coefficient and dual coefficient approaches were based on adjustment of the FAO 56 coefficients for local condition. AquaCrop was also used to estimate crop evapotranspiration using the dual coefficient approach. It was found that the dual coefficients, basal or transpiration coefficient K_{cb} and evaporation coefficient K_e, correctly depict the actual process. The effects of weather (rainfall and radiation) and crop phenology were correctly simulated in this method. However, single coefficient does not show the high evaporation component during the initial and late season stages. Generally, there is a strong agreement among different estimation methods except that the dual coefficient approach had better estimate during the initial and late season stages. The evapotranspiration estimated using different approaches was as follows: soil water balance (ET_a = 313 mm), single crop coefficient (ET_c = 332 mm), dual coefficient approach (ET_c = 366 mm with E of 79 mm and T of 288 mm), AquaCrop (ET_c = 382 mm with E of 139 mm and T of 243 mm). Evapotranspiration estimated using soil water balance method is actual evapotranspiration ET_a, while other methods estimate potential (maximum) evapotranspiration. Accordingly, ET estimated using rootzone water balance is lower than the ET estimated using the other methods. The single coefficient approach resulted in the lowest ET_c as it is not taking into account the evaporation spikes after rainfall during the initial and late season stages.

7. Acknowledgments

The senior author was research fellow at EH Graham Centre for Agricultural Innovation during this study. We also would like to thank David Luckett, Raymond Cowley, Peter Heffernan, David Roberts, and Peter Deane for professional and technical assistance.

8. References

Allen R.G., Pereira L.S., Raes D., Smith M. 1998. Crop evapotranspiration: guidelines for computing crop water requirements, FAO Irrigation and Drainage Paper 56., 300 p.

Allen R.G., Pereira L.S., Smith M., Raes D., Wright J.L. 2005. FAO-56 dual crop coefficient method for estimating evaporation from soil and application extensions. J Irrig Drain Eng ASCE, 131(1):2–13

Blaney, H.F. and Criddle, W.D. 1950. Determining water requirements in irrigated areas from climatological and irrigation data. USDA Soil Conserv. Serv. SCS-TP96. 44 pp.

Bonder, G., Loiskandl, W., Kaul, H.P. 2007. Cover crop evapotranspiration under semiarid conditions using FAO dual coefficient method with water stress compensation. Agric. Water Manag., 93 : 85-98.

Dragoni , D., Lakso, A.N., Piccioano, R.M. 2004. Transpiration of an apple orchard in a cool humid climate: measurement and modeling, Acta Horticulturae, 664:175-180.

Hawkins, R. H., Hjelmfelt, A. T., and Zevenbergen, A. W. 1985. Runoff probability, storm depth, and curve numbers. J. Irrig. Drain. Eng., 111(4): 330–340.

Hillel, D. 1997. Small scale irrigation for arid zones: Principles and options, Development monograph No. 2 , FAO, Rome.

Hillel, D. 1998. Environmental soil physics. Academic press. 771 pp. Elsevier (USA).

Monteith, J.L. 1981. Evaporation and surface temperature. Quart. J. Roy. Meteorol. Soc., 107:1-27.

Penman, H. L. 1948. "Natural evaporation from open water, bare soil and grass." Proc. Roy. Soc. London, A193, 120-146.

Penman, H.L. 1956. Estimating evaporation. Trans. Amer. Geoph. Union, 37:43-50.

Raes, D. 2009. *ET_o Calculator*: a software program to calculate evapotranspiration from a reference surface. FAO Land Water Division. Digital Media Service No 36.

Raes, D., Steduto, P., Hsiao, T.C., Fereres, E., 2009. AquaCrop—The FAO crop model to simulate yield response to water: II. Main algorithms and soft ware description. Agron. J. 101:438–447.

Ritchie, J.T., 1972. Model for predicting evaporation from a row crop with incomplete cover. Water Resour. Res. 8, 1204–1213.

Riverina Development Australia, RDA (2011). Riverina - Food basket of Australia. Industry and Investment , NSW Government. accessed 30 July 2011.

Smith, M. 1992. CROPWAT, a computer program for irrigation planning and management. FAO Irrigation and Drainage Paper 46, FAO, Rome.

Steduto, P., Hsiao, T.C., Raes, D., Fereres, E., 2009. AquaCrop—the FAO crop model to simulate yield response to water. I. Concepts. Agron. J. 101:426–437.

Stern, H., de Hoedt, G., Ernst, J., 2000. Objective classification of Australian climates. Bureau of meteorology, Melbourne.

Suleiman A.A., Tojo Soler, C.M., Hoogenboom, G. 2007. Evaluation of FAO-56 crop coefficient procedures for deficit irrigation management of cotton in a humid climate. Agric. Water Maneg., 91:33-42.

Thornthwaite, C.W. 1948. An approach toward a rational classification of climate. Geograph. Rev., 38:55-94.

4

Hargreaves and Other Reduced-Set Methods for Calculating Evapotranspiration

Shakib Shahidian[1], Ricardo Serralheiro[1], João Serrano[1], José Teixeira[2], Naim Haie[3] and Francisco Santos[1]
[1]University of Évora/ICAAM
[2]Instituto Superior de Agronomia
[3]Universidade do Minho
Portugal

1. Introduction

Globally, irrigation is the main user of fresh water, and with the growing scarcity of this essential natural resource, it is becoming increasingly important to maximize efficiency of water usage. This implies proper management of irrigation and control of application depths in order to apply water effectively according to crop needs. Daily calculation of the Reference Potential Evapotranspiration (ETo) is an important tool in determining the water needs of different crops. The United Nations Food and Agriculture Organization (FAO) has adopted the Penman-Monteith method as a global standard for estimating ETo from four meteorological data (temperature, wind speed, radiation and relative humidity), with details presented in the Irrigation and Drainage Paper no. 56 (Allen et al., 1998), referred to hereafter as PM:

$$ET_o = \frac{0.408\Delta(R_n - G) + \gamma \frac{900}{T+273} u_2 (e_s - e_a)}{\Delta + \gamma(1 + 0.34u_2)} \quad (1)$$

where:
R_n – net radiation at crop surface [MJ m^{-2} day^{-1}],
G – soil heat flux density [MJ m^{-2} day^{-1}],
T – air temperature at 2 m height [°C],
u_2 – wind speed at 2 m height [m s^{-1}],
e_s – saturation vapor pressure [kPa],
e_a – actual vapor pressure [kPa],
e_s-e_a – saturation vapor pressure deficit [kPa],
Δ – slope vapor pressure curve [kPa °C^{-1}],
γ – psychrometric constant [kPa °C^{-1}],
The PM model uses a hypothetical green grass reference surface that is actively growing and is adequately watered with an assumed height of 0.12m, with a surface resistance of 70s m^{-1} and an albedo of 0.23 (Allen et al., 1998) which closely resemble evapotranspiration from an extensive surface of green grass cover of uniform height, completely shading the ground

and with no water shortage. This methodology is generally considered as the most reliable, in a wide range of climates and locations, because it is based on physical principles and considers the main climatic factors, which affect evapotranspiration.

Need for reduced-set methods

The main limitation to generalized application of this methodology in irrigation practice is the time and cost involved in daily acquisition and processing of the necessary meteorological data. Additionally, the number of meteorological stations where all these parameters are observed is limited, in many areas of the globe. The number of stations where *reliable* data for these parameters exist is an even smaller subset.

There are also concerns about the accuracy of the observed meteorological parameters (Droogers and Allen, 2002), since the actual instruments, specifically pyranometers (solar radiation) and hygrometers (relative humidity), are often subject to stability errors. It is common to see a drift, of as much as 10 percent, in pyranometers (Samani, 2000, 1998). Henggeler et al. (1996) have observed that hygrometers loose about 1 percent in accuracy per installed month. There are also issues related to the proper irrigation and maintenance of the reference grass, at the weather stations. Jensen et al. (1997) observed that many weather stations are often not irrigated or inadequately irrigated, during the summer months, and thus the use of relative humidity and air temperature from these stations could introduce a bias in the computed values for *ETo*. Additionally, they observed that the measured values of solar radiation, *Rs*, are not always reliable or available and that wind data are quite site specific, unavailable, or of questionable reliability. Thus, they recommend the use of *ETo* equations that require fewer variables. These authors compared various methods, including FAO Penman Monteith, PM, and Hargreaves and Samani, HS, with lysimeter data and noted r^2 values of 0.94-0.97, with monthly SEE values of 0.30-0.34mm. Based on these data they concluded that the differences in *ETo* values, calculated by the different methods, are minor when compared with the uncertainties in estimating actual crop evapotranspiration from ETo. Additionally, these equations can be more easily used in adaptive or smart irrigation controllers that adjust the application depth according to the daily *ETo* demand (Shahidian et al., 2009).

This has created interest and has encouraged development of practical methods, based on a single or a reduced number of weather parameters for computing *ETo*. These models are usually classified according to the weather parameters that play the dominant role in the model. Generally these classifications include the *temperature-based models* such as Thornthwaite (1948); Blaney-Criddle (1950) and Hargreaves and Samani (1982); The *radiation models* which are based on solar radiation, such as Priestly-Taylor (1972) and Makkink (1957); and the *combination models* which are based on the energy balance and mass transfer principles and include the Penman (1948), modified Penman (Doorenbos and Pruitt, 1977) and FAO PM (Allen et al., 1998).

Objectives and methods

The objective of this chapter is to review the underlying principles and the genesis of these methodologies and provide some insight into their applicability in various climates and regions. To obtain a global view of the applicability of the reduced-set equations, each equation is presented together with a review of the published studies on its regional calibration as well as its application under different climates.

The main approach for evaluation and calibration of the reduced-set equations has been to use the PM methodology or lysimeter measurements as the benchmark for assessing their performance. Usually a linear regression equation, established with PM *ETo* values or lysimeter readings plotted as the dependent variable and values from the reduced-set equation plotted as the independent variable. The intercept, *a*, and calibration slope, *b*, of the best fit regression line, are then used as regional calibration coefficients:

$$ET_0PM = a + b(ET_0Equation) \tag{2}$$

The quality of the fit between the two methodologies is usually presented in terms of the coefficient of determination, r^2, which is the ratio of the explained variance to the total variance or through the Root Mean Square Error, *RMSE*:

$$RMSE = \sqrt{\frac{1}{n}\sum_{i=1}^{n}\left(ETo_{yi} - ETo_{PM}\right)^2} \tag{3}$$

and the mean Bias error:

$$MBE = \frac{1}{n}\sum_{i=1}^{n}\left(ETo_{yi} - ETo_{PM}\right) \tag{4}$$

where n is the number of estimates and ETo_{yi} is the estimated values from the reduced-set equation.

2. Temperature based equations

Temperature is probably the easiest, most widely available and most reliable climate parameter. The assumption that temperature is an indicator of the evaporative power of the atmosphere is the basis for temperature-based methods, such as the Hargreaves-Samani. These methods are useful when there are no data on the other meteorological parameters. However, some authors (McKenny and Rosenberg, 1993, Jabloun and Sahli, 2007) consider that the obtained estimates are generally less reliable than those which also take into account other climatic factors.

Mohan and Araumugam (1995) and Nandagiri and Kovoor (2006) carried out a multivariate analysis of the importance of various meteorological parameters in evapotranspiration. They concluded that temperature related variables are the most crucial required inputs for obtaining ETo estimates, comparable to those from the PM method across all types of climates. However, while wind speed is considered to be an important variable in arid climate, the number of sunshine hours is considered to be the more dominant variable in sub-humid and humid climates.

2.1 The Hargreaves- Samani methodology

Hargreaves, using grass evapotranspiration data from a precision lysimeter and weather data from Davis, California, over a period of eight years, observed, through regressions, that for five-day time steps, 94% of the variance in measured *ET* can be explained through average temperature and global solar radiation, *Rs*. As a result, in 1975, he published an equation for predicting *ETo* based only on these two parameters:

$$ET_o = 0.0135\, R_s(T + 17.8) \quad (5)$$

where *Rs* is in units of water evaporation, in mm day^{-1}, and *T* in °C. Subsequent attempts to use wind velocity, U_2, and relative humidity, *RH,* to improve the results were not encouraging so these parameters have been left out (Hargreaves and Allen, 2003).
The clearness index, or the fraction of the extraterrestrial radiation that actually passes through the clouds and reaches the earth's surface, is the main energy source for evapotranspiration, and later studies by Hargreaves and Samani (1982) show that it can be estimated by the difference between the maximum, T_{max}, and the minimum, T_{min} daily temperatures. Under clear skies the atmosphere is transparent to incoming solar radiation so the T_{max} is high, while night temperatures are low due to the outgoing longwave radiation (Allen et al., 1998). On the other hand, under cloudy conditions, T_{max} is lower, since part of the incoming solar radiation never reaches the earth, while night temperatures are relatively higher, as the clouds limit heat loss by outgoing longwave radiation. Based on this principle, Hargreaves and Samani (1982) recommended a simple equation to estimate solar radiation using the temperature difference, ΔT:

$$R_s / R_a = K_T\, (T_{\max} - T_{\min})^{0.5} \quad (6)$$

where *Ra* is the extraterrestial radiation in mm day^{-1}, and can be obtained from tables (Samani, 2000) or calculated (Allen et al., 1998). The empirical coefficient, K_T was initially fixed at 0.17 for Salt Lake City and other semi-arid regions, and later Hargreaves (1994) recommended the use of 0.162 for interior regions where land mass dominates, and 0.190 for coastal regions, where air masses are influenced by a nearby water body. It can be assumed that this equation accounts for the effect of cloudiness and humidity on the solar radiation at a location (Samani, 2000). The clearness index (*Rs/Ra*) ranges from 0.75 on a clear day to 0.25 on a day with dense clouds.
Based on equations (5) and (6), Hargreaves and Samani (1985) developed a simplified equation requiring only temperature, day of year and latitude for calculating *ETo*:

$$ET_o = 0.0135\, K_T\, (T + 17.78)(T_{mzx} - T_{\min})^{0.5} R_a \quad (7)$$

Since K_T usually assumes the value of 0.17, sometimes the 0.0135 K_T coefficient is replaced by 0.0023. The equation can also be used with *Ra* in MJ m^{-2} day^{-1}, by multiplying the right hand side by 0.408.
This method (designated as HS in this chapter) has produced good results, because at least 80 percent of *ETo* can be explained by temperature and solar radiation (Jensen, 1985) and ΔT is related to humidity and cloudiness (Samani and Pessarakli, 1986). Thus, although this equation only needs a daily measurement of maximum and minimum temperatures, and is presented here as a temperature-based method, it effectively incorporates measurement of radiation, albeit indirectly. As will be seen later, the ability of the methodology to account for both temperature and radiation provides it with great resilience in diverse climates around the world.
Sepashkhah and Razzaghi (2009) used lysimeters to compare the Thornthwaithe and the HS in semi-arid regions of Iran and concluded that a calibrated HS method was the most accurate method. Jensen et al.(1997) compared this and other *ETo* calculation methods and concluded that the differences in *ETo* values computed by the different methods are not larger than those introduced as a result of measuring and recording weather variables or the uncertainties

associated with estimating crop evapotranspiration from *ETo*. López-Urrea et al. (2006) compared seven *ETo* equations in arid southern Spain with Lysimeter data, and observed daily RMSE values between 0.67 for FAO PM and 2.39 for FAO Blaney-Criddle. They also observed that the Hargreaves equation was the second best after PM, with an RMSE of only 0.88.

Since the HS method was originally calibrated for the semi-arid conditions of California, and does not explicitly account for relative humidity, it has been observed that it can overestimate *ETo* in humid regions such as Southeastern US (Lu et al. 2005), North Carolina (Amatya et al. 1995), or Serbia (Trajkovic, 2007).

In Brasil, Reis et al. (2007) studied three regions of the Espírito Santo State: The north with a moderately humid climate, the south with a sub-humid climate, and the mountains with a humid climate (Table 1). The HS equation overestimated *ETo* in all three regions by as much as 32%, but the performance of the HS equation improved progressively as the climate became drier. Only further south, at a latitude of 24° S, and in a warm temperate climate did HS provide good agreement with PM, though still with a small overestimation. Borges and Mendiondo (2007) obtained an r^2 of 0.997 for HS when compared to PM, when using a calibrated α of 0.0022 (Sept-April) and 0.0020 for the rest of the year.

On the other hand, in dry regions such as Mahshad, Iran and Jodhpur, India, the HS equation tends to underestimate *ETo* by as much as 24% (Rahimkoob, 2008; Nandagiri and Kovoor, 2006). Rahimkoob (2008) studied the *ETo* estimates obtained from the HS equation in the very dry south of Iran. His data indicate that the HS equation fails to calculate *ETo* values above 9 mm day^{-1}, even when the PM reaches values of more than 13 mm day^{-1} (Fig. 1).

Wind removes saturated air from the boundary layer and thus increases evapotranspiration (Brutsaert, 1991). Since most of the reduced-set equations do not explicitly account for wind speed, it is natural for the calibration slope to be influenced by this parameter. Itensifu et al. (2003) carried out a major study using weather data from 49 diverse sites in the United States. They obtained ratios ranging from 0.805 to 1.242 between HS and PM and concluded that the HS equation has difficulty in accounting for the effects of high winds and high vapor pressure deficits, typical of the Great Plains region. They also observed that the HS equation tends to overestimate *ETo* when mean daily *ETo* is relatively low, as in most sites in the eastern region of the US, and to underestimate when *ETo* is relatively high, as in the lower Midwest of the US. As will be seen later, this seems to be a common issue with most of the reduced set evapotranspiration equations (see section 4.3, Fig. 7).

For the Mkoji sub-catchment of the Great Ruaha River in Tanzania, Igbadun et al. (2006) calculated the monthly *ETo* values of three very distinct areas of the catchment: the humid Upper Mkoji with an altitude of 1700m, the middle Mkoji with an average altitude of 1100 m, and the semi-arid lower Mkoji with an altitude of 900m. Their data indicate a strong relation between the monthly average wind speed and the performance of the HS equation as measured by the slope of the calibration equation (PM/HS ratio). Although the three areas have distinct climates, the HS equation clearly underestimated ETo for wind speed values below 2-2.3 ms^{-1}, and overestimated it for higher wind speed values (Fig. 2).

Trajkovic, et al. (2005) studied the HS equation in seven locations in continental Europe with different altitudes (42-433m) with RH ranging from 55 to 71%, representative of the distinct climates of Serbia. Their data show that despite the different altitudes and climatic conditions, wind speed was the major determinant for the calibration of the HS equation (Fig. 3). The results from these works indicate that wind is the main factor affecting the calibration of the HS equation and that the equation should be calibrated in areas with very high or low wind speeds.

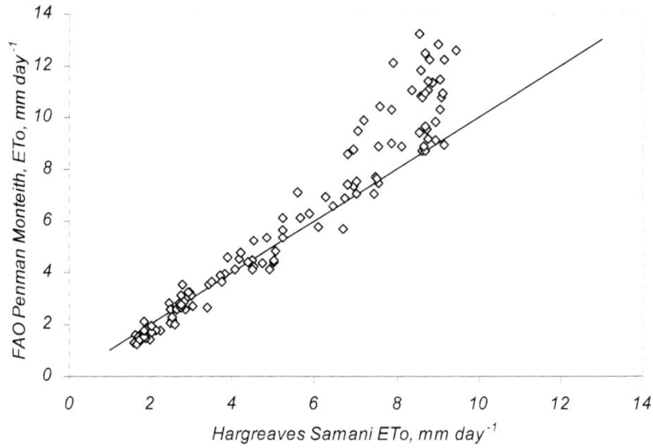

Fig. 1. Relation between *ETo* calculated with the HS equation and the PM for the dry conditions of Abadan, Iran. The Hargreaves Samani equation fails to calculate *ETo* values above 9 mm day-1 (data kindly provided by Rahimkoob)

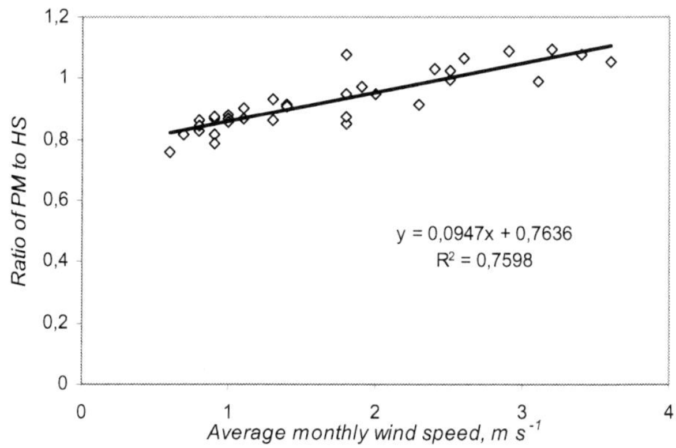

Fig. 2. Correlation between average wind speed and the calibration slope in distinct climates of the Great Ruana River in Tanzania (based on the original data from Igbadun et al. 2006).

Jabloun and Sahli (2008) studied eight stations in the semi-arid Tunisia and concluded that in inland stations, HS tends to overestimate *ETo* due to high ΔT values. In the coastal station of Tunis, HS underestimated *ETo* values, which they attributed to an underestimation of *Rs*. Various attempts have been made to improve the accuracy of the HS equation through incorporation of additional measured parameters, such as rainfall (Droogers and Allen, 2002) and altitude (Allen, 1995). These methodologies have had limited global application, probably because *ETo* is influenced by a combination of different parameters, and although in a certain region there appears to be a good correlation between the calibration slope and a certain parameter, this might not be so in a different climate.

The alternative is to use regional calibration, in which, based on the climatic characteristics of the region, the *ETo* calculated by the HS equation is adjusted to account for the combined

effect of the dominant climate parameters, and thus accuracy of the equations is improved (Teixeira et al., 2008). Table 1 presents a compilation of most of the published studies on the regional calibration of the HS equation. This compilation contains 33 published works covering 21 countries with all types of climatic conditions according to the Koppen classification. Whenever various stations from a similar climate were studied, only parameters from one representative station are presented. In some studies, HS and PM were calibrated against a third methodology (such as Pan A) and thus no direct calibration parameters for the PM/HS regression were provided. In these cases, a linear regression was obtained by plotting the PM calibration equation as the dependent variable and the HS calibration equation as the independent variable. The parameters of the resulting regression equation are then presented as the PM-HS calibration parameters.
In order to contextualize the information and allow for extension of the results to other regions with a similar climate, the locations are grouped according to Koppen climate classification. These calibration coefficients can be used in the area where they were obtained or can be extrapolated for areas with similar conditions where no actual calibration has been carried out yet.

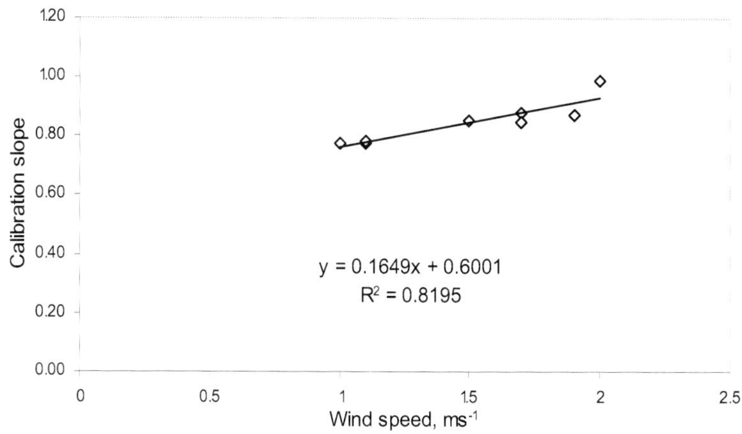

Fig. 3. Correlation between wind speed and the calibration slope for seven different locations in Serbia, representing the diverse local climates (original data from Trajkovic, 2005).

2.2 The Thornthwaite method

Thornthwaite (1948) devised a methodology to estimate *ETo* for short vegetation with an adequate water supply in certain parts of the USA. The procedure uses the mean air temperature and number of hours of daylight, and is thus classified as a temperature based method. Monthly *ETo* can be estimated according to Thornthwaite (1948) by the following equation:

$$Et_0 = ET_0sc\left(\frac{N}{12}\right)\left(\frac{dm}{30}\right) \tag{8}$$

Country	Station	latitude	Altitude	Koppen	Rainfall	RH	U2	Regression adjustment		R2	RMSE	Source
		m	m	classification	mm	%	ms^{-1}	intercept a	slope b			
Arid												
Desert												
China, NW	Shandan Heihe R.	38°90' N	1483	BWk	250	40	1.98	0.5431	1.148			Zhao et al. 2005
China, NW	Minle	38°80' N	2271	BWk	100			-0.32	1.065			Zhao et al. 2005
US	Aquila	33°56' N	655	BWh	195	35.3	3.2	0.0378	1.3155			Alexandris, 2006
Steppe												
India	Jodhpur	26°18' N	224	BSh	402	38.9	2.1	-0.3827	1.1924			Nandagiri and Kovoor, 2006
India	Heydarabad	17°32' N	545	BSh	820	65.6	2.8	-1.97	1.48			Nandagiri and Kovoor, 2006
Syria	Tel Hadya	36°01' N	293	BSh	231	57.4	2.82		1.04	0.91		Stockle, 2004
Iran	Shiraz	30°07'N	1650	BSh	306	36.4	2.49	0.41	0.82			Razzaghi and Sepahskah, 2009
Iran	Shiraz	30°07' N	1650	BSh	305.6	36.4	2.49		1.13			Sepashkah and Razzaghi, 2009
México	Progreso (Yucatán)	21°17' N	2	BSh	511			-0.26	1.012	0.78		Bautista et al 2009
Dry summer												
Spain	Daroca (NE Spain)	41°07' N	779	Bsk	364	66.5	1.08	-0.203	0.93			Mártinez-Cob andTejero-Juste, 2004
Spain	Zaragoza (NE Spain)	41°43' N	225	Bsk	353	73.7	2.43	-0.012	0.99			Mártinez-Cob andTejero-Juste, 2004
Spain	Cordoba, inland	37°52' N	117	Bsk	696	63.3	1.6		1.06			Gavilán et al, 2008
Bolivia	Patacamaya and Oruro	17°15'S	3749	Bsk	375	57.4	1.2	0.8622	0.6422			Garcia et al 2004
Spain	Albacete	39°14' N	695	Bsk	283	68.7	1.08	0.34*	1.14*			Lopéz-Urra et al 2005
Spain	Cordoba, inland	37°51' N	110	Bsk	696	63.3	1.6	-1.49	1.3			Berengena and Gavilan, 2005
Tanzania	Lower Mkoji	7°80'	900	Bsh	520			-0.0027	0.9092			Igbadun et al
Mesothermal												
Mediterranean												
Spain	Malaga (Andalucia) Coast	36°40' N	7	Csa	531	68.1	1.9		0.962			Vanderlinden et al., 2004
Spain	Sevilla (Andalucia) interior	37°125' N	31	Csa	473	67.8	0.93		1.165			Vanderlinden et al., 2004
Spain	La Mojonera, coast	37°45' N	142	Csa	272	62.3	1.9		1.27			Gavilán et al, 2008
Portugal, S	Evora	38°55' N	246	Csa	627	63.3	4.3		0.866			Santos and Maia, 2007
US	Davis	38°32' N	18.3	Csa	458	63.3	2.62	-0.844	1.245			Alexandris, 2006
Portugal	Elvas	38°60' N	202	Csa	506	58.2	1.97	-0.08	1.04			Teixeira et al. 2008
Spain	Niebla (Andalucia)	37°21' N	52	Csa	702	65.3	1.3		1.035	0.93		Gavilán et al., 2008
Spain	Vejer Frontera (Andalucia)	36° 17' N	24	Csa	571	69.4	2.9		1.404			Gavilán et al., 2008
Greece	Athens	38°23' N	100	Csa	371	61.8	1.87	0.264	0.781			Alexandris, 2006
USA	Prosser, WA	46°15' N	380	Csb	994	69.7	1.62		1.02	0.98		Stockle, 2004
Spain	Lleida	41°42' N	221	Csb	601	68.8	0.97		1.1	0.95		Stockle, 2004
Dry winter												
Tanzania	Middle Mkoji	8°30'	1070	Cwa	800			-0.4	0.955			Igbadun et al, 2006
Brasil	Douradas, Mato G. Sul	22°16'S	452	Cwa	1603	73.8	1.74	1.73	0.67	0.7		Fietz, 2004
Brasil	S. Mantiqueira, MG		1500	Cwb	2150			0.153	1.16			Pereira et al. 2009
fully humid												
Netherlands	Haarweg	51°58' N	9	Cfb	778	87.3	2.41		1.02	0.91		Stockle, 2004
US	Louisiana, inland	31° N	low land	Cfa	1500	92	0.82	-0.28	1.05			Fontenot, 2004
US	Lousisana, coastal	29° N	low land	Cfa	1500	88.7	0.6	-0.17	0.87			Fontenot, 2004
US	North Carolina, Plymouth	35°52'	6	Cfa	1299	80.2	4.9	0.03	0.83		1.23	Amatya et al. 1995
Brasil	Palotina, Paraná	24°18'S	310	Cfa	1700	73.8	1.74	-108	1			Syperreck, 2006
Brasil	Jacupiranga river, SP	24°29'S	52	Cfa	1879	91.5	0.97	-0.365	1.042			Borges and Mendiondo, 2007

Values in grey are annual averages obtained from Climwat data base.
When calibration parameters of the HS vsFAO PM were not directly provided, linear regression equations were established with FAO-56 PM daily ET0 estimates as the dependent variable and daily ET0 values estimated by HS as an independent variable. The parameters of the regression equation were then presented as the calibration parameters.

Country	Station	latitude m	Altitude m	Classification Koppen	Rainfall mm	RH %	U2 ms^{-1}	Regression adjustment intercept a	slope b	R2	RMSE	Source
Microthermal												
Fully humid												
Serbia	Kragujevac	44º00' N	190	Dfa		75%	1		0.78		0.451	Trajkovic, 2005
Serbia	Belgrade	44º45' N	132	Dfa	684	69%	1.7		0.99			Trajkovic, 2005
Cro., Ser. Bos.	Zagreb, Sarajevo, etc.	42.6- 46.1	42-630	Dfb		68-76	1.0-1.9		0.424 [3]			Trajkovic, 2007
Canada	Southern Ontario, Drumbo	43º16' N	310	Dfb		79%	1.5		0.74	0.7	0.704	Sentelhas et al. 2010
Canada	Southern Ontario, Harrow	42º12' N	190	Dfb		73%	2.2		0.94	0.64	0.704	Sentelhas et al. 2010
Dry winter												
China	Tibete plateau- Yushu	33º06' N	3681	Dwb	200	45.4	0.83	0.347	0.883	0.91	0.622	Ye et al. 2009
Polar												
Bulgaria	Trace plain, Plovdiv	42º 25' N	160	ET	492				1.11			Popova et al, 2006
Switzerland	Changins	46º24'N	416	ET	904	73	2.5	-0.31	1.12	0.99		Xu and Singh, 2002
Tropical												
Winter dry												
México	Mérida (Yucatán)	20º56' N	15	Aw	11.74			0.1754	1.021	0.78		Bautista et al 2009
Tanzania	Upper Mkoji	9º00'	1700	Aw	1070	77.5	1.23	0.006	0.987			Igbadun et al
India	Kharagpur			Aw				-2.64	1.561			Kashyap and Panda, 2001
India	Bangalore	13º00' N	921	Aw	940	66	1.9	-0.1063	1.0244			Nandagiri and Kovoor, 2006
Nigeria	Abeokuta	7º10' S	62	Aw	1506	92	2.12	-1.41	0.938			Adeboye, 2009
Nigeria	Abeokuta	7º10' S	63	Aw	1506	92	2.12	0.0025 [1]	16.8 [2]			Adeboye, 2009
Brasil	Goiânia, GO	16º28' S	823	Aw	1785	87.9	0.82	0.6923	0.3811	0.47		Oliveira et al. 2005
Brasil	Sooretama (South Espirito Santo	19º22'S	75	Am		75.9	3.34	-2.62	1.572			Reis et al, 2007
Summer dry												
Brasil	Campina Grande	7º14'S	550	As'	700	80	1.38	-0.488	0.893			Henrique, 2006
Fully humid												
Brasil	North Rio de Janeiro	21º19'S	13	Af	1172.9	73.1	0.3	-0.76	1			Mendonça et al 2003
Philipines	Los Banos	14º13' N	41	Af	1987	83.3	1.35		0.96	0.65		Stockle, 2004

* compared with lysimeter values

Values in grey are annual averages obtained from Climwat data base.

(1) (2) (3) Respectively, the K'T, d and e of regionally calibrated HS equation, according to Equation 7 in the text.

When calibration parameters of the HS vsFAO PM were not directly provided, linear regression equations were established with FAO-56 PM daily ET0 estimates as the dependent variable and daily ET0 values estimated by HS as an independent variable. The parameters of the regression equation were then presented as the calibration parameters.

Table 1. Regional calibration for the Hargreaves Samani equation compiled from published works

Where N is the maximum number of sunny hours as a function of the month and latitude and dm is the number of days per month. ETo_{sc} is the gross evapotranspiration (without corrections) and can be calculated as:

$$Et_0sc = 16\left(\frac{10T_a}{I}\right)a \tag{9}$$

where T_a is the mean daily temperature (°C), a is an exponent as a function of the annual index: a = 0.49239 + 1792 × 10^{-5} I - 771 × 10^{-7} I^2 + 675 × 10^{-9} I^3; and I is the annual heat index obtained form the monthly heat indecies:

$$I = \sum_{m=1}^{12}\left(\frac{T_m}{5}\right)1.514 \tag{10}$$

Bautista et al. (2009) found that the precision of the Thorntwaite methodology improved during the winter months in Mexico. Garcia et al. (2004) observed that under the dry and arid conditions of the Bolivian highlands the Thornthwaite equation strongly underestimates ETo because the equation does not consider the saturation deficit of the air (Stanhill, 1961; Pruitt, 1964; Pruitt and Doorenbos, 1977). Additionally, at high altitudes, the Thornthwaite equation also underestimates the effect of radiation, because the equation is calibrated for temperate low altitude climates. Studies in Brazil have shown that the underestimation of ETo produced by temperature-based equations under arid conditions, may be reduced by using the daily thermal amplitude instead of the mean temperature (Paes de Camargo, 2000) as in the case of the Hargreaves–Samani equation.

Gonzalez et al. (2009) studied the Thorthwaite method in the Bolivian Amazon. They observed that the Thornthwaite method underestimates evapotranspiration at all the three stations studied. This is expected, considering that normally this method leads to underestimations in humid areas (Jensen et al., 1990).

2.3 Blaney-Criddle method

The FAO Temperature Methodology recommended by Doorenbos and Pruitt (1977) is based on the Blaney-Criddle method (Blaney and Criddle, 1950), introducing a correction factor based on estimates of humidity, sunshine and wind.

$$ET_o = \alpha + \beta\left[p(0.46T + 8.13)\right] \tag{11}$$

where α and β are calibration parameters and p is the mean annual percentage of daytime hours. Values for α can be calculated using the daily RH_{min} and n/N as follows:

$$\alpha = 0.043RH_{\min} - \left(\frac{n}{N}\right) - 1.41 \tag{12}$$

$$\frac{n}{N} = 2(Rs / Ra) - 0.5 \tag{13}$$

For windy South Nebraska, Irmak et al. (2008) compared 12 different ET methodologies and found that the Blaney-Criddle method was the best temperature method and it had an RMSE value (0.64 mm d^{-1}) which was similar to some of the combination methods. The obtained estimates were good and were within 3% of the ASCE-PM *ETo* with a high r^2 of 0.94. The estimates were consistent with no large under or over estimations for the majority of the dataset. They attributed this to the fact that, unlike most of the other temperature methods, this method takes into account humidity and wind speed in addition to air temperature.

Lee et al. (2004) compared various *ETo* calculation methods in the West Coast of Malaysia and concluded that the Blaney-Criddle method was the best, among the reduced-set equations, for estimating ET in the region. They also observed that HS gave the highest estimates followed by the Priestly-Taylor equation. Similarly, in the humid Goiânia region of Brazil, Oliveira et al. (2005) observed that the Blaney-Criddle method produced the best results, next to the full PM equation.

Various studies indicate that the Blaney-Criddle equation might show some bias under arid conditions. For semi-arid conditions of Iran, Dehghani Sanij et al. (2004) found the Blaney-Criddle and the Makkink method to overestimate ETo during the growing season. Lopéz-Urrea et al. (2006) compared seven different methods for calculating ETo in the semiarid regions of Spain and observed that the Blaney-Criddle method significantly over-estimated average daily ETo.

For arid conditions of Iran, Fard et al. (2009) compared nine different methodologies with lysimeter data and observed that the Turc and the Blaney-Criddle methods showed very close agreement with the lysimeter data, while PM showed moderate agreement with the lysimeter data. The other methods showed bias, systematically over estimating the lysimeter data (Fig. 4).

Although recognizing the historical value of the Blaney-Criddle method and its validity, the FAO Expert Commission on Revision of FAO Methodologies for Crop Water Requirements (Smith et al. 1992) did not recommend the method further, in view of difficulties in estimating humidity, sunshine and wind parameters in remote areas. Nevertheless, they emphasized the value of the method for areas having only the mean daily temperature, and where appropriate correction factors can be found.

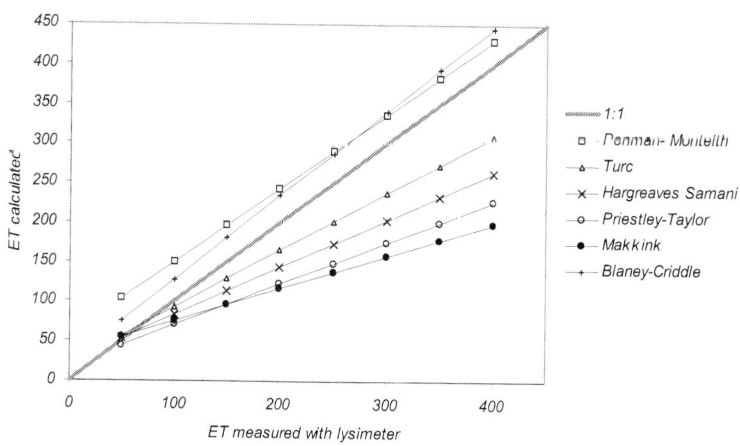

Fig. 4. Comparision of six ET methods with lysimeter data for Isfahan (adapted from Fard et al., 2009).

2.4 Reduced-set PM

The PM methodology has provisions for application in data-short situations (Allen et al. 1998), including the use of temperature data alone. The reduced-set PM equation requiring only the measured maximum and minimum temperatures uses estimates of solar radiation, relative humidity, and wind speed. Solar radiation, *Rs*, MJ m^{-2} d^{-1} can be estimated using equation 3 (Hargreaves and Samani, 1985) or using averages from nearby stations. For island locations *Rs* can be estimated as (Allen et al. 1998):

$$R_s = 0.7 R_a - b \tag{14}$$

where b is an empirical constant with a value of 4 MJ m^{-2} d^{-1} . Relative humidity can be estimated by assuming that the dewpoint temperature is approximately equal to T_{min} (Allen 1996; Allen et al. 1998) which is usually experienced at sunrise. In this case, e_a can be calculated as:

$$e_a = e^o(T_{\min}) = 0.611 \exp\left[\frac{17.27 T_{\min}}{T_{\min} + 237.3}\right] \tag{15}$$

where $e^o(T_{min})$ is the vapour pressure at the minimum temperature, expressed in mbar. For wind speed, Allen et al. (1998) recommend using average wind speed data from nearby locations or using a wind speed of 2 m s^{-1}, since, they consider, the impact of wind speed on the *ETo* results is relatively small, except in arid and windy areas. The soil heat flux density, *G*, for monthly periods can be estimated as:

$$G_i = 0.07(T_{i+1} - T_{i-1}) \tag{16}$$

where G_i is the soil heat flux density in month *I* in MJ m^{-2} d^{-1}; and T_{i+1} and T_{i-1} are the mean air temperatures in the previous and following months, respectively.

Allen (1995) evaluated the reduced-set PM (using only *Tmax* and *Tmin*) and HS using the mean annual monthly data from the 3,000 stations in the FAO CLIMWAT data base, with the full PM serving as the comparative basis. He found little difference in the mean monthly *ETo* between the two methods. Wright et al. (2000) found similar results in Kimberly, and 75 years of data from California (Hargreaves and Allen, 2003). Other data generally indicate that the reduced-set PM performs better in humid areas (Popova, 2005, Pereira et al., 2003), while HS performs better in dry climates (Temesgen et al. 2005, Jabloun et al. 2008).

Trajkovic (2005) compared the reduced-set PM, Hargreaves, and Thornthwaite temperature-based methods with the full PM in Serbia and found that the reduced-set PM estimates were better than those produced from the Hargreaves and Thornthwaite equations. Popova et al. (2006) found the reduced-set PM to provide more accurate results compared to the Hargreaves equation, which tended to overestimate reference evapotranspiration in the Trace plain in south Bulgaria. Jabloun and Sahli (2008) also found the Hargreaves equation to overestimate reference evapotranspiration in Tunisia and found the reduced-set PM equation to provide better estimates. Nevertheless, the reduced-set PM can produce poor results in areas where wind speed is significantly different from 2 ms^{-1} (Trajkovic, 2005).

3. Radiation based methods

It is known that water loss from a crop is related to the incident solar energy, and thus it is possible to develop a simple model that relates solar radiation to evapotranspiration.

Various models have been developed, over the years, for relating the measured net global radiation to the estimated reference evapotranspiration; such as the Priestley-Taylor method (1972), the Makkink method (1957), the Turc radiation method (1961), and the Jensen and Haise method (1965).
Irmak et al. (2008) compared 11 ET models and studied the relevance of their complexity for direct prediction of hourly, daily and seasonal scales. They concluded that radiation is the dominant driver of evaporative losses, over seasonal time scales, and that other meteorological variables, such as temperature and wind speed, gained importance in daily and hourly calculations.

3.1 The Priestley-Taylor method

The Priestley-Taylor method (Priestley and Taylor, 1972; De Bruin, 1983) is a simplified form of the Penman equation, that only needs net radiation and temperature to calculate *ETo*. This simplification is based on the fact that *ETo* is more dependant on radiation than on relative humidity and wind. The Priestly-Taylor method is basically the radiation driven part of the Penman Equation, multiplied by a coefficient, and can be expressed as:

$$ET_o = \alpha \frac{\Delta(R_n - G)}{\Delta + \gamma} + \beta \qquad (17)$$

where α and β are calibration factors, assuming values of 1.26 and 0, respectively. This model was calibrated for Switzerland (Xu and Singh, 1998) and values of 0.98 and 0.94 were obtained for α and β, respectively. In the Priestley-Taylor equation, evapotranspiration is proportional to net radiation, while in the Makkink equation (section 3.2), it is proportional to short-wave radiation.
Van Kraalingen and Stol (1997) found that application of the Priestly-Taylor equation during the Dutch winter months was not possible because it is based on net radiation. Since net radiation is often negative in the winter, it predicts dew formation, whereas the actual ET is positive. The situation would be different for a humid climate such as the Philippines, or in a semi-arid climate such as Israel, where the equation should compare well with PM.
Irmak et al. (2003) calibrated the Priestly-Taylor method against the FAO PM method using 15 years of climate data (1980–1994) in humid Florida, United States. The monthly values of the calibration coefficient (Fig. 5) show a considerable seasonal variation, aside from the natural difference in annual values. In general, the calibration coefficients are lower in winter months indicating that the Priestley and Taylor method underestimates *ETo*, and they are higher than 1.0 during the summer months, indicating that the method overestimates during the summer months. The long-term average lowest calibration values were obtained in January and December (0.70) and the highest values in July (1.10). These results indicate the importance of developing monthly calibration coefficients for regional use based on historic records. For the semi-arid conditions of southern Portugal, the authors also observed that the Priestley-Taylor method over-estimates daily ETo during the summer months (Shahidian et al., 2007).
Shuttleworth and Calder (1979) showed that Priestley-Taylor significantly underestimates wet forest evaporation, but also overestimates dry forest transpiration by as much as 20%. Berengena and Gavilán (2005) found that the Priestley–Taylor equation shows a considerable tendency to underestimate ETo, on average 23%, under convective conditions.

They concluded that the Priestly-Taylor equation is very sensitive to advection, and local calibration does not ensure an acceptable level of accuracy.

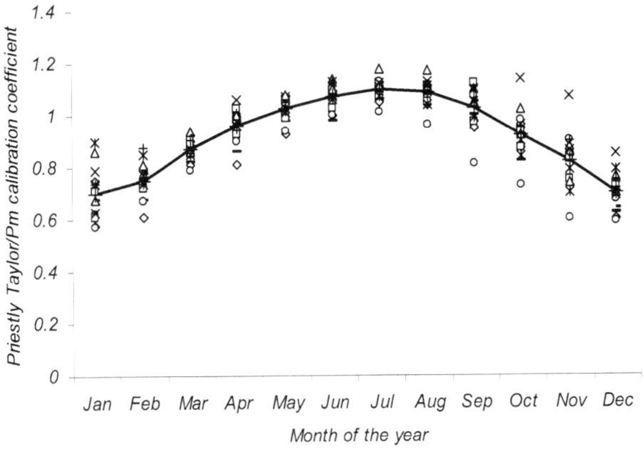

Fig. 5. Average monthly calibration coefficient for the Priestly-Taylor equation against PM for humid southern United States (based on data from Irmak et al. 2003).

3.2 The Makkink method

The Makkink method can be seen as a simplified form of the Priestley-Taylor method and was developed for grass lands in Holland. The difference is that the Makkink method uses incoming short-wave radiation *Rs* and temperature, instead of using net radiation, *Rn*, and temperature. This is possible, because on average, there is a constant ratio of 50% between net radiation and short wave radiation. The equation can be expressed as:

$$Et_o = \alpha \frac{\Delta}{\Delta + \gamma} \frac{R_s}{2,45} + \beta \tag{18}$$

where α is usually 0.61, and β is -0.012. Doorenbos and Pruitt (1975) proposed the FAO Radiation method based on the Makkink equation (1957), introducing a correction factor based on estimates for wind and humidity conditions to compensate for advective conditions. This radiation method has been proven valid, in particular under humid conditions, but can differ systematically from the PM reference method under special conditions, such as during dry months (Bruin and Lablands, 1998).

It has also been observed that it is difficult to use this radiation based method during winter months: Van Kraalingen and Stol (1997) found that application of the Makkink equation in Dutch winter months was not possible, though the Makkink equation did not produce negative values for ET, as was the case with the Priestley-Taylor method. Bruin and Lablans (1998) also concluded that there is no relationship between Makkink and PM in the winter months, December and January, since Makkink's method has no physical meaning, in this period.

It is reasonable to expect the Makkink and the Priestley-Taylor equations to compare well with the Penman's method, since in all these approaches the radiation terms are dominant and radiation is the main driving force for evaporation in short vegetation.

ET models tend to perform best in climates in which they were designed. A study by Amayta et al. (1995) showed that while the Makkink model generally performed well in North Carolina, the model underestimated *ETo* in the peak months of summer. Yet, the Makkink model shows excellent results in Western Europe where it was designed, both in comparison to PM as well as to the measured ETo data (Bruin and Lablans 1998, Xu and Singh 2000, Bruin and Stricker 2000, Barnett et al., 1998).

3.3 The Turc method

Also known as the Turc-Radiation equation, this method was presented by Turc in 1961, using data from the humid climate of Western Europe (France). This method only uses two parameters, average daily radiation and temperature and for RH>50% can be expressed as:

$$ET_p = \alpha\left((23{,}9001R_s) + 50\right)\left(\frac{T}{T+15}\right) \tag{19}$$

And for RH < 50% as:

$$ET_p = \alpha\left((23{,}9001R_s) + 50\right)\left(\frac{T}{T+15}\right)\left(1+\left(\frac{50-RH}{70}\right)\right) \tag{20}$$

Where α is 0.01333 and *Rs* is expressed in MJ m^{-2} day^{-1}.

Yoder et al. (2005) compared six different ET equations in humid southeast United States, and found the Turc equation to be second best only to the full PM. Jensen et al. (1990) analyzed the properties of twenty different methods against carefully selected lysimeter data from eleven stations, located worldwide in different climates. They observed that the Turc method compared very favorably with combination methods at the humid lysimeter locations. The Turc method was ranked second when only humid locations were considered, with only the Penman-Monteith method performing better. Trajkovic and Stojnic (2007) compared the Turc method with full PM in 52 European sites and found a SEE (Standard Error of Estimate) of between 0.10 and 0.37 mm d^{-1}. They also found that the reliability of the Turc method depends on the wind speed (Fig 6). The Turc method overestimated PM ETo in windless locations and generally underestimated ETo in windy locations.

Amatya et al. (1995) compared 5 different ETo methodologies in North Carolina and concluded that the Turc and the Priestley-Taylor methods were generally the best in estimating ETo. They observed that all other radiation methods and the temperature based Thorntwaite method underestimated the annual ET by as much as 16%.

Kashyap and Panda (2001) compared 10 different methods with lysimeter data in the sub humid Kharagupur region of India and observed that the Turc method had a deviation of only 2.72% from lysimeter values, followed by Blaney-Criddle with a 3.16% and Priestly Taylor with a 6.28% deviation (Fig. 7). The Kashyap and Panda data are also important because they show that under sub humid conditions, most of the equations, including the PM, tend to overestimate when evapotranspiration is low, and underestimate when it is high.

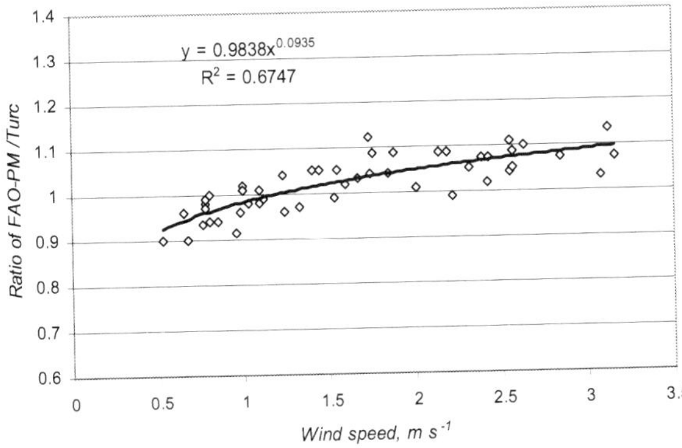

Fig. 6. Effect of wind on the ratio of evapotranspiration calculated with the FAO PM and the Turc methods (based on data from Trajkovic and Stojnic (2007), using average annual values).

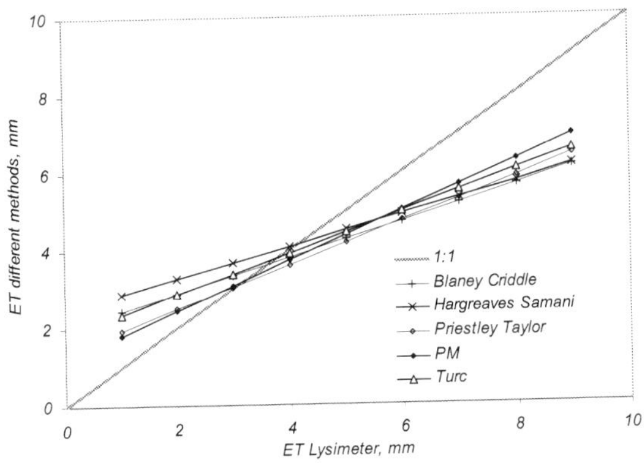

Fig. 7. Comparison of various ETo methods with Lysimeter readings in the sub-humid region of Kharagpur, India (adapted from Kashyap and Panda, 2001).

For Florida, Martinez and Thepadia (2010) compared the reduced-set PM equation with various temperature and radiation based equations and concluded that in the absence of regionally calibrated methods, the Turc equation has the least error and bias when using measured maximum and minimum temperatures. They also observed that the reduced-set PM and Hargreaves equations overestimate ET.

Fontenote (2004) studied the accuracy of seven evapotranspiraiton models for estimating grass reference ET in Louisiana. He observed that, statewide and in the coastal region, the Turc model was the most accurate daily model with a MAE of 0.26mm day^{-1}. Inland, the Blaney-Criddle performed best with a MAE of 0.31mm day^{-1} (Fig. 8).

Hence, it can be safely concluded that the Turc model can be expected to perform well in warm, humid climates such as those found in North Carolina (Amatya et al., 1995), India (George et al., 2002), and Florida (Irmak et al., 2003; Martinez and Thepadia, 2010).

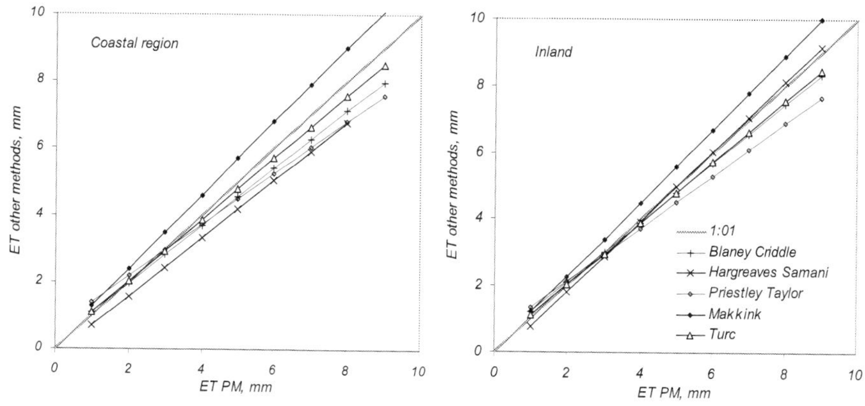

Fig. 8. Comparison of five ET methods with PM in two different regions of Louisiana (Adapted from Fontenote, 1999).

3.4 The Jensen and Haise method

This method was derived for the drier parts of the United States and is based on 3,000 observations of ET. Jensen and Haise used 35 years of measured evapotranspiration and solar radiation to derive the equation, based on the assumption that net radiation is more closely related to ET than other variables such as air temperature and humidity (Jensen and Haise, 1965). The equation can be expressed as:

$$ET = C_t\left(T - T_x\right)R_s \tag{21}$$

The original study of Jensen and Haise provides a calculation procedure to obtain *Rs* from the cloudiness, *Cl*, and the solar and sky radiation flux on cloudless days. The temperature Constant, C_t, and the intercept of the temperature exis, T_x, can be calculated as follows:

$$C_t = \frac{1}{\left[\left(45 - \frac{h}{137}\right) + \left(\frac{365}{e^0\left(T_{\max}\right) - e^0\left(T_{\min i}\right)}\right)\right]} \tag{22}$$

and

$$T_x = -2.5 - 0.14\left(e^0\left(T_{\max}\right) - e^0\left(T_{\min}\right)\right) - h/500 \tag{23}$$

where h is the altitude of the location in m, *Rs* is solar radiation (MJ m^{-2} d^{-1}); e^oT_{max} and e^oT_{min} are vapour pressures of the month with the mean maximum temperature and the month with the mean minimum temperature, respectively, expressed in mbar.

For the humid and rainy Rio Grande watershed in Brazil, Pereira et al. (2009) compared 10 different equations and concluded that the methods based on solar radiation are more accurate than those based only on air temperature, with the Jensen and Haise method presenting the smallest MBE, and thus being the method most recommended for this region.

4. Conclusions

Both temperature and radiation can be used successfully to calculate daily *ETo* values with relative accuracy. All the equations can be used for areas that have a climate that is similar to the one for which the equations were originally developed; while most of the equations can be used with some confidence for areas with moderate conditions of humidity and wind speed.

Regional calibration, especially if including monthly calibration coefficients, is important in decreasing the bias of the ETo estimates. Wind speed can greatly influence the results obtained with reduced-set equations, since wind removes the boundary layer from the leaf surface and can significantly increase evapotranspiration. Relative Humidity is another important factor that can affect the results.

Globally, it is observed that the Turc equation is highly recommended for humid or semi-humid areas, where it can produce very good results even without calibration, while the Thornthwaite equation tends to underestimate *ETo*.

The Priestley-Taylor and the Makkinik equations should not be used in the winter months in locations with high latitude, such as northern Europe.

Both the Hargreaves and the reduced-set Panman-Monteith can be effectively used with only temperature measurements, although the results can be improved if wind speed is taken into consideration.

The use of the reduced-set equations can be very important in actual irrigation management, since the error involved in using these equations can be much smaller than that resulting from using data from a weather station located many miles away.

5. References

Allen RG (1993) Evaluation of a temperature difference method for computing grass reference evapotranspiration. Report submitted to the Water Resources Develop. and Man. Serv., Land and Water Develop. Div., FAO, Rome. 49 p.

Allen RG, Pereira LS, Raes D, Smith M (1998) Crop evapotranspiration: Guidelines for computing crop requirements. Irrigation and Drainage Paper No. 56, FAO, Rome, Italy.

Amatya DM, Skaggs RW, Gregory JD(1995) Comparison of Methods for Estimating REF-ET. Journal of Irrigation and Drainage Engineering 121:427-435.

Bautista F, Bautista D, Delgado-Carranza (2009) Calibration of the equations of Hargreaves and Thornthwaite to estimate the potential evapotranspiration in semi-arid and sub-humid tropical climates for regional applications. Atmósfera 22(4): 331-348

Berengena J, Gavilán P (2005) Reference evapotranspiration estimation in a highly advective semiarid environment. J. Irrig. Drain. Eng. ASCE 131 (2):147–163.

Borges AC, Mendiondo EM (2007) Comparação entre equações empíricas para estimativa da evapotranspiração de referência na Bacia do Rio Jacupiranga. Revista Brasileira de Engenharia Agrícola e Ambiental 11(3): 293–300.

Blaney, HF, Criddle, WD (1950). Determining water requirements in irrigated áreas from climatological and irrigation data. In ISDA Soil Conserv. Serv., SCS-TP-96,

Blaney HF, Criddle WD. (1962). Determining Consumptive Use and Irrigation Water Requirements. USDA Technical Bulletin 1275, US Department of Agriculture, Beltsvill

Bruin, HAR Lablans, WN, (1998) Reference crop evapotranspiration determined with a modified Makkink equation. Hydrological Processes

Brutsaert, W (1991) Evapotration into the atmosphere. D. Reidel Publishing Company.

DehghaniSanij H, Yamamoto T, Rasiah V (2004) Assessment of evapotranspiration estimation models for use in semi-arid environments. Agricultural Water Management 64: 91–106

Hargreaves, G.H. (1994). Simplified coefficients for estimating monthly solar radiation in North America and Europe. Departmental Paper, Dept. of Biol. And Irrig. Engrg., Utah State University, Logan, Utah

Doorenbos J., Pruit WO. (1977) Guidelines for predicting crop water requirements. FAO irrigation and drainagem paper, 24

Droogers P, Allen, RG (2002) Estimating reference evapotranspiration under inaccurate data conditions. *Irrig. Drain. Syst.* 16(1): 33-45.

Fontenot Rl (2004) An evaluation of reference evapotranspiration models in Louisiana, MSc thesis, Louisiana State University, August.

Garcia M, Raes D, Allen R, Herbas C, (2004) Dynamics of reference evapotranspiration in the Bolivian highlands (Altiplano). Agric. Forest Meteorol. 125: 67–82.

George BA, Reddy BRS, Raghuvanshi NS, Wallender WW (2002) Decision support system for estimating reference evapotranspiration. J. Irrig. Drain. Eng. 128(1): 1–10.

Hargreaves GH, Samni ZA. (1982) Estimation of potential evapotranspiration. Journal of Irrigation and Drainage Division, Proceedings of the American Society of Civil Engineers 108: 223-230

Hargreaves GH, Samani ZA. (1985) Reference crop evapotranspiration from temperature. Appl Engine Agric. 1(2):96–99.

Hargreaves GH, Allen RG (2003) History and Evaluation of Hargreaves Evapotranspiration Equation. Journal of Irrigation and Drainage Engineering129(1): 53-63.

Henggeler J.C., Z. Samani, M.S. Flynn, J.W. Zeitler (1996) Evaluation of various evapotranspiration equations for Texas and New Mexico. Proceeding of Irrigation Association International Conference, San Antonio, Texas

Igbadun H, Mahoo H, Tarimo A, Salim B (2006) Performance of Two Temperature-Based Reference Evapotranspiration Models in the Mkoji Sub-Catchment in Tanzania. Agricultural Engineering International: the CIGR Ejournal. Manuscript LW 05 008. Vol. VIII. March,

Irmak S, Irmak A, Allen RG, Jones JW (2003) Solar and Net Radiation-Based Equations to Estimate Reference Evapotranspiration in Humid Climates. Journal of irrigation and drainage engineering. September/October

Irmak S, Istanbulluoglu, E, Irmak A. (2008) An Evaluation of Evapotranspiration model complexity against performance in comparison with Bowen Ration Energy Balance measurements. Transactions of the ASABE 51(4):1295-1310

Jabloun M, Sahli A (2007) Ajustement de l'e´quation de Hargreaves-Samani aux conditions climatiques de 23 stations climatologiques Tunisiennes. Bulletin Technique no. 2. Laboratoire de Bioclimatologie, Institut National Agronomique de Tunisie, 21 p.

Jabloun M, Sahli A (2008) Evaluation of FAO-56 methodology for estimating reference evapotranspiration using limited climatic data. Application to Tunisia. Agricultural Water Management 95: 707-715.

Jensen ME, Haise HR (1963) Estimating evapotranspiration from solar radiation. Journal of Irrigation and Drainage Division, Proc. Amer. Soc. Civil Eng. 89:15–41.

Jensen DT, Hargreaves GH, Temesgen B, Allen RG (1997) Computation of ETo under non ideal conditions. J. Irrig. Drain. Eng. ASCE 123 (5): 394–400.

Kashyap PS, Panda RK (2001) Evaluation of evapotranspiration estimation methods and development of crop-coefficients for potato crop in a sub-humid region. Agricultural Water Management 50: 9-25

López-Urrea R, Martín de Santa Olalla F, Fabeiro C, Moratalla A (2006) Testing evapotranspiration equations using lysimeter observations in a semiarid climate. Agricultural Water Management 85: 15–26.

Lu, J, Sun G, McNulty S, Amatya DM (2005) A Comparison of Six Potential Evapotranspiration Methods for Regional Use in the Southeastern United States. Journal of the American Water Resources Association (JAWRA) 41(3):621-633.

Makkink GF. (1957) Testing the Penman formula by means of lysimeters. Journal of the Institution of Water Engineers 11: 277-28

Martinez, A.M., Thepadia, M. (2010) Estimating Reference Evapotranspiration with Minimum Data in Florida. J. Irrig. Drain. Eng. 136(7): 494-501

McKenney, M. S. and N. J. Rosenberg, (1993) Sensitivity of some potential evapotranspiration estimation methods to climate change. - *Agricultural and Forest Meteorology* 64, 81-110.

Mohan S, Arumugam N (1996) Relative importance of meteorological variables in evapotranspiration: Factor analysis approach. Water Resour. Manage. 10, 1–20.

Nandagiri L, Kovoor GM (2006) Performance Evaluation of Reference Evapotranspiration Equations across a Range of Indian Climates. Journal of Irrigation and Drainage Engineering 132(3) . DOI: 10.1061/(ASCE)0733-9437(2006)132:3(238)

Oliveira RZ, Oliveira LFC, Wehr TR, Borges LB, Bonomo R (2005) Comparative study of estimative models for reference evapotranspiration for the region of Goiânia, Go. Biosci. J. Uberlândia, 21(3):19-27.

Paes de Camargo, A. and Paes de Camargo, M.(2000) Numa revisão analítica da evapotranspiração potencial Bragantia. Campinas 59 2, pp. 125–137.

Penman, H.L. (1948) Natural evaporation from open water, bare soil, and grass. Proc. Roy. Soc. London A193:120-146.

Pereira DR, Yanagi SNM, Mello CR, Silva AM, Silva LA (2009) Performance of the reference evapotranspiration estimating methods for the Mantiqueira range region, MG, Brazil. Ciência Rural, Santa Maria 39(9):2488-2493.

Priestley CHB, Taylor RJ (1972) On the assessment of the surface heat flux and evaporation using large-scale parameters. Monthly Weather Review 100: 81–92

Popova Z, Kercheva M, Pereira LS (2006) Validation of the FAO methodology for computing ETo with limited data. Application to south Bulgaria. Irrig Drain. 55:201–215.

Rahimikhoob AR (2008) Comparative study of Hargreaves's and artificial neural network's methodologies in estimating reference evapotranspiration in a semiarid environment. Irrig Sci 26:253–259.

Reis EF, Bragança R, Garcia GO, Pezzopane JEM, Tagliaferre C (2007) Comparative study of the estimate of evaporate transpiration regarding the three locality state of Espirito Santo during the dry period. IDESIA (Chile) 25(3) 75-84

Samani Z (2000) Estimating solar radiation and evapotranspiration using minimum climatological data. J Irrig Drain Engin. 126(4):265–267.

Samani ZA, Pessarakli M (1986) Estimating potential crop evapotranspiration with minimum data in Arizona. Trans. ASAE (29): 522–524.

Sepaskhah, AR, Razzaghi, FH (2009) Evaluation of the adjusted Thornthwaite and Hargreaves-Samani methods for estimation of daily evapotranspiration in a semi-arid region of Iran. Archives of Agronomy and Soil Science, 55: 1, 51- 6

Sentelhas C, Gillespie TJ, Santos EA (2010) Evaluation of FAO Penman-Monteith and alternative methods for estimating reference evapotranspiration with missing data in Southern Ontario, Canada. Agricultural Water Management 97: 635–644.

Shahidian, S., Serralheiro, R.P., Teixeira, J.L., Santos, F.L., Oliveira, M.R.G., Costa, J.L., Toureiro, C.. Haie, N. (2007) Desenvolvimento dum sistema de rega automático, autónomo e adaptativo. I Congreso Ibérico de Agroingeneria.

Shahidian S. , Serralheiro R. , Teixeira J.L., Santos F.L., Oliveira M.R., Costa J., Toureiro C., Haie N., Machado R. (2009) Drip Irrigation using a PLC based Adaptive Irrigation System WSEAS Transactions on Environment and Development, Vol 2- Feb.

Shuttleworth, W.J., and I.R. Calder. (1979) Has the Priestley-Taylor equation any relevance to the forest evaporation? Journal of Applied Meteorology, 18: 639-646.

Smith, M, R.G. Allen, J.L. Monteith, L.S. Pereira, A. Perrier, and W.O. Pruitt. (1992) Report on the expert consultation on procedures for revision of FAO guidelines for prediction of crop water requirements. Land and Water Development Division, United Nations Food and Agriculture Service, Rome, Italy

Stanhill, G., 1961. A comparison of methods of calculating potential evapotranspiration from climatic data. Isr. J. Agric. Res.: Bet-Dagan (11), 159–171.

Teixeira JL, Shahidian S, Rolim J (2008) Regional analysis and calibration for the South of Portugal of a simple evapotranspiration model for use in an autonomous landscape irrigation controller.

Thornthwaite CW (1948) An approach toward a rational classification of climate. Geograph Rev.38:55–94.

Trajkovic S. (2005) Temperature-based approaches for estimating reference evapotranspiration. J Irrig Drain Engineer. 131(4):316–323

Trajkovic S (2007) Hargreaves versus Penman-Monteith under Humid Conditions Journal of Irrigation and Drainage Engineering, Vol. 133, No. 1, February 1.

Trajkovic, S, Stojnic, V. (2007) Effect of wind speed on accuracy of Turc Method in a humid climate. Facta Universitatis. 5(2):107-113

Xu, CY, Singh, VP (2000) Evaluation and Generalization of Radiation-based Methods for Calculating Evaporation, *Hydrolog. Processes* 14: 339–349.

Xu CY, Singh VP (2002) Cross Comparison of Empirical Equations for Calculating Potential evapotranspiration with data from Switzerland. Water Resources Management 16: 197-219.

5

Fuzzy-Probabilistic Calculations of Evapotranspiration

Boris Faybishenko
Lawrence Berkeley National Laboratory
Berkeley, CA
USA

1. Introduction

Evaluation of evapotranspiration uncertainty is needed for proper decision-making in the fields of water resources and climatic predictions (Buttafuoco et al., 2010; Or and Hanks, 1992; Zhu et al., 2007). However, in spite of the recent progress in soil-water and climatic uncertainty quantification, using stochastic simulations, the estimates of potential (reference) evapotranspiration (E_0) and actual evapotranspiration (*ET*) using different methods/models, with input parameters presented as PDFs or fuzzy numbers, is a somewhat overlooked aspect of water-balance uncertainty evaluation (Kingston et al., 2009). One of the reasons for using a combination of different methods/models and presenting the final results as fuzzy numbers is that the selection of the model is often based on vague, inconsistent, incomplete, or subjective information. Such information would be insufficient for constructing a single reliable model with probability distributions, which, in turn, would limit the application of conventional stochastic methods.

Several alternative approaches for modeling complex systems with uncertain models and parameters have been developed over the past ~50 years, based on fuzzy set theory and possibility theory (Zadeh, 1978; 1986; Dubois & Prade, 1994; Yager & Kelman, 1996). Some of these approaches include the blending of fuzzy-interval analysis with probabilistic methods (Ferson & Ginzburg, 1995; Ferson, 2002; Ferson et al., 2003). This type of analysis has recently been applied to hydrological research, risk assessment, and sustainable water-resource management under uncertainty (Chang, 2005), as well as to calculations of E_0, ET, and infiltration (Faybishenko, 2010).

The objectives of this chapter are to illustrate the application of a combination of probability and possibility conceptual-mathematical approaches—using fuzzy-probabilistic models—for predictions of potential evapotranspiration (E_0) and actual evapotranspiration (*ET*) and their uncertainties, and to compare the results of calculations with field evapotranspiration measurements.

As a case study, statistics based on monthly and annual climatic data from the Hanford site, Washington, USA, are used as input parameters into calculations of potential evapotranspiration, using the Bair-Robertson, Blaney-Criddle, Caprio, Hargreaves, Hamon, Jensen-Haise, Linacre, Makkink, Penman, Penman-Monteith, Priestly-Taylor, Thornthwaite, and Turc equations. These results are then used for calculations of evapotranspiration based on the modified Budyko (1974) model. Probabilistic calculations are performed using Monte

Carlo and p-box approaches, and fuzzy-probabilistic and fuzzy simulations are conducted using the RAMAS Risk Calc code. Note that this work is a further extension of this author's recently published work (Faybishenko, 2007, 2010).

The structure of this chapter is as follows: Section 2 includes a review of semi-empirical equations describing potential evapotranspiration, and a modified Budyko's model for evaluating evapotranspiration. Section 3 includes a discussion of two types of uncertainties—epistemic and aleatory uncertainties—involved in assessing evapotranspiration, and a general approach to fuzzy-probabilistic simulations by means of combining possibility and probability approaches. Section 4 presents a summary of input parameters and the results of E_o and ET calculations for the Hanford site, and Section 5 provides conclusions.

2. Calculating potential evapotranspiration and evapotranspiration

2.1 Equations for calculations of potential evapotranspiration

The potential (reference) evapotranspiration E_o is defined as evapotranspiration from a hypothetical 12 cm grass reference crop under well-watered conditions, with a fixed surface resistance of 70 s m^{-1} and an albedo of 0.23 (Allen et al., 1998). Note that this subsection includes a general description of equations used for calculations of potential evapotranspiration; it does not provide an analysis of the various advantages and disadvantages in applying these equations, which are given in other publications (for example, Allen et al., 1998; Allen & Pruitt, 1986; Batchelor, 1984; Maulé et al., 2006; Sumner & Jacobs, 2005; Walter et al., 2002).

The two forms of Baier-Robertson equations (Baier, 1971; Baier & Robertson, 1965) are given by:

$$E_o = 0.157T_{max} + 0.158\,(T_{max} - T_{min}) + 0.109R_a - 5.39 \tag{1}$$

$$E_o = -0.0039T_{max} + 0.1844(T_{max} - T_{min}) + 0.1136\,R_a + 2.811(e_s - e_a) - 4.0 \tag{2}$$

where E_o= daily evapotranspiration (mm day^{-1}); T_{max} = the maximum daily air temperature, ^oC; T_{min}= minimum temperature, ^oC; R_a = extraterrestrial radiation (MJ m^{-2} day^{-1}) (ASCE 2005), e_s = saturation vapor pressure (kPa), and e_a = mean actual vapor pressure (kPa). Equation (1) takes into account the effect of temperature, and Equation (2) takes into account the effects of temperature and relative humidity.

The Blaney-Criddle equation (Allen & Pruitt, 1986) is used to calculate evapotranspiration for a reference crop, which is assumed to be actively growing green grass of 8–15 cm height:

$$E_o = p\,(0.46 \cdot T_{mean} + 8) \tag{3}$$

where E_o is the reference (monthly averaged) evapotranspiration (mm day^{-1}), T_{mean} is the mean daily temperature (°C) given as T_{mean} = $(T_{max} + T_{min})/2$, and p is the mean daily percentage of annual daytime hours.

The Caprio (1974) equation for calculating the potential evapotranspiration is given by

$$E_o = 6.1 \cdot 10^{-6}\,R_s\,[(1.8 \cdot T_{mean}) + 1.0] \tag{4}$$

where E_o = mean daily potential evapotranspiration (mm day^{-1}); R_s = daily global (total) solar radiation (kJ m^{-2} day^{-1}); and T_{mean} = mean daily air temperature (°C).

The Hansen (1984) equation is given by:

$$E_o = 0.7\,\Delta/\ (\Delta + \gamma)\ \cdot R_i/\lambda \tag{5}$$

where Δ = slope of the saturation vapor pressure vs. temperature curve, γ = psychrometric constant, R_i = global radiation, and λ = latent heat of water vaporization.
The Hargreaves equation (Hargreaves & Samani, 1985) is given by

$$E_o = 0.0023(T_{mean} + 17.8)(T_{max} - T_{min})^{0.5}\,R_a \tag{6}$$

where both E_o and R_a (extraterrestrial radiation) are in millimeters per day^{-1} (mm day^{-1}).
The Jensen and Haise (1963) equation is given by

$$E_o = R_s/2450\ [(0.025\ T_{mean}) + 0.08] \tag{7}$$

where E_o = monthly mean of daily potential evapotranspiration (mm day^{-1}); R_s = monthly mean of daily global (total) solar radiation (kJ m^{-2} day^{-1}); and T_{mean} = monthly mean temperature.
The Linacre (1977) equation is given by:

$$E_o = [500T_m\ /\ (100\text{-}L) + 15(T\text{-}Td)]\ /\ (80\text{-}T) \tag{8}$$

where E_o is in mm day^{-1}, T_m = temperature adjusted for elevation, $T_m = T + 0.006\text{h}$ (°C), h = elevation (m), T_d = dew point temperature (°C), and L = latitude (°).
The Makkink (1957) model is given by

$$E_o = 0.61\ \Delta/\ (\Delta + \gamma)\ R_s/2.45 - 0.12 \tag{9}$$

where R_s = solar radiation (MJ m^{-2} day^{-1}), and Δ and γ are the parameters defined above.
The Penman (1963) equation is given by

$$E_o = mR_n + \gamma\ 6.43(1+0.536\ u_2)\ \delta e\ /\ \lambda_v\ (m + \gamma) \tag{10}$$

where Δ = slope of the saturation vapor pressure curve (kPa K^{-1}), R_n = net irradiance (MJ m^{-2} day^{-1}), ρ_a = density of air (kg m^{-3}), c_p = heat capacity of air (J kg^{-1} K^{-1}), δe = vapor pressure deficit (Pa), λ_v = latent heat of vaporization (J kg^{-1}), γ = psychrometric constant (Pa K^{-1}), and E_o is in units of kg/(m^{2}s).
The general form of the Penman-Monteith equation (Allen et al., 1998) is given by

$$E_o = [0.408\ \Delta\ (R_n - G) +\ C_n\,\gamma\ /(T{+}273)\ u_2\ (e_s\text{-}e_a)]\ /\ [\Delta + \gamma\ (1{+}C_d\,u_2)] \tag{11}$$

where E_o is the standardized reference crop evapotranspiration (in mm day^{-1}) for a short (0.12 m, with values C_n=900 and C_d=0.34) reference crop or a tall (0.5 m, with values C_n=1600 and C_d=0.38) reference crop, R_n = net radiation at the crop surface (MJ m^{-2} day^{-1}), G = soil heat flux density (MJ m^{-2} day^{-1}), T = air temperature at 2 m height (°C), u_2 = wind speed at 2 m height (m s^{-1}), e_s = saturation vapor pressure (kPa), e_a = actual vapor pressure (kPa), (e_s - e_a) = saturation vapor pressure deficit (kPa), Δ = slope of the vapor pressure curve (kPa °C^{-1}), and γ = psychrometric constant (kPa °C^{-1}).
The Priestley-Taylor (1972) equation is given by

$$E_o = \alpha\ 1/\lambda\ \Delta\ (R_n - G)\ /\ (\Delta + \gamma) \tag{12}$$

where λ = latent heat of vaporization (MJ kg^{-1}), R_n = net radiation (MJ m^{-2} day^{-1}), G = soil heat flux (MJ m^{-2} day^{-1}), Δ= slope of the saturation vapor pressure-temperature relationship (kPa °C^{-1}), γ = psychrometric constant (kPa °C^{-1}), and α = 1.26. Eichinger et al. (1996) showed that α=1.26 is practically constant for all typically observed atmospheric conditions and relatively insensitive to small changes in atmospheric parameters. (On the other hand, Sumner and Jacobs [2005] showed that α is a function of the green-leaf area index [LAI] and solar radiation.)

The Thornthwaite (1948) equation is given by

$$E_o = 1.6\,(L/12)\,(N/30)\,(10\,T_{\mathrm{mean}\,(i)}/I)^{\alpha} \tag{13}$$

where E_o is the estimated potential evapotranspiration (cm/month), $T_{\mathrm{mean}\,(i)}$ = average monthly (i) temperature (°C); if $T_{\mathrm{mean}\,(i)} < 0$, $Eo = 0$ of the month (i) being calculated, N = number of days in the month, L = average day length (hours) of the month being calculated, and I = heat index given by

$$I = \sum_{i=1}^{12}\left(\frac{T_{\mathrm{mean}(i)}}{5}\right)^{1.514}$$

and $\alpha = (6.75\cdot10^{-7})\,I^3 - (7.71\cdot10^{-5})\,I^2 + (1.792\cdot10^{-2})I + 0.49239$

The Turc (1963) equation is given by

$$E_o = (0.0239\cdot R_s + 50)\,[0.4/30\cdot T_{\mathrm{mean}}/(T_{\mathrm{mean}} + 15.0)] \tag{14}$$

where E_o = mean daily potential evapotranspiration (mm/day); R_s = daily global (total) solar radiation (kJ/m^2/day); T_{mean} = mean daily air temperature (°C).

2.2 Modified Budyko's equation for evaluating evapotranspiration

For regional-scale, long-term water-balance calculations within arid and semi-arid areas, we can reasonably assume that (1) soil water storage does not change, (2) lateral water motion within the shallow subsurface is negligible, (3) the surface-water runoff and runon for regional-scale calculations simply cancel each other out, and (4) ET is determined as a function of the aridity index, ϕ: $ET=f(\phi)$, where $\phi = E_o/P$, which is the ratio of potential evapotranspiration, E_o, to precipitation, P (Arora 2002).

Budyko's (1974) empirical formula for the relationship between the ratio of ET/P and the aridity index was developed using the data from a number of catchments around the world, and is given by:

$$ET/P = \{\phi \tanh(1/\phi)\,[1 - \exp(-\phi)]\}^{0.5} \tag{15}$$

Equation (1) can also be given as a simple exponential expression (Faybishenko, 2010):

$$ET/P = a[1-\exp(-b\phi)] \tag{16}$$

with coefficients a =0.9946 and b =1.1493. The correlation coefficient between the calculations using (15) and (16) is R=0.999. Application of the modified Budyko's equation, given by an exponential function (2) with the ϕ value in single term, will simplify further calculations of ET.

3. Types of uncertainties in calculating evapotranspiration and simulation approaches

3.1 Epistemic and aleatory uncertainties

The uncertainties involved in predictions of evapotranspiration, as a component of soil-water balance, can generally be categorized into two groups—*aleatory* and *epistemic* uncertainties. Aleatory uncertainty arises because of the natural, inherent variability of soil and meteorological parameters, caused by the subsurface heterogeneity and variability of meteorological parameters. If sufficient information is available, probability density functions (PDFs) of input parameters can be used for stochastic simulations to assess aleatory evapotranspiration uncertainty. In the event of a lack of reliable experimental data, fuzzy numbers can be used for fuzzy or fuzzy-probabilistic calculations of the aleatory evapotranspiration uncertainty (Faybishenko 2010).

Epistemic uncertainty arises because of a lack of knowledge or poor understanding, ambiguous, conflicting, or insufficient experimental data needed to characterize coupled-physics phenomena and processes, as well as to select or derive appropriate conceptual-mathematical models and their parameters. This type of uncertainty is also referred to as subjective or reducible uncertainty, because it can be reduced as new information becomes available, and by using various models for uncertainty evaluation. Generally, variability, imprecise measurements, and errors are distinct features of uncertainty; however, they are very difficult, if not impossible, to distinguish (Ferson & Ginzburg, 1995).

In this chapter the author will consider the effect of aleatory uncertainty on evapotranspiration calculations by assigning the probability distributions of input meteorological parameters, and the effect of epistemic uncertainty is considered by using different evapotranspiration models.

3.2 Simulation approaches

3.2.1 Probability approach

A common approach for assessing uncertainty is based on Monte Carlo simulations, using PDFs describing model parameters. Another probability-based approach to the specification of uncertain parameters is based on the application of probability boxes (Ferson, 2002; Ferson et al., 2003). The probability box (p-box) approach is used to impose bounds on a cumulative distribution function (CDF), expressing different sources of uncertainty. This method provides an envelope of distribution functions that bounds all possible dependencies. An uncertain variable x expressed with a probability distribution, as shown in Figure 1a, can be represented as a variable that is bounded by a p-box $[\overline{F}, \underline{F}]$, with the right curve $\underline{F}\ (x)$ bounding the higher values of x and the lower probability of x, and the left curve $\overline{F}\ (x)$ bounding the lower values and the higher probability of x. With better or sufficiently abundant empirical information, the p-box bounds are usually narrower, and the results of predictions come close to a PDF from traditional probability theory.

3.2.2 Possibility approach

In the event of imprecise, vague, inconsistent, incomplete, or subjective information about models and input parameters, the uncertainty is captured using *fuzzy modeling theory*, or *possibility theory*, introduced by Zadeh (1978). For the past 50 years or so, possibility theory has successfully been applied to describe such systems as complex, large-scale engineering systems, social and economic systems, management systems, medical diagnostic processes, human perception, and others. The term *fuzziness* is, in general, used in possibility theory to

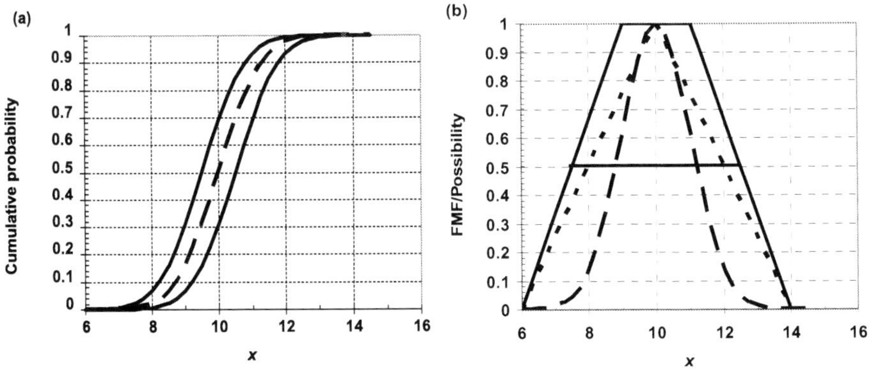

Fig. 1. Graphical illustration of uncertain numbers: (a) Cumulative normal distribution function (dashed line), with mean=10 and standard deviation σ=1, and a p-box—left bound with mean=9.5 and σ=0.9, and right bound with mean=10.5 and σ=1.1; and (b) Fuzzy trapezoidal (solid line) number, plotted using Eq. (17) with a=6, b=9, c=11, and d=14. Interval [b,c]=[9, 11] corresponds to FMF=1. Triangular (short dashes) and Gaussian (long dashes) fuzzy numbers are also shown. Figure (b) also shows an α-cut=0.5 (thick horizontal line) through the trapezoidal fuzzy number (Faybishenko 2010).

describe objects or processes that cannot be given precise definition or precisely measured. *Fuzziness* identifies a class (set) of objects with nonsharp (i.e., fuzzy) boundaries, which may result from imprecision in the meaning of a concept, model, or measurements used to characterize and model the system. Fuzzification implies replacing a set of crisp (i.e., precise) numbers with a set of fuzzy numbers, using fuzzy membership functions based on the results of measurements and perception-based information (Zadeh 1978). A fuzzy number is a quantity whose value is imprecise, rather than exact (as is the case of a single-valued number). Any fuzzy number can be thought of as a function whose domain is a specified set of real numbers. Each numerical value in the domain is assigned a specific "grade of membership," with 0 representing the smallest possible grade (full nonmembership), and 1 representing the largest possible grade (full membership). The grade of membership is also called the degree of possibility and is expressed using fuzzy membership functions (FMFs). In other words, a fuzzy number is a fuzzy subset of the domain of real numbers, which is an alternative approach to expressing uncertainty.

Several types of FMFs are commonly used to define fuzzy numbers: triangular, trapezoidal, Gaussian, sigmoid, bell-curve, Pi-, S-, and Z-shaped curves. As an illustration, Figure 1b shows a trapezoidal fuzzy number given by

$$f(x) = \begin{Bmatrix} 0, x \le a \\ \dfrac{x-a}{b-a}, a \le x \le b \\ 1, b \le x \le c \\ \dfrac{d-x}{d-c}, c \le x \le d \\ 0, d \le x \end{Bmatrix}, \quad (17)$$

where coefficients a, b, c, and d are used to define the shape of the trapezoidal FMF. When $a= b$, the trapezoidal number becomes a triangular fuzzy number.

Figure 1b also illustrates one of the most important attributes of fuzzy numbers, which is the notion of an α-cut. The α-cut interval is a crisp interval, limited by a pair of real numbers. An α-cut of 0 of the fuzzy variable represents the widest range of uncertainty of the variable, and an α-cut value of 1 represents the narrowest range of uncertainty of the variable.

Possibility theory is generally applicable for evaluating all kinds of uncertainty, regardless of its source or nature. It is based on the application of both hard data and the subjective (perception-based) interpretation of data. Fuzzy approaches provide a distribution characterizing the results of all possible magnitudes, rather than just specifying upper or lower bounds. Fuzzy methods can be combined with calculations of PDFs, interval numbers, or p-boxes, using the RAMAS Risk Calc code (Ferson 2002). In this paper, the RAMAS Risk Calc code is used to assess the following characteristic parameters of the fuzzy numbers and p-boxes:

- Mean—an interval between the means of the lower (left) and upper (right) bounds of the uncertain number x.
- Core—the most possible value(s) of the uncertain number x, i.e., value(s) with a possibility of one, or for which the probability can be any value between zero and one.
- Iqrange—an interval guaranteed to enclose the interquartile range (with endpoints at the 25th and 75th percentiles) of the underlying distribution.
- Breadth of uncertainty—for fuzzy numbers, given by the area under the membership function; for p-boxes, given by the area between the upper and lower bounds. The uncertainty decreases as the breadth of uncertainty decreases.

When fuzzy measures serve as upper bounds on probability measures, one could expect to obtain a conservative (bounding) prediction of system behavior. Therefore, fuzzy calculations may overestimate uncertainty. For example, the application of fuzzy methods is not optimal (i.e., it overestimates uncertainty) when sufficient data are available to construct reliable PDFs needed to perform a Monte Carlo analysis.

In a recent paper (Faybishenko 2010), this author demonstrated the application of the fuzzy-probabilistic method using a hybrid approach, with direct calculations, when some quantities can be represented by fuzzy numbers and other quantities by probability distributions and interval numbers (Kaufmann and Gupta 1985; Ferson 2002; Guyonnet et al. 2003; Cooper et al. 2006). In this paper, the author combines (aggregates) the results of Monte Carlo calculations with multiple E_o models by means of fuzzy numbers and p-boxes, using the RAMAS Risk Calc software (Ferson 2002).

4. Hanford case study

4.1 Input parameters and modeling scenarios for the Hanford Site

The Hanford Site in Southeastern Washington State is one of the largest environmental cleanup sites in the USA, comprising 1,450 km^2 of semiarid desert. Located north of Richland, Washington, the Hanford Site is bordered on the east by the Columbia River and on the south by the Yakima River, which joins the Columbia River near Richland, in the Pasco Basin, one of the structural and topographic basins of the Columbia Plateau. The areal topography is gently rolling and covered with unconsolidated materials, which are sufficiently thick to mask the surface irregularities of the underlying material. Areas adjacent to the Hanford Site are primarily agricultural lands.

Meteorological parameters used to assign model input parameters were taken from the Hanford Meteorological Station (HMS—see http://hms.pnl.gov/), located at the center of the Hanford Site just outside the northeast corner of the 200 West Area, as well as from publications (DOE, 1996; Hoitink et al., 2002; Neitzel, 1996.) At the Hanford Site, the E_0 is estimated to be from 1,400 to 1,611 mm/yr (Ward et al. 2005), and the ET is estimated to be 160 mm/yr (Figure 2). A comparison of field estimates with the results of calculations performed in this paper is shown in Section 4.2. Calculations are performed using the temperature and precipitation time-series data representing a period of active soil-water balance (i.e., with no freezing) from March through October for the years 1990-2007. A set of meteorological parameters is summarized in Table 1, which are then used to develop the input PDFs and fuzzy numbers shown in Figure 3.

Several modeling scenarios were developed (Table 2) to assess how the application of different models for input parameters affects the uncertainty of E_0 and ET calculations. For the sake of simulation simplicity, the input parameters are assumed to be independent variables. Scenarios 0 to 8, described in detail in Faybishenko (2010), are based on the application of a single Penman model for E_0 calculations, with annual average values of input parameters. Scenario 0 was modeled using input PDFs by means of Monte Carlo simulations, using RiskAMP Monte Carlo Add-In Library version 2.10 for Excel. Scenarios 1 through 8 were simulated by means of the RAMAS Risk Calc code. Scenario 1 was simulated using input PDFs, and the results are given as p-box numbers. Scenarios 2 through 6 were simulated applying both PDFs and fuzzy number inputs, corresponding to α-cuts from 0 to 1). Scenarios 7 and 8 were simulated using only fuzzy numbers. The calculation results of Scenarios 0 through 8 are compared in this chapter with newly calculated Scenarios 9 and 10, which are based on Monte Carlo calculations by means of all E_0 models, described in Section 2, and then bounding the resulting PDFs by a trapezoidal fuzzy number (Scenario 9) and the p-box (Scenario 10).

Type of data	Parameters		Wind speed (km/hr)	Relative humidity (%)		Albedo	Solar radiation (Ly/day)	Annual precipitation (mm/yr)	Temperature (°C)	
				Max	Min				Max	Min
PDFs	Mean		15.07	80.2	33.3	0.21	332.55	185	33.41	2.87
	Standard Deviation		0.92	4.01	1.66	0.021	16.63	55.62	1.08	1.11
Trapezoidal FMFs	α= 0	Min	12.31	68.17	28.29	0.15	282.66	46.0	30.17	0.0
		Max	17.84	92.23	38.31	0.27	382.44	324.1	36.65	6.17
	α=1	Min	14.61	78.2	32.47	0.22	324.24	157.2	32.87	2.32
		Max	15.53	82.2	34.14	0.27	382.44	212.8	33.95	3.42

Table 1. Meteorological parameters from the Hanford Meteorological Station used for E_0 calculations for all scenarios (the data sources are given in the text).

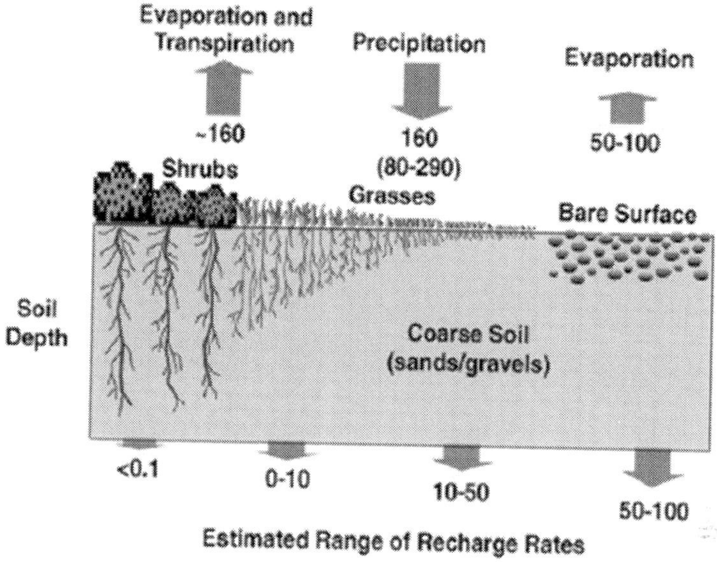

Fig. 2. Estimated water balance *ET* and recharge/infiltration at the Hanford site (Gee et al, 2007).

Scenarios	Input parameters						Output parameters
	Wind speed	Humidity	Albedo	Solar radiation	Precipitation	Temperature	
0	PDF	PDF	PDF	PDF	PDF	PDF	PDF
1	PDF	PDF	PDF	PDF	PDF	PDF	p-box
2	Fuzzy	PDF	PDF	PDF	PDF	PDF	Hybrid
3	Fuzzy	Fuzzy	PDF	PDF	PDF	PDF	Hybrid
4	Fuzzy	Fuzzy	Fuzzy	PDF	PDF	PDF	Hybrid
5	Fuzzy	Fuzzy	Fuzzy	Fuzzy	PDF	PDF	Hybrid
6	Fuzzy	Fuzzy	Fuzzy	Fuzzy	Fuzzy	PDF	Hybrid
7[1)]	Fuzzy	Fuzzy	Fuzzy	Fuzzy	Fuzzy	Fuzzy	Fuzzy
8[2)]	Fuzzy	Fuzzy	Fuzzy	Fuzzy	Fuzzy	Fuzzy	Fuzzy
9[3)]	PDF	PDF	PDF	PDF	PDF	PDF	Fuzzy
10[3)]	PDF	PDF	PDF	PDF	PDF	PDF	p-box

Notes:
[1)] In Scenario 7, all FMFs are trapezoidal.
[2)] In Scenario 8, all FMFs are triangular: the mean values of parameters, which are given in Table 1, are used for α=1; and the minimum and maximum values of parameters, given in Table 1 for trapezoidal FMFs (Scenario 7), are also used for α =0 of triangular FMFs in Scenario 8.
[3)] In Scenarios 9 and 10, input parameters are monthly averaged.

Table 2. Scenarios of input and output parameters used for water-balance calculations (Scenarios 0, and 1-8 are from Faybishenko, 2010).

4.2 Results and comparison with field data

4.2.1 Potential evapotranspiration (E_o)

Figure 4a shows cumulative distributions of E_o from different models, along with an aggregated p-box, and Figure 4b shows the corresponding FMFs (calculated as normalized PDFs) of E_o from different models, along with an aggregated trapezoidal fuzzy E_o. These figures illustrate that the Baier-Robertson (Eq. 1), Blaney-Criddle (Eq. 3), Hargreaves (Eq. 6), Penman (Eq. 10), Penman-Monteith (Eq. 11) (for tall plants), and Priestly-Taylor (Eq. 12) models provide the best match with field data, while the Makkink (Eq. 9) and Thornthwaite (Eq. 13) models significantly underestimate the E_o, and the Linacre (Eq. 8) and Baier-Robertson (Eq. 2) models greatly overestimate E_o.

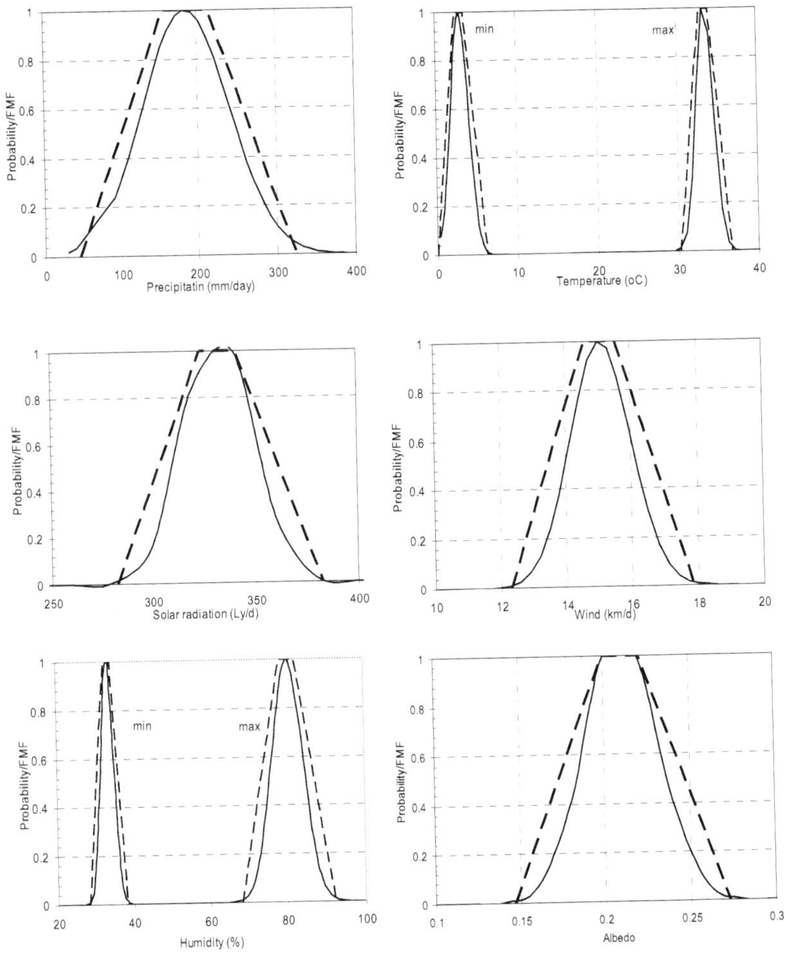

Fig. 3. Input PDFs (solid lines) and fuzzy numbers (dashed lines) used for calculations (Faybishenko, 2010).

Figure 5a demonstrates that the E_o mean from Monte Carlo simulations is within the mean ranges from the p-box (Scenario 1) and fuzzy-probabilistic scenarios (Scenarios 2-6). It also corresponds to a midcore of the fuzzy scenario with trapezoidal FMFs (Scenario 7), the core of the fuzzy scenario with triangular FMFs (Scenario 8), and the centroid values of the fuzzy E_o of Scenario 9, as well as a p-box of Scenario 10.

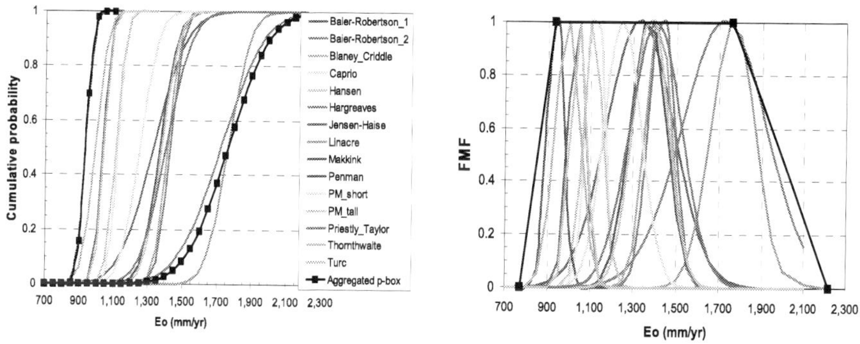

Fig. 4. (a) Cumulative probability of potential evapotranspiration calculated using different E_o formulae; an aggregated p-box, which is shown by a black line with solid squares: normal distribution with the left/minimum curve—mean=933, var=1070, and the right /max curve—mean=1763, var=35755; and (b) corresponding fuzzy numbers (calculated from normalized PDFs); an aggregated trapezoidal fuzzy number is shown by a black line—Eq. (17) with a=772, b=933, c=1763, and d=2222. (all numbers of E_o are in mm/yr)

The range of means from the p-box and fuzzy-probabilistic calculations for α=1 is practically the same, indicating that including fuzziness within the input parameters does not change the range of most possible E_o values. Figure 5a shows that the core uncertainty of the trapezoidal FMFs (Scenario 7) is the same as the uncertainty of means for fuzzy-probabilistic calculations for α =1. Obviously, the output uncertainty decreases for the input triangular FMFs (Scenario 8), because these FMFs resemble more tightly the PDFs used in other scenarios. Figure 5a also illustrates that a relatively narrow range of field estimates of E_o—from 1,400 to 1,611 mm/yr for the Hanford site (Ward 2005)—is well within the calculated uncertainty of E_o values. Note from Figure 5a that the uncertainty ranges from p-box, hybrid, and fuzzy calculations significantly exceed those from Monte Carlo simulations for a single Penman model, but are practically the same as those from calculations using multiple E_o models.

Characteristic parameters (Figures 5a) and the breadth of uncertainty (Figure 6a) of E_o calculated from multiple models—Scenarios 9 and 10—are in a good agreement with field measurements and other calculation scenarios.

4.2.2 Evapotranspiration (*ET*)

Figure 5b shows that the mean *ET* of ~184 mm/yr from Monte Carlo simulations (Scenario 0) is practically the same as the *ET* means for Scenarios 1 through 5 and the core value for Scenario 8. The greater *ET* uncertainty for Scenario 6 (precipitation is simulated using a fuzzy number) can be explained by the relatively large precipitation range for α=0—from 46 to 324 mm/yr. At the same time, the means of *ET* values for α =1 range within relatively narrow limits, as the precipitation for α =1 changes from 157.2 to 212.8 mm/yr (see Table 1).

The breadth of uncertainty of ET (Figure 6b) is practically the same for Scenarios 1 through 5, increase for Scenarios 6, 7, and 8 in the account of calculations using a fuzzy precipitation, and then decrease for Scenarios 9 and 10 using multiple E_o models. A smaller range of ET uncertainty calculated using multiple E_o models can be explained by the fact that the Budyko curve asymptotically reaches the limit of ET/P=1 for high values of the aridity index, which are typical for the semi-arid climatic conditions of the Hanford site.

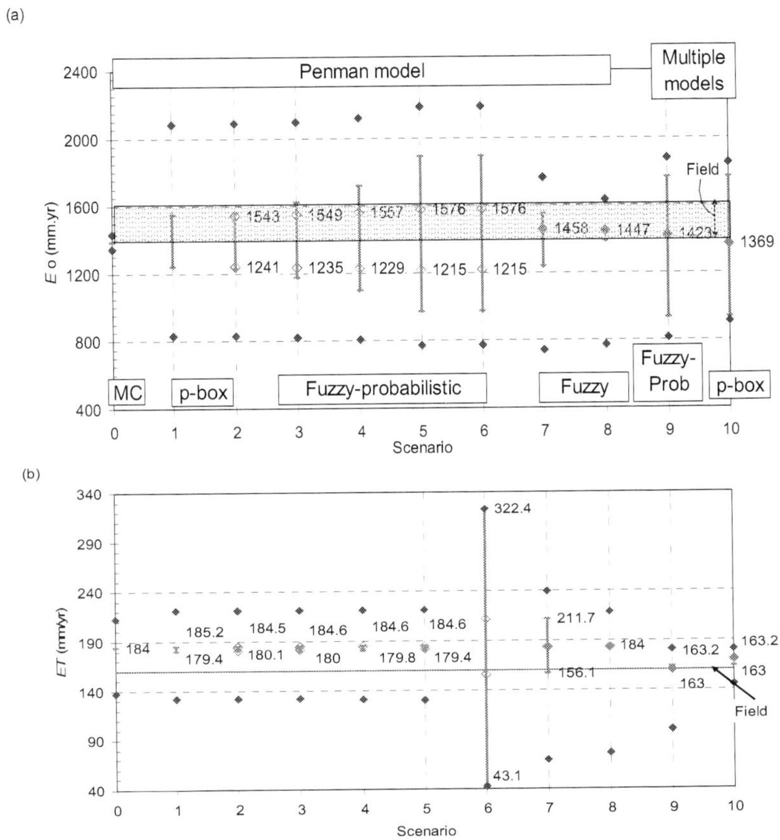

Fig. 5. Results of calculations of E_o (a) and ET (b) and comparison with field measurements. Red vertical lines are the mean intervals (Scenarios 1-6, and 10) and core intervals (Scenarios 7, 8, and 9), the blue diamonds indicate the interquartile ranges with endpoints at the 25th and 75th percentiles of the underlying distribution. Red open diamonds for Scenarios 2-6 indicate the mean intervals for the hybrid level=10 (Faybishenko 2010), and red solid diamonds for Scenarios 7-10 indicate centroid values. The height of a shaded area in figure *a* indicates the range of E_o from field measurements. (Results of calculations of Scenarios 0-8 are from Faybishenko, 2010.)

The calculated means for Scenarios 0, 1–5, and 8 exceed the field estimates of ET of 160 mm/yr (Gee et al., 1992; 2007) by 22 to 24 mm/yr. This difference can be explained by Gee et al. using a lower value of annual precipitation (160 mm/yr for the period prior to 1990) in their calculations, while our calculations are based on using a greater mean annual precipitation (185 mm/yr), averaged for the years from 1990 to 2007. The field-based data are within the ET uncertainty range for Scenarios 6 and 7, since the precipitation range is wider for these scenarios. Calculations using multiple E_o models generated the ET values (Scenarios 9 and 10), which are practically the same as those from field measurements.

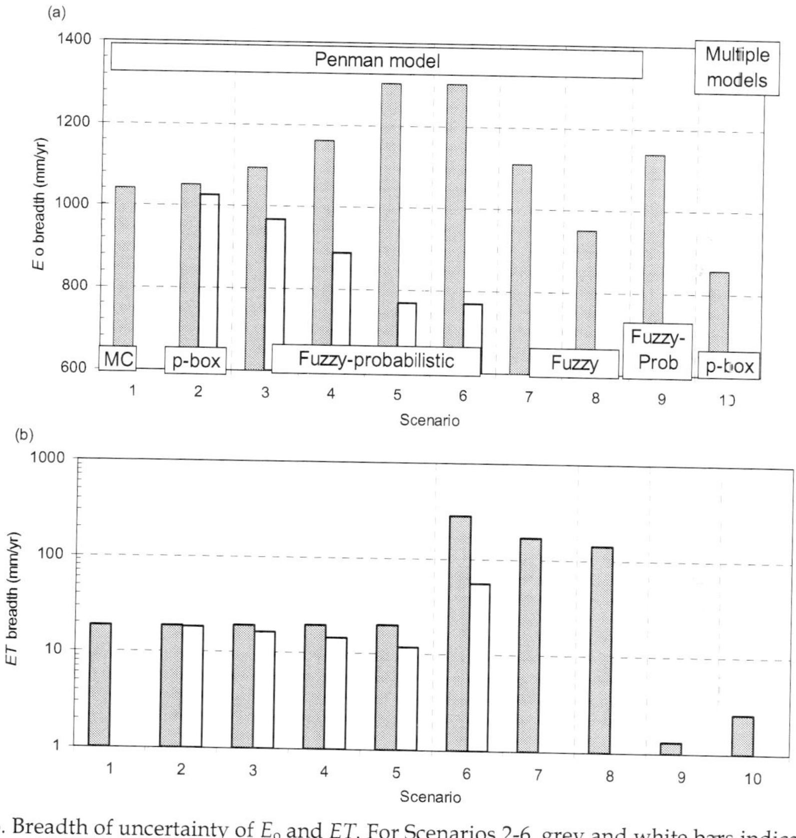

Fig. 6. Breadth of uncertainty of E_o and ET. For Scenarios 2-6, grey and white bars indicate the maximum and minimum uncertainty, correspondingly. (Results of calculations of Scenarios 0-8 are from Faybishenko, 2010.)

5. Conclusions

The objectives of this chapter are to illustrate the application of a fuzzy-probabilistic approach for predictions of E_o and ET, and to compare the results of calculations with those from field measurements at the Hanford site. Using historical monthly averaged data from the Hanford Meteorological Station, this author employed Monte-Carlo simulations to assess the frequency distribution and statistics of input parameters for these models, which are then used as input into probabilistic simulations. The effect of aleatory uncertainty on calculations of evapotranspiration is assessed by assigning the probability distributions of input meteorological parameters, and the combined effect of aleatory and epistemic (model) uncertainty is then expressed by means of aggregating the results of calculations using a p-box and fuzzy numbers. To illustrate the application of these approaches, the potential evapotranspiration is calculated using the Bair-Robertson, Blaney-Criddle, Caprio, Hargreaves-Samani, Hamon, Jensen-Haise, Linacre, Makkink, Priestly-Taylor, Penman, Penman-Monteith, Thornthwaite, and Turc models, and evapotranspiration is then determined based on the modified Budyko (1974) model. Probabilistic and fuzzy-probabilistic calculations using multiple E_0 models generate the E_0 and ET results, which are well within the range of field measurements and the application of a single Penman model. The Baier-Robertson, Blaney-Criddle, Hargreaves, Penman, Penman-Monteith, and Priestly-Taylor models provide the best match with field data.

6. Acknowledgment

This work was partially supported by the Director, Office of Science, Office of Biological and Environmental Remediation Sciences of the U.S. Department of Energy, and the DOE EM-32 Office of Soil and Groundwater Remediation (ASCEM project) under Contract No. DE-AC02-05CH11231 to Lawrence Berkeley National Laboratory.

7. References

Allen, R.G. & Pruitt, W.O. (1986), Rational Use of the FAO Blaney-Criddle Formula, *Journal of Irrigation and Drainage Engineering*, Vol. 112, No. 2, pp. 139-155, doi 10.1061/(ASCE)0733-9437(1986)112:2(139)

Allen, R.G.; Pereira L.S.; Raes, D. & Smith, M. (1998). Crop evapotranspiration - Guidelines for computing crop water requirements - FAO Irrigation and drainage paper 56.

Arora V.K. (2002). The use of the aridity index to assess climate change effect on annual runoff, *J. of Hydrology*, Vol. 265, pp. 164–177.

ASCE (2005). The ASCE Standardized Reference Evapotranspiration Equation. Edited by R. G. Allen, I.A. Walter, R.; Elliott, T. Howell, D. Itenfisu, and M. Jensen. New York, NY: American Society of Civil Engineers.

Baier, W. & Robertson, G.W. 1965. Estimation of latent evaporation from simple weather observations. Can. J. Soil Sci. 45, pp. 276-284.

Baier, W. (1971). Evaluation of latent evaporation estimates and their conversion to potential evaporation. *Can. J. of Plant Sciences* 51, pp. 255-266.

Budyko, M.I. (1974).*Climate and Life*, Academic, San Diego, Calif., 508 pp.

Buttafuoco, G.; Caloiero, T.; & Coscarelli, R. (2010). Spatial uncertainty assessment in modelling reference evapotranspiration at regional scale, *Hydrol. Earth Syst. Sci.*, 14, pp. 2319-2327, doi:10.5194/hess-14-2319-2010, 2010.

Caprio, J.M. (1974). The Solar Thermal Unit Concept in Problems Related to Plant Development and Potential Evapotranspiration. In: H. Lieth (Editor), *Phenology and Seasonality Modeling. Ecological Studies.* Springer Verlag, New York, pp. 353-364.

Chang N-B (2005). Sustainable water resources management under uncertainty, *Stochast Environ Res and Risk Assess*, 19, pp. 97–98.

Cooper J.A.; Ferson, S.; & Ginzburg, L. (2006). Hybrid processing of stochastic and subjective uncertainty data, *Risk Analysis*, Vol. 16, No. 6, pp. 785–791.

DOE (1996). Final Environmental Impact Statement for the Tank Waste Remediation System, Hanford Site, Richland, Washington, DOE/EIS-0189. Available from http://www.globalsecurity.org/wmd/library/report/enviro/eis-0189/app_i_3.htm

Dubois, D. & Prade, H. (1994). Possibility theory and data fusion in poorly informed environments. *Control Engineering Practice* 2(5), pp. 811-823.

Eichinger W.E.; Parlange, M.B. & Stricker H. (1996). On the Concept of Equilibrium Evaporation and the Value of the Priestley-Taylor Coefficient, *Water Resour. Res.*, 32(1): 161-164, doi:10.1029/95WR02920.

Faybishenko, B. (2007). Climatic Forecasting of Net Infiltration at Yucca Mountain Using Analogue Meteorological Data, *Vadose Zone Journal*, 6: 77–92.

Faybishenko, B. (2010), Fuzzy-probabilistic calculations of water-balance uncertainty, *Stochastic Environmental Research and Risk Assessment*, Vol. 24, No. 6, pp. 939–952.

Ferson, S. (2002). RAMAS Risk Calc 4.0 Software: Risk assessment with uncertain numbers, CRC Press.

Ferson, S. & Ginzburg, L. (1995) Hybrid arithmetic. *Proceedings of the 1995 Joint ISUMA/ NAFIPS Conference*, IEEE Computer Society Press, Los Alamitos, California, pp. 619-623.

Ferson, S.; Kreinovich, V.; Ginzburg, L; Myers, D.S. & Sentz, K. (2003). Constructing probability boxes and Dempster-Shafer structures, SAND REPORT, SAND2002-4015.

Gee, G.W.; Fayer, M.J.; Rockhold, M.L. & Campbell, M.D. (1992). Variations in recharge at the Hanford Site. *Northwest Sci.* 66, pp. 237–250.

Gee, G.W.; Oostrom, M.; Freshley, M.D.; Rockhold, M.L. & Zachara, J.M. (2007). Hanford site vadose zone studies: An overview, *Vadose Zone Journal* Vol. 6, pp 899-905.

Guyonne, D.; Dubois, D. ; Bourgine, B.; Fargier, H.; Côme, B. & Chilès, J.-P. (2003). Hybrid method for addressing uncertainty in risk assessments. *Journal of Environmental Engineering* 129: 68-78.

Hansen, S. (1984). Estimation of Potential and Actual Evapotranspiration, *Nordic Hydrology*, 15, 1984, pp. 205-212, Paper presented at the Nordic Hydrological Conference (Nyborg, Denmark, August 1984).

Hargreaves, G.H. & Z.A. Samani (1985). Reference crop evapotranspiration from temperature. *Transaction of ASAE* 1(2), pp. 96-99.

Hoitink, D.J.; Burk, K.W ; Ramsdell, Jr, J.V; & Shaw, W.J. (2003). Hanford Site Climatological Data Summary 2002 with Historical Data . PNNL-14242, Pacific Northwest National Laboratory, Richland, WA.

Jensen, M.E. & Haise, H.R. (1963). Estimating evapotranspiration from solar radiation. *J. Irrig. Drainage Div. ASCE*, 89: 15-41.

Kaufmann, A. & Gupta, M.M. (1985). *Introduction to Fuzzy Arithmetic*, New York: Van Nostrand Reinhold.

Kingston, D.G.; Todd, M.C.; Taylor, R.G.; Thompson, J.R. & Arnell N.W. (2009). Uncertainty in the estimation of potential evapotranspiration under climate change, *Geophysical Research Letters*, Vol. 36, L20403. doi:10.1029/2009GL040267

Linacre, E. T. (1977). A simple formula for estimating evaporation rates in various climates, using temperature data alone. *Agric. Meteorol.*, 18, pp. 409--424.

Makkink, G. F. (1957). Testing the Penman formula by means of lysimiters, *J. Institute of Water Engineering*, 11, pp. 277-288.

Maulé C.; Helgason, W.; McGin, S. & Cutforth, H. (2006). Estimation of standardized reference evapotranspiration on the Canadian Prairies using simple models with limited weather data. *Canadian Biosystems Engineering* 48, pp. 1.1 - 1.11.

Neitzel, D.A. (1996) Hanford Site National Environmental Policy Act (NEPA) Characterization. PNL-6415, Rev. 8. Pacific Northwest National Laboratory. Richland, Washington.

Or D. & Hanks, R.J. (1992). Spatial and temporal soil water estimation considering soil variability and evapotranspiration uncertainty, *Water Resour. Res.* Vol. 28, No. 3, pp. 803-814. doi:10.1029/91WR02585

Penman H.L (1963). *Vegetation and hydrology*. Tech. Comm. No. 53, Commonwealth Bureau of Soils, Harpenden, England. 125 pp.

Priestley, C.H.B. & Taylor, R.J. (1972). On the assessment of surface heat flux and evaporation using large-scale parameters. *Mon. Weather Rev.* 100(2), pp. 81–92.

Sumner D.M. & Jacobs, J.M. (2005). Utility of Penman–Monteith, Priestley–Taylor, reference evapotranspiration, and pan evaporation methods to estimate pasture evapotranspiration, *Journal of Hydrology*, 308, pp. 81–104.

Thornthwaite, C.W. (1948). An approach toward a rational classification of climate. *Geogr. Rev.* 38, pp. 55–94.

Turc, L. (1963). Evaluation des besoins en eau d'irrigation, évapotranspiration potentielle, formulation simplifié et mise à jour. *Ann. Agron.*, 12: 13-49.

Walter, I.A.; Allen, R.G.; Elliott, R.; Itenfisu, D.; Brown, P.; Jensen, M.E.; Mecham, B.; Howell, T.A.; Snyder, R.L.; Eching, S.; Spofford, T.; Hattendorf, M.; Martin, D.; Cuenca, R.H. & Wright, J.L. (2002). The ASCE standardized reference evapotranspiration equation. *Rep. Task Com. on Standardized Reference Evapotranspiration July 9, 2002, EWRI-Am. Soc. Civil. Engr.*, Reston, VA, 57 pp. /w six Appendices. http://www.kimberly.uidaho.edu/water/asceewri/main.pdf.

Ward, A.L.; Freeman, E.J.; White, M.D. & Zhang, Z.F. (2005). STOMP: Subsurface Transport Over Multiple Phases, Version 1.0, Addendum: Sparse Vegetation Evapotranspiration Model for the Water-Air-Energy Operational Mode, PNNL-15465.

Yager. R. & Kelman, A. (1996). Fusion of fuzzy information with considerations for compatibility, partial aggregation, and reinforcement. *International Journal of Approximate Reasoning*, 15(2), pp. 93-122.

Zadeh, L. (1978). Fuzzy sets as a basis for a theory of possibility. *Fuzzy Sets and Systems*, 1, pp. 3-28.

Zadeh, L.A. (1986). A Simple view of the Dempster-Shafer theory of evidence and its implication for the rule of combination. *The AI Magazine* 7, pp. 85-90.

Zhu J.; Young, M.H. & Cablk, M.E. (2007) Uncertainty Analysis of Estimates of Ground-Water Discharge by Evapotranspiration for the BARCAS Study Area, DHS Publication No. 41234.

6

Using Soil Moisture Data to Estimate Evapotranspiration and Development of a Physically Based Root Water Uptake Model

Nirjhar Shah[1], Mark Ross[2] and Ken Trout[2]
[1]AMEC Inc. Lakeland, FL
[2]Univ. of South Florida. Tampa, FL
USA

1. Introduction

In humid regions such as west-central Florida, evapotranspiration (ET) is estimated to be 70% of precipitation on an average annual basis (Bidlake et al. 1993; Knowles 1996; Sumner 2001). ET is traditionally inferred from values of potential ET (PET) or reference ET (Doorenabos and Pruitt 1977). PET data are more readily available and can be computed from either pan evaporation or from energy budget methods (Penman 1948; Thornthwaite 1948; Monteith 1965; Priestly and Taylor 1972, etc.). The above methodology though simple, suffer from the fact that meteorological data collected in the field for PET are mostly under non-potential conditions, rendering ET estimates as erroneous (Brutsaert 1982; Sumner 2006). Lysimeters can be used to determine ET from mass balance, however, for shallow water table environments, they are found to give erroneous readings due to air entrapment (Fayer and Hillel 1986), as well as fluctuating water table (Yang et al. 2000). Remote sensing techniques such as, satellite-derived feedback model and Surface Energy Balance Algorithm (SEBAL) as reviewed by Kite and Droogers (2000) and remotely sensed Normalized Difference Vegetation Index (NDVI) as used by Mo et al. (2004) are especially useful for large scale studies. However, in the case of highly heterogeneous landscapes , the resolution of ET may become problematic owing to the coarse resolution of the data (Nachabe et al. 2005). The energy budget or eddy correlation methodologies are also limited to computing net ET and cannot resolve ET contribution from different sources. For shallow water table environments, continuous soil moisture measurements and water table estimation have been found to accurately determine ET (Nachabe et al. 2005; Fares and Alva 2000). Past studies, e.g., Robock et al. (2000), Mahmood and Hubbard (2003), and Nachabe et al. (2005), have clearly shown that soil moisture monitoring can be successfully used to determine ET from a hydrologic balance. The approach used herein involves use of soil moisture and water table data measurements. Using point measurement of soil moisture and water table observations from an individual monitoring well ET values can be accurately determined. Additionally, if similar measurements of soil moisture content and water table are available from a set of wells along a flow transect , other components of water budgets and attempts to comprehensively resolve other components of the water budget at the study site.

The following section describes a particular configuration of the instruments, development of a methodology, and an example case study where the authors have successfully applied measurement of soil moisture and water table in the past to estimate and model ET at the study site. The authors also used the soil moisture dataset to compute actual root water uptake for two different land-covers (grassed and forested). The new methodology of estimating ET is based on an eco-hydrological framework that includes plant physiological characteristics. The new methodology is shown to provide a much better representation of the ET process with varying antecedent conditions for a given land-cover as compared to traditional hydrological models.

2. Study site

The study site for gathering field data and using it for ET estimation and vadose zone process modeling was located in the sub basin of Long Flat Creek, a tributary of the Alafia River, adjacent to the Tampa bay regional reservoir, in Lithia, Florida. **Figure 1** shows the regional and aerial view of the site location. Two sets of monitoring well transects were installed on the west side of Long Flat Creek. One set of wells designated as PS-39, PS-40, PS-41, PS-42, and PS-43 ran from east to west while the other set consisting of two wells was roughly parallel to the stream (Long Flat Creek), running in the North South direction. The wells were designated as USF-1 and USF-3.

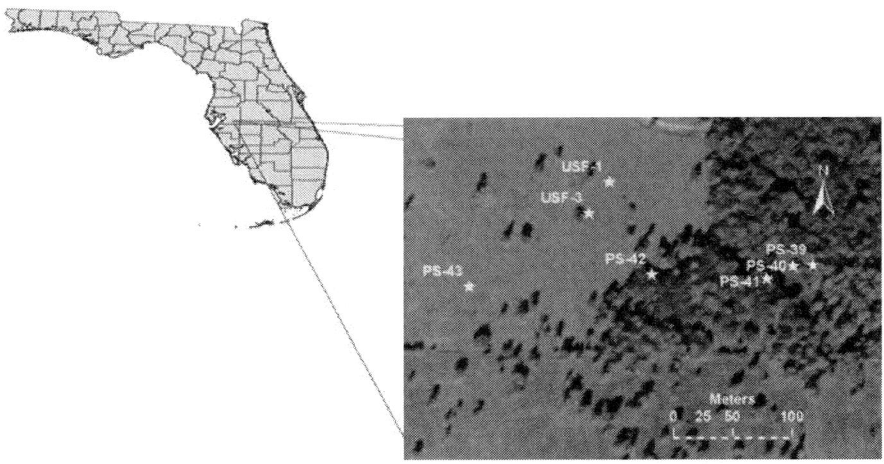

Fig. 1. Location of the study site in Hillsborough County, Florida

The topography of the area slopes towards the stream with PS-43 being located at roughly the highest point for both transects. The vegetation varied from un-grazed Bahia pasture grass in the upland areas (in proximity of PS-43, USF-1, and USF-3), to alluvial wetland forest comprised of slash pine and hardwood trees near the stream. The area close to PS-42 is characterized as a mixed (grassed and forested) zone. Horizontal distance between the wells is approximately 16, 22, 96, 153 m from PS-39 to PS-43, with PS-39 being approximately 6 m from the creek. The horizontal distance between USF-1 and USF-3 was 33 m. All wells were surveyed and land surface elevations were determined with respect to National Geodetic Vertical Datum 1929 (NGVD).

The data captured from this configuration was used both for point estimation as well as transect modeling, however , for this particular chapter, only point estimation o: ET and and point data set will be used to develop conceptualizations of vadose zone processes will be discussed. For details regarding transect modeling to generate water budget estimates refer to Shah (2007).

3. Instrumentation

For measurement of water table at a particular location a monitoring well instrumented with submersible pressure transducer (manufactured by Instrumentation Northwest, Kirkland, WA) 0-34 kPa (0-5 psi), accurate to 0.034 kPa (0.005 psi) was installed. Adjacent to each well, an EnviroSMART® soil moisture probe (Sentek Pty. Ltd., Adelaide, Australia) carrying eight sensors was installed (see **Figure 2**). The soil moisture sensors allowed measurement of volumetric moisture content along a vertical profile at different depths from land surface. The sensors were deployed at 10, 20, 30, 50, 70, 90, 110, 150 cm from the land surface. The sensors work on the principle of frequency domain reflectometery (FDR) to convert electrical capacitance shift to volumetric water content ranging from oven dryness to saturation with a resolution of 0.1% (Buss 1993). Default factory calibration equations were used for calibrating these sensors. Fares and Alva (2000) and Morgan et al. (1999) found no significant difference in the values of observed recorded water content from the sensors when compared with the manually measured values. Two tipping bucket and two manual rain gages were also installed to record the amount of precipitation.

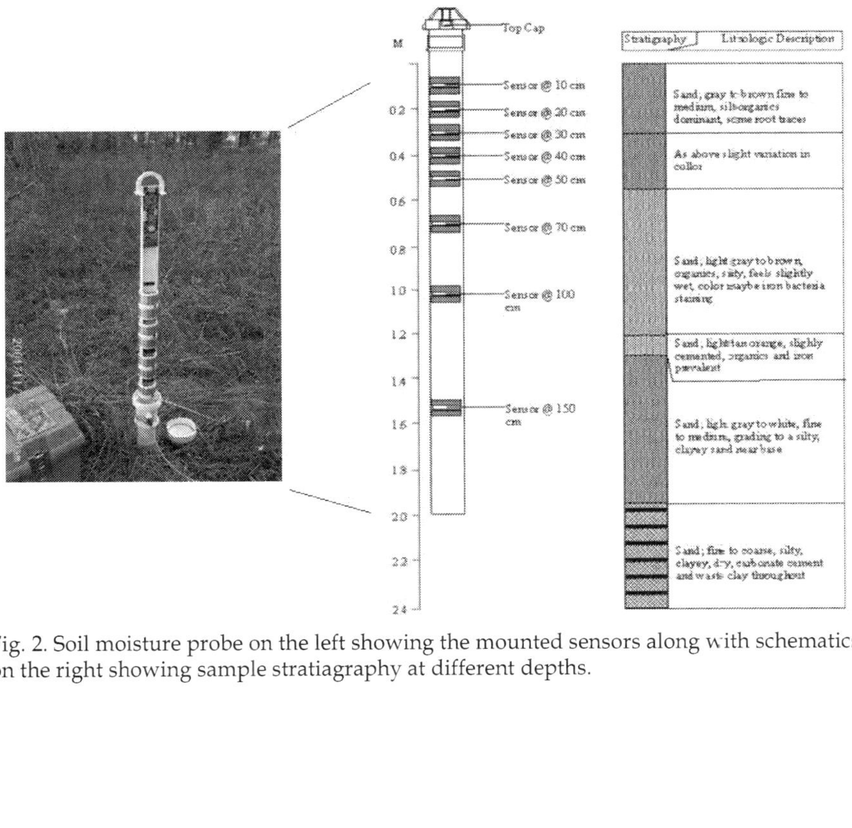

Fig. 2. Soil moisture probe on the left showing the mounted sensors along with schematics on the right showing sample stratiagraphy at different depths.

4. Point estimation of evapotranspiration using soil moisture data

At any given well location variation in total soil moisture on non-rainy days can be due to (a) subsurface flow from or to the one dimensional soil column (0 - 155 cm below land surface) over which soil moisture is measured, and (b) evapotranspiration from this soil column. Mathematically

$$\frac{\partial TSM}{\partial t} = Q - ET \tag{1}$$

where t is time [T], Q is subsurface flow rate [LT^{-1}], and ET is evapotranspiration rate [LT^{-1}]. TSM is total soil moisture, determined as below

$$TSM = \int_{\varsigma} \theta \, dz \tag{2}$$

where θ[L^3L^{-3}] is the measured water content, z [L] is the depth below land surface ζ[L] is the depth of monitored soil column (155 cm). The values in the square brackets (for all the variables) represent the dimensions (instead of units) e.g. L is length, T is time.

The negative sign in front of ET in **Equation 1** indicates that ET depletes the TSM in the column. The subsurface flow rate can be either positive or negative. In a groundwater discharge area, the subsurface flow rate, Q, is positive because it acts to replenish the TSM in the soil column (Freeze and Cherry, 1979). Thus, this flow rate is negative in a groundwater recharge area. **Figure 3** illustrates the role of subsurface flow in replenishing or depleting total soil moisture in the column. An inherent assumption in this approach is that the deepest sensor is below the water table which allows accounting for all the soil moisture in the vadose zone. Hence, monitoring of water table is critical to make sure that the water table is shallower than the bottom most sensor. To estimate both ET and Q in **Equation 1**, it was important to decouple these fluxes. In this model the subsurface flow rate was estimated from the diurnal fluctuation in TSM. Assuming ET is effectively zero between midnight and 0400 h, Q can be easily calculated from **Equation 3** using:

$$Q = \frac{TSM_{0400h} - TSM_{midnight}}{4} \tag{3}$$

where TSM_{0400h} and $TSM_{midnight}$ are total soil moisture measured at 0400 h and midnight, respectively. The denominator in **Equation 3** is 4 h, corresponding to the time difference between the two TSM measurements. The assumption of negligible ET between midnight and 0400h is not new, but was adopted in the early works of White (1932) and Meyboom (1967) in analyzing diurnal water table fluctuation. It is a reasonable assumption to make at night when sunlight is absent.

Taking Q as constant for a 24h period (White 1932; Meyboom, 1967), the ET consumption in any single day was calculated from the following equation

$$ET = TSM_j - TSM_{j+1} + 24 \times Q \tag{4}$$

where TSM_j is the total soil moisture at midnight on day j, and TSM_{j+1} is the total soil moisture 24h later (midnight the following day). Q is multiplied by 24 as the **Equation 4** provides daily ET values. **Figure 4** show a sample observations for 5 day period showing the evolution of TSM in a groundwater discharge and recharge area respectively. Also marked on the graphs are different quantities calculated to determine *ET* from the observations.

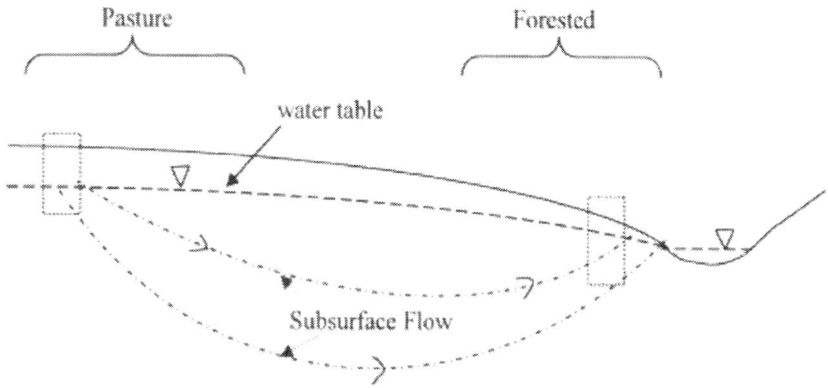

Fig. 3. Total soil moisture is estimated in two soil columns. The first is in a groundwater recharge area (pasture), and the second is in a groundwater discharge area (forested). In a groundwater discharge area, subsurface flow acts to replenish the total soil.

Equation 1 applies for dry periods only, because it does not account for the contribution of interception storage to ET on rainy days. Also, the changes in soil moisture on rainy days can occur due to other processes like infiltration, upstream runoff infiltration (as will be discussed later) etc. The results obtained from the above model were averaged based on the land cover of each well and are presented as ET values for grass or forested land cover. The values for the grassed land cover were also compared against ET values derived from pan evaporation measurements.

The ET estimates from the data collected at the study site using the above methodology are shown in **Figure 5**. **Figure 5** shows variability in the values of *ET* for a period of about a year and half. It can be seen from **Figure 5** that the method was successful in capturing spatial variability in the ET rates based on the changes in the land cover, as the ET rate of forested (alluvial wetland forest) land cover was found to be always higher than that of the grassland (in this case un-grazed Bahia grass). In addition to spatial variability, the method seemed to capture well the temporal variability in ET. The temporal variability for this particular analysis existed at two time scales, a short-term daily variation associated with daily changes in atmospheric conditions (e.g. local cloud cover, wind speed etc.) and a long-term, seasonal, climatic variation. The short-term variation tends to be less systematic and is demonstrated in **Figure 5** by the range marks. The seasonal variation is more systematic and pronounced and is clearly captured by the method.

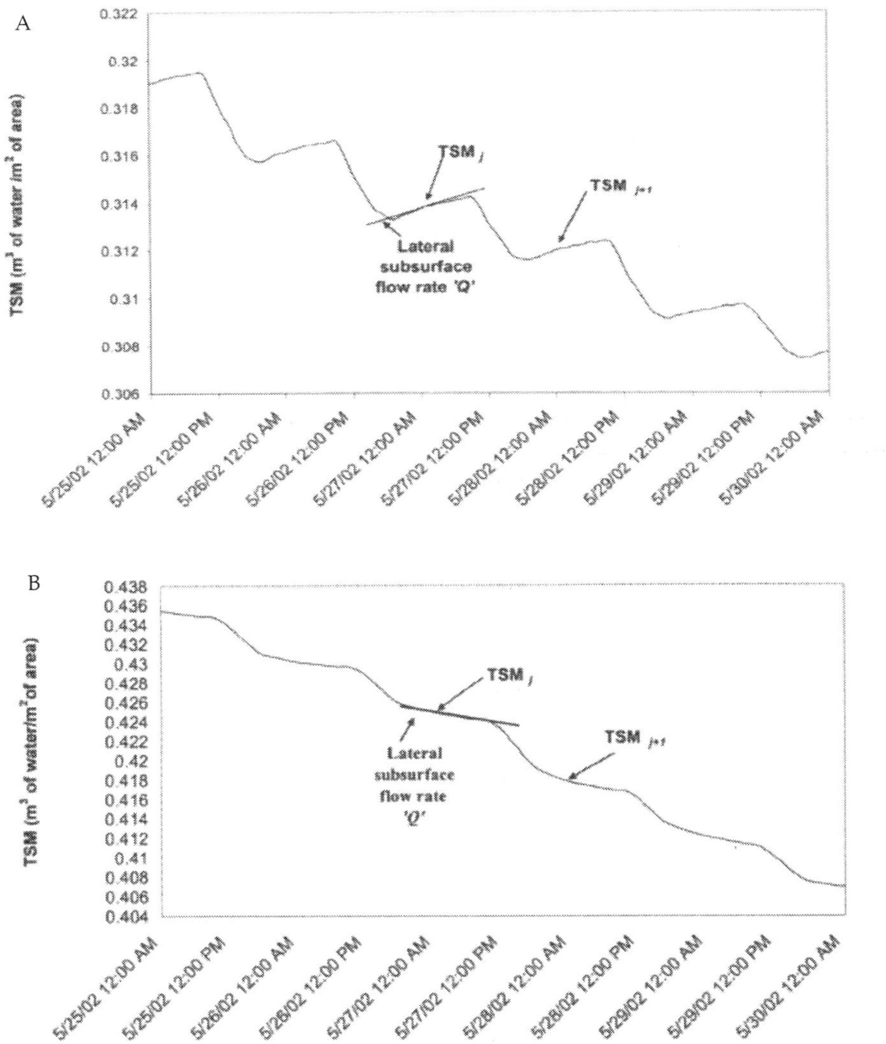

Fig. 4. Total soil moisture versus time in the (a) groundwater discharge area and (b) ground water recharge area. The subsurface flux is the positive slope of the line between midnight and 4 AM.

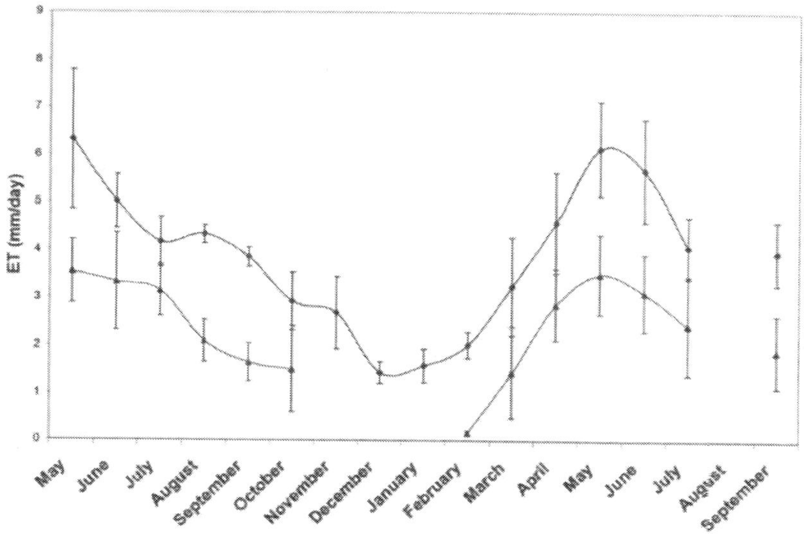

Fig. 5. Monthly average of evapotranspiration (ET) daily values in forested (diamonds) and pasture (triangles) areas. The gap in the graph represents a period of missing data. Standard deviations of daily values are also shown in the range limits.

To assess the reasonableness of the methodology, the estimated ET values for pasture were compared with ET estimated from the evaporation pan. The measured pan evaporation was multiplied by a pan coefficient for pasture to estimate ET for this vegetation cover. A monthly variable crop coefficient was adopted (Doorenbos and Pruitt, 1977) to account for changes associated with seasonal plant phenology (see **Table 1**). The consumptive water use or the crop evapotranspiration is calculated as:

$$ET_C = E_P \times K_C \quad (5)$$

where E_P is the measured pan evaporation, K_C is a pan coefficient for pastureland, and ET_C is the estimated evapotranspiration $[LT^{-1}]$ (mm/d) by the pan evaporation method. **Figure 6** compares the ET estimated by both the evaporation pan and moisture sensors for pasture. Although the two methods are fundamentally different, on average, estimated *ET* agreed well with an R^2 coefficient of 0.78. This supported the validity of the soil moisture methodology, which further captured the daily variability of ET ranging from a low of 0.3 mm/d to a maximum of 4.9 mm/d. The differences between the two methods can be attributed to fundamental discrepancies. The pan results are based on atmospheric potential with crude average monthly coefficients while the TSM approach inherently incorporates plant physiology and actual moisture limitations. Indeed, both methods suffer from limitations. The pan coefficient is generic and does not account for regional variation in vegetation phenology or other local influences such as soil texture and fertility. Similarly, the accuracy of the soil moisture method proposed in this study depends on the number of sensors used in monitoring total moisture in the soil column.

Month		Coefficient
January		0.4
February		0.45
March		0.55
April		0.64
May		0.7
June		0.7
July		0.7
August		0.7
September		0.7
October		0.6
November		0.5
December		0.5

Table 1. Pan coefficients used to obtain pasture evapotranspiration for different months.

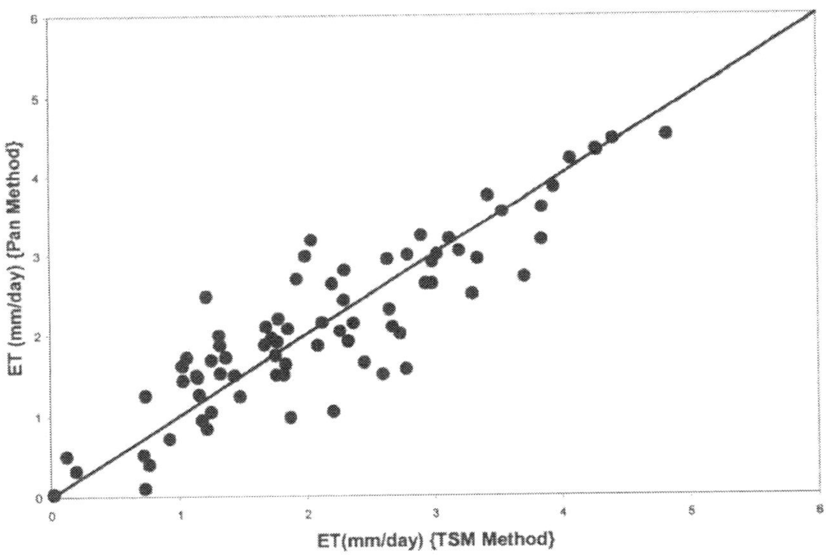

Fig. 6. Evapotranspiration estimates for pasture by the pan and point scale model. Data points represent the daily values of ET from both techniques.

5. Development of root water uptake model

The preceding sections described a novel data collection approach that can be used to measure ET (and other water budget components). The measured values can be subsequently used to develop modeling parameters or validate modeling results for areas which are similar to the study site in terms of climatic and land-cover conditions. The next step is the development of a generic modeling framework to accurately determine ET.

Transpiration by its very nature is a process that is primarily based on plant physiology and the better one can determine root water uptake the more accurate will be the estimation of actual transpiration and, therefore ET. Traditionally used models and concepts, however, make over simplifying assumptions about plants (Shah et al. 2007), hence casting doubt on the model results. What needs to be done is to try and combine land cover characteristics in the root water uptake models to produce more reliable results. With this intent in mind, recently, a new branch of study called "Eco-Hydrology" has been initiated. The aim of eco-hydrology is to encourage the interdisciplinary work on ecology and hydrology with an objective of improving hydrological modeling capabilities.

Soil moisture datasets (as described in Section 2) can be used to provide insight into the process of root water uptake which can then be combined with plant characteristics to develop a more physically based ET model. The next sections describe how the soil moisture dataset has been used by the authors to estimate vertical distribution of root-water uptake for two land-cover classes (shallow rooted and deep rooted) and how the results were then used to develop a land-cover based modeling framework.

5.1 Traditional root water uptake models

The governing equation for soil moisture dynamics in the unsaturated soil zone is the Richards's equation (Richards 1931). Richards's equation is derived from Darcy's law and the continuity equation. What follows is a brief description of Richards's equation and how can it incorporates root water uptake. For more detailed information about formulation of Richards's equation, including its derivation in three dimensions, the readers are directed to any text book on soil physics e.g. Hillel (1998).

Due to ease of measurement and conceptualization, energy of water (E) is represented in terms of height of liquid column and is called the hydraulic head (h). It is defined as the total energy of water per unit weight. Mathematically hydraulic head, h, can be represented as

$$h = \frac{E}{\rho_W g} \tag{6}$$

where ρ_W is the density of water and g is the acceleration due to gravity. The flow of water always occurs along decreasing head. In soil physics the fundamental equation used to model the flow of water along a head gradient is known as Darcy Law (Hillel 1998). Mathematically the equation can be written as

$$q = K\frac{\Delta h}{l} \tag{7}$$

where q [$L^3L^{-2}T^{-1}$] is known as the specific discharge and is defined as the flow per unit cross-sectional area, K[LT^{-1}] is termed as the hydraulic conductivity, which indicates ease of flow, Δh [L] is the head difference between the points of interest and l[L] is the distance between them. Darcy's Law is analogous to Ohm's law with head gradient being analogous to the potential difference and current being analogous to specific discharge and hydraulic conductivity being similar to the conductance of a wire.

The second component of Richards's equation is the continuity equation. Continuity equation is based on the law of mass conservation, and for any given volume it states that

net increase in storage in the given volume is inflow minus sum of outflow and any sink present in the volume of soil. Mathematically, it is this sink term that allows the modeling of water extracted from the given volume of soil.
In one dimension for flow occurring in the vertical direction (z axis is positive downwards) Richards's equation can be written as

$$\left(\frac{\partial \theta}{\partial t}\right)=\left(\frac{\partial}{\partial z}K\left(\frac{\partial h}{\partial z}+1\right)\right)-S \tag{8}$$

where θ is the water content, defined as the ratio of volume of water present and total volume of the soil element , t is time, S represents the sink term while other terms are as defined before.
If flow in X and Y directions is also considered , Richards's equation in three dimensions can be derived. Solution of Equation 8 can theoretically provide the spatial and temporal variability of moisture in the soil. However, due to high degree of non linearity of the equation no analytical solution exists for Richards's equation and numerical techniques are used to solve it. For a numerical solution of Richards's equation two essential properties that need to be defined a-priori are (a) relationship between soil water content and hydraulic head, also known as, soil moisture retention curves, and (b) a model that relate hydraulic head to root water uptake. Details about the soil moisture retention curves and numerical techniques used to solve Richards's equation can be found in Simunek et al. (2005). While much literature and field data exist describing the soil moisture retention curves, relatively less information exists about root water uptake models. The root water uptake models generally used, especially, on a watershed scale, are mostly empirical and lack any field verification.
The most common approach used to model root water uptake is to define a sink term S as a function of hydraulic head using the following equation

$$S(h)=\alpha(h)S_p \tag{9}$$

where $S(h)[L^3L^{-3}T^{-1}]$ is the actual root water uptake (RWU) from roots subjected to hydraulic or capillary pressure head *'h'*. On the right hand side of the equation S_p $[L^3L^{-3}T^{-1}]$ is the maximum (also known as potential) uptake of water by the roots. The $a(h)$ is a root water uptake stress response function, with its values varying between 0 and 1.
The idea behind conceptualization of **Equation 9** is based on three basic assumptions. The first assumption being , as the soil becomes dryer the amount of water that can be extracted will decrease proportionally. Secondly, the amount of water extracted by the roots is affected by the ambient climatic conditions. Drier and hotter conditions result in more water loss from surface of leaves, hence, initiating more water extraction from the soil. The third and final assumption is that the uptake of water from a particular section of a root is directly proportional to the amount of roots present in that section.
The root water stress response function (α) is a result of the first assumption. Two models commonly used to define α are the Feddes model (Feddes et al. 1978) and the van Genuchten model (van Genuchten 1987). **Figure 7** (a and b, respectively) show the variation of α with decreasing hydraulic head which is same as decreasing water content or increasing soil dryness. Both models for α are empirical and do not involve any plant physiology to define the thresholds for the water stress response function. An interesting

contrast, due to empiricism that is clearly evident is the value of α during saturated conditions. While the Feddes model predict the value of a to decrease to zero van Genuchten model predicts totally opposite with α rising to become unity under saturated conditions.

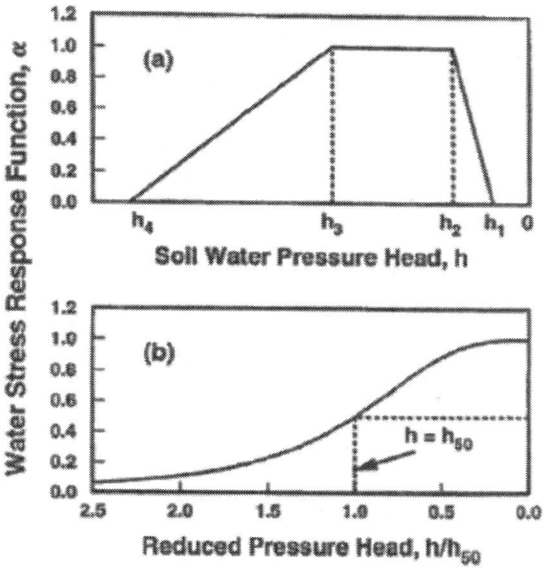

Fig. 7. Water stress response function as given conceptualized by (a) Feddes et al. 1978 and (b) van Genuchten (1980) [Adapted from Simunek et al. 2005].

Recently couple of different models (Li et al. 2001, Li et al. 2006) have been presented to overcome the empiricism in α. However these models are more a result of observation fitting and fail to bring in the plant physiology, which is what causes the changes in the water uptake rate due variation in soil moisture conditions.

Combining the second and the third assumptions in **Equation 9** results in the definition of S_p. S_p for any section of roots is defined as the product of root fraction in that section and the maximum possible water loss by the plant which is also known as the potential evapotranspiration. Potential evapotranspiration is a function of ambient atmospheric conditions and standard models like Penman-Monteith (Allen et al. 1998) are used to calculate the potential evapotranspiration rate. The problem with this definition of Sp is that for any given value of potential evapotranspiration, limiting the value of root water uptake by the root-fraction restrict the amount of water that can be extracted from a particular section. In other words, the amount of water extracted by a particular section of root is directly proportional to the amount of roots present and ignores the amount of ambient soil moisture present. This as will discussed later using field data is a significant limitation especially during dry period when the top soil with maximum roots get dry while the deep soil layer with lesser root mass still have soil moisture available for extraction.

5.2 Use of soil moisture data to estimate root water uptake

For the current analysis, the soil moisture data as described in Section 2 are used. Soil moisture and water-table data from well locations PS-43 and PS-40 were used to determine root water uptake from forested versus grassed land cover. The well PS-43 is referred to as Site A while PS-40 will be called Site B. Hourly averaged data at four hour time step were used for the analysis.

Extensive soil investigations including in-situ and laboratory analysis were performed for the study site. The soil in the study area is primarily sandy marine sediments with high permeability in the surface and subsurface layers. Detailed information about soil and site characteristics can be found in Said et al. (2005), and Trout and Ross (2004). Data for period of record January 2003 to December 2003 were used in this analysis.

van Genuchten (1980) proposed a model relating the water content and hydraulic conductivity with the suction head (soil suction pressure) represented by the following equations

$$S_e = \frac{\theta - \theta_r}{\theta_s - \theta_r} \tag{10}$$

$$h(\theta) = \frac{(S_e^{\frac{1}{m}} - 1)^{\frac{1}{n}}}{\phi} \tag{11}$$

$$K(h) = \begin{cases} K_S S_e^l [1 - (1 - S_e^{1/m})^m & h < 0 \\ K_S & h \geq 0 \end{cases} \tag{12}$$

where m = 1 – 1/n for n > 1, S_e [-] is the normalized water content, varying between 0 and 1. θ is the observed water content, while θ_r and θ_s are the residual and saturated water content values respectively K_S [LT^{-1}] is the hydraulic conductivity when the soil matrix is saturated, l[-] is the pore connectivity parameter assumed to be 0.5 as an average for most soils (Mualem, 1976), and ϕ[L^{-1}], n[-] and m[-] are the van Genuchten empirical parameters. Negative values of hydraulic head (suction head) indicate the water content in the soil matrix is less than saturated while the positive value indicate saturated conditions. From the **Equations 11** and **12**, it is clear that for each type of soil five parameters, namely, K_S, n, ϕ, θ_r and θ_s have to be determined to uniquely define relationship of hydraulic conductivity and water content with soil suction head.

Figure 8 shows the schematics of the vertical soil column which is monitored using eight soil moisture sensors and a pressure transducer measuring the water table elevation, at each of the two locations. Shown also in **Figure 8** is the zone of influence of each sensor along with the elevation of water table and arrows showing possible flow directions. For the purpose of defining moisture retention and hydraulic conductivity curves, each section is treated as a different soil layer and independently parameterized. Hence, for each of the two locations for this particular study eight soil cores from depths corresponding to the zone of influence of each sensor were taken and analyzed (see Shah, 2007 for more details). **Table 2(a)** and **(b)** shows the parameters values that were obtained following the all the soil tests.

Sensor Location Below Land Surface (cm)	$\theta_S(\%)$	$\theta_R(\%)$	Φ (cm^{-1})	n(-)	K_S(cm/hr)
10	35	3	0.03	1.85	4.212
20	35	3	0.07	1.70	2.520
30	32	3	0.07	1.70	2.520
50	34	3	0.03	1.60	0.803
70	31	3	0.03	1.60	0.005
90	32	3	0.05	1.90	0.005
110	32	3	0.05	1.80	0.005
150	30	3	0.05	1.80	0.001

(a)

Sensor Location Below Land Surface (cm)	$\theta_S(\%)$	$\theta_R(\%)$	Φ (cm^{-1})	n(-)	K_S(cm/hr)
10	38	3	0.02	1.35	0.0100
20	34	3	0.03	1.35	0.0100
30	31	3	0.03	1.35	0.0100
50	31	3	0.07	1.90	0.0100
70	31	3	0.2	2.20	0.0100
90	31	3	0.2	2.20	0.0004
110	33	3	0.2	2.20	0.0004
150	35	3	0.2	2.10	0.0012

(b)

Table 2. Soil parameters for study locations in (a) Grassland and (b) Forested area.

Once the soil parameterization is complete root water uptake from each section can be calculated. For any given soil layer in the vertical soil column (**Figure 8**), above the observed water table, observed water content and **Equation 11** can be used to calculate the hydraulic head. For soil layers below the water table hydraulic head is same as the depth of soil layer below the water table due to assumption of hydrostatic pressure. Similarly using **Equation 12** hydraulic conductivity can be calculated. Hence, at any instant in time hydraulic head in each of the eight soil layers can be calculated. To determine total head, gravity head, which is the height of the soil layer above a common datum, has to be added to the hydraulic head.

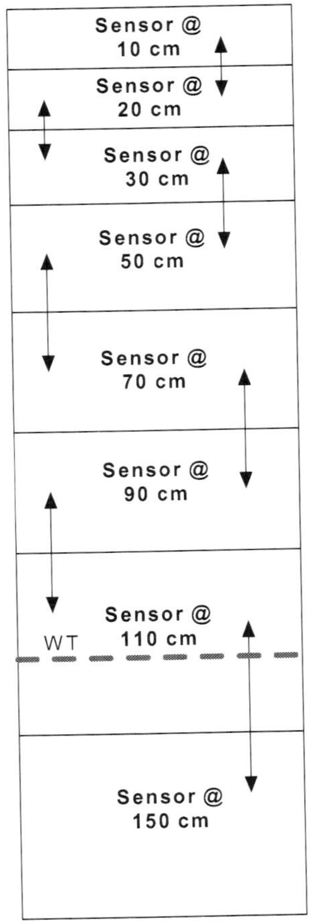

Fig. 8. Schematics of the vertical soil column with location of the soil moisture sensors and water table.

To quantify flow across each soil layer, Darcy's Law (**Equation 7**) is used. Average head values between two consecutive time steps are used to determine the head difference. Also, flow across different soil layers is assumed to be occurring between the midpoints of one layer to another, hence, to determine the head gradient ($\Delta h/l$) the distance between the midpoints of each soil layer is used. The last component needed to solve Darcy's Law is the value of hydraulic conductivity. For flow occurring between layers of different hydraulic conductivities equivalent hydraulic conductivity is calculated by taking harmonic means of the hydraulic conductivities of both the layers (Freeze and Cherry 1979). Hence for each time step harmonically averaged hydraulic conductivity values (**Equation 13**) were used to calculate the flow across soil layers.

$$K_{eq} = \frac{2K_1K_2}{K_1 + K_2} \tag{13a}$$

where K_1 [LT^{-1}]and K_2 [LT^{-1}]are the two hydraulic conductivity values for any two adjacent soil layers and K_{eq} [LT^{-1}]is the equivalent hydraulic conductivity for flow occurring between those two layers.

Figure 9 shows a typical flow layer with inflow and outflow marked. Now using **simple mass balance** changes in water content at two consecutive time steps can be attributed to net inflow minus the root water uptake (assuming no other sink is present). Equation 6.9 can hence be used to determine root water uptake from any given soil layer

$$RWU = (\theta^t - \theta^{t+1}) - (q_{out} - q_{in}) \tag{13b}$$

Using the described methodology one can determine the root water uptake from each soil layer at both study locations (site A and site B).Time step for calculation of the root water uptake was set as four hours and the root water uptake values obtained were summed up to get a daily value for each soil layer.

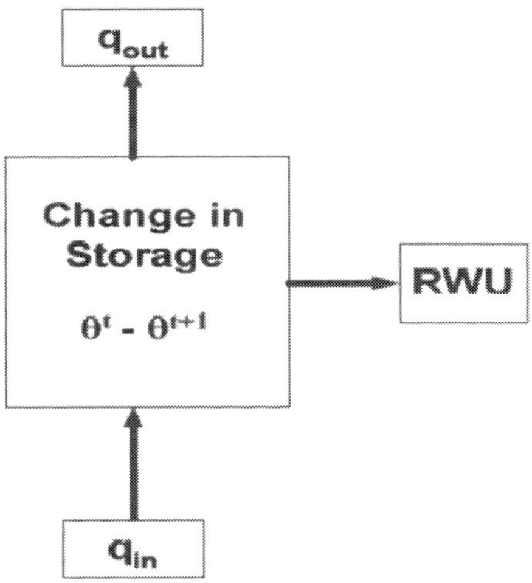

Fig. 9. Schematics of a section of vertical soil column showing fluxes and change in storage.

Using the above methodology root water uptake was calculated from each section of roots for tree and grass land cover from January to December 2003 at a daily time step. **Figure 10** (a and b) shows the variation of root water uptake for a representative period from May 1st to May 15th 2003, This particular period was selected as the conditions were dry and their was no rainfall. Graphs in **Figure 10** (a and b) show the root water uptake variation from section corresponding to each section. Also plotted on the graphs is the normalized water content, which also gives an indication, of water lost from the section.

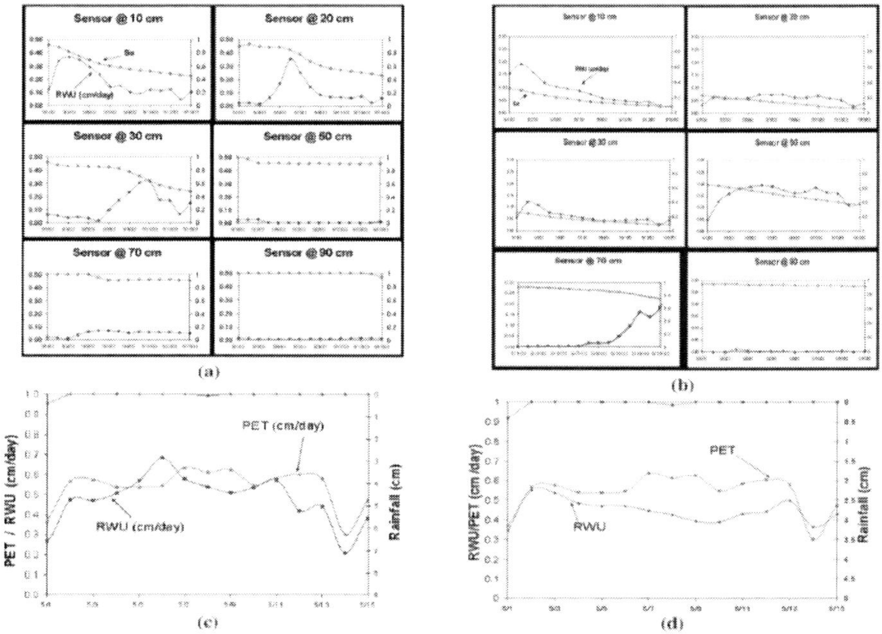

Fig. 10. Root water uptake from sections of soil corresponding to each sensor on the soil moisture instrument for (a, c) Grass land and (b, d) Forest land cover

Figure 10(a) shows the root water uptake from grassed site while panel of graphs in **Figure 10(b)** plots RWU from the forested area. From **Figure 10** (a and b) it can be seen that in both the cases of grass and forest the root water uptake varies with water content and as the top layers starts to get dry, the water uptake from the lower layer increases so as to keep the root water uptake constant clearly indicating that the compensation do take place and hence the models need to account for it. Another important point to note is that in **Figure 10(a)** root water uptake from top three sensors is accounts for the almost all the water uptake while in **Figure 10(b)** the contribution from fourth and fifth sensor is also significant. Also, as will be shown later, in case of forested land cover, root water uptake is observed from the sections that are even deeper than 70 cm below land surface. This is expected owing to the differences in the root system of both land cover types. While grasses have shallow roots, forest trees tend to put their roots deeper into the soil to meet their high water consumptive use.

Figure 10(**c** and **d**) show the values of PET plotted along with the observed values of root water uptake. On comparing the grass versus forested graphs it is evident while the grass is still evapotranspiring at values close to PET root water uptake from forested land covers is occurring at less than potential. This behavior can be explained by the fact that water content in the grassed region (as shown by the normalized water content graph, Se) is greater than that of the forest and even though the 70 cm sensor shows significant contribution the uptake is still not sufficient to meet the potential demand.

Figure 11 shows an interesting scenario when a rainfall event occurs right after a long dry stretch that caused the upper soil layers to dry out. **Figure 11(a)** shows the root water uptake profile on 5/18/2003 for forested land cover with maximum water being taken from section of soil profile corresponding to 70 cm below the land surface. A rainfall event of 1inch took place on 5/19/2003. As can be clearly seen in **Figure 11(b)** the maximum water uptake shifts right back up to 10 cm below the land surface, clearly showing that the ambient water content directly and quickly affects the root water uptake distribution. **Figure 11(c)** which shows the snapshot on 5/20/2003 a day after the rainfall where the root water uptake starts redistributing and shifting toward deeper wetter layers. In fact this behavior was observed for all the data analyzed for the period of record for both the grass and forested land covers. With roots taking water from deeper wetter layers and as soon as the shallower layer becomes wet the uptakes shifts to the top layers. **Figure 12** (a and b) show a long duration of record spanning 2 months (starting October to end November), with the whiter shade indicating higher root water uptake. From both the figures it is evident that water uptake significantly shifts in lieu of drier soil layers especially in case of forest land cover (**Figure 12(b)**), while in case of grass uptake is primarily concentrated in the top layers.

As a quick summary the results indicate that

a. Assuming RWU as directly proportional to root density may not be a good approximation.
b. Plants adjust to seek out water over the root zone
c. In case of wet conditions preferential RWU from upper soil horizons may take place
d. In case of low ET demands the distribution on ET was found to be occurring as per the root distribution, assuming an exponential root distribution

Hence, traditionally used models are not adequate, to model this behavior. Changes in regard to the modeling techniques as well as conceptualizations, hence, need to occur. Plant physiology is one area that needs to be looked into to see what plant properties affect the water uptake and how can they be modeled mathematically. The next section discusses a modeling framework based on plant root characteristics which can be employed to model the aforesaid observations.

5.3 Incorporation of plant physiology in modeling root water uptake

Any framework to model root water uptake dynamically, will have to explicitly account for all the four points listed above. The dynamic model should be able to adjust the uptake pattern based on root density as well as available water across the root zone. The model should use physically based parameters so as to remove empiricism from the formulation of the equations. For a given distribution of water content along the root zone (observed or modeled) knowledge of root distribution as well as hydraulic characteristics of roots is hence essential to develop a physically based root water uptake model. The following two sections will describe how root distributions can be modeled as well as how do roots need to be characterized to model uptake from root's perspective.

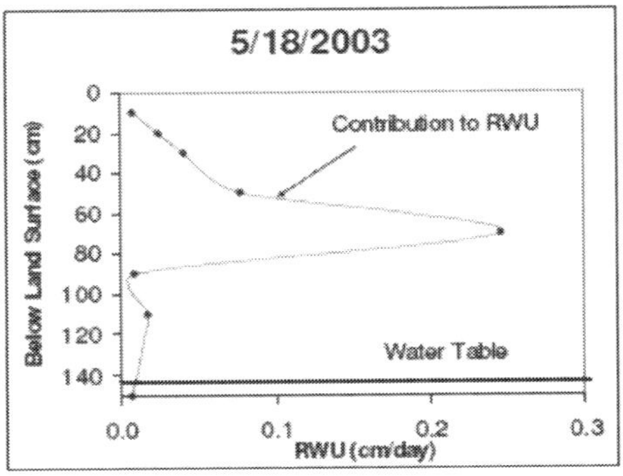

(a)

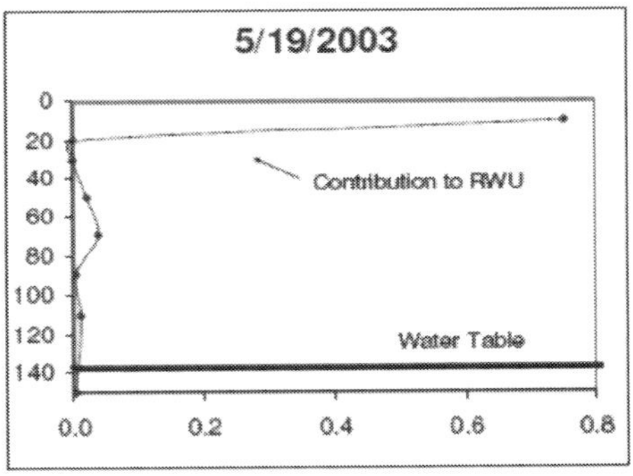

(b)

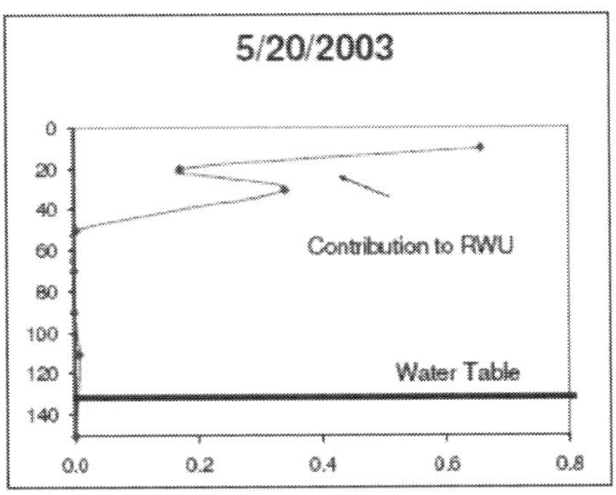

(c)

Fig. 11. Root water uptake variation due to a one inch rainfall even on 5/19/2003.

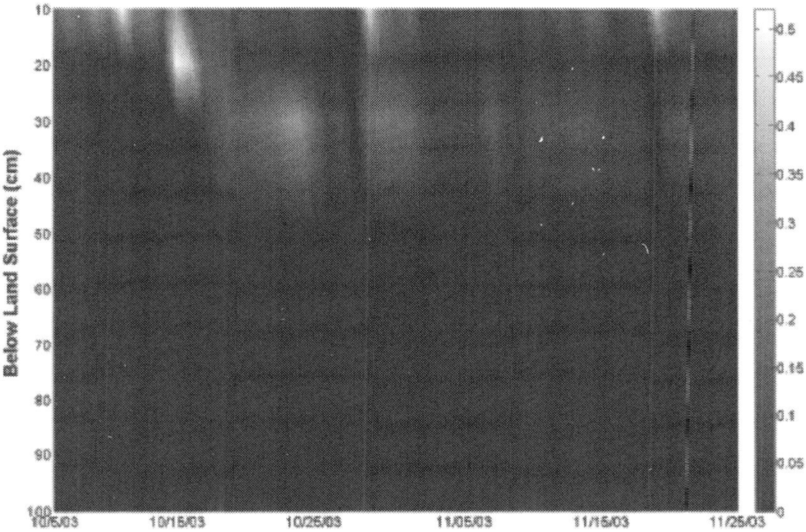

(a)

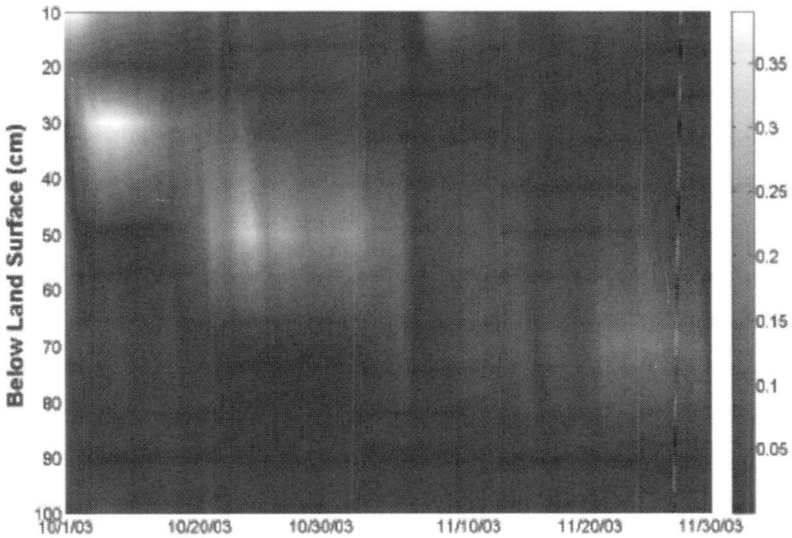

(b)

Fig. 12. Daily root water uptake variation for two October and November 2003 for (a) grass land cover and (b) forested land cover.

5.3.1 Root distribution

Schenk and Jackson (2002) expanded an earlier work of Jackson et al. (1996) to develop a global root database having 475 observed root profiles from different geographic regions of the world. It was found that by varying parameter values the root distribution model given by Gale and Grigal (1987) can be used with sufficient accuracy to describe the observed root distributions. **Equation 14** describes the root distribution model.

$$Y = 1 - \gamma^{d} \tag{14}$$

where Y is the cumulative fraction of roots from the surface to depth d, and γ is a numerical index of rooting distribution which depends on vegetation type. **Figure 13** shows the observed distribution (shown by data points) versus the fitted distribution using **Equation 14** for different vegetation types. The figure clearly indicates the goodness of fit of the above model. Hence, for a given type of vegetation a suitable γ can be used to describe the root distribution.

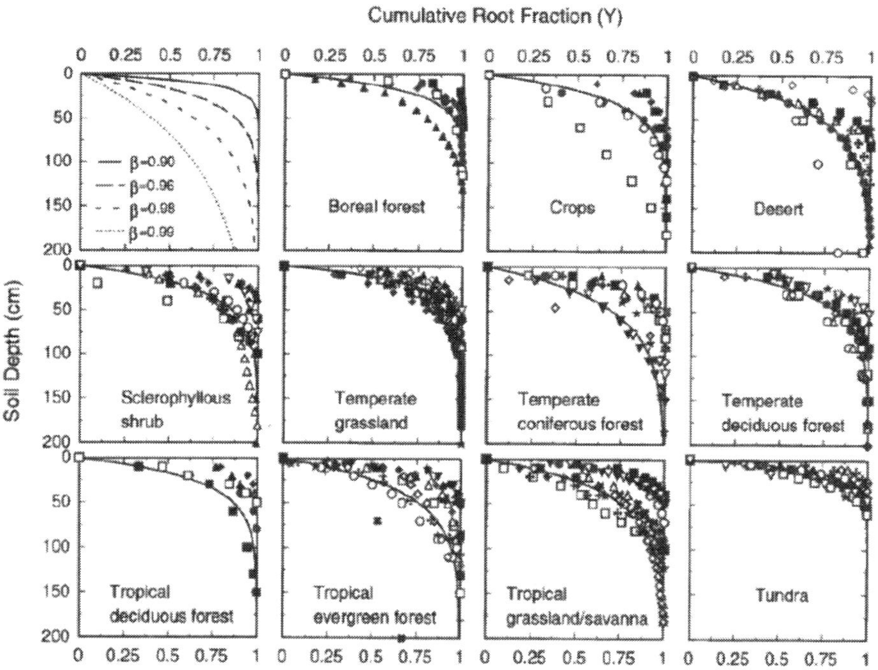

Fig. 13. Observed and Fitted Root Distribution for different type of land covers. [Adapted from Jackson et al. 1996]

5.3.2 Hydraulic characterization of roots

Hydraulically, soil and xylem are similar as they both show a decrease in hydraulic conductivity with reduction in soil moisture (increase in soil suction). For xylem the relationship between hydraulic conductivity and soil suction pressure is called 'vulnerability curve' (Sperry et al. 2003) (see **Figure 14**). The curves are drawn as a percentage loss in conductivity rather than absolute value of conductivity due to the ease of determination of former. Tyree et al (1994) and Hacke et al (2000) have described methods for determination of vulnerability curves for different types of vegetation.

Commonly, the stems and/or root segments are spun to generate negative xylem pressure (as a result of centrifugal force) which results in loss of hydraulic conductivity due to air seeding into the xylem vessels (Pammenter and Willigen 1998). This loss of hydraulic conductivity is plotted against the xylem pressure to get the desired vulnerability curve.

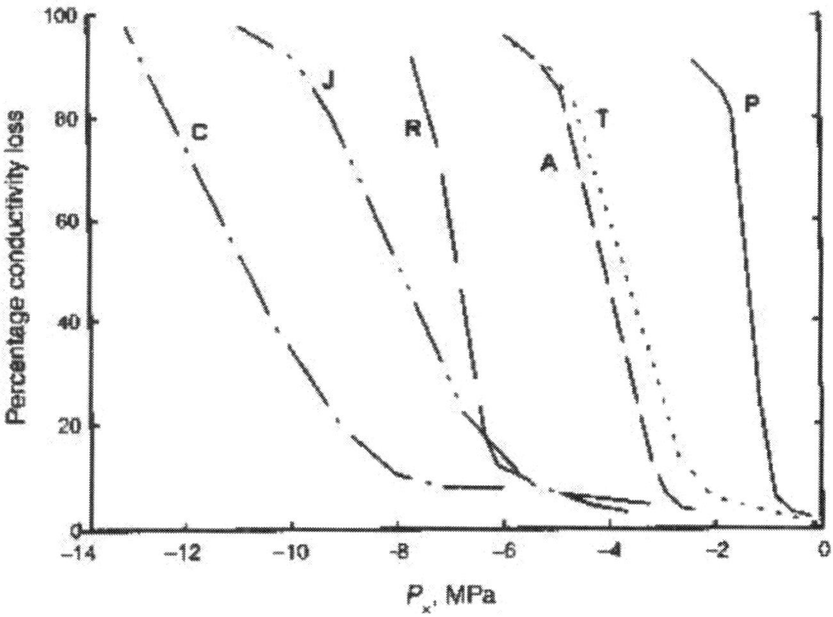

Fig. 14. Vulnerability curves for various species. [Adapted from Tyree, 1999]

For different plant species the vulnerability curve follows an S-Shape function, see **Figure 14** (Tyree 1999). In **Figure 14**, y-axis is percentage loss of hydraulic conductivity induced by the xylem pressure potential Px, shown on the x-axis. C= Ceanothus megacarpus, J = Juniperus virginiana, R = Rhizphora mangel, A = Acer saccharum, T= Thuja occidentalis, P = Populus deltoids.

Pammenter and Willigen (1998) derived an equation to model the vulnerability curve by parametrizing the equation for different plant species. **Equation 15** describes the model mathematically.

$$PLC = \frac{100}{1 + e^{a.(P - P_{50PLC})}} \tag{15}$$

where PLC denotes the percentage loss of conductivity P_{50PLC} denotes the negative pressure causing 50% loss in the hydraulic conductivity of xylems, P represents the negative pressure and a is a plant based parameter. **Figure 15** shows the model plotted against the data points for different plants. Oliveras et al. (2003) and references cited therein have parameterize the model for different type of pine and oak trees and found the model to be successful in modeling the vulnerability characteristics of xylem.

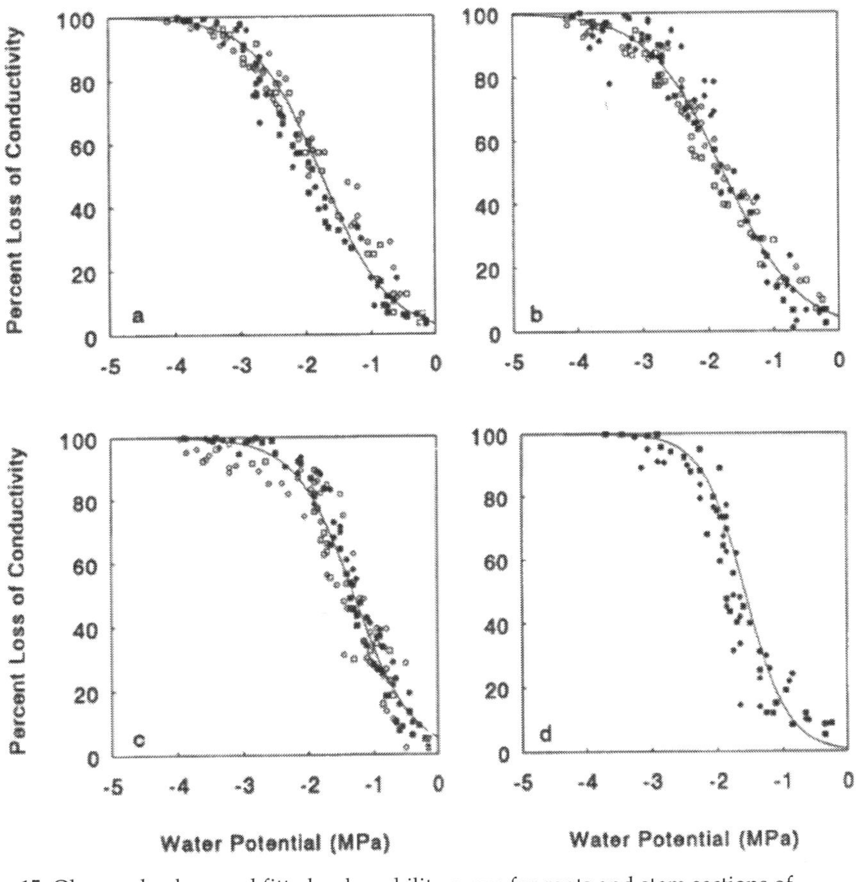

Fig. 15. Observed values and fitted vulnerability curve for roots and stem sections of different Eucylaptus trees. [Adapted from Pammenter and Willigen, 1998].

The knowledge of hydraulic conductivity loss can be used analogous to the water stress response function a (**Equation 9**) by scaling PLC from 0 to 1 and converting the suction pressure to water head. The advantage of using vulnerability curves instead of Feddes or van Genuchten model is that vulnerability curves are based on xylem hydraulics and hence can be physically characterized for each plant species.

5.3.3 Development of a physically based root water uptake model

The current model development is based on model conceptualization proposed by Jarvis (1989) however the parameters for the current model are physically defined and include plant physiological characteristics.

For a given land cover type **Equation 14** and **15** can be parameterize to determine the root fraction for any given segment in root zone and percentage loss of conductivity for a given soil suction pressure. For consistency of representation percentage loss of conductivity will be hence forth represented by α (scaled between 0 and 1 similar to **Equation 9**) and will be called stress index.

For any section of root zone, for example i^{th} section, root fraction can be written as R_i and stress index, determined from vulnerability curve and ambient soil moisture condition, can be written as a_i. Average stress level $\bar{\alpha}$ over the root zone can be defined as the

$$\bar{\alpha} = \sum_{i=1}^{n} R_i \alpha_i \tag{16}$$

where n represents the number of soil layers and the other symbols are as previously defined. Thus, as can be seen from **Equation 16** the average stress level $\bar{\alpha}$ combines the effect of both the root distribution and the available water content (via vulnerability curve).

As shown in **Figure 12(b)** if there is available moisture in the root zone, plant can transpire at potential by increasing the uptake from the lower wetter section of the roots. In terms of modeling it can be conceptualized that above a certain critical average stress level ($\bar{\alpha}_C$) plants can transpire at potential and below $\bar{\alpha}_C$ the value of total evapotranspiration decreases. The decrease in the ET value can be modeled linearly as shown by Li et al (2001). The graph of average stress level versus ET (expresses as a ratio with potential ET rate) can hence be plotted as shown in **Figure 16**. In **Figure 16**, ET_a is the actual ET out of the soil column while ET_p is the potential value of ET. **Figure 16** can be used to determine the value of actual ET for any given average stress level.

Once the actual ET value is known, the contribution from individual sections can be modeled depending on the weighted stress index using the relationship defined by

$$S_i = \left(\frac{E_a}{\Delta Z_i}\right)\left(\frac{R_i \alpha_i}{\bar{\alpha}}\right) \tag{17}$$

where S_i defined as the water uptake from the ith *section*, ΔZi is the depth of ith section and other symbols are as previously defined

Jarvis (1989) used empirical values to simulate the behavior of the above function and **Figure 17** shows the result of root water uptake obtained from his simulation. The values next to each curve in **Figure 17** represent the day after the start of simulation and actual ET rate as expressed in mm/day. On comparison with **Figure 12**, the model successfully reproduced the shift in root water uptake pattern with the uptake being close to potential value (ET_P = 5.0 mm/d) for about a month from the start of simulation. The decline in ET rate occurred long after the start of the simulation in accordance with the observed values. The model was successful not only in simulating peak but also in the observed magnitude of the root water uptake.

From the above analysis it can be concluded that the root water uptake is just not directly proportional to the distribution of the roots but also depends on the ambient water content. Under dry conditions roots can easily take water from deeper wetter soil layers.

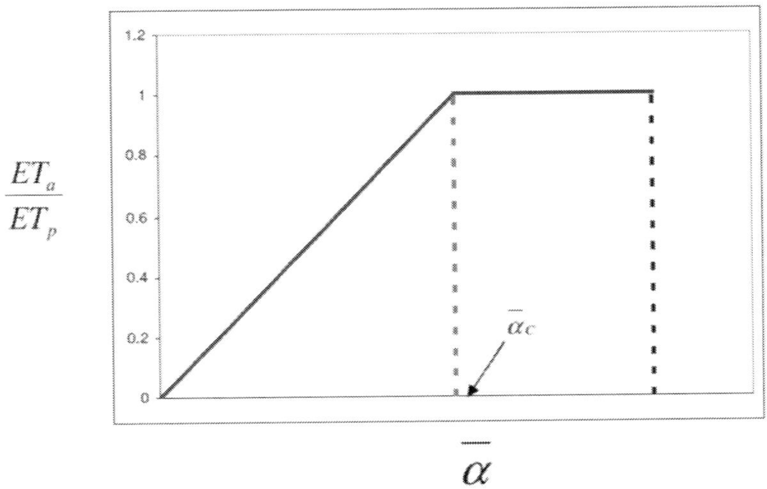

Fig. 16. Variation of ratio of actual to potential ET with location of the critical stress level.

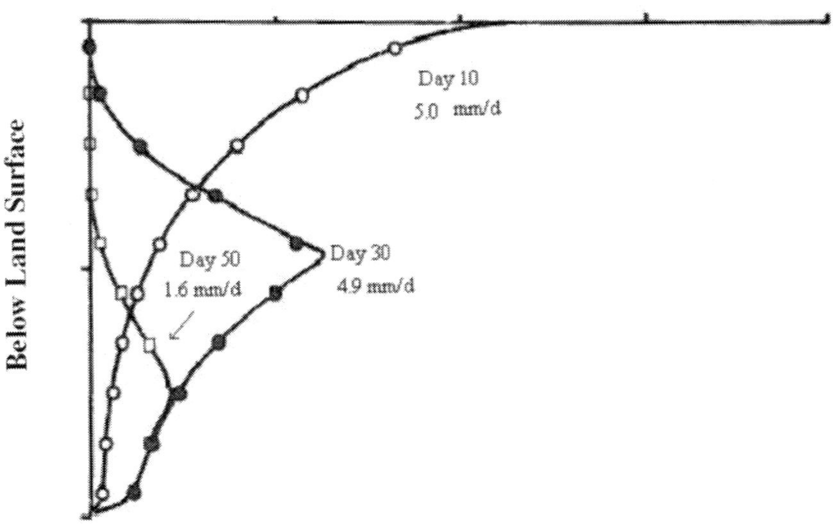

Fig. 17. Variation in the vertical distribution of root water uptake, at different times. [Adapted from Jarvis (1989)]

The methodology described here involves initial laboratory analyses to determine the hydraulic characteristics of the plant. However, once a particular plant specie is characterized then the parameters can be use for that specie elsewhere under similar conditions. The approach shows that eco-hydrological framework has great potential for improving predictive hydrological modeling.

6. Conclusion

The chapter described a method of data collection for soil moisture and water table that can be used for estimation of evapotranspiration. Also described in the chapter is the use of vertical soil moisture measurements to compute the root water uptake in the vadose zone and use that uptake to validate a root water uptake model based on plant physiology based root water uptake model. As evaporation takes place primarily from the first few centimeters (under normal conditions) of the soil profile and the biggest component of the ET is the root water uptake. Hence to improve our estimates of ET, which constitutes ~70% of the rainfall, the estimation and modeling of root water uptake needs to be improved. Eco-hydrology provides one such avenue where plant physiology can be incorporated to better represent the water loss. Also, hydrological model incorporating plant physiology can be modified easily in future to be used to predict land-cover changes due to changes in rainfall pattern or other climatic variables.

7. References

Allen RG, Pereira LS, Raes D, Smith M.1998. Crop evapotranspiration—guidelines for computing crop water requirements.FAO Irrigation & Drainage Paper 56. FAO, Rome

Bidlake, W. R., W.M.Woodham, and M.A.Lopez. 1993. Evapotranspiration from areas of native vegetation in Wets-Central Florida: U.S. Geological Survey open file report 93-415, 35p.

Brutsaert, W.1982. Evaporation into the Atmosphere: Theory, History, and Applications. Kluwer Academic Publishers, Boston, MA.

Doorenbos, J., and W.O.Pruitt. 1977. Crop Water Requirements. FAO Irrigation and drainage paper 24. Food and agricultural organization of the United Nations, Rome.

Fares, A. and A.K. Alva. 2000. Evaluating the capacitance probes for optimal irrigation of citrus through soil moisture monitoring in an Entisol profile. Irrigation. Science 19:57–64.

Fayer,M.J. and D.Hillel.1986. Air Encapsulation I - Measurement in a field soil. Soil Science Society of America Journal. 50:568-572.

Feddes,R.A., P.J.Kowalik, and H.Zaradny. 1978. Simulation of field water use and crop yield. New York: John Wiley & Sons.

Freeze,R. and J.Cherry. 1979. Groundwater. Prentice Hall, Old Tappan, NJ.

Hacke.U.G., J.S.Sperry, and J.Pittermann. 2000. Drought Experience and Cavitation Resistance in Six Shrubs from the Great Basin, Utah. Basic Applied Ecology 1:31-41.

Hillel,D. 1998. Environmental soil physics. Academic Press, New York, NY

Jackson, R.B., J.Canadell, J.R.Ehleringer, H.A.Mooney, O.E.Sala, and E.D.Schulze. 1996. A global analysis of root distributions for terrestrial biomes. Oecologia 108:389-411.

Jarvis.N.J. 1989. A Simple Empirical Model of Root Water Uptake. Journal of Hydrology.107:57-72.

Kite, G.W., and P. Droogers. 2000. Comparing evapotranspiration estimates from satellites, hydrological models and field data. Journal of Hydrology 229:3–18.

Knowles, L., Jr. 1996. Estimation of evapotranspiration in the Rainbow Springs and Silver Springs basin in north-central Florida. Water resources investigation report. 96-4024. USGS, Reston, VA.

Li, K.Y., R.De jong, and J.B. Boisvert. 2001. An exponential root-water-uptake model with water stress compensation. Journal of hydrology 252:189-204.

Li,K.Y., R.De Jong, and M.T.Coe. 2006. Root water uptake based upon a new water stress reduction and an asymptotic root distribution function. Earth Interactions 10 (paper 14):1-22.

Mahmood, R. and K.G. Hubbard. 2003. Simulating sensitivity of soil moisture and evapotranspiration under heterogeneous soils and land uses. Journal of Hydrology. 280:72–90.

Meyboom, P. 1967. Ground water studies in the Assiniboine river drainage basin: II. Hydrologic characteristics of phreatophytic vegetation in south-central Saskatchewan. Geological Survey of Canada Bulletin 139, no.64.

Monteith, J. L. 1965. Evaporation and environment. *In* G.E.Fogg (ed). The state and movement of water in living organisms. Symposium of the Society of Experimental Biology: San Diego, California, Academic Press, New York, p.205-234

Morgan,K.T., L.R.Parsona, T.A. Wheaton, D.J.Pitts and T.A.Oberza. 1999. Field calibration of a capacitance water content probe in fine sand soils. Soil Science Society of America Journal 63: 987-989.

Mo, X., S. Liu, Z. Lin, and W. Zhao. 2004. Simulating temporal and spatial variation of evapotranspiration over the Lushi basin. Journal of Hydrology 285:125–142.

Mualem, Y. 1976. A new model predicting the hydraulic conductivity of unsaturated porous media. Water Resources Research 12(3):513-522.

Nachabe, M., N.Shah, M.Ross, and J.Vomacka. 2005. Evapotranspiration of two vegetation covers in a shallow water table environment. Soil Science Society of America Journal 69:492-499.

Oliveras,I., J.Martinez-Vilalta, T.Jimenez-Ortiz, M.J Lledo, A.Escarre, and J.Pinol. 2003. Hydraulic Properties of Pinus Halepensis, Pinus Pinea, and Tetraclinis Articulata in a Dune Ecosystem of Eastern Spain. Plant Ecology 169:131-141.

Pammenter.N.W. and C.V.Willigen. 1998. A Mathematical and Statistical Analysis of the Curves Illustrating Vulnerability of Xylem to Cavitation. Tree Physiology 18:589-593.

Priestley, C.H.B., and Taylor, R.J. 1972. On the assessment of surface heat flux and evaporation using large-scale parameters. Monthly Weather Review 100(2): 81-92.

Richards.L.A .1931. Capillary conduction of liquids through porous mediums, Journal of Applied Physics, 1(5), 318-333.

Said , A., M.Nachabe, M.Ross, and J.Vomacka. 2005. Methodology for estimating specific yield in shallow water environment using continuous soil moisture data. ASCE Journal of Irrigation and Drainage Engineering 131, no.6:533-538.

Schenk, H.J. and R. B. Jackson. 2002. Rooting Depths, Lateral Root Spreads and Below-Ground/Above-Ground Allometries of Plants in Water-Limited Ecosystems. The Journal of Ecology 90(3):480-494.

Shah,N. 2007. Vadose Zone Processes Affecting Water Table Fluctuations - Conceptualization and Modeling Considerations. PhD Disseration, University of South Florida, Tampa, Fl, 233 pp.

Shah, N., M. Ross, and A. Said. 2007. Vadose Zone Evapotranspiration Distribution Using One-Dimensional Analysis and Conceptualization for Integrated Modeling. Proceedings of ASCE EWRI conference, May 14th -May 19th 2007, Tampa.

Simunek, J., M. Th. van Genuchten and M. Sejna. 2005. The HYDRUS-1D software package for simulating the movement of water, heat, and multiple solutes in variably saturated media, version 3.0, HYDRUS software series 1. Department of Environmental Sciences, University of California Riverside, Riverside, California, USA, 270 pp.

Sperry, J.S., V.Stiller, and U.G.Hacke. 2003 Xylem Hydraulics and the Soil-Plant-Atmosphere Continuum: Oppurtunities and Unresolved Issues. Agronomy Journal 95:1362-1370.

Sumner, D.M. 2001. Evapotranspiration from a cypress and pine forest subjected to natural fires, Volusia County, Florida, 1998-99. Water Resources Investigations Report 01-4245. USGS, Reston, VA.

Sumner, D. 2006.Adequacy of selected evapotranspiration approximations for hydrological simulation. Journal of the American Water Resources Association. 42(3):699- 711.

Thornthwaite, C.W. 1948. An approach toward a rational classification of climate. Geographic Review 38:55-94.

Trout, K., and M.Ross. 2004. Intensive hydrologic data collection in as small watershed in West-Central Florida. Hydrological Science and Technology 21(1-4):187-197.

Tyree, M.T. S.Yang, P.Cruiziat, and, B.Sinclair. 1994. Novel Methods of Measuring Hydraulic Conductivity of Tree Root Systems and Interpretation Using AMAIZED. Plant Physilogy 104:189-199

van Genuchten, M.Th. 1980. A closed-form equation for predicting the hydraulic conductivity of unsaturated soils. Soil Science Society of America Journal 44:892-898.

van Genuchten, M. Th.1987. A numerical model for water and solute movement in and below the root zone. Research report No 121, U.S. Salinity laboratory, USDA, ARS, Riverside, California, 221pp.

Yang, J., B. Li, and S. Liu. 2000. A large weighing lysimeter for evapotranspiration and soil water-groundwater exchange studies. Hydrological Processes 14:1887–1897.

White,W.N. 1932.A method of estimating ground-water supplies based on discharge by plants and evaporation from soil: Results of investigation in Escalante Valley, Utah. Water-Supply Paper 659-A.

7

Impact of Irrigation on Hydrologic Change in Highly Cultivated Basin

Tadanobu Nakayama[1,2]
[1]*National Institute for Environmental Studies (NIES)*
16-2 Onogawa, Tsukuba, Ibaraki
[2]*Centre for Ecology & Hydrology (CEH)*
Crowmarsh Gifford, Wallingford, Oxfordshire
[1]*Japan*
[2]*United Kingdom*

1. Introduction

With the development of regional economies, the water use environment in the Yellow River Basin, China, has changed greatly (Fig. 1). The river is well known for its high sediment content, frequent floods, unique channel characteristics in the downstream (where the river bed lies above the surrounding land), and limited water resources. This region is heavily irrigated, and combinations of increased food demand and declining water availability are creating substantial pressures. Some research emphasized human activities such as irrigation water withdrawals dominate annual streamflow changes in the downstream in addition to climate change (Tang et al., 2008a). The North China Plain (NCP), located in the downstream area of the Yellow River, is one of the most important grain cropping areas in China, where water resources are also the key to agricultural development, and the demand for groundwater has been increasing. Groundwater has declined dramatically over the previous half century due to over-pumping and drought, and the area of saline-alkaline land has expanded (Brown and Halweil, 1998; Shimada, 2000; Chen et al., 2003b; Nakayama et al., 2006).

Since the completion of a large-scale irrigation project in 1969, noticeable cessation of flow has been observed in the Yellow River (Yang et al., 1998; Fu et al., 2004) resulting from intense competition between water supply and demand, which has occurred increasingly often. The ratio of irrigation water use (defined as the ratio of the annual gross use for irrigation relative to the annual natural runoff) having increased continuously from 21% to 68% during the last 50 years, indicating that the current water shortage is closely related to irrigation development (Yang et al., 2004a). This shortage also reduces the water renewal time (Liu et al., 2003) and renewability of water resources (Xia et al., 2004). This has been accompanied by a decrease in precipitation in most parts of the basin (Tang et al., 2008b). To ensure sustainable water resource use, it is also important to understand the contributions of human intervention to climate change in this basin (Xu et al., 2002), in addition to clarifying the rather complex and diverse water system in the highly cultivated region.

The objective of this research is to clarify the impact of irrigation on the hydrologic change in the Yellow River Basin, an arid to semi-arid environment with intensive cultivation.

Combination of the National Integrated Catchment-based Eco-hydrology (NICE) model (Nakayama, 2008a, 2008b, 2009, 2010, 2011a, 2011b; Nakayama and Fujita, 2010; Nakayama and Hashimoto, 2011; Nakayama and Watanabe, 2004, 2006, 2008a, 2008b, 2008c; Nakayama et al., 2006, 2007, 2010, 2011) with complex components such as irrigation, urban water use, and dam/canal systems has led to the improvement in the model, which simulates the balance of both water budget and energy in the entire basin with a resolution of 10 km. The simulated results also evaluates the complex hydrological processes of river dry-up, agricultural/urban water use, groundwater pumping, and dam/canal effects, and to reveal the impact of irrigation on both surface water and groundwater in the basin. This approach will help to clarify how the substantial pressures of combinations of increased food demand and declining water availability can be overcome, and how effective decisions can be made regarding sustainable development under sound socio-economic conditions in the basin.

2. Material and methods

2.1 Coupling of process-based model with complex irrigation procedures

Previously, the author developed the process-based NICE model, which includes surface-unsaturated-saturated water processes and assimilates land-surface processes describing the variations of LAI (leaf area index) and FPAR (fraction of photosynthetically active radiation) from satellite data (Fig. 2) (Nakayama, 2008a, 2008b, 2009, 2010, 2011a, 2011b; Nakayama and Fujita, 2010; Nakayama and Hashimoto, 2011; Nakayama and Watanabe, 2004, 2006, 2008a, 2008b, 2008c; Nakayama et al., 2006, 2007, 2010, 2011). The unsaturated layer divides canopy into two layers, and soil into three layers in the vertical dimension in the SiB2 (Simple Biosphere model 2) (Sellers et al., 1996). About the saturated layer, the NICE solves three-dimensional groundwater flow for both unconfined and confined aquifers. The hillslope hydrology can be expressed by the two-layer surface runoff model including freezing/thawing processes. The NICE connects each sub-model by considering water/heat fluxes: gradient of hydraulic potentials between the deepest unsaturated layer and the groundwater, effective precipitation, and seepage between river and groundwater.

In an agricultural field, NICE is coupled with DSSAT (Decision Support Systems for Agro-technology Transfer) (Ritchie et al., 1998), in which automatic irrigation mode supplies crop water requirement, assuming that average available water in the top layer falls below soil moisture at field capacity for cultivated fields (Nakayama et al., 2006). The model includes different functions of representative crops (wheat, maize, soybean, and rice) and simulates automatically dynamic growth processes. Potential evaporation is calculated on Priestley and Taylor equation (Priestley and Taylor, 1972), and plant growth is based on biomass formulation, which is limited by various reduction factors like light, temperature, water, and nutrient, et al. (Nakayama et al., 2006; Nakayama and Watanabe, 2008b; Nakayama, 2011a).

In this study, the NICE was coupled with complex sub-systems in irrigation and dam/canal in order to develop coupled human and natural systems and to analyze impact of irrigation on hydrologic change in highly cultivated basin. The return flow was evaluated from surface drainage and from groundwater, whereas previous studies had considered only surface drainage (Liu et al., 2003; Xia et al., 2004; Yang et al., 2004a). The gross loss of river water to irrigation includes losses via canals and leakage into groundwater in the field, and can be estimated as the difference between intake from, and return to the river.

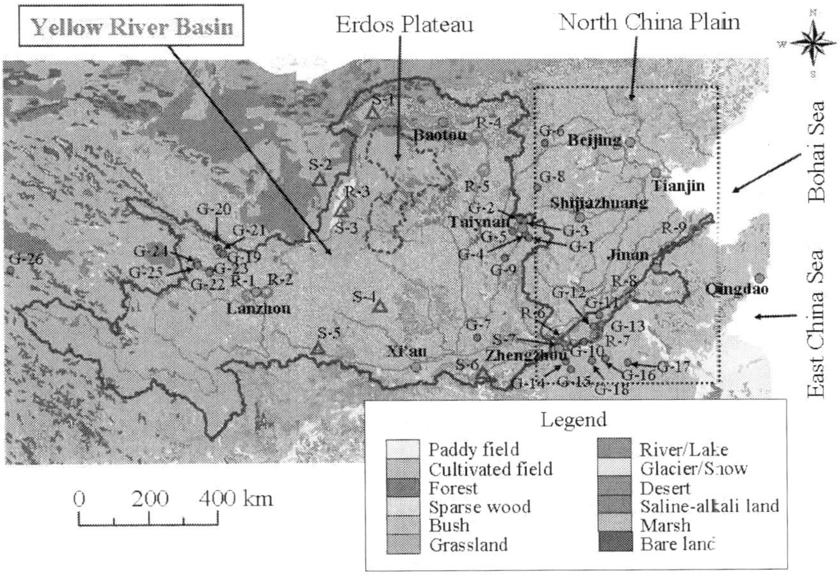

Fig. 1. Land cover in the study area of the Yellow River Basin in China. Bold black line shows the boundary of the basin. Black dotted line is the border of the North China Plain (NCP), which includes the downstream of the Yellow River Basin. Verification data are also plotted in this figure: river discharge (open red circle), soil moisture (open brown triangle), and groundwater level (green dot).

Irrigation withdrawals in the basin account for about 90% of total surface abstraction and 60% of groundwater withdrawal (Chen et al., 2003a). The model was improved for application to irrigated fields where water is withdrawn from both groundwater and river, and therefore the river dry-up process can be reproduced well. As the initial conditions, the ratios of river to aquifer irrigation were set at constant values. In the calibration procedure, these values were changed from initial conditions in order to reproduce the observation data as closely as possible after repeated trial and error (Oreskes et al., 1994). A validation procedure was then conducted in order to confirm the simulation under the same set of parameters, which resulted into reproducing reasonably the observed values. Spring/winter wheat, summer maize, and summer rice were automatically simulated in sequence analysis mode in succession by inputting previous point data for each crop type (Wang et al., 2001; Liu et al., 2002; Tao et al., 2006) and spatial distribution data (Chinese Academy of Sciences, 1988; Fang et al., 2006). The deficit water in the irrigated fields was automatically withdrawn and supplied from the river or the aquifer in the model in order to satisfy the observed hydrologic variables like soil moisture, river discharge, groundwater level, LAI, evapotranspiration, and crop coefficient. So, NICE simulates drought impact and includes the effect of water stress implicitly. Details are given in the previous researches (Nakayama et al., 2006; Nakayama and Watanabe, 2008b; Nakayama, 2011a, 2011b).

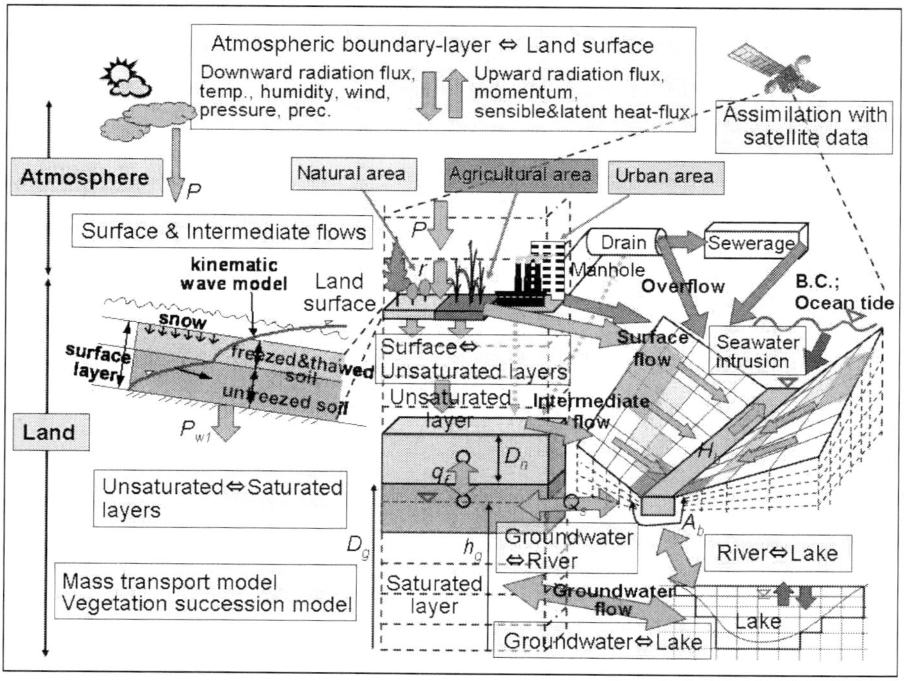

Fig. 2. National Integrated Catchment-based Eco-hydrology (NICE) model.

Another important characteristics of the study area is that there are many dams and canals to meet the huge demand for agricultural, industrial, and domestic water use (Ren et al., 2002) (Fig. 1), and exist six large dams on the main river (Yang et al., 2004a). Because there are few available data on discharge control at most of these dams, the model uses a constant ratio of dam inflow to outflow, which is a simpler approach than that of the storage-runoff function model (Sato et al., 2008). There are also many complex canals in the three large irrigation zones (Qingtongxia in Ningxia Hui, Hetao in Inner Mongolia, and Weisan in Shandong Province), in addition to the NCP, making it very difficult to evaluate the flow dynamics there. Because it is impossible to obtain the observed discharge and data related to the control of the weir/gate at every canal, it is effective to estimate the flow dynamics only in main canals as the first approximation when attempting to evaluate the hydrologic cycle in the entire basin in the same way as (Nakayama et al., 2006; Nakayama, 2011a). Therefore, NICE simulates the discharge only in a main canal assuming that this is defined as the difference in hydraulic potentials at both junctions similar to the stream junction model (Nakayama and Watanabe, 2008b). The dynamic wave effect is also important for the simulation of meandering rivers and smaller slopes, because the backwater effect is predominant (Nakayama and Watanabe, 2004). When a river flow is very low and almost zero at some point in the simulation, the dynamic wave theory requires a lot more computation time and sometimes becomes unstable. Therefore, the model applies a threshold water level of 1 mm to ensure simulation stability and to include the dry-up process. The model also includes the seepage process which is decided by some parameters such as hydraulic conductivity of the river bed, cross-sectional area of the groundwater section, and river bed thickness. Details are described in Nakayama (2011b).

2.2 Model input data and running the simulation

Six-hour reanalysed data for downward radiation, precipitation, atmospheric pressure, air temperature, air humidity, wind speed at a reference level, FPAR, and LAI were input into the model after interpolation of ISLSCP (International Satellite Land Surface Climatology Project) data with a resolution of 1° x 1° (Sellers et al., 1996) in inverse proportion to the distance back-calculated in each grid. Because the ISLSCP precipitation data had the least reliability and underestimated the observed values at peak times, rain gauge daily precipitation data collected at 3,352 meteorological stations throughout the study area were used to correct the ISLSCP precipitation data. Mean elevation of each 10-km grid cell was calculated from the spatial average of a global digital elevation model (DEM; GTOPO30) with a horizontal grid spacing of 30 arc-seconds (~1 km) (USGS, 1996). Digital land cover data produced by the Chinese Academy of Sciences (CAS) based on Landsat TM data from the early 1990s (Liu, 1996) were categorized for the simulation (Fig. 1). Vegetation class and soil texture were categorized and digitized into 1-km mesh data by using 1:4,000,000 and 1:1,000,000 vegetation and soil maps of China (Chinese Academy of Sciences, 1988, 2003). The author's previous research showed that these finer-resolution products are highly correlated with the ISLSCP (Nakayama, 2011b). The geological structure was divided into four types on the basis of hydraulic conductivity, the specific storage of porous material, and specific yield by scanning and digitizing the geological material (Geological Atlas of China, 2002) and core-sampling data at some points (Zhu, 1992).

The irrigation area was calculated from the GIS data based on Landsat TM data from the early 1990s (Liu, 1996), and the calculated value agrees well with the previous results from that period (Yang et al., 2004a) (Table 1), as described in Nakayama (2011b). Most of the irrigated fields are distributed in the middle and lower regions of the Yellow River mainstream and in the NCP (Fig. 3). The agricultural areas in the upper regions and Erdos Plateau are dominated by dryland fields. Spring/winter wheat was predominant in the upper and middle of the arid and semi-arid regions, and double cropping of winter wheat and summer maize was usually practised in the middle and downstream and in the NCP's relatively warm and humid environment (Wang et al., 2001; Liu et al., 2002; Fang et al., 2006; Nakayama et al., 2006; Tao et al., 2006). The averaged water use during 1987-1988 at the main cities in the Yellow River Basin and the NCP (Hebei Department of Water Conservancy, 1987-1988; Yellow River Conservancy Commission, 2002) was directly input to the model. In the 1990s, return flow was as much as 35% of withdrawal in the upper and 25% in the middle, but close to 0% in the downstream (Chen et al., 2003a; Cai and Rosegrant, 2004). The return flows at Qingtongxia and Hetao irrigation zones are 59% and 25% of withdrawal, whereas that at Weisan irrigation zone is close to 0%, because the river bed is above the level of the plain (Chen et al., 2003a; Cai and Rosegrant, 2004). This information was also input into the model.

At the upstream boundaries, a reflecting condition on the hydraulic head was used assuming that there is no inflow from the mountains in the opposite direction (Nakayama and Watanabe, 2004). At the eastern sea boundary, a constant head was set at 0 m. The hydraulic head values parallel to the observed ground level were input as initial conditions for the groundwater sub-model. As initial conditions, the ratios of river and aquifer irrigation were set at the same constant values as in the above section. In river grids decided by digital river network from 1:50,000 and 1:100,000 topographic maps (CAS, 1982), inflows or outflows from the riverbeds were simulated at each time step depending on the difference in the hydraulic heads of groundwater and river. The simulation area covered 3,000 km by 1,000 km with a grid spacing of 10 km, covering the entire Yellow River Basin and the NCP. The vertical layer was discretized in thickness with depth, with each layer increased in thickness by a factor of 1.1 (Nakayama, 2011b; Nakayama and Watanabe, 2008b; Nakayama et al., 2006). The upper layer was set at 2 m depth, and the 20th layer was defined as an elevation of -500 m from the sea surface. Simulations were performed with a time step of 6 h for two years during 1987–1988 after 6 months of warm-up period until equilibrium. The author first calibrated the simulated values including irrigation water use in 1987 against previous results, and then validated them in 1988. Previously observed data about river discharge (9 points; Yellow River Conservancy Commission, 1987-1988), soil moisture (7 points of the Global Soil Moisture Data Bank; Entin et al., 2000; Robock et al., 2000), and groundwater level (26 points; China Institute for Geo-Environmental Monitoring, 2003) were also used for the verification of the model (Fig. 1 and Table 2) in addition to values published in the literature (Clapp and Hornberger, 1978; Rawls et al., 1982). Details are described in Nakayama (2011b).

Reaches[a]	Irrigation area (x 10^4 ha)	
	GIS database (Liu 1996)	Previous research (Yang et al. 2004a)
Above LZ	46.8	39.5
LZ - TDG	342.1	344.1
TDG - LM	43.0	53.4
LM - SMX	295.6	281.3
SMX - HYK	59.2	60.6
Below HYK	160.7	155.0
Sum	947.3	933.9

[a]Abbreviation in the following; LZ, Lanzhou (R-1);
TDG, Toudaoguai (R-4); LM, Longmen; SMX, Sanmenxia;
HYK, Huayuankou (R-6).

Table 1. Comparison of irrigation area in the simulated condition with that in the previous research.

No.	Point Name	Type	Lat.	Lon.	Elev.(m)
R-1	Lanzhou	River Discharge	35°55.8'	103°19.8'	1794.0
R-2	Lanzhou	River Discharge	36°4.2'	103°49.2'	1622.0
R-3	Qingtongxia	River Discharge	38°21.0'	106°24.0'	1096.0
R-4	Toudaoguai	River Discharge	40°16.2'	111°4.2'	861.0
R-5	Hequ	River Discharge	39°22.2'	111°9.0'	861.0
R-6	Huayuankou	River Discharge	34°55.2'	113°39.0'	104.0
R-7	Lankao	River Discharge	34°55.2'	114°42.0'	73.0
R-8	Juancheng	River Discharge	35°55.8'	115°54.0'	1.0
R-9	Lijin	River Discharge	37°31.2'	118°18.0'	1.0
S-1	Bameng	Soil Moisture	40°46.2'	107°24.0'	1059.0
S-2	Xilingaole	Soil Moisture	39°4.8'	105°22.8'	1238.0
S-3	Yongning	Soil Moisture	38°15.0'	106°13.8'	1130.0
S-4	Xifengzhen	Soil Moisture	35°43.8'	107°37.8'	1435.0
S-5	Tianshui	Soil Moisture	34°34.8'	105°45.0'	1196.0
S-6	Lushi	Soil Moisture	34°0.0'	111°1.2'	675.0
S-7	Zhengzhou	Soil Moisture	34°49.2'	113°40.2	99.0
G-1	Shanxi-1	Groundwater Level	37°44.0'	112°34.2	772.51
G-2	Shanxi-2	Groundwater Level	38°0.7'	112°25.8	831.10
G-3	Shanxi-3	Groundwater Level	37°58.0'	112°29.6'	788.96
G-4	Shanxi-4	Groundwater Level	37°47.3'	112°31.3'	779.03
G-5	Shanxi-5	Groundwater Level	37°43.2'	111°57.5'	780.51
G-6	Shanxi-6	Groundwater Level	40°3.4'	113°17.4'	1059.73
G-7	Shanxi-7	Groundwater Level	34°56.9'	110°45.1'	346.73
G-8	Shanxi-8	Groundwater Level	38°47.1'	112°44.0'	823.18
G-9	Shanxi-9	Groundwater Level	37°7.6'	111°54.2'	733.24
G-10	Henan-1	Groundwater Level	34°48.2'	114°18.2'	73.40
G-11	Henan-2	Groundwater Level	35°42.0'	115°1.3'	52.20
G-12	Henan-3	Groundwater Level	35°31.0'	115°1.0'	53.30
G-13	Henan-4	Groundwater Level	35°31.0'	115°12.5'	54.00
G-14	Henan-5	Groundwater Level	34°1.5'	113°50.8'	66.80
G-15	Henan-6	Groundwater Level	33°49.0'	113°56.3'	63.91
G-16	Henan-7	Groundwater Level	34°4.1'	115°18.2'	47.20
G-17	Henan-8	Groundwater Level	33°55.9'	116°22.3'	31.90
G-18	Henan-9	Groundwater Level	34°48.2'	114°18.2'	52.60
G-19	Qinghai-1	Groundwater Level	36°35.5'	101°44.7'	2321.17
G-20	Qinghai-2	Groundwater Level	37°0.0'	101°37.9'	2506.94
G-21	Qinghai-3	Groundwater Level	36°58.5'	101°38.9'	2474.63
G-22	Qinghai-4	Groundwater Level	36°34.3'	101°43.9'	2346.37
G-23	Qinghai-5	Groundwater Level	36°32.2'	101°40.5'	2451.32
G-24	Qinghai-6	Groundwater Level	36°43.1'	101°30.3'	2425.32
G-25	Qinghai-7	Groundwater Level	36°41.9'	101°30.9'	2383.98
G-26	Qinghai-8	Groundwater Level	36°14.8'	94°46.6'	3007.81

Table 2. Lists of observation stations for calibration and validation shown in Fig. 1.

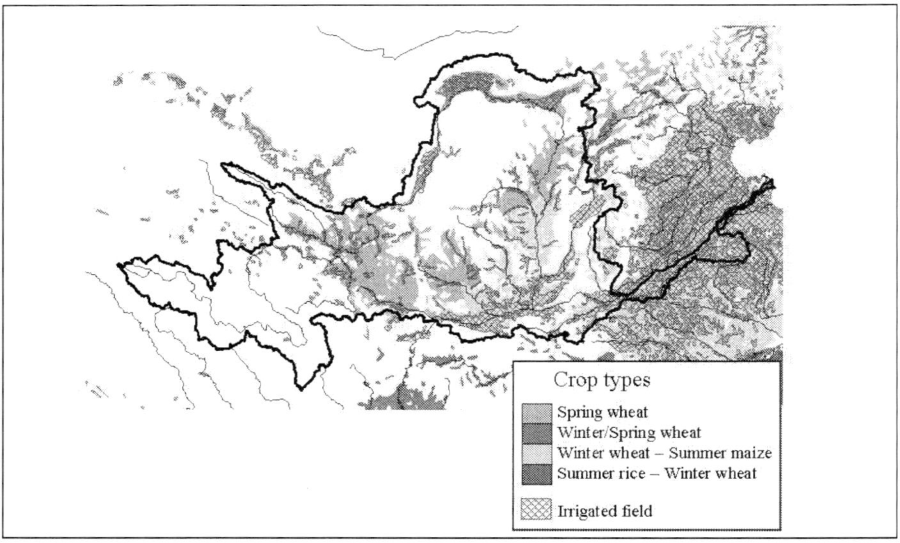

Fig. 3. Crop types in the agricultural areas. Irrigation areas are also overshaded. Irrigated fields cover most of the NCP for double cropping of winter wheat and summer maize in addition to three large irrigation zones.

3. Result and discussion

3.1 Verification of hydrologic cycle in the basin

The irrigation water use simulated in 1987 was firstly calibrated against previous results (Cai and Rosegrant, 2004; Liu and Xia, 2004; Yang et al., 2004a; Cai, 2006), showing a close agreement with the results of Cai and Rosegrant (2004). Then, the simulated value in 1988 was validated with the previous researches (Table 3), which indicates that there was reasonable agreement with each other and that irrigation water use is higher in large irrigation zones (LZ-TDG), along the Wei and Fen rivers (LM-SMX) in the middle, and in the downstream (below HYK). The results also show a high correlation between irrigation area and water use: r^2 = 0.986 (Chen et al., 2003a) and r^2 = 0.826 (Liu and Xia, 2004). Details of calibration and validation procedures are described in Nakayama (2011b).

The actual ET simulated by NICE reproduces reasonably the general trend estimated by integrated AVHRR NDVI data (Sun et al., 2004), which may give a good support on the predictive skill of the model (Fig. 4a-b). Although there are some discrepancies particularly for the lowest ET area (EP < 200 mm/year) mainly because of the banded colour figures, the simulated result reproduces the characteristics that the value is lowest in the downstream area of middle and on the Erdos Plateau—less than 200-300 mm per year (except in the irrigated area)—where vegetation is dominated by desert and soil is dominated by sand, and increases gradually towards the south-east. The simulated result also indicates that this spatial heterogeneity is related to human interventions and the resultant water stress by spring/winter cultivation in the upper/middle areas (Chen et al., 2003a; Tao et al., 2006), and winter wheat and summer maize cultivations in the middle/downstream (including the

Wei and Fen tributaries) and the NCP (Wang et al., 2001; Liu et al., 2002: Nakayama et al., 2006). Although the satellite-derived data are effective for grasping the spatial distribution of actual ET, there are some inefficiencies with regard to underestimation in sparsely vegetated regions (Inner Mongolia and Shaanxi Province) and overestimation in densely vegetated or irrigated regions (source area and Henan Province), as suggested by previous research (Sun et al., 2004; Zhou et al., 2007), which the simulation overcomes and improves mainly due to the inclusion of drought impact in the model. Details are described in Nakayama (2011b).

The model also simulated effect of irrigation on evapotranspiration at rotation between winter wheat and summer maize in the downstream of Yellow River (Fig. 4c). Because more water is withdrawn during winter-wheat period due to small rainfall in the north, the irrigation in this period affects greatly the increase in evapotranspiration. The simulated result indicates that the evapotranspiration increases predominantly during the seasons of grain filling and harvest of winter wheat with the effect of irrigation. In particular, most of the irrigation is withdrawn from aquifer in the NCP because surface water is seriously limited there (Nakayama, 2011b; Nakayama et al., 2006). This over-irrigation also affects the hydrologic change such as river discharge, soil moisture, and groundwater level in addition to evapotranspiration, as described in the following.

Reaches[a]	Irrigation water use (x 10^9 m^3)				
	Simulated value (1988)[b]	Cai and Rosegrant 2004 (2000)[b]	Liu and Xia 2004 (1990s)[b]	Yang et al. 2004a (1990s)[b]	Cai 2006 (1988-1992)[b]
Above LZ	1.5	2.9	13.2		12.4
LZ - TDG	6.8	12.2			
TDG - LM	1.1	1.0		18.9	
LM - SMX	10.4	7.3	6.0		4.8
SMX - HYK	2.0	2.4			
Below HYK	8.4	10.6	10.8	9.5	11.2
Sum	30.2	36.4	30.0	28.5	28.4

[a]Abbreviation in the following; LZ, Lanzhou (R-1); TDG, Toudaoguai (R-4); LM, Longmen; SMX, Sanmenxia; HYK, Huayuankou (R-6).
[b]Value in parenthesis shows the target year in the simulation and the literatures.

Table 3. Validation of irrigation water use simulated by the model with that in the previous research.

The model could simulate reasonably the spatial distribution of irrigation water use after the comparison with a previous study based on the Penman-Monteith method and the crop coefficient (Fang et al., 2006) not only in reach level but also in the spatial distribution, as described in Nakayama (2011b). In particular, simulated ratios of river to total irrigation

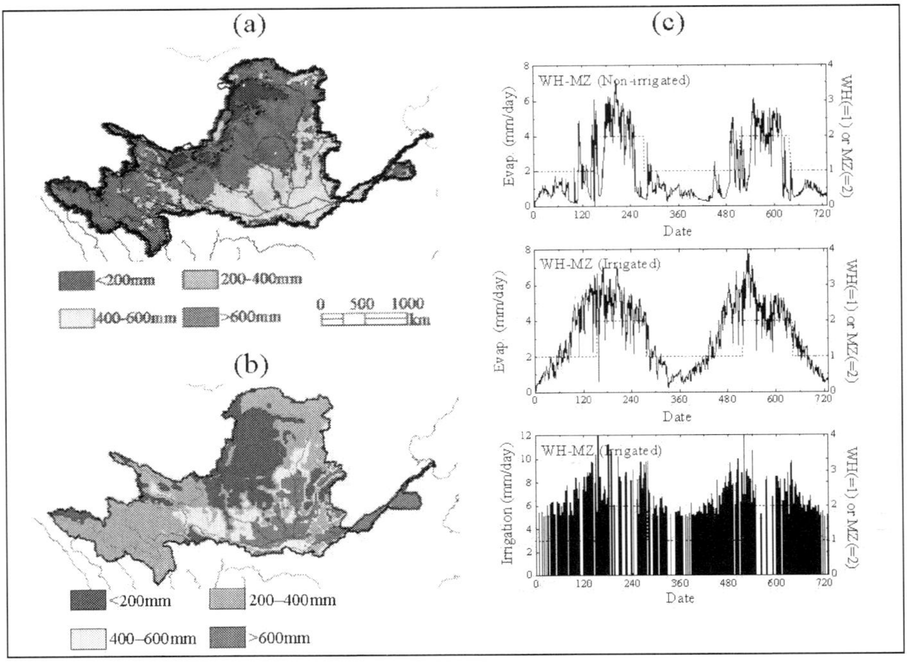

Fig. 4. Annual-averaged spatial distribution of evapotranspiration in 1987; (a) previous research; (b) simulated result; and (c) simulated value about impact of irrigation on evapotranspiration at rotation between winter wheat and summer maize. In Fig. 4c, right axis (dotted line) shows a period of each crop (WH; wheat, and MZ; maize).

(=river + aquifer) showed great variation and spatial heterogeneity in the basin. Fig. 5 shows the effect of over-irrigation on the decrease in river discharge on the downstream. The model reproduces reasonably the observed discharge for a low flow, and sometimes dry-up in the downstream (Yellow River Conservancy Commission, 1987-1988) with relatively high correlation r^2 and Nash-Sutcliffe criterion (NS; Nash and Sutcliffe, 1970) because the model includes the irrigation procedure and dynamic wave effect (Nakayama and Watanabe, 2004) in the model (Fig. 5b). The discharge decreases seriously in the downstream area mainly because of the water withdrawal for agriculture, which is more than 90% of the total withdrawal (Cai, 2006). At the downstream point at Lijin (R-9 in Table 2), the river discharge dries out during the spring mainly because most of the water is used for the irrigation of winter wheat in correspondence with the great increase in evapotranspiration shown in Fig. 4c. The model also indicated that the effect of groundwater irrigation is predominant in the downstream (data not shown), mainly on account of intensified water-use conflicts between upstream and downstream, and between various sectors like agriculture, municipality, and industry (Brown and Halweil, 1998; Nakayama, 2011a, 2011b; Nakayama et al., 2006). The smaller change in groundwater level in the upper was largely attributable to its unsuitability for crop production and the higher dependence of irrigation on surface water, as described previously (Yellow River Conservancy Commission, 2002).

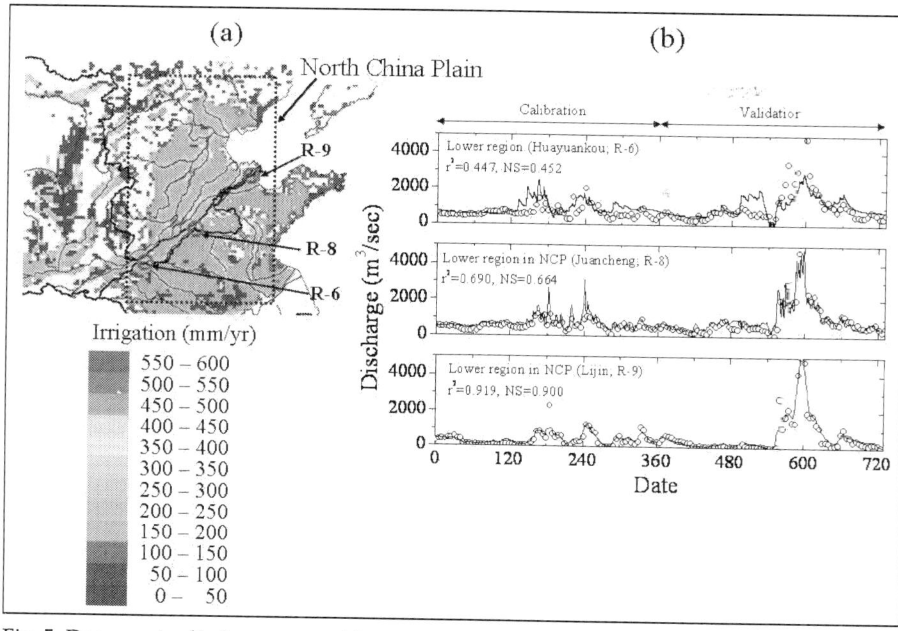

Fig. 5. Decrease in discharge caused by over-irrigation in the downstream region; (a) simulated result of river irrigation in 1987; (b) river discharge at the downstream. In Fig. 5b, solid line is the simulated result with irrigation, and circle is the observed value.

The simulated groundwater levels and soil moisture contents were calibrated and validated against observed data (Entin et al., 2000; Robock et al., 2000; China Institute for Geo-Environmental Monitoring, 2003) shown in Fig. 1 and Table 2 (data not shown). Although the correlation of groundwater level relative to the surface was not as good (r^2 = 0.401) as that of the absolute groundwater level (r^2 = 0.983) and the simulated value showed a tendency to overestimate the observed value in the calibration procedure for 1987, the simulation reproduced well the general distribution (BIAS = -21.2%, RMSE = 5.6 m, RRMSE = -0.468, MSSS = 0.356) (Nakayama, 2011b). This disagreement was due to the difference in surface elevation on the point-scale and mesh-scale (scale dependence), and the resolution of the groundwater flow model (changes in elevation from 0 m to 3000–4000 m in the basin). Because the simulated level is the hydraulic head in an aquifer, it might take a larger value than the land surface, particularly for a grid cell near or on the river. Another reason is that the irrigation water use simulated by the model might be underestimated because automatic irrigation supplied the water requirement for crops in order to satisfy the observed soil moisture, river discharge, groundwater level, LAI, evapotranspiration, and crop coefficient, which was theoretically pumped up from the river or the aquifer in the model. In reality, it has a possibility that farmers might use more irrigation water than the theoretical water requirement for crops if possible though there were not enough statistical or observed data to support it. The simulated water level decreases rapidly around the source area, indicating that there are many springs in this region. It is very low in the downstream (below sea level in some regions) because of the low elevation and overexploitation, as is the case in the NCP (Nakayama, 2011a, 2011b; Nakayama et al., 2006). The soil moisture is higher in the source area and in the paddy-dominated Qingtongxia Irrigation Zone (data not shown), corresponding closely with the distribution of the groundwater level. Details are described in Nakayama (2011b).

3.2 Impact of irrigation on hydrologic changes

Scenario analysis of conversion from unirrigated to irrigated run predicted the hydrologic changes (Fig. 6). The predicted result without irrigation generally overestimates the observed river discharge (Fig. 6a) and this effect is more prominent in the middle and downstream, as supported by reports that the difference between natural and observed runoff is larger downstream (Ren et al., 2002; Fu et al., 2004; Liu and Zheng, 2004). The difference between simulations considering and not considering irrigation strongly supports previous studies from the point of view that the influence of human interventions on river runoff has increased downstream over the last five decades (Chen et al., 2003a; Liu and Xia, 2004; Yang et al., 2004a; Cai, 2006; Tang et al., 2007) (Table 3), as also represented by the decline of water renewal times (Liu et al., 2003) and water resource renewability (Xia et al., 2004). This difference is greatly affected by complex irrigation procedures of various crops, which are roughly represented by spring/winter wheat in the upper-middle, and double cropping of winter wheat and summer maize in the middle-downstream regions (Wang et al., 2001; Liu et al., 2002; Fang et al., 2006; Nakayama et al., 2006; Tao et al., 2006).

Because there is some time lag between periods of increase in irrigation and decrease in runoff, the river discharge does not necessarily decrease in the winter and sometimes decreases in the summer. Further, the discharge sometimes increases slightly in the flood season, which indicates that the precipitation in irrigated fields sometimes responds quickly to flood drainage. Although both r^2 and NS have relatively low values across the basin (max: r^2 = 0.447, NS = 0.452), the simulated results with irrigation reproduce these characteristics better, and the statistics for MV (mean value), SD (standard deviation), and CV (coefficient of variation; CV = SD/MV) generally agree better with the observed values, as also supported by the better reproduction of other components of the hydrologic cycle, such as annual ET (Fig. 4) (data not shown in the case without irrigation) and irrigation water use (Fig. 5a, Table 3). The simulated result considering irrigation also reproduces the observed data for a low flow, and sometimes dry-up in the same way as Fig. 5b (Zhang et al., 1990; Yang et al., 1998; Ren et al., 2002), being attributable to inclusion of the dynamic wave effect in NICE, which other previous NICE series were unable to reproduce. Furthermore, the model improves the reproduction of river discharge in the basin in comparison with previous research (Yang and Musiake, 2003), where the ratio of absolute error to the mean was more than 60% at Huayuankou hydrological station, one of the worst such cases on a major river in Asia. The major reason for this disagreement is artificial water regulation such as reservoirs, water intakes, and diversions, which the model generally includes in addition to the extreme annual variation in flood seasons (Nakayama and Watanabe, 2008b).

Scenario analysis also predicts the groundwater level change and indicates that the effect of groundwater over-irrigation is predominant in the middle and downstream (Fig. 6b), where surface water is seriously limited, as shown in Fig. 5b and described in section 2.1 (Yellow River Conservancy Commission, 2002). The predicted result indicates a serious situation of water shortage in the downstream region and the NCP where groundwater level degrades over a wide area (Brown and Halweil). The result also implies that the model accounts for intensified water-use conflicts between upstream and downstream areas, and between agriculture, municipal, and industrial sectors (Brown and Halweil, 1998; Shimada, 2000; Chen et al., 2003b; Nakayama et al., 2006). These analyses of the impact of human intervention on hydrologic changes present strong indicatives of the seriousness of the situation, and imply the need for further correct estimation and appropriate measures against such irrigation loss and the low irrigation efficiency described previously (Wang et al., 2001). Details are described in Nakayama (2011b).

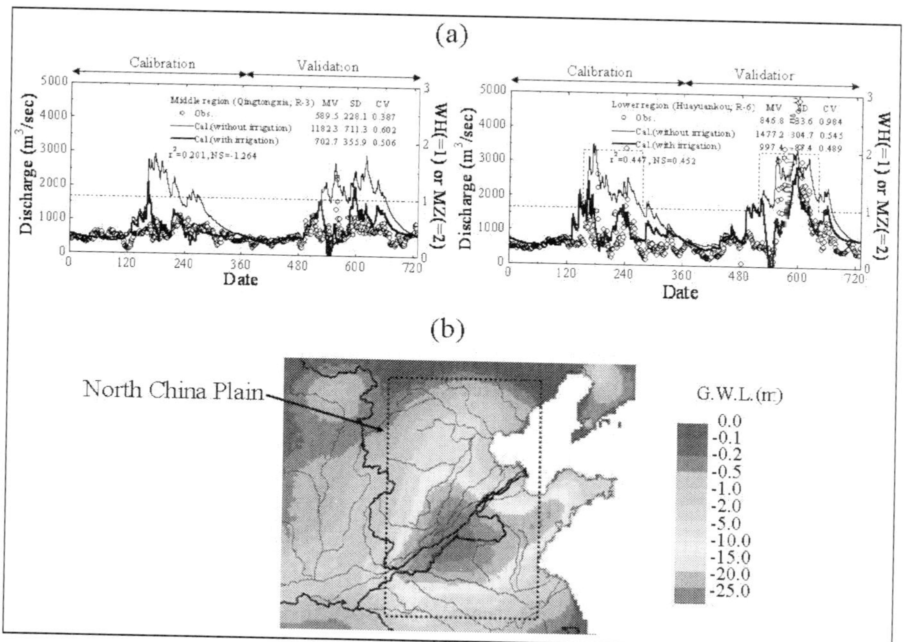

Fig. 6. Scenario analysis of conversion from unirrigated to irrigated run; (a) prediction of river discharge at the upper-middle (R-3; Qingtongxia) and the lower (R-6; Huayuankou) in Fig. 1 and Table 2; (b) groundwater level change in the middle-downstream regions. In Fig. 6a, circles show observation data; solid line is the simulated result without irrigation effect; bold line is the simulated result with irrigation. Right axis (dotted line) shows a period of each crop (WH; wheat, and MZ; maize) in the same way as Fig. 4c.

3.3 Discussion

Water scarcity and resource depletion in the downstream and the NCP, referred to as the 'bread basket' of China, is becoming more severe every year against increased crop production based on irrigation water, in addition to the expansion of municipal and industrial usage (Nakayama, 2011a; Nakayama et al., 2006). The simulated result shows the discharge was affected greatly by the rapid development of cities and industries and the increase in farmland irrigation (Fig. 6), which is closely related to severe groundwater degradation owing to the high clay content of the surface soil (Nakayama et al., 2006;

Nakayama, 2011a, 2011b). Because the dry-up of river reaches and groundwater exhaustion have been very severe so far (Chen et al., 2003b; Xia et al., 2004; Yang et al., 2004a), it is urgently necessary to perform effective control of water diversions (Liu and Xia, 2004; Liu and Zheng, 2004); the results simulated by NICE can be taken as strong indicatives of the seriousness of the situation. There are some reasons for the gap between irrigation water use (Fig. 5a) and groundwater level distribution (Fig. 6b). Firstly, the simulated levels have a temporally averaged distribution, and it takes some time for water levels to reach equilibrium after the boundary conditions have changed. This in turn affects the replenishment of groundwater from adjacent regions in addition to the heterogeneity of three-dimensional groundwater flow. Secondly, irrigation water is drawn not only from groundwater but also from river, and the ratio of river to total irrigation changes spatio-temporally in the basin (Fig. 5b); more river irrigation is drawn in the upper, and most of the irrigation depends on groundwater in the downstream, particularly in the NCP. This effect is clearly evident in comparison with the simulated results and the degradation value in the downstream is smaller that that in Fig. 6b.

Though the simulation reproduced reasonably hydrologic cycle such as evapotranspiration (Fig. 4), irrigation water use (Fig. 5a and Table 3), groundwater level, and river discharge (Fig. 5b and Fig. 6a), there were some discrepancies due to very complex and inaccurate nature of water withdrawal in the basin. In particular, the model achieved a relatively reasonable agreement though the model tried to calibrate and validate irrigation water use during only two years against other studies focusing on irrigation during long period (Fig. 5a and Table 3), which might lead to a substantial bias on model parameters. Because the objective of this study is primarily to evaluate the complex hydrological processes and reveal the impact of irrigation on hydrologic cycle in the basin through the verification during a fixed period, it is a future work to run model for the long period in the next step. At the same time, it will be of importance to derive better estimates of water demand in agricultural and urban areas during the long period by using more detailed statistical data, GIS data, and satellite data in longer period. Although the geological structure in the model included the general characteristics of several aquifers by reference to previous literature (Geological Atlas of China, 2002), the detailed structure of each aquifer layer was simplified as much as possible (Nakayama and Watanabe, 2008b; Nakayama et al., 2006). It will be necessary to obtain more precise data for the complex channel geometry of both natural and artificial rivers, soil properties, and geological structure. The spatial and temporal resolution used in the simulation also requires further improvement in order to overcome the problem of scale dependence and to improve verification and future reliability (Nakayama, 2011b).

Simulated results about the impact of irrigation on evapotranspiration change show a heterogeneous distribution (Fig. 4a-b). In particular, the irrigation of winter wheat increases greatly evapotranspiration, which is supplied by the limited water resources of river discharge and groundwater there (Fig. 5). This implies that energy supply is abundant relative to the water supply and the hydrological process is more sensitive to precipitation in the north, whereas the water supply is abundant relative to the energy supply and sun duration has a more significant impact in the south (Cong et al., 2010). The NICE is effective to provide better evaluation of hydrological trends in longer period including 'evaporation paradox' (Roderick and Farquhar, 2002; Cong et al., 2010) together with observation networks because the model does not need the crop coefficient (depending on a growing stage and a kind of crop) for the calculation of actual evaporation and simulates it directly without detailed site-specific information or empirical relation to calculate effective precipitation (Nakayama, 2011a; Nakayama et al., 2006). It is further necessary to clarify feedback and inter-relationship between micro, regional, and global scales; Linkage with global-scale dynamic vegetation model including two-way interactions between seasonal crop growth and atmospheric variability (Bondeau et al., 2007; Oleson et al., 2008); From stochastic to deterministic processes towards relationship between seedling establishment, mortality, and regeneration, and growth process based on carbon balance (Bugmann et al., 1996); From CERES-DSSAT to generic (hybrid) crop model by combinations of growth-development functions and mechanistic formulation of photosynthesis and respiration (Yang et al., 2004b); Improvement of nutrient fixation in seedlings, growth rate parameter, and stress factor, etc. for longer time-scale (Hendrickson et al., 1990). These future works might make a great contribution to the construction of powerful strategy for climate change problems in global scale.

Importance is that authority for water management in the basin is delineated by water source (surface water or groundwater) in addition to topographic boundaries (basin) and integrated water management concepts. In China, surface water and groundwater are managed by different authorities; the Ministry of Water Resources is responsible for surface water, while groundwater is considered a mineral resource and is administered by the Ministry of Minerals. In order to manage water resources effectively, any change in water accounting procedures may need to be negotiated through agreements brokered at relatively high levels of government, because surface water and groundwater are physically closely related to each other. Furthermore, the future development of irrigated and unirrigated fields and the associated crop production would affect greatly hydrologic change and usable irrigation water from river and aquifer, and vice versa (Nakayama, 2011b). The changes seen in this water resource are also related to climate change because groundwater storage moderates basin responses and climate feedback through evapotranspiration (Maxwell and Kollet, 2008). This is also related to a necessity of further evaluation about the evaporation paradox as described in the above. Although the groundwater level has decreased rapidly mainly due to overexploitation in the middle and downstream (Nakayama et al., 2006; Nakayama, 2011a, 2011b), regions where the land surface energy budget is very sensitive to groundwater storage are dominated by a critical water level (Kollet and Maxwell, 2008). The predicted hydrologic change indicates heterogeneous vulnerability of water resources and implies the associated impact on climate change (Fig. 6).

Basin responses will also be accelerated by an ambitious project to divert water from the Changjiang to the Yellow River, so-called, the South-to-North Water Transfer Project (SNWTP) (Rich, 1983; Yang and Zehnder, 2001). It can be estimated that the degradation of crop productivity may become severe, because most of the irrigation is dependent on vulnerable water resources (McVicar et al., 2002). Further research is necessary to examine the optimum amount of water that can be transferred, the effective management of the Three Gorges Dam (TGD) in the Changjiang River, the overall economic and social consequences of both projects, and their environmental assessment. It will be further necessary to obtain more observed and statistical data relating to water level, soil and water temperatures, water quality, and various phenological characteristics and crop productivity of spring/winter wheat and summer maize, in addition to satellite data of higher spatiotemporal resolution describing the seasonal and spatial vegetation phenology more accurately. The linear relationship between evapotranspiration and biomass production, which is very conservative and physiologically determined, is also valuable for further evaluation of the relationship between changes in water use and crop production by coupling with the numerical simulation and the satellite data analysis. Furthermore, it is powerful to develop a more realistic mechanism for sub-models, and to predict future hydrologic cycle and associated climate change using the model in order to achieve sustainable development under sound socio-economic conditions.

4. Conclusion

This study coupled National Integrated Catchment-based Eco-hydrology (NICE) model series with complex sub-models involving various factors, and clarified the importance of and diverse water system in the highly cultivated Yellow River Basin, including hydrological processes such as river dry-up, groundwater deterioration, agricultural water use, et al. The model includes different functions of representative crops (wheat, maize, soybean, and rice) and simulates automatically dynamic growth processes and biomass formulation. The model reproduced reasonably evapotranspiration, irrigation water use, groundwater level, and river discharge during spring/winter wheat and summer maize cultivations. Scenario analysis predicted the impact of irrigation on both surface water and groundwater, which had previously been difficult to evaluate. The simulated discharge with irrigation was improved in terms of mean value, standard deviation, and coefficient of variation. Because this region has experienced substantial river dry-up and groundwater degradation at the end of the 20th century, this approach would help to overcome substantial pressures of increasing food demand and declining water availability, and to decide on appropriate measures for whole water resources management to achieve sustainable development under sound socio-economic conditions.

5. Acknowledgment

The author thanks Dr. Y. Yang, Shijiazhuang Institute of Agricultural Modernization of the Chinese Academy of Sciences (CAS), China, and Dr. M. Watanabe, Keio University, Japan, for valuable comments about the study area. Some of the simulations in this study were run on an NEC SX-6 supercomputer at the Center for Global Environmental Research (CGER), NIES. The support of the Asia Pacific Environmental Innovation Strategy (APEIS) Project and the Environmental Technology Development Fund from the Japanese Ministry of Environment is also acknowledged.

6. References

Bondeau, A., Smith, P.C., Zaehle, S., Schaphoff, S., Lucht, W., Cramer, W., Gerten, D., Lotze-Campen, H., Muller, C., Reichstein, M. & Smith, B. (2007) Modelling the role of agriculture for the 20th century global terrestrial carbon balance. *Global Change Biol.*, Vol.13, pp.679-706, doi: 10.1111/j.1365-2486.2006.01305.x, ISSN 1354-1013

Brown, L.R. & Halweil, B. (1998). China's water shortage could shake world food security. *World Watch*, July/August, Vol.11(4), pp.10-18

Bugmann, H.K.M., Yan, X., Sykes, M.T., Martin, P., Linder, M., Desanker, P.V. & Cumming, S.G. (1996) A comparison of forest gap models: model structure and behaviour. *Climatic Change*, Vol.34, pp.289–313, ISSN 0165-0009

Cai, X. & Rosegrant, M.W. (2004). Optional water development strategies for the Yellow River Basin: Balancing agricultural and ecological water demands. *Water Resour. Res.*, Vol.40, W08S04, doi: 10.1029/2003WR002488, ISSN 0043-1397

Cai, X. (2006). *Water stress, water transfer and social equity in Northern China: Implication for policy reforms.* Human Development Report 2006, UNEP, Available from http://hdr.undp.org/en/reports/global/hdr2006/papers/cai ximing.pdf

Chen, J., He, D. & Cui, S. (2003a). The response of river water quality and quantity to the development of irrigated agriculture in the last 4 decades in the Yellow River Basin, China. *Water Resour. Res.*, Vol.39(3), 1047, doi: 10.1029/2001WR001234. ISSN 0043-1397

Chen, J.Y., Tang, C.Y., Shen, Y.J., Sakura, Y., Kondoh, A. & Shimada, J. (2003b). Use of water balance calculation and tritium to examine the dropdown of groundwater table in the piedmont of the North China Plain (NCP). *Environ. Geol.*, Vol.44, pp.564-571, ISSN 0943-0105

Chen, Y.M., Guo, G.S., Wang, G.X., Kang, S.Z., Luo, H.B. & Zhang, D.Z. (1995). *Main crop water requirement and irrigation of China.* Hydrologic and Electronic Press, Beijing, 73-102

China Institute for Geo-Environmental Monitoring (CIGEM) (2003). *China Geological Envoronment Infonet, Database of groundwater observation in the People's Republic of China*, Available from http://www.cigem.gov.cn

Chinese Academy of Sciences (CAS) (1982). *Topographic maps of 1:50,000 and 1:100,000*

Chinese Academy of Sciences (CAS) (1988). *Administrative division coding system of the People's Republic of China*, Beijing

Chinese Academy of Sciences (CAS) (2003). *China soil database*, Available from http://www.soil.csdb.cn

Clapp, R.B. & Hornberger, G.M. (1978). Empirical equations for some soil hydraulic properties. *Water Resour. Res.*, Vol.14, 601-604, ISSN 0043-1397

Cong, Z., Zhao, J., Yang, D. & Ni, G. (2010) Understanding the hydrological trends of river basins in China. *J. Hydrol.*, Vol.388, pp.350-356, doi: 10.1016/j.jhydrol.2010.05.013, ISSN 0022-1694

Doll, P. & Siebert, S. (2002). Global modeling of irrigation water requirements. *Water Resour. Res.*, Vol.38, 8-1 – 8-10, ISSN 0043-1397

Entin, J.K., Robock, A., Vinnikov, K.Y., Hollinger, S.E., Liu, S. & Namkhai, A. (2000). Temporal and spatial scales of observed soil moisture variations in the extratropics. *J. Geophys. Res.*, Vol.105(D9), pp.11865-11877, ISSN 0148-0227

Fang, W., Imura, H. & Shi, F. (2006). Wheat irrigation water requirement variability (2001-2030) in the Yellow River Basin under HADCM3 GCM scenarios. *Jpn. J. Environ. Sci.*, Vol.19(1), pp.3-14

Fu, G., Chen, S., Liu, C. & Shepard, D. (2004). Hydro-climatic trends of the Yellow River basin for the last 50 years. *Climatic Change*, Vol.65, pp.149-178, ISSN 0165-0009

Geological Atlas of China (2002). Geological Publisher, Beijing, China (in Chinese)

Godwin, D.C. & Jones, C.A. (1991). Nitrogen dynamics in soil-plant systems, In: *Modeling plant and soil systems*, Hanks, R.J. & Ritchie, J.T. (Eds.), 287-321, Agronomy 31, American Society of Agronomy, Madison, Wisconsin, USA

Hebei Department of Water Conservancy (1987). *Hebei year book of water conservancy for 1987* (in Chinese)

Hebei Department of Water Conservancy (1988). *Hebei year book of water conservancy for 1988* (in Chinese)

Hendrickson, O.Q., Fogal, W.H. & Burgess, D. (1990) Growth and resistance to herbivory in N2-fixing alders. *Can. J. Bot.*, Vol.69, pp.1919–1926, ISSN 0008-4026

Kollet, S.J., Maxwell, R.M., 2008. Capturing the influence of groundwater dynamics on land surface processes using an integrated, distributed watershed model. *Water Resour. Res.*, Vol.44, W02402, doi: 10.1029/2007WR006004, ISSN 0043-1397

Lee, T.M. (1996). Hydrogeologic controls on the groundwater interactions with an acidic lake in karst terrain, Lake Barco, Florida. *Water Resour. Res.*, Vol.32, 831-844, ISSN 0043-1397

Liu, C., Zhang, X. & Zhang, Y. (2002). Determination of daily evapotranspiration of winter wheat and corn by large-scale weighting lysimeter and micro-lysimeter. *Agr. Forest. Meteorol.*, Vol.111, pp.109-120, ISSN 0168-1923

Liu, C. & Xia, J. (2004). Water problems and hydrological research in the Yellow River and the Huai and Hai River basins of China. *Hydrol. Process.*, Vol.18, pp.2197-2210, doi: 10.1002/hyp.5524, ISSN 0885-6087

Liu, C. & Zheng, H. (2004). Changes in components of the hydrological cycle in the Yellow River basin during the second half of the 20th century. *Hydrol. Process.*, Vol.18, pp.2337-2345, doi: 10.1002/hyp.5534, ISSN 0885-6087

Liu, J.Y. (1996). *Macro-scale survey and dynamic study of natural resources and environment of China by remote sensing*, Chinese Science and Technology Publisher, Beijing, China (in Chinese)

Liu, L., Yang, Z. & Shen, Z. (2003). Estimation of water renewal times for the middle and lower sections of the Yellow River. *Hydrol. Process.*, Vol.17, pp.1941-1950, doi: 10.1002/hyp.1219, ISSN 0885-6087

Maxwell, R.M. & Kollet, S.J. (2008). Interdependence of groundwater dynamics and land-energy feedbacks under climate change. *Nat. Geosci.*, Vol.1, pp.665-669, doi: 10.1038/ngeo315, ISSN 1752-0894

McVicar, T.R., Zhang, G.L., Bradford, A.S., Wang, H.X., Dawes, W.R., Zhang, L. & Li, L. (2002). Monitoring regional agricultural water use efficiency for Hebei Province on the North China Plain. *Aust. J. Agric. Res.*, Vol.53, pp.55-76, ISSN 0004-9409

Nakayama, T. & Watanabe, M. (2004). Simulation of drying phenomena associated with vegetation change caused by invasion of alder (Alnus japonica) in Kushiro Mire. *Water Resour. Res.*, Vol.40, W08402, doi: 10.1029/2004WR003174, ISSN 0043-1397

Nakayama, T. & Watanabe, M. (2006). Simulation of spring snowmelt runoff by considering micro-topography and phase changes in soil layer. *Hydrol. Earth Syst. Sci. Discuss.*, Vol.3, pp.2101-2144, ISSN 1027-5606

Nakayama, T., Yang, Y., Watanabe, M. & Zhang, X. (2006). Simulation of groundwater dynamics in North China Plain by coupled hydrology and agricultural models. *Hydrol. Process.*, Vol.20(16), pp.3441-3466, doi: 10.1002/hyp.6142, ISSN 0885-6087

Nakayama, T., Watanabe, M., Tanji, K. & Morioka, T. (2007). Effect of underground urban structures on eutrophic coastal environment. *Sci. Total Environ.*, Vol.373(1), pp.270-288, doi: 10.1016/j.scitotenv.2006.11.033, ISSN 0048-9697

Nakayama, T. (2008a). Factors controlling vegetation succession in Kushiro Mire. *Ecol. Model.*, Vol.215, pp.225-236, doi: 10.1016/j.ecolmodel.2008.02.017, ISSN 0304-3800

Nakayama, T. (2008b). Shrinkage of shrub forest and recovery of mire ecosystem by river restoration in northern Japan. *Forest Ecol. Manag.*, Vol.256, pp.1927-1938, doi: 10.1016/j.foreco.2008.07.017, ISSN 0378-1127

Nakayama, T. & Watanabe, M. (2008a). Missing role of groundwater in water and nutrient cycles in the shallow eutrophic Lake Kasumigaura, Japan. *Hydrol. Process.*, Vol.22, pp.1150-1172, doi: 10.1002/hyp.6684, ISSN 0885-6087

Nakayama, T. & Watanabe, M. (2008b). Role of flood storage ability of lakes in the Changjiang River catchment. *Global Planet. Change*, Vol.63, pp.9-22, doi: 10.1016/j.gloplacha.2008.04.002, ISSN 0921-8181

Nakayama, T. & Watanabe, M. (2008c). Modelling the hydrologic cycle in a shallow eutrophic lake. *Verh. Internat. Verein. Limnol.*, Vol.30

Nakayama, T. (2009). Simulation of Ecosystem Degradation and its Application for Effective Policy-Making in Regional Scale, In: *River Pollution Research Progress*, Mattia N. Gallo & Marco H. Ferrari (Eds.), 1-89, Nova Science Publishers, Inc., ISBN 978-1-60456-643-7, New York

Nakayama, T. (2010). Simulation of hydrologic and geomorphic changes affecting a shrinking mire. *River Res. Appl.*, Vol.26(3), pp.305-321, doi: 10.1002/rra.1253, ISSN 1535-1459

Nakayama, T. & Fujita, T. (2010). Cooling effect of water-holding pavements made of new materials on water and heat budgets in urban areas. *Landscape Urban Plan.*, Vol.96, pp.57-67, doi: 10.1016/j.landurbplan.2010.02.003, ISSN 0169-2046

Nakayama, T., Sun, Y. & Geng, Y. (2010). Simulation of water resource and its relation to urban activity in Dalian City, Northern China. *Global Planet. Change*, Vol.73, pp.172-185, doi: 10.1016/j.gloplacha.2010.06.001, ISSN 0921-8181

Nakayama, T. (2011a). Simulation of complicated and diverse water system accompanied by human intervention in the North China Plain. *Hydrol. Process.*, Vol.25, pp.2679-2693 doi: 10.1002/hyp.8009, ISSN 0885-6087

Nakayama, T. (2011b). Simulation of the effect of irrigation on the hydrologic cycle in the highly cultivated Yellow River Basin. *Agr. Forest Meteorol.*, Vol.151, pp.314-327, doi: 10.1016/j.agrformet.2010.11.006, ISSN 0168-1923

Nakayama, T. & Hashimoto, S. (2011). Analysis of the ability of water resources to reduce the urban heat island in the Tokyo megalopolis. *Environ. Pollut.*, Vol.159, pp.2164-2173, doi: 10.1016/j.envpol.2010.11.016, ISSN 0269-7491

Nakayama, T., Hashimoto, S. & Hamano, H. (2011). Multi-scaled analysis of hydrothermal dynamics in Japanese megalopolis by using integrated approach. *Hydrol. Process.* (in press), ISSN 0885-6087

Nash, J.E. & Sutcliffe, J.V. (1970). Riverflow forecasting through conceptual model. *J. Hydrol.*, Vol.10, pp.282-290, ISSN 0022-1694

Oleson, K.W., Niu, G.-Y., Yang, Z.-L., Lawrence, D.M., Thornton, P.E., Lawrence, P.J., Stockli, R., Dickinson, R.E., Bonan, G.B., Levis, S., Dai, A. & Qian, T. (2008)

Improvements to the Community Land Model and their impact on the hydrological cycle. *J. Geophys. Res.*, Vol.113, G01021, doi: 10.1029/2007JG000563, ISSN 0148-0227

Oreskes, N., Shrader-Frechette, K. & Belitz, K. (1994). Verification, validation, and confirmation of numerical models in the earth sciences. *Science*, Vol.263, pp.641-646, ISSN 0036-8075

Priestley C.H.B. & Taylor, R.J. (1972). On the assessment of surface heat flux and evaporation using large-scale parameters. *Mon. Weather Rev.*, Vol.100, pp.81-92, ISSN 0027-0644

Rawls, W.J., Brakensiek, D.L. & Saxton, K.E. (1982). Estimation of soil water properties. *Trans. ASAE*, Vol.25, pp.1316-1320

Ren, L., Wang, M., Li, C. & Zhang, W. (2002). Impacts of human activity on river runoff in the northern area of China. *J. Hydrol.*, Vol.261, pp.204-217, ISSN 0022-1694

Rich, V. (1983). Yangtze to cross Yellow River. *Nature*, Vol.305, pp.568, ISSN 0028-0836

Ritchie, J.T., Singh, U., Godwin, D.C. & Bowen, W.T. (1998). Cereal growth, development and yield, In: *Understanding Options for Agricultural Production*, Tsuji, G.Y., Hoogenboom, G. & Thornton, P.K. (Eds.), 79-98, Kluwer, ISBN 0-7923-4833-8, Great Britain

Robock, A., Konstantin, Y.V., Govindarajalu, S., Jared, K.E., Steven, E.H., Nina, A.S., Suxia, L. & Namkhai, A. (2000). The global soil moisture data bank. *Bull. Am. Meteorol. Soc.*, Vol.81, pp.1281-1299, Available from http://climate.envsci.rutgers.edu/soil_moisture/

Roderick, M.L. & Farquhar, G.D. (2002) The cause of decreased pan evaporation over the past 50 years. *Science*, Vol.298(15), pp.1410-1411, ISSN 0036-8075

Sato, Y., Ma, X., Xu, J., Matsuoka, M., Zheng, H., Liu, C. & Fukushima, Y. (2008). Analysis of long-term water balance in the source area of the Yellow River basin. *Hydrol. Process.*, Vol.22, pp.1618-1929, doi: 10.1002/hyp.6730, ISSN 0885-6087

Sellers, P.J., Randall, D.A., Collatz, G.J., Berry, J.A., Field, C.B., Dazlich, D.A., Zhang, C., Collelo, G.D. & Bounoua, L. (1996). A revised land surface prameterization (SiB2) for atomospheric GCMs. Part I : Model formulation. *J. Climate*, Vol.9, pp.676-705, ISSN 0894-8755

Shimada, J. (2000). Proposals for the groundwater preservation toward 21st century through the view point of hydrological cycle. *J. Japan Assoc. Hydrol. Sci.*, Vol.30, pp.63-72 (in Japanese)

Sun, R., Gao, X., Liu, C.M. & Li, X.W. (2004). Evapotranspiration estimation in the Yellow River Basin, China using integrated NDVI data. *Int. J. Remote Sens.*, Vol.25, pp.2523-2534, ISSN 0143-1161

Tang, Q., Oki, T., Kanae, S. & Hu, H. (2007). The influence of precipitation variability and partial irrigation within grid cells on a hydrological simulation. *J. Hydrometeorol.*, Vol.8, pp.499-512, doi: 10.1175/JHM589.1, ISSN 1525-755X

Tang, Q., Oki, T., Kanae, S. & Hu, H. (2008a). Hydrological cycles change in the Yellow River basin during the last half of the twentieth century. *J. Climate*, Vol.21, pp.1790-1806, doi: 10.1175/2007JCLI1854.1, ISSN 0894-8755

Tang, Q., Oki, T., Kanae, S. & Hu, H. (2008b). A spatial analysis of hydro-climatic and vegetation condition trends in the Yellow River basin. *Hydrol. Process.*, Vol.22, pp.451-458, doi: 10.1002/hyp.6624, ISSN 0885-6087

Tao, F., Yokozawa, M., Xu, Y., Hayashi, Y. & Zhang, Z. (2006). Climate changes and trends in phenology and yields of field crops in China, 1981-2000. *Agr. Forest Meteorol.*, Vol.138, pp.82-92, ISSN 0168-1923

U.S. Geological Survey (USGS) (1996). *GTOPO30 Global 30 Arc Second Elevation Data Set*, USGS, Available from http://www1.gsi.go.jp/geowww/globalmap-gsi/gtopo30/gtopo30.html

Wang, H., Zhang, L., Dawes, W.R. & Liu, C. (2001). Improving water use efficiency of irrigated crops in the North China Plain - measurements and modeling. *Agr. Forest. Meteorol.*, Vol.48, pp.151-167, ISSN 0168-1923

Xia, J., Wang, Z., Wang, G. & Tan, G. (2004). The renewability of water resources and its quantification in the Yellow River basin, China. *Hydrol. Process.*, Vol.18, pp.2327-2336, doi: 10.1002/hyp.5532, ISSN 0885-6087

Xu, Z.X., Takeuchi, K., Ishidaira, H. & Zhang, X.W. (2002). Sustainability analysis for Yellow River Water Resources using the system dynamics approach. *Water Resour. Manag.*, Vol.16, pp.239-261, ISSN 0920-4741

Yang, Z.S., Milliman, J.D., Galler, J., Liu, J.P. & Sun, X.G. (1998). Yellow River's water and sediment discharge decreasing steadily. *EOS*, Vol.79(48), pp.589-592, ISSN 0096-3941

Yang, H. & Zehnder, A. (2001). China's regional water scarcity and implications for grain supply and trade. *Environ. Plann. A*, Vol.33, pp.79-95

Yang, D. & Musiake, K. (2003). A continental scale hydrological model using the distributed approach and its application to Asia. *Hydrol. Process.*, Vol.17, pp 2855-2869, doi: 10.1002/hyp.1438, ISSN 0885-6087

Yang, D., Li, C., Hu, H., Lei, Z., Yang, S., Kusuda, T., Koike, T. & Musiake, K. (2004a). Analysis of water resources variability in the Yellow River of China during the last half century using historical data. *Water Resour. Res.*, Vol.40, W06502, doi: 10.1029/2003WR002763, ISSN 0043-1397

Yang, H.S., Dobermann, A., Lindquist, J.L., Walters, D.T., Arkebauer, T.J. & Cassman, K.G. (2004b) Hybrid-maize–a maize simulation model that combines two crop modeling approaches. *Field Crop. Res.*, Vol.87, pp.131-154, ISSN 0378-4290

Yellow River Conservancy Commission (1987). *Annual report of discharge and sediment in Yellow River*, Interior report of the committee (in Chinese)

Yellow River Conservancy Commission (1988). *Annual report of discharge and sediment in Yellow River*, Interior report of the committee (in Chinese)

Yellow River Conservancy Commission (2002). *Yellow River water resources bulletins*, Available from http://www.yrcc.gov.cn/ (in Chinese)

Zhang, J., Huang, W.W. & Shi, M.C. (1990). Huanghe (Yellow River) and its estuary: sediment origin, transport and deposition. *J. Hydrol.*, Vol.120, pp.203-223, ISSN 0022-1694

Zhou, M.C., Ishidaira, H. & Takeuchi, K. (2007). Estimation of potential evapotranspiration over the Yellow River basin: reference crop evaporation or Shuttleworth-Wallance?. *Hydrol. Process.*, Vol.21, pp.1860-1874, doi: 10.1002/hyp.6339, ISSN 0885-6087

Zhu, Y. (1992). *Comprehensive hydro-geological evaluation of the Huang-Huai-Hai Plain*, Geological Publishing House of China, 277p., Beijing, China (in Chinese)

8

Estimation of Evapotranspiration Using Soil Water Balance Modelling

Zoubeida Kebaili Bargaoui
Tunis El Manar University
Tunisia

1. Introduction

Assessing evapotranspiration is a key issue for natural vegetation and crop survey. It is a very important step to achieve the soil water budget and for deriving drought awareness indices. It is also a basis for calculating soil-atmosphere Carbon flux. Hence, models of evapotranspiration, as part of land surface models, are assumed as key parts of hydrological and atmospheric general circulation models (Johnson et al., 1993). Under particular climate (represented by energy limiting evapotranspiration rate corresponding to potential evapotranspiration) and soil vegetation complex, evapotranspiration is controlled by soil moisture dynamics. Although radiative balance approaches are worth noting for evapotranspiration evaluation, according to Hofius (2008), the soil water balance seems the best method for determining evapotranspiration from land over limited periods of time. This chapter aims to discuss methods of computing and updating evapotranspiration rates using soil water balance representations.

At large scale, Budyko (1974) proposed calculating annual evapotranspiration from data of meteorological stations using one single parameter w_0 representing a critical soil water storage. Using a statistical description of the sequences of wet and dry days. Eagleson (1978 a) developed an average annual water balance equation in terms of 23 variables including soil, climate and vegetation parameters with the assumption of a homogeneous soil-atmosphere column using Richards (1931) equation. On the other hand, the daily bucket with bottom hole model (BBH) proposed by Kobayashi et al. (2001) was introduced based on Manabe model (1969) involving one single layer bucket but including gravity drainage (leakage) as well as capillary rise. Vrugt et al. (2004) concluded that the daily Bucket model and the 3-D model (MODHMS) based on Richards equation have similar results. Also, Kalma & Boulet (1998) compared simulation results of the rainfall runoff hydrological model VIC which assumes a bucket representation including spatial variability of soil parameters to the one dimensional physically based model SiSPAT (Braud et al. , 1995). Using soil moisture profile data for calibration, they conclude that catchment's scale wetness index for very dry and very wet periods are misrepresented by SiSPAT while captured by VIC. Analyzing VIC parameter identifiability using streamflow data, DeMaria et al. (2007) concluded that soil parameters sensitivity was more strongly dictated by climatic gradients than by changes in soil properties especially for dry environments. Also, studying the measurements of soil moisture of sandy soils under semi-arid conditions, Ceballos et al. (2002) outlined the dependence of soil moisture time series on intra annual rainfall variability. Kobayachi et al. (2001) adjusted soil humidity profiles measurements for model calibration while Vrugt et al. (2004) suggested that effective soil hydraulic properties are poorly identifiable using drainage discharge data.

The aim of the chapter is to provide a review of evapotranspiration soil water balance models. A large variety of models is available. It is worth noting that they do differ with respect to their structure involving empirical as well as conceptual and physically based models. Also, they generally refer to soil properties as important drivers. Thus, the chapter will first focus on the description of the water balance equation for a column of soil-atmosphere (one dimensional vertical equation) (section 2). Also, the unsaturated hydrodynamic properties of soils as well as some analytical solutions of the water balance equation are reviewed in section 2. In section 3, key parameterizations generally adopted to compute actual evapotranspiration will be reported. Hence, several soil water balance models developed for large spatial and time scales assuming the piecewise linear form are outlined. In section 4, it is focused on rainfall-runoff models running on smaller space scales with emphasizing on their evapotranspiration components and on calibration methods. Three case studies are also presented and discussed in section 4. Finally, the conclusions are drawn in section 5.

2. The one dimensional vertical soil water balance equation

As pointed out by Rodriguez-Iturbe (2000) the soil moisture balance equation (mass conservation equation) is "likely to be the fundamental equation in hydrology". Considering large spatial scales, Sutcliffe (2004) might agree with this assumption. In section 2.1 we first focus on the presentation of the equation relating relative soil moisture content to the water balance components: infiltration into the soil, evapotranspiration and leakage. Then water loss through vegetation is addressed. Finally, infiltration models are discussed in section 2.2.

2.1 Water balance

For a control volume composed by a vertical soil column, the land surface, and the corresponding atmospheric column, and under solar radiation and precipitation as forcing variables, this equation relates relative soil moisture content s to infiltration into the soil $I(s,t)$, evapotranspiration $E(s,t)$ and leakage $L(s,t)$.

$$nZ_a\, \delta s/\delta t = I(s,t) - E(s,t) - L(s,t) \quad (1a)$$

Where t is time, n is soil effective porosity (the ratio of volume of voids to the total soil matrix volume); and Z_a is the active depth of soil.

Soil moisture exchanges as well as surface heat exchanges depend on physical soil properties and vegetation (through albedo , soil emissivity, canopy conductance) as well as atmosphere properties (turbulent temperature and water vapour transfer coefficients, aerodynamic conductance in presence of vegetation) and weather conditions (solar radiation, air temperature, air humidity, cloud cover, wind speed). Soil moisture measurements require sampling soil moisture content by digging or soil augering and determining soil moisture by drying samples in ovens and measuring weight losses; also, in situ use of tensiometry, neutron scattering, gamma ray attenuation, soil electrical conductivity analysis, are of common practice (Gardner et al. (2001) ; Sutcliffe, 2004; Jeffrey et al. (2004)).

The basis of soil water movement has been experimentally proposed by Darcy in 1856 and expresses the average flow velocity in a porous media in steady-state flow conditions of groundwater. Darcy introduced the notion of hydraulic conductivity. Boussinesq in 1904 introduced the notion of specific yield so as to represent the drainage from the unsaturated zone to the flow in the water table. The specific yield is the flux per unit area draining for a unit fall in water table height. Richards (1931) proposed a theory of water movement in the unsaturated homogeneous bare soil represented by a semi infinite homogeneous column:

$$\delta\theta/\delta t = \delta/\delta z\ [\ K\ \delta\psi/\delta z - K(\theta)] \tag{1b}$$

Where t is time; θ is volumetric water content (which is the ratio between soil moisture volume and the total soil matrix volume cm^3cm^{-3}); z is the vertical coordinate (z>0 downward from surface); K is hydraulic conductivity (cms^{-1}); ψ is the soil water matrix potential. Both K and ψ are function of the volumetric water content. Richards equation assumes that the effect of air on water flow is negligible. If accounting for the slope surface, it comes:

$$\delta\theta/\delta t = \delta/\delta z\ [K\ \delta\psi/\delta\theta\ \delta\theta/\delta z] - \delta\ K/\delta\theta\ \ \delta\theta/\delta z]\ \cos\beta \tag{2}$$

Where β is surface slope angle and cos is the cosinus function. We notice that the term [K δψ/δz - K(θ)] represents the vertical moisture flux. In particular, as reported by Youngs (1988) the soil-water diffusivity parameter D has been proposed by Childs and Collis-George (1950) as key soil-water property controlling the water movement.

$$D(\theta) = K(\theta)\ \delta\psi/\delta\theta \tag{3}$$

Thus, the Richards equation is often written as following:

$$\delta\theta/\delta t = \delta/\delta z\ [\ D(\theta)\ \delta\theta/\delta z\ ] - \delta\ K(\theta)/\delta z] \tag{4}$$

Eq. (4) is generally completed by source and sink terms to take into account the occurrence of precipitation infiltrating into the soil $I_{nf}(\theta,z_0)$ where z_0 is the vertical coordinate at the surface and vegetation uptake of soil moisture $g_r(\theta,z)$,. Vegetation uptake (transpiration) depends on vegetation characteristics (species, roots, leaf area, and transfer coefficients) and on the potential rate of evapotranspiration E_0 which characterizes the climate. Consequently, Eq. (4) becomes:

$$\delta\theta/\delta t = \delta/\delta z\ [\ D(\theta)\delta\theta/\delta z - K(\theta)] - g_r(\theta,z) + I_{nf}(\theta,z_0) \tag{5}$$

Youngs (1988) noticed that near the soil surface where temperature gradients are important Richards equation may be inadequate. We find in Raats (2001) an important review of evapotranspiration models and analytical and numerical solutions of Richards equation. However, it should be noticed that after Feddes et al. (2001) "in case of catchments with complex sloping terrain and groundwater tables, a vertical domain model has to be coupled with either a process or a statistically based scheme that incorporates lateral water transfer". So, a key task in the soil water balance model evaluation is the estimation of $I_{nf}(\theta,z_0)$ and $g_r(\theta,z)$. Both depend on the distribution of soil moisture. We focus here on vegetation uptake (or transpiration) $g_r(\theta,z)$ which is regulated by stomata and is driven by atmospheric demand. Based on an Ohm's law analogy which was primary proposed by Honert in 1948 as outlined by Eagleson (1978 b), the conceptual model of local transpiration uptake u(z,t)= $g_r(\theta,z)$ as volume of water per area per time is expressed as (Guswa, 2005)

$$u(z,t)=\Delta z\ (\psi(z,t) - \psi_p)\ /[\ R_1(\theta\ (z,t))+R_2] \qquad (6)$$

ψ soil moisture potential (bars), ψ_p leaf moisture potential (bars); R_1 (s cm^{-1}) a resistance to moisture flow in soil; it depends on soil and root characteristics and is function of the volumetric water content; R_2 (s cm^{-1}) is vegetation resistance to moisture flow; Δz is soil depth. It is worth noting that $\psi_p > \psi*$ where $\psi*$ is the wilting point potential; In Ceballos et al. (2002) the wilting point is taken as the soil-moisture content at a soil-water potential of -1500 kPa.

Estimations of air and canopy resistances R_1 and R_2 often use semi-empirical models based on meteorological data such as wind speed as explanatory variables (Monteith (1965); Villalobos et al., 2000). Jackson et al. (2000) pointed out the role of the Hydraulic Lift process which is the movement of water through roots from wetter, deeper soil layers into drier, shallower layers along a gradient in ψ. On the basis of such redistribution at depth, Guswa (2005) introduced a parameter to represent the minimum fraction of roots that must be wetted to the field capacity in order to meet the potential rate of transpiration. The field capacity is defined as the saturation for which gravity drainage becomes negligible relative to potential transpiration (Guswa, 2005). The potential matrix at field capacity is assumed equal to 330 hPa (330 cm) (Nachabe, 1998). The resulting u(z,t) function is strongly non linear versus the average root moisture with a relative insensitivity to changes in moisture when moisture is high and sensitivity to changes in moisture when the moisture is near the wilting point conditions. We also emphasize the Perrochet model (Perrochet, 1987) which links transpiration to potential evapotranspiration E_0 through:

$$g_r(\theta,z,t) = \alpha(\theta)r(z)\ E_0(t) \qquad (7)$$

Where r(z) (cm^{-1}) is a root density function which depends both on vegetation type and climatic conditions, $\alpha(\theta)$ is the root efficiency function. Both r(z) and $\alpha(\theta)$ represent macroscopic properties of the root soil system; they depend on layer thickness and root distribution . Lai and Katul (2000) and Laio (2006) reported some models assigned to r(z) which are linear or non linear. As out pointed by Laio (2006), models generally assume that vegetation uptake at a certain depth depends only on the local soil moisture. It is noticeable that in Feddes et al. (2001), a decrease of uptake is assumed when the soil moisture exceeds a certain limit and transpiration ceases for soil moisture values above a limit related to oxygen deficiency.

2.2 Review of models for hydrodynamic properties of soils

Many functional forms are proposed to describe soil properties evolution as a function of the volumetric water content (Clapp et al. , 1978). They are called retention curves or pedo transfer functions. We first present the main functional forms adopted to describe hydraulic parameters (section 2.2.1). Then, we report some solutions of Richards equation (section 2.2.2).

2.2.1 Functional forms of soil properties

According to Raats (2001), four classes of models are distinguishable for representing soil hydraulic parameters. Among them the linear form with D as constant and K linear with θ and the function Delta type as proposed by Green Ampt $D= \frac{1}{2}\ s^2\ (\theta_1 - \theta_0)^{-1}\ \delta(\theta_1 - \theta_0)$ where s is the degree of saturation (which is the ratio between soil moisture volume and voids volume; s=1 in case of saturation) and θ_1; θ_0 parameters. Also power law functions for $\psi(\theta)$ and $K(\theta)$ are proposed by Brooks and Corey (1964) on the basis of experimental observations while Gardner (1958) assumes exponential functions. The power type model proposed by Brooks & Corey (1964) are the most often adopted forms in rainfall-runoff transformation models. The Brooks and Corey model for K and ψ is written as:

$$K(s) = K(1)\, s^{c'} ; \psi(s) = \psi(1)\, s^{-1/m} \tag{8}$$

where m is a pore size index and c' a pore disconnectedness index (Eagleson 1978 a,b); After Eagleson (1978a, b), c' is linked to m with c'=(2+3m)/m. In Eq. (8), K(1) is hydraulic conductivity at saturation (for s=1); $\psi(1)$ is the bubbling pressure head which represents matrix potential at saturation. During dewatering of a sample, it corresponds to the suction at which gas is first drawn from the sample; As a result, Brooks and Corey (*BC*) model for diffusivity is derived as:

$$D(\theta) = s^d\, \psi(1)\ K(1) / (nm) \tag{9}$$

where n is effective soil porosity; and d=(c'-1- (1/m)). Let's consider the intrinsic permeability k which is a soil property. (K and k are related by $K = k\, \rho_w / \mu$ where μ dynamic viscosity of water; ρ_w specific weight of pore water). After Eagleson (1978 a, b), three parameters involved in pedo transfer functions may be considered as independent parameters: n, c' and k(1) where k(1) is intrinsic permeability at saturation.
On the other hand, Gardner (1958) model assumed the exponential form for the hydraulic conductivity parameter (Eq. 10):

$$K(\psi) = K_S\, e^{-a'\psi} \tag{10}$$

Where K_S saturated hydraulic conductivity at soil surface; a' pore size distribution parameter. Also, in Gardner (1958) model, the degree of saturation and the soil moisture potential are linked according to Eq. (11). The power function introduces a parameter l which is a factor linked to soil matrix tortuosity (l= 0.5 is recommended for different types of soils);.

$$s(\psi) = [e^{-0.5\, a'\psi}\,(1+ 0.5\, a'\, \psi\,)]^{2/(l+2)} \tag{11}$$

Van Genutchen model (1980) is another kind of power law model but it is highly non linear

$$K(\psi) = K_S\, s^{\,\lambda+1}\, [\, 1- (1- s^{\,(\lambda+1)/\lambda})^{\lambda/(\lambda+1)}]^2 \tag{12}$$

$$\begin{aligned} s(\psi) &= [1+ (\psi(1)/\psi))^{-(\lambda+1)}\,]^{-\,\lambda/(\lambda+1)} && \text{for } \psi \le \psi(1); \\ s&=1 && \text{for } \psi < \psi(1) \end{aligned} \tag{13}$$

In Eq. (12) and (13) λ is a parameter to be calibrated. Calibration is generally performed on the basis of the comparison of computed and observed retention curves.
In order to determine K_S one way is to adopt Cosby et al. (1984) model (Eq. 14).

$$\mathrm{Log}(K_S) = -0.\ 6 + (\ 0.0126\ S_{\%} - 0.0064\ C_{\%}) \tag{14}$$

Where $S_{\%}$ and $C_{\%}$ stand for soil percents of sand and clay. Also, we may find tabulated values of K_S (in m/day) according to soil texture and structure properties in FAO (1980). On the other hand, soil field capacity S_{FC} plays a key role in many soil water budget models. In Ceballos et al. (2002) the field capacity was considered as "the content in humidity corresponding to the inflection point of the retention curve before it reached a trend parallel to the soil water potential axis". In Guswa (2005), it is defined as the saturation for which gravity drainage becomes negligible relative to potential transpiration. As pointed out by Liao (2006) who agreed with Nachabe (1998), there is an "intrinsic subjectivity in the definition of field capacity". Nevertheless, many semi-empirical models are offered in the literature for S_{FC} estimation as a function of soil properties (Nachabe, 1988). In Cosby (1984), S_{FC} expressed as a degree of saturation is assumed s:

$$S_{FC} = 50.1 + (-0.142\, S_{\%} - 0.037\, C_{\%}) \tag{15}$$

On the other hand, according to Cosby (1984) and Saxton et al. (1986) S_{FC} may be derived as:

$$S_{FC}= (20/A')^{1/B'} \tag{16}$$

where
$A'=100*\exp(a_1+a_2C_{\%}+a_3S_{\%}^2+a_4S_{\%}^2C_{\%})$; $B'=a_5+a_6C_{\%}^2+a_7S_{\%}^2+a_8S_{\%}^2C_{\%}$; $a_1=$ - 4,396; $a_2 =$ - 0,0715; a_3 = - 0,000488; $a_4 =$ -0,00004285; $a_5 =$ -0,00222; $a_6 =$ -0,00222 ; $a_7 =$ -0,00003484; $a_8 =$ -0,00003484
Recently, this model was adopted by Zhan et al. (2008) to estimate actual evapotranspiration in eastern China using soil texture information. Also, soil characteristics such as S_{FC} may be obtained from Rawls & Brakensiek (1989) according to soil classification (Soil Survey Division Staff, 1998). Nasta et al. (2009) proposed a method taking advantage of the similarity between shapes of the particle-size distribution and the soil water retention function and adopted a log-Normal Probability Density Function to represent the matrix pressure head function retention curve.

2.2.2 Review of analytical solutions of the movement equation

Two well-known solutions of Richards equation are reported here (Green &Ampt model (1911), Philip model (1957)) as well as a more recent solution proposed by Zhao and Liu (1995). These solutions are widely adopted in rainfall-runoff models to derive infiltration.
In the Green &Ampt method (1911), it is assumed that infiltration capacity f from a ponded surface is:

$$f = K_{av}\,(1 + \Delta\psi\, \Delta\theta\, F^{-1}) \tag{17}$$

K_{av} average saturated hydraulic conductivity ; $\Delta\psi$ difference in average matrix potential before and after wetting; $\Delta\theta$ difference in average soil water content before and after wetting; F the cumulative infiltration for a rainfall event (with f = dF/dt).
In the Philip (1957) solution, it is assumed that the gravity term is negligible so that $\delta K(\theta)/\delta z]\approx 0$. A time series development considers the soil water profile of the form:

$$z(\theta,t) = f_1(\theta)\, t^{1/2} + f_2(\theta)\, t + f_3(\theta)\, t^{3/2} + \ldots \tag{18}$$

Where f_1, f_2, … are functions of θ. Hence, the cumulative infiltration $\Omega_f(t)$ is:

$$\Omega_f(t)= S\, t^{1/2} + (A_2 + K_S)\, t + A_3\, t^{3/2} + \ldots \tag{19}$$

Where S soil sorptivity, K_S is saturated hydraulic conductivity of the soil and A_1, A_2, … are parameters. Philip suggested adopting a truncation that results in:

$$\Omega_f(t) = S\,t^{1/2} + K_S / n'\,t \tag{20}$$

Where n' is a factor $0.3 < n' < 0.7$. It is worth noting that the soil sorptivity S depends on initial water content. So it has to be adjusted for each rainfall event. This is usually performed by comparing observed and simulated cumulative infiltration. For further discussion of Philip model, the reader may profitably refer to Youngs (1988).
Another model of infiltration is worth noting. It is the model of Zhao and Liu (1995) which introduced the fraction of area under the infiltration capacity:

$$i(t) = i_{max}\left[1 - (1 - A(t))^{1/b''}\right] \tag{21}$$

Where i(t) is infiltration capacity at time t. Its maximum value is i_{max}. A(t) is the fraction of area for which the infiltration capacity is less than i(t) and b″ is the infiltration shape parameter. As out pointed by DeMaria et al. (2007), the parameter b″ plays a key role. Effectively, an increase in b″ results in a decrease in infiltration.

3. Review of various parameterizations of actual evapotranspiration

Many early works on radiative balance combination methods for estimating latent heat using Penman - Monteith method (Monteith, 1965) were coupled with empirical models for representing the conductance of the soil-plant system (the conductance is the inverse function of the resistance). Based on observational evidence, these works have assumed a linear piecewise relation between volumetric soil moisture and actual evapotranspiration. Thus, several water balance models have been developed for large spatial and time scales assuming this piecewise linear form beginning from the work of Budyko in 1956 as pointed out by Manabe (1969)), Budyko (1974), Eagleson (1978 a, b), Entekhabi & Eagleson (1989) and Milly (1993). In fact, soil water models for computing actual evapotranspiration differ according to the time and space scales and the number of soil layers adopted as well as the degree of schematization of the water and energy balances. Moreover, specific canopy interception schemes, pedo transfer sub-models and runoff sub-models often distinguish between actual evapotranspiration schemes. Also, models differ by the consideration of mixed bare soil and vegetation surface conditions or by differencing between vegetation and soil cover. In the former, there is a separation between bare soil evapotranspiration and vegetation transpiration as distinct terms in the computation of evapotranspiration. In the following, we first present a brief review of land surface models which fully couple energy and mass transfers (section 3.1). Then, we make a general presentation of soil water balance models based on the actualisation of soil water storage in the upper soil zone assuming homogeneous soil (section 3.2).Further, it is focused on the estimation of long term actual evapotranspiration using approximation of the solution of the water balance model (section 3.3). In section 3.4, large scale soil water balance models (bucket schematization) are outlined with much more details. Finally a discussion is performed in section 3.5.

3.1 Review of land surface models

In Soil-Vegetation-Atmosphere-Transfer (SVAT) models or land surface models, energy and mass transfers are fully coupled solving both the energy balance (net radiation equation, soil heat fluxes, sensible heat fluxes, and latent heat fluxes) in addition to water movement equations. Usually this is achieved using small time scales (as for example one hour time increment). The specificity of SVAT models is to describe properly the role of vegetation in the evolution of water and energy budgets. This is achieved by assigning land type and soil information to each model grid square and by considering the physiology of plant uptake. Many SVAT models have been developed in the last 25 years. We may find in Dickinson and al. (1986) perhaps one of the first comprehensive SVAT models which was addressed to be used for General circulation modelling and climate modelling. It was called BATS (Biosphere-Atmosphere Transfer Scheme). It was able to compute surface temperature in response to solar radiation, water budget terms (soil moisture, evapotranspiration and, runoff), plant water budget (interception and transpiration) and foliage temperature. ISBA model (Noilhan et Mahfouf, 1996) was further developed in France and belongs to "simple models with mono layer energy balance combined with a bulk soil description" (after Olioso et al. (2002)). An example of using ISBA scheme is presented in Olioso et al. (2002). The following variables are considered: surface temperature, mean surface temperature, soil volumetric moisture at the ground surface, total soil moisture, canopy interception reservoir. The soil volumetric moisture at the ground surface is adopted to compute the soil evaporation while the total soil moisture is used to compute transpiration. The total latent heat is assumed as a weighted average between soil evaporation and transpiration using a weight coefficient depending on the degree of canopy cover. Canopy albedo and emissivity, vegetation Leaf area index LAI, stomatal resistance, turbulent heat and transfer coefficients are parameters of the energy balance equations. It is worth noting that soil parameters in temperature and moisture are computed using soil classification databases. Without loss of generality we briefly present the two layers water movement model adopted by Montaldo et al. (2001)

$$\delta\theta_g / \delta t = C_1 / (\rho_w d_1) \, [P_g - E_g] - C_2/\tau \, [\theta_g - \theta_{geq}] \qquad 0 \le \theta_g \le \theta_s \tag{22}$$

$$\delta\theta_2 / \delta t = C_1 / (\rho_w d_2) \, [P_g - E_g - E_{tr} - q_2] \qquad 0 \le \theta_2 \le \theta_s \tag{23}$$

d_1 and d_2 depth of near surface and root zone soil layers; ρ_w density of the water; θ_g and θ_2 volumetric water contents of near surface and root zone soil layers; θ_{geq} equilibrium surface volumetric soil moisture content ideally describing a reference soil moisture for which gravity balances capillary forces such that no flow crosses the bottom of the near surface zone of depth d_1; P_g precipitation infiltrating into the soil; E_g bare soil evaporation rate at the surface; E_{tr} transpiration rate from the root zone of depth d_2; q_2 rate of drainage out of the bottom of the root zone; It is assumed to be equal to the hydraulic conductivity of the root zone at $\theta=\theta_2$. τ ; C_1 and C_2 are parameters. In this model, the rescaling of the root zone soil moisture θ_2 seems to be highly recommended in order to achieve adequate prediction of θ_g in comparison to observations (Montaldo et al. (2001)). Using an assimilation procedure, Montaldo et al. (2001) achieved overcoming misspecification of K_S of two orders magnitude in the simulation of θ_2.

According to Franks et al. (1997), the calibration of SVAT schemes requires a large number of parameters. Also, field experimentations needed to calibrate these parameters are rather important. Moreover up scaling procedures are to be implemented. Boulet and al. (2000) argued that "detailed SVAT models especially when they exhibit small time and space steps are difficult to use for the investigation of the spatial and temporal variability of land surface fluxes".

3.2 Review of average long term evapotranspiration or "regional" evapotranspiration models

Considering the soil water balance at monthly time scale, Budyko (1974) introduced one single parameter which is a critical soil water storage w_0 corresponding to 1 m homogeneous soil depth. According to Budyko (1974), w_0 is a regional parameter seasonally constant and essentially depending on the climate-vegetation complex. The main assumption is that monthly actual evapotranspiration starts from zero and is a piecewise linear function of the degree of saturation expressed as the ratio w/w_0 where w is the actual soil water storage. Either, for $w \geq w_0$ actual evapotranspiration is assumed at potential value E_0.

Average annual water balance equation is also developed in Eagleson (1978 a) in terms of 23 variables (six for soil, six for climate and one for vegetation) with the assumption of a homogeneous soil-atmosphere column using Richards equation. Further, the behaviour of soil moisture in the upper soil zone (1 m deep or root zone) is expressed in terms of the following three independent soil parameters: effective porosity n, pore disconnectedness index c′ and saturated hydraulic conductivity at soil surface K_S while storm and inter storm net soil moisture flux are coupled to storm and inter storm Probability Density Functions. The average annual evapotranspiration E_m is finally expressed as :

$$E_m = J(E_e, M_v, k_v)\,(E_{pa} - E_{ra}) \tag{24}$$

J(.) evapotranspiration function; E_{pa} average annual potential evapotranspiration; E_{ra} average annual surface retention; E_e exfiltration parameter as function of initial degree of saturation s_0; k_v plant coefficient. It is approximately equal to effective transpiring leaf surface per unit of vegetated land surface; M_v vegetation fraction of surface.

Further, Milly (1993) developed similar probabilistic approach for soil water storage dynamics based on Manabe model (Manabe, 1969). A key assumption is that the soil is of high infiltration capacity. The model adopts the so-called water holding capacity W_0, which is a storage capacity parameter allowing the definition of the state "reservoir is full". For well developed vegetation, W_0 is interpreted as the difference between the volumetric moisture contents θ_f of the soil at field capacity and the wilting point θ_w ($W_0=\theta_f-\theta_w$). Furthermore, Milly (1994) adopted seasonally Poisson and exponential Probability Density Functions, together with seasonality of evapotranspiration forcing. To take into account horizontal large length scales, the spatial variability of water holding capacity W_0 was introduced, adopting a Gamma Probability Density Function with mean W_{m0}. In total, the model involved only seven parameters: a dryness index EDI = P / ETP, the mean holding capacity of soil W_{m0} and a shape parameter of the Gamma distribution,, mean storm arrival rate, and one measure of seasonality for respectively annual precipitation, potential evapotranspiration and storm arrival rate. Performing a comparison with observed annual runoff in US, it was found that the geographical distribution of calculated runoff shares at least qualitatively the large scale features of observed maps. In effect, 88% of the variance of grid runoff and 85% of the variance of grid evapotranspiration is reproduced by this model. However, it is outlined that the model presents failures within areas with elevation. Average annual precipitation and runoff over 73 large basins worldwide were also studied by (Milly and Dunne, 2002). Using precipitation and net radiation as independent variables, they compared observed mean runoff amounts to those computed by Turc-Pike and Budyko models. In northern Europe, they found a tendency for underestimation of observed evapotranspiration.

3.3 Empirical model for estimating regional evapotranspiration

Combining the water balance to the radiative balance at monthly scale, Budyko proposed an asymptotic solution in which R_n stands for average annual net radiation (which is the net energy exchange with the atmosphere equal to net radiation - sensible heat flux - latent heat flux), P average annual precipitation, E_m average (long term) annual evapotranspiration, ϕ a function expressed in Eq. (26).

$$E_m / P = \phi (R_n/P) \tag{25}$$

$$\phi (x) = [x (\tanh(x^{-1})) (1 - \cosh(x) + \sinh(x))]^{1/2} \tag{26}$$

Where tanh(.) stands for hyperbolic tangent, cosh(.) hyperbolic cosines, sinh(.) hyperbolic sinus

According to Shiklomavov (1989) and Budyko (1974), Ol'dekop was the first to propose in 1911 an empirical formulation of the relationship between climate characteristics and water balance terms (rainfall and runoff) assuming the concept of « maximum probable evaporation» E_{max} and using the ratio P / E_{max}. According to Milly (1994), works of Budyko in 1948 resulted, on the basis of dimensional analysis, to propose the ratio R_n/P as radiative index of aridity. Conversely, the function ϕ (Eq. 26) was empirical and was derived assuming that in arid climate E_m approaches P while it approaches R_n under humid climate.Budyko model was validated using 1200 watersheds world wild computing E_m as the difference between average long term annual observed rainfall and annual observed runoff. Model accuracy is reflected by the fact that the ratio E_m /P is simulated within a relative error of 10% (Budyko, 1974). However, larger discrepancy values are found for basins with important orography. Choudhury (1999) proposed to adopt Eq. (27) to derive ϕ :

$$\phi (x) = (1+x^{-\nu})^{-1/\nu} \tag{27}$$

where ν is a parameter depending of the basin characteristics. Milly et Dunne (2002) reported that ν=2.1 closely approximates Budyko model, while ν=2 corresponds to Turc-Pike model. According to Choudhury (1999), the more the basin area is large, the more ν is small and smaller is E_m. ν=2.6 is recommended for micro-basins while ν=1.8 for large basins. According to Milly et Dunne (2002), it was found that for a large interval of watershed areas, ν=1.5 to 2.6.

Another approximation of Budyko model is the Hsuen Chun (1988) model (H.C.) introducing the ratio $ID_{etp} = E_0/P$ and an empirical parameter k'.

$$E_m = E_0 [ID_{etp}^{k'} / (1+ ID_{etp}^{k'})]^{1/k'} \tag{28}$$

After Hsuen Chun (1988) the value *k'*=2.2 reproduces Budyko model results. According to Pinol et al. (1991), the adjusted values of k' are in the interval 1.03 <k'< 2.40. Also, they noticed that k' depends on the type of vegetation cover. After Donohue et al. (2007), Eq. (28) may be adopted for basins with area < 1000 Km^2 and series of at least 5 year length.

3.4 Modeling of actual evapotranspiration for long time series and large scale applications

Simple soil water balance models based on bucket schematization have been developed to fulfil the need to simulate long time series of water balance outputs allowing the calculation of actual evapotranspiration. We focus the review on the Manabe model (1969), the

Rodriguez-Iturbe et al. (1999) model and the Bottom hole bucket model of Kobayachi et al. (2001).

3.4.1 Manabe bucket model

In fact, the single layer single bucket model of Manabe (1969) takes a central place in large scale water budget modelling. It was proposed as part of the climate and ocean circulation model. This conceptual model runs at the monthly scale and adopts the field capacity S_{FC} as key parameter. Also, it assumes an effective parameter W_k representing a fraction of the field capacity ($W_k = 0.75* S_{FC}$). Here we notice that the field capacity S_{FC} is now expressed as a water content. The climatic forcing is represented by the potential evapotranspiration E_0. Let w be the actual soil water content. The actual evapotranspiration E_a is expressed as a linear piecewise function:

For $w \geq W_k$ $E_a = E_0$

For $w < W_k$ $E_a = E_0*(w/W_k)$

On the other hand, the surface runoff R_s component in Manabe model depends on the actual soil moisture content in comparison to the field capacity as well as on the precipitation forcing compared to the potential evapotranspiration uptake. Let Δw the change in soil water content. Thus, surface runoff is assumed as following:

For $w = S_{FC}$ and $P > E_0$; $\Delta w=0$ and $R_s = P - E_0$

For $w < S_{FC}$; $\Delta w = P - E_a$; $R_s = 0$

Another well-known model is FAO-56 model (Allen et al. (1998)). In fact, it is based on Manabe soil water budget. However, it takes into account the water stress through an empirical coefficient K'_s. First of all, in FAO-56 model, it is important to outline that the potential evapotranspiration is replaced by a reference evapotranspiration E_r computed using Penman-Montheith model with respect to a reference grass corresponding to an albedo value equal 0.23. Then, a seasonal crop coefficient K_c is introduced. The parameter K_c depends on both the crop type and the vegetative stage. Default K_c values are reported in (Allen et al. (1998)) for various crop types. This crop coefficient corresponds to ideal soil moisture conditions related to no water stress conditions and to good biological conditions. In real conditions, K_c is corrected by a correction coefficient K'_s ($0<K'_s<1$) such that the product $K_c K'_s$ includes the vegetation type as well as the water stress conditions. So actual evapotranspiration is written as:

$$E_a = K_c K'_s E_r \tag{29}$$

According to Biggs et al. (2008) mild stress conditions would correspond to K'_s of 0.8 and moderate stress conditions to K'_s of 0.6. Based on the findings that default K_c values underestimate lysimeter experiments K_c values, Biggs et al. (2008) built a non linear regression relationships between the product ($K_c K'_s$) and the ratio of seasonal precipitation to potential evapotranspiration for various crop types. To that purpose they fitted a Beta Probability Density Function to the correction factor K'_s. They adopted lysimeter observations to fit this modified FAO-56 model.. The model explained (49–90%) of the variance in actual evapotranspiration, depending on the crop type.

3.4.2 Rodriguez-Iturbe model

In Rodriguez-Iturbe et al. (1999), the point of departure is infiltration into the soil which is expressed as function of the existing soil moisture which is reported in terms of saturation (corresponding to $s = w/nZ_a$ where Za is effective depth of soil and n soil effective porosity). Soil drainage varies according to a power law although it is approximated by two linear segments. Consequently, it is assumed that soil drainage occurs for s exceeding a threshold value s_1, going from zero for $s=s_1$ to K_S for saturated condition (s=1) where K_S is the saturated hydraulic conductivity of the soil. Moreover, a saturation threshold s* is assumed to reduce evapotranspiration in case of water stress. Its value depends on the type of vegetation. Thus, for $s \le s^*$, the evapotranspiration is computed as the potential rate scaled by the ratio s/s* while the evapotranspiration is at potential value for s> s*.

$$E_a(s)=E_0\, s/s^* \qquad \text{For } s \le s^* \tag{30}$$

$$E_a(s)=E_0 \qquad \text{For } s>s^* \tag{31}$$

Milly (2001) model corresponds to the case $s^* \rightarrow 0$ and $K_S \rightarrow$ infinity. According to Milly (2001), the introduction of the threshold parameter s* is much recommended especially under arid conditions. In the case where no distinction is made between forested and bare soil areas, Rodriguez-Iturbe et al. (1999) pointed out that s* is considerably lower than the field capacity S_{FC} conversely to Manabe model which corresponds to $s^* = 0.75\ S_{FC}$. Laio (2006) adopted a generalized form of Rodriguez-Iturbe et al. (1999) model by accounting for the reduction of evapotranspiration in case of water stress by introducing the soil moisture at wilting point s_w. He represented s* as a soil moisture level above which plant stomata are completely opened (Eq. 32 and Eq. 33).

$$E_a(s)=E_0\, (s-s_w)/(s^*-s_w) \qquad \text{For } s \le s^* \tag{32}$$

$$E_a(s)=E_0 \qquad \text{For } S_{FC} > s > s^* \tag{33}$$

On the other hand, Rodriguez-Iturbe et al. (1999) model the leakage component is represented by the exponential decay Gardner model. This model was also adopted by Guswa et al. (2002). Leakage component is assumed as exponential decay function of the effective degree of soil saturation, as well as soil characteristics (saturated hydraulic conductivity, drainage curve parameter and field capacity).

3.4.3 Bottom hole bucket model

The daily bucket with bottom hole model (BBH) proposed by Kobayashi et al. (2001) is also based on Manabe model involving one layer bucket but including gravity drainage (leakage) as well as capillary rise. Kobayashi et al. (2001) outlined that the soil moisture dynamics is better simulated by BBH than by Bucket (Manabe) model. Kobayashi et al. (2007) developed a new version of BBH named BBH-B including a second soil layer in order to take into account for the variability of the soil profile when the root zone is rather deep (1 m or more).

In the following, we focus on BBH model where forcing variables are precipitation P and potential evapotranspiration E_0. The actual evapotranspiration is assumed as:

$$\begin{aligned} E_a &= M'\, E_0 \qquad \text{For } s \le s^* \\ E_a &= E_0 \qquad \text{For } s > s^* \end{aligned} \tag{34}$$

Where M' is a water stress factor updated at each time step and expressed as:

$$M'=\mathrm{Min}\ (1,w/(\sigma W_{max})) \qquad \text{For } s\leq s^* \tag{35}$$

σ: parameter representing the resistance of vegetation to evapotranspiration; $W_{max}=nZ_a$ where W_{max}: total water-holding capacity (mm); Z_a: thickness of active soil layer (mm); n: effective soil porosity.
Percolation and capillary rise term Gd(t) is assumed according to exponential function.

$$Gd(t)=\exp\ ((w(t)-a)/b)-c \tag{36}$$

Where a: parameter related to the field capacity (mm); b: parameter representing the decay of soil moisture (mm); c: parameter representing the daily maximal capillary rise (mm). On the other hand, daily surface runoff Rs(t) is expressed as:

$$Rs(t)=\mathrm{Max}\ [P(t)-(W_{BC}-W(t))-E_a(t)-Gd(t),\ 0] \tag{37}$$

Where $W_{BC}=\eta\ W_{max}$; η : parameter representing the moisture retaining capacity ($0<\eta<1$). According to Kobayachi and al. (2001) the parameter a (which corresponds here to a/Wmax) is "nearly equal to or somewhat smaller than the field capacity". After Teshima et al. (2006), parameter b is a measure of soil moisture recession that depends on hydraulic conductivity and thickness of active soil layer Z_a. In Iwanaga et al. (2005), a sensitivity analysis of BBH model applied to an irrigated area in semi-arid region suggests that error soil moisture is most sensitive to σ, η and c.

3.5 Discussion

According to the previous presentation and model comparison, bucket type models involves one parameter in Manabe model (W_k) up to six parameters in BBH (W_{max},a,b,c,σ, η). The minimum level of model complexity for bucket type models is discussed using a daily time step by Atkinson et al. (2002). These authors introduced the permanent wilting point θ_{pwp} to refine the bucket capacity $S_{bc}=(n-\theta_{pwp})Z_a$. Also, complexity is raised by the inclusion of a separation between transpiration and evaporation from bare soil. Hence a parameter which represents the fraction of basin area covered by forests is incorporated. A linear piecewise function is assumed similarly to Rodriguez-Iturbe et al. (1999) in both cases (bare soil areas and forest areas). They suppose that storage at field capacity S_{fc} is the bucket capacity S_{bc} scaled by a threshold storage parameter fc with S_{fc} = fc S_{bc} and fc $=(\theta_{fc}-\theta_{pwp})/(n-\theta_{pwp})$ where θ_{fc} is volumetric water content corresponding to field capacity. In addition, they assume that saturation excess runoff occurs when the storage exceeds S_{bc} and that subsurface runoff occurs when the storage exceeds S_{fc} with a piecewise non linear drainage function involving two recession parameters. These parameters are further calibrated using observed discharge recession curves while the other parameters are adapted from soil properties (via field data interpretation). Under wet, energy limited catchments authors conclude that the threshold storage parameter fc has a little control on runoff. Conversely, under drier catchments they conclude that the threshold storage parameter fc controls runoff volumes. Either, Kalma & Boulet (1998) compared simulation results of the hydrological model VIC which assumes a bucket representation including spatial variability of soil parameters to the one dimensional physically based model SiSPAT. Using soil moisture profile data for calibration, they conclude that catchment scale wetness index for very dry and very wet periods are misrepresented by SiSPAT while VIC model may better capture the water flux near and by the land surface. However, they outlined that

the difficulty of physical interpretation of the bucket VIC model parameters (maximum and minimum storage capacity) constitutes a major drawbacks of the bucket approach.
Guswa et al. (2002) also compared simulations of Richards (1D) and daily bucket model for African Savanna. They outlined that the differences between models outputs are mainly in the relationship between evapotranspiration and average root zone saturation, timing and intensity of transpiration as well as uptake separation between transpiration and evaporation. Vrugt et al. (2004) as well compared the daily Bucket model to a 3-D model (MODHMS) based on Richards equation while taking into account drainage observations. They concluded that Bucket model results are similar to MODHMS results. They also noticed that physical interpretation of MODHMS parameters is difficult since they represent effective properties. Moreover it is noticed that soil control on evapotranspiration is important in dry conditions. Besides, the introduction of a threshold parameter for evapotranspiration uptake is much recommended under arid conditions. Else, according to Rodriguez-Iturbe et al. (1999) under dry conditions, the spatial variation in soil properties has very little impact on the mean soil moisture. DeMaria et al. (2007) analyzed VIC parameter identifiability using stream flows data. Classifying four basins according to their climatic conditions (driest, dry, wet, wettest) they concluded that parameter sensitivity was more strongly dictated by climatic gradients than by changes in soil properties.

4. Rainfall runoff hydrological models

Soil water balance represents a key component of the structure of many Rainfall-runoff (R-R) models. Rainfall-runoff models are primarily tools for runoff prediction for water infrastructure sizing, water management and water quality management. On the basis of rainfall and temperature information, they aim to simulate the water balance at local and regional scales often adopting daily time step. In the majority of cases, model structure is a conceptual representation of the water balance, model parameters having to be adjusted using climatic and soil information as well as hydrological data, in order to match model outputs to observed outputs (Wagener et al., 2003). R-R models have two main components: a soil moisture-accounting module (also named production function) and a routine module (also named transfer function). In the former, the soil moisture status is up-dated while in the latter the runoff hydrograph is simulated. Models differ by the sub-models which are used for each hydrological process in both modules. The way of computing infiltration, evapotranspiration and leakage is of amount importance in the moisture-accounting module which simulates the soil moisture dynamics. It is worth noting that the Rainfall-Runoff Modelling Toolkit (RRMT), developed at Imperial College offers a generic modeling covering to the user to help him (her) to implement different lumped model structures to built his (her) own model (http://www3.imperial.ac.uk/ewre/research/software/toolkit). The system architecture of RRMT is composed by the production and transfer functions modules, and either an off-line data processing module, a visual analysis module and optimization tools module for calibration purposes (Wagener et al. 2001). In this section, we focus on evapotranspiration sub-models of two well-used R-R models (section 4.1). Then, we review the main steps of the calibration process required to estimate the model parameters (section 4.2). Finally three case studies are reported (section 4.3).

4.1 Evapotranspiration sub models

Despite the focus on runoff results in R-R modeling, evapotranspiration computation is a key part of R-R models. As an example, we emphasize the evapotranspiration sub-model of

GR4 model which is a parsimonious lumped model proposed by CEMAGREF (France) and running at the daily step with four parameters. A full model description is available in (Perrin et al., 2003). At each time step, a balance of daily rainfall and daily potential evapotranspiration is performed. Consequently, a net evapotranspiration capacity E_n and a net rainfall P_n are computed. If $P_n \neq 0$, a part P_s of P_n fills up the soil reservoir (so, P_s represents infiltration). It is noticeable that this quantity P_s depends on the actual soil moisture content w according to a non linear decreasing function of the w/x_1 where x_1 is the maximum capacity of the reservoir soil (which might represent the field capacity). On the other hand, if the net evapotranspiration capacity $E_n \neq 0$, actual evapotranspiration E_s is computed as a non linear increasing function of the water content involving the ratio w/x_1. Also, this function is parameterized through the ratio E_n/x_1 which refers to the characteristics of climate-soil complex. Furthermore, a leakage component is assumed with a power law function of the reservoir water content w.

$$\text{For } P \geq E_0; \quad P_n = P - E_0 \quad \text{and} \quad E_n = 0 \tag{38}$$

$$\text{For } P < E_0; \quad P_n = 0 \quad \text{and} \quad E_n = E_0 - P \tag{39}$$

$$E_s = w\,(2-(w/x_1))\tanh(E_n/x_1)/\{1+[(1-wx_1)\tanh(E_n/x_1)]\} \tag{40}$$

Where tanh(.) stands for hyperbolic tangent.

As second example, we underline the sub-models adopted in the *HBV* conceptual semi-distributed model proposed by the Swedish hydrological institute (Begstrōm, 1976). The fraction ΔQ of precipitation entering the soil reservoir is assumed as power law function of the ratio (w/FC) of reservoir water content w to a parameter FC representing soil field capacity in HBV model.

$$\Delta Q = P_e[1-(w/FC)^{\beta'}] \tag{41}$$

Where β' is a calibration parameter usually estimated by fitting observed and simulated runoff data. Also, P_e is effective precipitation. In addition, the actual evapotranspiration is a piecewise linear function. The control of actual evapotranspiration rates is performed using a parameter PWP representing a threshold water content. If w< PWP, the evapotranspiration uptake is a fraction of the potential evapotranspiration E_0 otherwise it is at potential rate.

$$E_a/E_0 = w/PWP \text{ for } w<PWP; \quad \text{and } E_a = E_0 \text{ for } w>PWP \tag{42}$$

4.2 Model calibration issues

As runoff has been for long time the main targeted response of rainfall-runoff modeling, rainfall-runoff models were often adjusted according to runoff observations. So far, observations from other control variables such as soil moisture content (Lamb et al., 1998), water table levels (Seibert, 2000) and either low flows (Dunne, 1999) have been adopted to enhance runoff predictions. Calibration of model parameters against runoff data is often performed using criteria such as bias and Root Mean Square Error (RMSE), which helps quantifying the discrepancy between observed discharges y_0 and simulated discharges y_i over a fixed time period with N observations.

$$\text{RMSE}=\left(\frac{1}{N}\sum_{i=1}^{i=N}\left(y_{si}-y_{oi}\right)^{2}\right)^{1/2} \tag{43}$$

The difficulty in the calibration process is that various parameter sets and even model structures might result in similarly good levels of performance, which constitutes a source of ambiguity as out pointed by Wagener et al. (2003) and many other authors before them (see the literature review of Wagener et al. (2003)). Also, it is noticeable that this problem of ability of various model structures and model parameters to perform equal quality with respect to matching observations is not dependent of the calibration process itself. In other words, the use of a performing optimisation tool does not prevent the problem. Another question is related to the single versus multi objective optimization. Wagener et al. (2003) reported that "single objective function is sufficient to identify only between three and five parameters" while lumped R-R models usually adopt far superior number of parameters. Multi-objective approach of calibration using additional output variables such as water table levels or soil moisture observations has been introduced to deal with the problem. Yet, inadequate model structure may be responsible of mismatching between observed and simulated outputs, as related by Boyle et al. (2000).

4.3 Case studies

Three case studies are presented in this section. In the first case, we propose a method for calibrating the empirical parameter k' of Hsuen Chun (1988) (Eq. 28). In the second case, we propose as example of calibrating HBV model using both runoff data and regional evapotranspiration information. In the third case, calibration of BBH model is performed using both runoff data and regional evapotranspiration information.

4.3.1 Fitting empirical models of regional evapotranspiration

This case study is presented in Bargaoui et al. (2008) and Bargaoui & Houcine (2010). It is aimed to calibrate the H.C. model using climatic, rainfall and runoff data from gauged watersheds. Monthly temperature and solar radiation data as well as annual rainfall and runoff data from various locations in Tunisia listed in Table 1 are adopted to calibrate the parameter k' of the empirical Hsuen Chen model (Eq. 28). To this end, 18 rainfall stations and 20 river discharge stations are considered, as well as 8 meteorological stations (Table 1). On the other hand, the potential evapotranspiration E_0 is computed at monthly scale using Turc formula.

$$E_0 = 0.4\, T_m\, [(R_g/N_j)+50]\ /\ [Rg+15] \tag{44}$$

T_m : monthly average temperature in (°C); R_g : global solar radiation (cal.cm^{-2} month^{-1}); N_j : number of days by month

For each river basin, simulated average (long term) annual evapotranspiration is computed using Eq. (28). Then, simulated mean annual runoff is computed as the difference between observed mean annual precipitation and simulated average annual evapotranspiration. The fitting of annual simulated runoff to annual observed runoff using the 20 river discharge stations results in k'= 1.5. The good adequacy of the model is well reflected in the plot of average simulated versus average observed annual runoff (Fig. 1).

River discharge stations			Rainfall stations			Meteorolcgical stations		
Stations	Latitude	Longitude	Stations	Latitude	Longitude	Stations	Latitude	Longitude
Jebel Antra	36°57′18″	9°27′45″	Ouchtata	36°57′53″	8°60′1″	Sfax	34°43′0″	10°41′0 ″
Joumine Mateur	37°2′19″	9°40′56″	Cherfech	36°57′0″	10°3′13″	Tunis	36°51′0″	10°20′0″
Zouara	36°54′15″	9°7′1″	Tabarka	36°56′59″	8°44′50″	Tabarka	36°57′0″	8°45′0″
Barbara	36°40′32″	8°32′56″	El Kef	36°10′53″	8°42′57″	Bizerte	37°14′0″	9°52′0″
Rarai sup.	36°27′36″	8°21′20″	Mellègue	36°7′16″	8°30′2″	Jendouba	36°29′0″	8°48′0″
Mellegue K13	36°7′1″	8°29′52″	Tajerouine	36°27′32″	9°14′57″	El Kef	36°8′0″	8°42′0″
Mellegue Rmel	36°1′1″	8°37′14″	Mejez El Bab	36°39′3″	9°36′17″	Kairouan	35°4′0″	10°4′0″
Haffouz	35°37′58″	9°39′33″	Tunis	36°47′23″	10°10′23″	Siliana	36°4′0″	9°22′0″
Merguellil Skhira	35°44′24″	9°23′3″	Feriana	34°56′49″	8°34′29″			
Chaffar	34°33′49″	10°29′14″	Jendouba	36°30′14″	8°46′52″			
Joumine Tine	36°58′3″	9°43′2″	Sejnane BV	37°3′35″	9°14′46″			
Miliane, Tuburbo Majus	36°23′39″	9°54′43″	Ksour	36°45′22″	9°28′27″			
M′khachbia aval	36°43′22″	9°24′24″	Sers	36°4′19″	9°1′25″			
Haidra Sidi Abdelhak	35°56′59″	8°16′22″	Ghardimaou	36°27′2″	8°25′58″			
Medjerda Jendouba	36°30′40″	8°46′7″	Bou Salem	36°36′30″	8°57′57″			
Sejnane	37°11′37″	9°30′16″	Merguellil H.	35°38′8″	9°40′36″			
Tessa Sidi Medien	36°16′44″	8°57′14″	Merguellil Skhira	35°44′24″	9°23′3″			
Rarai plaine	36°29′16″	8°32′18″	Chaffar PVF	34°40′0″	10°5′0″			
Ghezala-Ichkeul	37°4′35″	9°32′12″						
Douimis	37°12′50″	9°37′38″						

Table 1. Location of stations to calibrate H.C. model (after Bargaoui &Houcire, 2010)

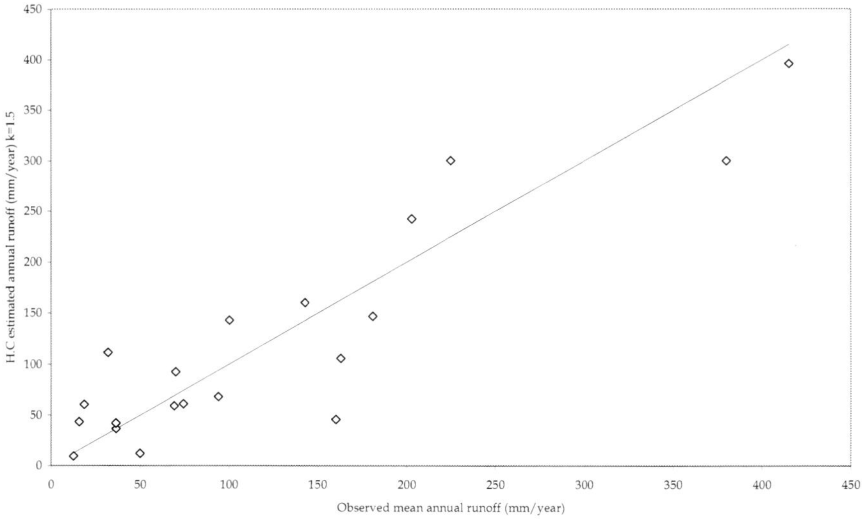

Fig. 1. Comparison of observed and simulated runoff for 20 river basins

4.3.2 Multicriteria calibration of HBV model using regional evapotranspiration information

This application is presented in Bargaoui et al. (2008). The idea is to use the information about the climatic regime as a driver for runoff prediction. Effectively, for a large number of basins with areas in the interval 50 à 1000 km², Wagener et al., (2007) suggested that there is a significant correlation between annual runoff and the ratio of forcing variables P/E_0. In the same way, we seek to use information about average (regional) actual evapotranspiration which is a bio-climatic indicator as means to improve accuracy of runoff predictions. To develop these ideas, the HBV rainfall-runoff model was adopted, coupled to a SCE-UA optimization tool. The calibration method adopts an objective function combining three criteria: minimisation of runoff root mean square error, minimisation of water budget simulation error, minimisation of the difference between mean annual simulated evapotranspiration E_a and regional E_m. The case study is a mountainous watershed of Wadi Sejnane (Tunisia). Mean daily runoff observations from September 1964 to August 1969 are available for a hydrometric station controlling an area of 378 km². Average basin annual rainfall is 931 mm/year. Over 8 years of rainfall observations, the minimum value of the series is 628 mm/year while the maximum value is 1141 mm/year denoting an important rainfall inter annual variability. Mean annual discharge is 2.43 m³/s. Average evapotranspiration computed using HC model (Eq. 28) with k'=1.5 results in E_m=643 mm/year. To calibrate the HBV model parameters, we adopt the period 1964/1967 for calibration and the period 1967/1969 for validation. The minimization of the objective function is performed using SCE-UA algorithm (Duan et al., 1994) in order to adjust 10 parameters (while 7 other HBV parameters have been set constant because they were found insensitive). First, the Nash coefficient of mean daily discharges is chosen as objective function F_0=$Nash_R$. The resulting value F_0=0.81 is quite good. However, for the validation period the ensuing optimal parameter set results in very poor fitting with a negative value of the Nash coefficient ($Nash_R$ = -0.084). Consequently, the objective function was modified to F_1 integrating the average model error (bias) of runoff output. Hence,

$$F_1 = Nash_R - w' \, ER_{RA} \quad (45)$$

Where ER_{RA} is the absolute relative error with respect to annual discharge. The weight coefficient w' = 0.1 is adopted according to Lindström and al. (1997) and helps aggregate the two criteria $Nash_R$ and ER_{RA}. In fact, the adoption of ER_{RA} aims to consider climatic zonality during the calibration process. Resulting optimal solution corresponds to $Nash_R$ =0.81 and ER_{RA} = 5%, which is believed good performance. It is worth noting that this modification of the objective function greatly improved $Nash_R$ also for the validation period ($Nash_R$ =0.55). The mean annual simulated evapotranspiration using HBV model is equal to 728 mm/ year while the H.C. model with k'=1.5 results in 643mm/year. To try to overcome such overestimation, it was proposed to directly include the information about evapotranspiration by adopting a new objective function F_2.

$$F_2 = Nash_R - 0{,}1\, ER_{RA} - 0{,}1\, ER_{ETRG} \quad (46)$$

Where ER_{ETRG} is the absolute relative error with respect to mean annual evapotranspiration (simulated by HBV versus estimated by H.C with k'=1.5). The resulting runoff Nash is a little smaller ($Nash_R$ =0.79) than for F_1, but a real improvement is obtained during the validation period ($Nash_R$ = 0.68). Fig. 2 reports HBV estimated annual evapotranspiration obtained with the optimal HBV solution (squares) versus annual rainfall. Comparatively, we also report annual evapotranspiration as evaluated using H.C model with k'=1.5 (interrupted line). Effect of year to year rainfall fluctuation on HBV estimations is well seen in the graph.

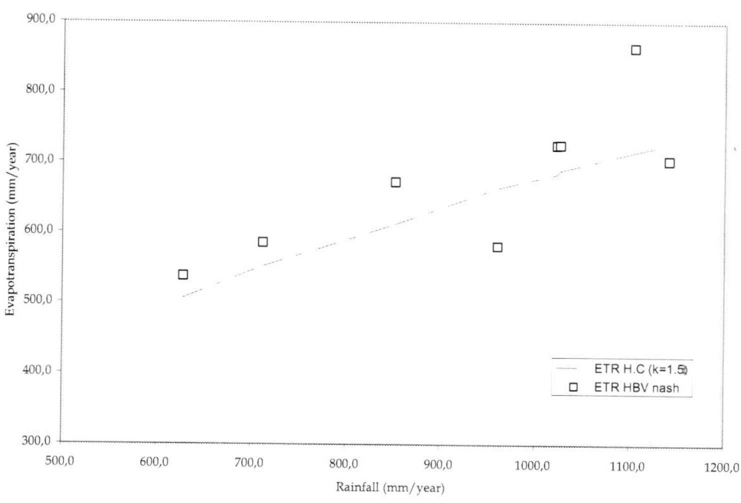

Fig. 2. Comparison of evapotranspiration estimates from HBV and HC models in relation with rainfall

4.3.3 Multicriteria calibration of BBH model using regional evapotranspiration information

In the third application it is aimed to compare BBH model results using the decadal time step. A part of this case study is presented in Bargaoui & Houcine (2011) using monthly data for model evaluation. Here will report results of decadal evaluations. Data are from the Wadi Chaffar watershed (250 km^2) situated under arid climate, South Tunisia. Vegetation cover comprises mainly olives. Meteorological data (solar radiation, air temperature and humidity, sky cloudiness, wind speed and Piche evaporation) are available from September 1989 to August 1999 for computing the daily reference evapotranspiration E_0 according to Allen et al. (1998). E_0 is multiplied by the crop coefficient K_c of olives trees to obtain daily potential evapotranspiration (Allen et al., 1998). Daily average basin rainfalls are available from September 1985 to August 1999. Stream discharge data are available for the basin outlet at the daily time step from September 1985 to August 1999. In the period September 1985 to August 1989, meteorological data are missing and the used E_0 values are the daily long term average computed for September 1989- August 1999. The H.C. model results in an average annual evapotranspiration E_m = 213 mm/year (Bargaoui & Houcine, 2010). BBH model inputs are precipitation and potential evapotranspiration and seven parameters are to be calibrated. To reduce the number of calibrated parameters, we first fix the thickness of active soil layer Z_a (in mm) and the effective soil porosity n (unit less). Also, we undertake a reformulation of leakage component L(s) by using the model of Guswa et al. (2002) where

$$L(s) = K_S \frac{e^{B(s-S_{FC})} - 1}{e^{B(1-S_{FC})} - 1} \tag{47}$$

where s is the degree of saturation (unit less); K_S saturated hydraulic conductivity at soil surface (mm/day); B is the soil water retention curve shape parameter; S_{FC} (unit less) is the field capacity; $W_{max} = nZ_a$ (W_{max} is the total water-holding capacity in mm).
Coupling this expression with pedo-transfer functions it makes it possible after Bargaoui & Houcine (2010), to derive the parameters (a, b, c) as following using pedo-transfer parameters K_S , B and S_{FC}:

$$a = W_{max}\left[S_{FC} - \frac{1}{B} Ln\left(K_S \frac{1}{e^{B(1-S_{FC})} - 1}\right)\right] \tag{48}$$

$$b = W_{max}\frac{1}{B} \tag{49}$$

$$c = \left(\frac{1}{e^{B(1-S_{FC})} - 1}\right)K_S \tag{50}$$

In this case, the model by Rawls et al. (1982) is adopted for K_S estimation while S_{FC} is derived according to the Cosby (1984) and Saxton et al. (1986) models recently adopted by Zhan et al., (2008). Finally B = 9 is assumed in agreement with Rodriguez-Iturbe et al. (1999). The dominant soil type is considered to represent the soil characteristics. So, the value n=0.34 corresponding to a sandy soil was adopted; these assumptions result in K_S = 3634 mm/d and S_{FC}= 0.166. Also, after many trials the value Z_a= 0.5 m was adopted. The two remaining parameters σ and η (0< σ <1; 0< η <1) represent respectively the resistance of vegetation to evapotranspiration and the moisture retaining capacity. The problem is now to fit the parameters σ and η. They are adjusted using two different methods: i.e. using only observed runoff (method 1) and using both observed runoff and regional evapotranspiration information (method 2). Also BBH model has been completed adopting a , contributing area sub-model (Betson, 1964); Dunne et Black (1970). According to this assumption, runoff originates from a part of the watershed (contributing area) contrarily to the assumption of runoff occurring from the entire watershed. For a fixed day j, the contributing area CA_j is herein assumed linked to the soil moisture content according to Dickinson & Whiteley (1969). Additionally, a logistic Probability Density Function as a function of humidity index IH_j is adopted with parameters a_c and b_c (Eq. 51). It means that the mean contributing area is a_c and that the variance of the contributing area is $(b_c\pi)^2/3$. The humidity index takes account for the rainfall accumulated during the actual day and the IX previous days (Eq. 52).

$$CA_j = \frac{e^{((IH_j - a_c)/b_c)}}{(1 + e^{((IH_j - a_c)/b_c)})} \tag{51}$$

$$IH_j = W_{j-1} + \omega'' \sum_{l=0}^{IX} P_{j-l} \tag{52}$$

where ω'' is a fixed weight (ω'' =0.1). Then, two cases are considered: case (a) when the total basin area contributes to runoff at the basin outlet; case (b) when only a contributive area gives rise to runoff at the outlet.

After many trials and errors we assumed IX= 90 days, a_c = 20 and b_c = 10 in case (b). The model was calibrated for σ and η using daily hydro meteorological data (solar radiation, air temperature, air humidity, mean areal rainfall) as well as daily runoff records and also average annual evapotranspiration. The decadal, monthly and annual totals are adopted to evaluate model performance.

In each case (a) and (b), a first criterion based on the matching of decadal runoff (Eq. 53) is adopted to delineate adequate solutions for σ and η ($0< \sigma <1$; $0< \eta <1$). A supplementary criterion is based on the matching of long term annual evapotranspiration (Eq. 54).

$$C_y(\sigma,\eta) = \frac{1}{N}\sum_{i=1}^{N} \left|(y_{si} - y_{oi})/y_{oi}\right| \tag{53}$$

$$C_E(\sigma,\eta) = \frac{1}{N'}\sum_{i=1}^{N'} \left|(E_{si} - E_m)/E_m\right| \tag{54}$$

In Eq. (53), y_{oi} and y_{si} are respectively decadal observed and simulated volume runoff and N is the number of simulated decades. In Eq. (54), E_{si} is simulated annual evapotranspiration and N' is the number of simulated years.

For each pair of simulated (σ, η) ($0< \sigma <1$; $0< \eta <1$), candidate solutions verifying the criterion $C_y(\sigma, \eta) < \alpha$ (Eq. 53) with α=20% the Nash coefficient R_N is then evaluated. Pairs for which it is found that R_N>0.5, are thus selected. Also, introducing E_m for calibration method 2, the absolute value $C_E(\sigma, \eta)$ of the relative error between mean annual simulated evapotranspiration and E_m, is used through the additional selection criterion of (Eq. 54).

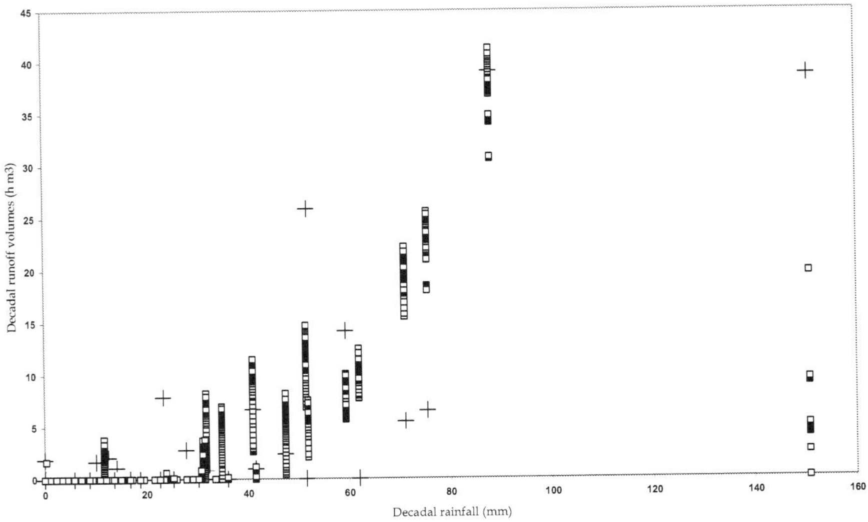

Fig. 3. Estimated decadal runoff versus decadal precipitation with the assumption of total watershed contributing to runoff (+ represent observed volumes and squares represent simulated volume for the selected pairs of (σ, η))

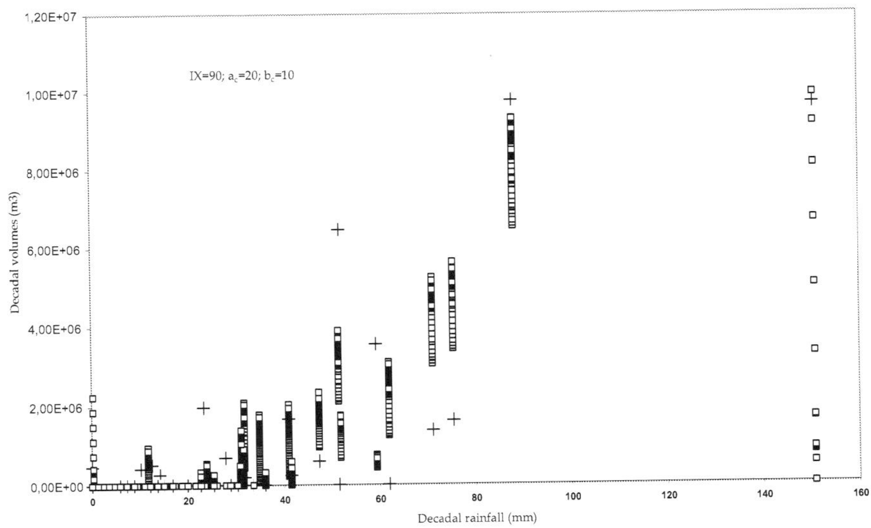

Fig. 4. Estimated decadal runoff versus decadal precipitation with the assumption of contributive area (+ represent observed volumes and squares represent simulated volume for the selected pairs of (σ, η)).

Pairs of simulated (σ, η) (0< σ <1; 0< η <1) which satisfy both C_y (σ, η) < 20%; R_N > 0.5 and C_E (σ, η) < α' with α'=30%are finally selected as adequate solutions.
Fig. 3 and 4 report model outputs for sets of (σ, η) fulfilling the above conditions under the assumptions of cases (a) and (b) in case where E_m information is included. Estimated decadal volumes (squares) for the selected pairs of (σ, η) are compared to observed decadal volumes (+) and are reported versus precipitation data. Fig. 3 is related to case (a) corresponding to the assumption of total watershed contributing to runoff. Fig. 4 is related to case (b) assuming a contributing area. The results suggest that the introduction of contributing area outcomes produce outputs which result in a better fitting of the rainfall-runoff evolution. In effect, in the case of total area contributing no solution is found able to simulate the most rainy decade,(squares are far from the symbol + for the Rainiest decade). Conversely, some solutions are found able to reproduce the most rainy decade if we consider contributing area scheme (some squares are located near the +). Also, evapotranspiration information has greatly reduced the interval of acceptable solutions. Effectively, selected solutions are such that 0.15 < σ <0.35 and 0.15 < η <0.25.

5. Conclusions

The simulation of evapotranspiration using the water balance equation is part of hydrological modelling (rainfall-runoff models) and is also important in the framework of global circulation models (Land surface models). A lot of models are now functioning and their formulation is based on different assumptions on soil characteristics in relation with soil moisture, transpiration schemes, as well as infiltration and runoff schemes.
Empirical models for estimating regional evapotranspiration are worth noting for estimating average long term evapotranspiration. They are generally based on climatic information (rainfall and potential evapotranspiration). They often require the adjustment of a single empirical parameter. Under particular climate and soil vegetation, evapotranspiration is controlled by soil moisture dynamics. Thus, Bucket type soil water budget models are worth noting for estimating time series of actual evapotranspiration at smaller time scales (daily to monthly). They involve from one parameter such as in the Manabe model (with parameter W_k) up to six parameters such as in BBH model (with parameters W_{max},a,b,c,σ, η). Parameters are linked to soil, climatic and vegetation characteristics. However, it is generally believed that the temporal variability of soil moisture series is mostly dependent on the rainfall variability especially under conditions of low precipitations. On the other hand, soil parameters such as field capacity, hydraulic conductivity at saturation and wilting point potential are key parameters controlling the evapotranspiration model outputs. One way to derive soil parameters is to adopt pedo transfer functions. Transpiration which corresponds to vegetation uptake is regulated by stomata and driven by atmospheric demand. It is widely represented by a linear piecewise function with parameters depending on vegetation characteristics. Thus, in computing evapotranspiration, a main assumption is the linear piecewise function of evapotranspiration in relation with potential evapotranspiration for taking account for soil water stress. Such an assumption is underlined in several rainfall runoff models (for example the two models GR4 and HBV studied here adopt such analytical form). Model adequacies introduce the question of the choice of the objective function as well as the output variables adopted for model

evaluation. In the case studies presented here, results suggest that the introduction of the information about average (long term) annual evapotranspiration may help improving the accuracy of the water balance simulation results. In effect the runoff Nash coefficient is found to be improved during the validation period in the case where long term evapotranspiration is taking account during the calibration process.

6. Annexe

6.1 Glossary

a: parameter related to the field capacity (mm)
a' : pore size distribution parameter
a_c: Logistic density distribution parameter
$\alpha(\theta)$: the root efficiency function.
A(t) : the fraction of area for which the infiltration capacity is less than i(t)
B : the soil water retention curve shape parameter;
b: parameter representing the decay of soil moisture (mm);
b″ is the infiltration shape parameter.
b_c: Logistic density distribution parameter
β: surface slope angle
β' : a calibration parameter in HBV model
c' : pore disconnectedness index
c: parameter representing the daily maximal capillary rise (mm)
$C_{\%}$: soil percent of clay
CA_j: : the contributing area
cos: the cosinus function
d_1 : depth of near surface soil layer
d_2 : depth of root zone soil layer;
D:soil-water diffusivity parameter
$\Delta\psi$: difference in average matrix potential before and after wetting
$\Delta\theta$: difference in average soil water content before and after wetting
Δw : the change in soil water content
Δz : soil depth.
C_1: parameter,
C_2 : parameter,
E(s,t): : evapotranspiration
E_a : actual evapotranspiration
EDI : dryness index
E_e exfiltration parameter as function of initial degree of saturation s_0
E_g : bare soil evaporation rate at the surface
E_m average annual evapotranspiration
E_n: net evapotranspiration capacity
E_r: reference evapotranspiration according to FAO model
E_{tr}: transpiration rate from the root zone of depth d_2
ER_{ETRG} : the absolute relative error with respect to mean annual evapotranspiration
E_{pa} average annual potential evapotranspiration

ER_{RA} : the absolute relative error with respect to annual discharge
E_{Turc} : monthly potential evapotranspiration (mm);
E_{ra} average annual surface retention
f : infiltration capacity
F cumulative infiltration for a rainfall event
fc: threshold storage parameter
FC: representing soil field capacity in HBV model
Gd(t): Daily percolation and capillary rise term
$g_r(\theta,z)$: vegetation uptake of soil moisture
I(s,t): infiltration into the soil
$I_{nf}(\theta,z_0)$: precipitation infiltrating into the soil
i(t): infiltration capacity at time t.
i_{max} : maximum value of infiltration capacity
Ω_f (t) : the cumulative infiltration
J(.): evapotranspiration function
k : intrinsic permeability
k(1): intrinsic permeability at saturation
K: hydraulic conductivity
K (1) hydraulic conductivity at saturation
k_v : plant coefficient
K_{av} : average saturated hydraulic conductivity
k′ : parameter of HC model
K_c : crop coefficient
K_S : the saturated hydraulic conductivity;
K'_s : correction coefficient of the crop coefficient
κ: shape parameter of the Gamma distribution
l: factor linked to soil matrix tortuosity
L(s,t) :leakage
LAI : Leaf area index
λ: mean storm arrival rate
M_v : vegetation fraction of surface.
μ : dynamic viscosity of water;
n: soil effective porosity
ν : parameter
θ: volumetric water content
θ_f :the volumetric moisture contents of the soil at field capacity
θ_w: the volumetric moisture contents at wilting point
θ_{pwp}: permanent wilting point
θ_g : volumetric water contents of near surface soil layer;
θ_s : saturated soil moisture content
θ_2 : volumetric water contents of root zone soil layer;
θ_{geq} : equilibrium surface volumetric soil moisture content
θ_1 : specific value of soil moisture content
θ_0 : specific value of soil moiqture content
N: number of observations

N_j : number of days by month
$Nash_R$: Nash coefficient of mean daily discharges
P : average annual precipitation
P_e : effective precipitation
PWP : parameter representing a threshold water content in HBV model.
P_g : precipitation infiltrating into the soil;
P_n: net rainfall
q_2 : rate of drainage out of the bottom of the root zone;
R_1 : (s cm^{-1}) a resistance to moisture flow in soil
R_2 : (s cm^{-1}) is vegetation resistance to moisture flow;
R_n : average annual net radiation
R_s : surface runoff
R_g : global solar radiation ($cal.cm^{-2}$ $month^{-1}$)
r(z) : a root density function (cm^{-1})
ρ_w : density of the water;
s: relative soil moisture content or degree of saturation
s*: saturation threshold
s_1 :threshold value of soil saturation
s_w: soil moisture at wilting point.
s_0: initial degree of saturation
S : sorptivity
$S_{\%}$: soil percent of sand
S_{bc}: bucket capacity
S_{FC} : soil field capacity
S_{fc}: storage at field capacity
σ : parameter representing the resistance of vegetation to evapotranspiration;
t: time
T_m : monthly average temperature in (°C);
u(z,t) : local transpiration uptake
w : the actual soil water storage
w_0 : critical soil water storage in Budyko model
W_0: water holding capacity
W_k : a fraction of the soil field capacity
W_{max} : total water-holding capacity (mm);
W_{m0}: mean water holding capacity
ω : a fixed weight
ψ : soil moisture potential (bars)
ψ_p : leaf moisture potential (bars)
ψ* : the wilting point potential
$\psi(1)$: the bubbling pressure head which represents matrix potential at saturation.
x_1 : maximum capacity of the reservoir soil
y_i: simulated discharges
y_0: observed discharges
z: the vertical coordinate (z>0 downward from surface)
Z_a: thickness of active soil layer (mm);
z_0 : the vertical coordinate at the surface

7. References

Allen RG, Pereira LS, Raes D, Smith M. (1998). Crop evapotranspiration. Guidelines for computing crop water requirements. *FAO Irrigation and drainage*. Paper No. 56. FAO, Rome.

Atkinson, S. E., R. A. Woods, and M. Sivapalan (2002), Climate and landscape controls on water balance model complexity over changing timescales, *Water Resour. Res.*, 38(12), 1314, doi:10.1029/2002WR001487.

Bargaoui Z., Dakhlaoui H, Houcine A. (2008). Modélisation pluie-débit et classification hydroclimatique. *Revue des sciences de l'eau*. 21 (2) : 233-245.

Bargaoui Z., Houcine A. (2010) Sensitivity to calibration data of simulated soil moisture related drought indices, *Revue Sécheresse*, 21(4), 1-7.

Bargaoui Z., Houcine A. (2011) Calibration of an evapotranspiration model using runoff records and regional evapotranspiration, *in Hydro-climatology: Variability and change IAHS Pub. 344*, 21-26.

Begström S. (1976). Development and application of a conceptual rainfall-runoff model for the scandanivain catchment , SMHI RH07, Norrköping.

Betson R. P. (1964). What is watershed runoff? *Journal of geophysical research*. 69(8). 1541-1552. In: Streamflow generation processes. Selection, Introduction and commentary by K. J. Beven. Benchmark Papers in Hydrology. IAHS Press 2006.

Biggs, T.W. , P. K. Mishra, H. Turral. (2008). Evapotranspiration and regional probabilities of soil moisture stress in rainfed crops, southern India, Agric. *Forest Meteorology*, doi:10.1016/j.agrformet.2008.05.012

Boulet G., Chehbouni A., Braud I., Vauclin M., Haverkamp R., Zammit C. (2000). A simple water and energy balance model designed for regionalization and remote sensing data utilization. *Agricultural and Forest Meteorology* 105 (2000) 117–132

Boyle, D.P., Gupta, H.V., Sorooshian, S., (2000). Toward improved calibration of hydrologic models: Combining the strengths of manual and automatic methods. Water R. Res. 36 (12), 3663–3674.

Braud I., Dantas Antonino A.C., Vauclin M., Thony J.-L., Ruelle P., A (1995). Simple Soil Plant Atmosphere Transfer model (SiSPAT) development and field verification, *J. Hydrol.* 166 (1995) 213–250.

Brooks, R.H., Corey, A.T., (1964). Hydraulic properties of porous media, *Hydrology paper*, 3. Colorado State University, Fort Collins.

Budyko, M. I. (1974) Climate and life. Academic Press. 508p.

Ceballos A., Martınez-Fernandez J.,, Santos F.& Alonso P. (2002). Soil-water behaviour of sandy soils under semi-arid conditions in the Duero Basin (Spain). *Journal of Arid Environments* (2002) 51: 501–519. doi:10.1006/jare.2002.0973, available online at http://www.idealibrary.com on

Childs, E.C. and Collis-George. N., (1950). The permeability of porous materials. *Proc. R. Soc.,London*, 201A: 392 405.

Choudhury B.J. (1999). Evaluation of an empirical equation for annual evaporation using field observations and results from a biophysical model. *J. Hydrol.* 216. 99-110.

Clapp, R. B., and G. M. Hornberger (1978). Empirical equations for some soil hydraulic properties, *Water Resour. Res.*, 14, 601–604.

Cosby (1984) A statistical exploration of the relationship of soil moisture characteristics to the physical properties of soils. *Water Resour. Res* 20(6). 682-690.

Cosby BJ, Hornberger GM, Clapp RB, Ginn TR. (1984). A Statistical Exploration of the Relationships of Soil Moisture Characteristics to the Physical Properties of Soils. *Water Resour. Res.* 20 (6) : 682-690.

DeMaria, E. M., Nijssen, B., Wagener, T. (2007) Monte Carlo sensitivity analysis of land surface parameters using the Variable Infiltration Capacity model, *Journal of geophysical research*, 112, D11113, doi:10.1029/2006JD007534, 2007

Dickinson., W.T., Whiteley, H. (1969) Watershed areas contributing to runoff, *Hydrological sciences Journal.* 12-26. Special issue: rainfall runoff modelling.

Dickinson R.E., Henderson-Sellers A., Kennedy P.J., Wilson M.F. (1986). Biosphere-Atmosphere Transfer Scheme (BATS) for NCAR Community Climate Model . In: Evaporation. Selection, Introduction and commentary by J. H. C. Gash and W.J. Shuttleworth. Benchmark Papers in Hydrology. ISSN 1993-4572. IAHS Press 2007.

Donohue R.J., Roderick M. L. and T.R. McVicar (2007). On the importance of including vegetation dynamics in Budyko hydrological model. *Hydrol. Earth Syst. Sci.*, 11, 983-995.

Duan, Q., Sorooshian, S. and Gupta, V. ,(1994). Optimal Use of the SCE-UA Global Optimisation Method for Calibrating Watershed Models. *Journal of Hydrology*, No. 158, 265-284.

Dunne, T., Black, R.D., (1970). Partial area contributions to storm runoff in a small New England watershed. *Water Resour. Res* 6, 1296–1311. In: Streamflow generation processes. Selection, Introduction and commentary by K. J. Beven. Benchmark Papers in Hydrology. IAHS Press 2006.

Dunne S. M. (1999). Imposing constraints on parameter values of a conceptual hydrological model using base flow response. *Hydrology and earth system sciences* 3: 271-284.

Eagleson PS. Climate, Soil, and Vegetation. (1978 a). Dynamics of the Annual Water Balance. *Water Resour. Res* 14 (5) : 749-764

Eagleson P. (1978 b). Climate, Soil, and Vegetation. The Expected Value of Annual Evapotranspiration. *Water Resour. Res* . 14, 5. 731- 739.

Entekhabi, D., Eagleson, P.S., (1989). Land surface fluxes hydrology parameterization for atmospheric general circulation models including subgrid variability. *J. Climate* 2 (8), 816–831.

FAO (1980). Drainage design factors. *Irrigation and Drainage* Paper No. 38. Rome. 52 pp

Feddes RA, Hoff H, Bruen M, Dawson T, de Rosnay P, Dirmeyer P,. (2001). Modeling root water uptake in hydrological and climate models. *Bull Amer Meteorol Soc.* ;82(12):2797–809.

Franks, S.W., Beven, K.J., Quinn, P.F., Wright, I.R., (1997). On the sensitivity of the soil-vegetation-atmosphere transfer (SVAT) schemes: equifinality and the problem of robust calibration. *Agricultural and Forest Meteorology* 86, 63–75.

Gardner, W . R., (1958). Somes teady-state solutions of the unsaturated moisture flow equation with application to evaporation from a water table, *Soil Sci.*, 85(4), 228-232, 1958.

Gardner , C. M. K., Robinson D. A., Blyth K. , Cooper J.D. (2001). Soil water content. In: Soil and environmental analysis: physical methods (Second Edn) (Ed. By K. A. Smith and C. E. Mullins), 1-64. Marcel Dekker Inc., New York. USA.

Green, W.H., Ampt, G.A., (1911). Studies on soil physics. I. Flow of air and water through soils. *J. Agric. Sci.* 4, 1–24.

Guswa A.J. (2005). Soil-moisture limits on plant uptake: An upscaled relationship for water-limited ecosystems. *Advances in Water Resources* 28. 543–552

Guswa, A.J., Celia, M, Rodriguez-Iturbe, I. (2002) Models of soil moisture dynamics in ecohydrology: A comparative study. *Water Resour. Res* ., 38 (9), 1156-1180.

Hsuen-Chun Y. (1988). A composite method for estimating annual actual evapotranspiration. *Hydrol. Sci. J.* ; 33 (4, 8): 345 - 356.

Hofius K. (2008). Evolving role of WMO in hydrology and water resources management. *WMO Bulletin* 57(3).147-151.

Iwanaga R, Kobayashi T, Wang W, He W, Teshima J, Cho H. (2005). Evaluating the Irrigation requirement at a Cornfield in the Yellow River Basin Based on the "Dynamic Field Capacity " *J. Japan Soc. Hydrol. & Water resour.*; 18: 664-674.

Jackson R. B. , J. S. Sperry, and T. E. Dawson, (2000). Root water uptake and transport: Using physiological processes in global predictions. *Trends Plant Sci.*, 5, 482–488.

Jeffrey P. Walker*, Garry R. Willgoose1, Jetse D. Kalma (2004) . In situ measurement of soil moisture: a comparison of techniques *Journal of Hydrology* 293. 85–99

Johnson K. D., Entekhabi D., Eagleson P.S. (1993). The implementation and validation of improved land surface hydrology in an atmospheric general circulation model. *Journal of Climate* 6. 1009-1026.

Kalma, J.D. , Boulet, G. (1998) Measurement and prediction of soil moisture in a medium-sized catchment. *Hydrological Sciences Journal*, 43(4), Special issue: onitoring and Modelling of Soil Moisture: Integration over Time and Space, 597-610.

Kobayashi T, Matsuda S, Nagai H, Teshima J. (2001). A bucket with a bottom hole (BBH) model of soil hydrology. In: Soil–Vegetation–Atmosphere Transfer Schemes and Large-Scale Hydrological Models (ed. Dolman, H. et al.), IAHS Publication. 270. 41-45.

Kobayashi T., Teshima J., Iwanaga R., Ikegami D., Yasutake D. , He W., Cho H.. (2007) An improvement in the BBH model for estimating evapotranspiration from cornfields in the upper Yellow river. *J. Agric. Meteorol.* 63(1). 1-10.

Lai C-T, Katul G. (2000).The dynamic role of root-water uptake in coupling potential to actual Transpiration. *Advances in Water Resources* 23 . 427±439

Laio F. (2006). A vertically extended stochastic model of soil moisture in the root zone. *Water Resour. Res* ., 42, W02406, doi:10.1029/2005WR004502,

Lamb R., K. Beven et S. Myrabo (1998) : Use of spatially distributed water table observations to constrain uncertainties in a rainfall-runoff model , *Advance in Water Resources* , Vol 22(4): 305-317.

Lindström G., B. Johanson, M. Person, M. Gardelin, S. Bergström, (1997). Development and test of the distributed HBV-96 hydrological model. *Journal of hydrology* 201 -272-288.

Noilhan, J. and J.-F. Mahfouf, 1996: The ISBA land surface parameterization scheme. *Global and Plan. Change*, 13, 145-159. http://www.cnrm.meteo.fr/isbadoc/pubs.html

Manabe, S. (1969) Climate and the ocean circulation, 1. The atmospheric circulation and the hydrology of the earth's surface. *Mon. Weath. Rev.* 97, 73Sept.774.

Milly PCD. (1993). An analytic solution of the stochastic storage problem applicable to soil water. *Water Resour. Res* . 29 (11) : 3755–3758.

Milly PCD. (1994). Climate, soil water storage, and the average annual water balance. *Water Resour. Res* . 30 (7) : 2143-2156.

Milly, P. C. D. (2001). A minimalist probabilistic description of root zone soil water, *Water Resour. Res* .., 37(3), 457–464.

Milly P. C. D and K. A. Dunne (2002). Macroscale water fluxes, 2, Water and energy supply control of their interannual variability. *Water Resour. Res* .38(10), 1206, doi:10.1029/2001 WR000760.

Montaldo Nicola Montaldo1 and John D. Albertson. Marco Mancini (2001). Robust simulation of root zone soil moisture with assimilation of surface soil moisture data. *Water Resour. Res* .37, 12, 2889–2900

Monteith, J.L., (1965). Evaporation and environment. Symp. Soc. Exp. Biol. XIX, 205–234. In: Evaporation. Selection, Introduction and commentary by J. H. C. Gash and W.J. Shuttleworth. Benchmark Papers in Hydrology. ISSN 1993-4572. IAHS Press 2007.

Nachabe, M. H. (1998), Refining the definition of field capacity in the literature, *J. Irrig. Drain. Eng.*, 124(4), 230- 232.

Nasta P., Kamai T., Chirico G. B., Hopmans J. W., Romano N., (2009). Scaling soil water retention functions using particle-size distribution. *Journal of Hydrology* 374 . 223

Olioso A., Braud I., Chanzy A., Courault D., Demarty J., Kergoat L. , Lewan E., Ottlé C., Prévot L., Zhao W.G., Calvet J. C., Cayrol P., Jongschaap R., Moulin S., Noilhan J., Wigneron J.P. (2002). SVAT modeling over the Alpilles-ReSeDA experiment: comparing SVAT models over wheat fields *Agronomie* 22 (2002) 651–668

Perrin Ch., Michel C., Andréassian V. (2003) .Improvement of a parsimonious model for streamflow simulation. *Journal of Hydrology* 279: 275–289

Perrochet P. (1987). Water uptake by plant roots — A simulation model, I. Conceptual model. *Journal of Hydrology*. Volume 95, Issues 1-2, 15 November 1987, Pages 55-61.

Philip, J.R., 1957. The theory of infiltration. *Soil Sci.* 1 (7), 83–85.

Pinol J., LLedo M.J. and Escarré A. (1991). Hydrological balance for two mediterranan forested catchments (Prades, Northeast Spain). *Hydrol. Sc. J.* 36(2)/4.

Raats P.A.C. (2001), Developments in soil-water physics since the mid 1960s. *Geoderma* 100. 355–387

Rawls W., Brakensiek D. et Saxton K. (1982) Estimation of soil water properties. *Trans. Am. Soc. Agric. Engrs* 25, 51-66

Rawls, W. J., and D. L. Brakensiek (1989), Estimation of soil water retention and hydraulic properties, in Unsaturated Flow in Hydrologic Modeling: Theory and Practice, edited by H. J. Morel-Seytoux, pp. 275- 300, Springer, New York.

Richards, L. A. (1931), Capillary conduction of liquids through porous medium, *Physics*, 1, 318- 333.

Rodriguez-Iturbe I, Proporato A, Ridolfi L, Isham V, Cox D. (1999). Probabilistic modeling of water balance at a point: the role of climate, soil and vegetation, *Proc. R. Soc. London* Ser. A, 455. 3789-3805.

Rodriguez-Iturbe I. (2000) Ecohydrology: A hydrologic perspective of climate-soil-vegetation dynamics. *Water Resour. Res* . 36, 1, 3-9

Saxton K.E., Rawls W.J., Romberger J.S., Papendick R.I. (1986) Estimating generalized soil-water characteristics from texture. Soil Sci. Soc. Am. J. 50, 1031-1036

Seibert J., Multi-criteria calibration of a conceptual runoff model using a genetic algorithm, Hydrol. Earth Syst. Sci., 4, 215- 224, 2000.

Shiklomavov I. A. (1989) Climate and water resources. *Hydrological Sciences - Journal - des Sciences Hydrologiques,* 34, 5, 10/1989

Soil Survey Division Staff (SSDS). (1998). Keys to Soil Taxonomy, eighth ed. US Department of Agriculture, Washington DC, USDA-NRCS.234

Sutcliffe J. V. (2004). Hydrology: a question of balance. *IAHS special publication* 7. , IAHS Press, 2004.

Teshima, J., Hirayama, Y., Kobayashi, T., Cho, H. (2006) Estimating evapotranspiration from a small area on a grass-covered slope using the BBH model of soil hydrology. *J. Agric. Meteorol.* , 62 (2), 65-74

Van Genuchten M.T. (1980). A closed-form equation for predicting the hydraulic conductivity of unsaturated soils, Soil Soc. Am. J. 44 892–898

Villalobos F.J., Orgaz F., Testi L., Fereres E. (2000). Measurement and modeling of evapotranspiration of olive (Olea europaea L.) orchards. *European Journal of Agronomy* 13 (2000) 155–163

Vrugt, J. A., Schoups, G., Hopmans, J. W., Young, C., Wallender, W. W., Harter, T., Bouten, W. (2004) Inverse modeling of large-scale spatially distributed vadose zone properties using global optimization. *Water Resour. Res* .40, W06503, doi :10.1029/2003WR002706

Wagener T., Boyle D. P., M. J. Lees, H. S. Wheater, H. V. Gupta and S. Sorooshian (2001). A framework for development and application of hydrological models; *Hydrology and Earth System Sciences*, 5(1), 13–26.

Wagener, T.* N. McIntyre, M. J. Lees, H. S. Wheater1 and H. V. Gupta. (2003). Towards reduced uncertainty in conceptual rainfall-runoff modelling: Dynamic identifiability analysis. *Hydrol. Process.* 17, 455–476.

Wagener T., M. Sivapalan, P. Troch and R. Woods (2007). Catchment classification and hydrologic similarity. *Geography compass*: 901-931

Youngs E. G. (1988) . Soil physics and hydrology. *Journ. of Hydrol.* 100, 411-451.

Zhan, Ch.-S., Xia, J., Chen, Z., Zuo, Q.-T. (2008) An integrated hydrological and meteorological approach for the simulation of terrestrial evapotranspiration. *Hydrological Sciences Journal.*

Zhao, R. J., and X. R. Liu (1995), The Xinjiang model, in Computer Models of Watershed Hydrology, edited by V. P. Singh, chap. 7, *Water Resour. Publ.*, Highlands Ranch, Colo.

9

Evapotranspiration of Grasslands and Pastures in North-Eastern Part of Poland

Daniel Szejba
Warsaw University of Life Sciences - SGGW
Poland

1. Introduction

The problem of plant water requirements and supply is of great importance to agricultural water management. It is crucial to determine and provide the water amount required in a certain region to support the plants assimilation function. The quantity of water required on a specific farm can be determined by analyzing the water balance, where precipitation and evapotranspiration are basic elements. Evapotranspiration data is also indispensable when mathematically modelling the water balance. The values of evapotranspiration can be obtained from lysimeter measurements. However, this measurement is labour intensive and also requires special equipment; thus, it is not widely applied. To address this problem, a number of methods of evapotranspiration estimation based on physical and empirical equations are available, where the quantity of evapotranspiration depends on other measured factors. Penman (1948) developed a method for determination of the potential evapotranspiration as a product of the crop coefficient for a certain crop in a certain development stage and the reference evapotranspiration (Łabędzki et al., 1996). Open water surface evaporation is the reference evapotranspiration used in this method. Currently, the method most widely applied in Poland for evapotranspiration estimation is a method called the "French Modified Penman method", which is a version of FAO Modified Penman method (Doorenbos & Pruitt, 1977), with the net radiation flux calculated by Podogrodzki (Roguski et al., 1988). Name of "Modified Penman method" is using in further part of this text. On the other hand, the Food and Agriculture Organization (FAO) recommends the Penman-Monteith method for evapotranspiration estimation (Allen et al., 1998). The aforementioned methods require relevant crop coefficients to estimate the potential evapotranspiration. Although crop coefficients for grasslands and pastures applicable to the modified Penman are available for Polish conditions (Roguski et al., 1988; Brandyk et al., 1996; Szuniewicz & Chrzanowski, 1996), the problem occurs when the potential evapotranspiration has to be calculated according to the FAO standards which require the Penman-Monteith method to be used. Both the methods (Modified Penman and Penman-Monteith) require meteorological data including: air temperature, humidity, cloudiness or sunshine and wind speed. If one or more of the required inputs are not available, then applying any of the two methods is difficult, perhaps even impossible. In such cases, the Thornthwaite method, developed in 1931, can be a viable alternative (Byczkowski, 1979; Skaags, 1980; Newman, 1981; Pereira & Pruitt, 2004). The Thornthwaite method is commonly used in the USA. This method requires only two basic climatic inputs that

determine the solar energy supply and are necessary to estimate the potential evapotranspiration: air temperature and day length.
There are two objectives of this chapter. The first objective is to determine the crop coefficient needed when estimating the potential evapotranspiration with the Penman-Monteith method. The second objective is a comparative analysis of the potential evapotranspiration estimates obtained from the Thornthwaite method and the crop coefficient approach with Penman-type formula as a reference evapotranspiration.

2. Reviewing the selected methods for evapotranspiration estimation: Modified Penman, Penman-Monteith and Thornthwaite

It can be assumed, that the amount of a farm plants evapotranspiration depends on such factors as atmosphere condition, plants development stage and soil moisture. The interdependence of these factors is complex and difficult to describe mathematically. This dependence can be expressed as a product of following functions:

$$ET = f_1(M) \cdot f_2(P) \cdot f_3(S) \tag{1}$$

where:
M - atmosphere factors,
P - plant factors,
S - soil moisture factors.
Groups of atmosphere factors can be formulated as a reference evapotranspiration (ET_0), which characterises meteorological conditions in the evapotranspiration process and describes evaporation ability in the atmosphere. This factor determines the intensity of evapotranspiration process in the case of unlimited access to a water source, that is deplete of soil water:

$$f_1(M) = ET_0 \tag{2}$$

$f_2(P)$ function describes the influence of plant parameters such as: plant species, development stage, mass of above ground and underground parts, leaf area index (LAI), growth dynamics, nutrients supply, yield and frequency of harvesting. A group of these parameters is expressed as a crop coefficient (k_c), which is empirically determined in independently by soil moisture conditions:

$$f_2(P) = k_c \tag{3}$$

$f_3(S)$ function describes the influence of soil moisture and the availability of soil water for plants (as a soil water potential) on evapotranspiration amount. With our knowledge of soil physics and plant physiology knowledge, it can be assumed that evapotranspiration during sufficient water supply does not depend or slightly depend on soil moisture (Łabędzki et al., 1996, as cited in: Kowalik, 1973; Salisbury & Ross, 1975; Feddes et al., 1978; Rewut, 1980; Olszta, 1981; Korohoda, 1985; Więckowski, 1985; Brandyk, 1990). Sufficient water supply does not limit evapotranspiration and plant yield is defined as a soil moisture range between optimum water content (when air content equals at least 8 - 10% in root zone) and refill point (pF 2.7 - 3.0). In other words, sufficient water supply means easily available water or readily available water (RAW). Evapotranspiration reductions has a place, when

RAW becomes consumed by plants. The deciding factor of evapotranspiration reduction amounts is the difference between actual soil moisture content and soil moisture content when the evapotranspiration process fades (wilting point). Thus, it can be showed in general (Łabędzki et al., 1996, as cited in: Olszta et al., 1990; Łabędzki & Kasperska, 1994; Łabędzki, 1995):

$$f_3(S) = k_s(\theta) \tag{4}$$

where:
$k_s(\theta)$ – soil coefficient as a function of soil moisture.
Summarizing, equation (1) can be noted as below, where ETa is called actual evapotranspiration:

$$ETa = ET_0 \cdot k_c \cdot k_s \tag{5}$$

In cases when sufficient water supply does not limiting evapotranspiration (k_s = 1), actual evapotranspiration (ETa) equals potential evapotranspiration (ETp):

$$ETp = ET_0 \cdot k_c \tag{6}$$

The problem becomes how to determine a reference evapotranspiration and a crop coefficient.

2.1 The reference evapotranspiration computing by the Modified Penman method

Penman (1948) estimated the evaporation from an open water surface, and than used that as a reference evaporation. This method requires measured climatic data on temperature, humidity, solar radiation and wind speed. Analyzing a range of lysimeter data worldwide, Doorenbos and Pruitt (1977) proposed the FAO Modified Penman method. These authors adopted the same approach as Penman to estimate reference evapotranspiration. They replaced Penman's open water evaporation with evapotranspiration from a reference crop. The reference crop was defined as "an extended surface of an 8 to 15 cm tall green grass cover of uniform height, actively growing, completely shading the ground, and not short of water". The reference evapotranspiration according to Modified Penman method commonly applied in Poland was calculated by the following algorithm. This algorithm was developed according to following literature: Roguski et al. (1988); Feddes & Lenselink (1994), Kowalik (1995), Kędziora (1999), Woś (1995), Łabędzki et al. (1996), Łabędzki (1997), Feddes et al. (1997) and van Dam et al. (1997). The parameters are as follows:
φ - latitude of meteorological station [°],
J – day number [-],
T – daily average air temperature [°C],
RH - daily average relative humidity [%],
h_i - anemometer level above ground level [m],
v_{hi} – average wind speed on 10 m level [m s^{-1}],
c – average daily cloudiness in 11 degree scale,
n – duration of direct sunshine [h],
R_a - solar radiation at the external atmosphere border [W m^{-2}],
α - albedo, in case of a crop equals to 0.23 [-],
γ - the psychrometric constant equals to 0.0655 [kPa K^{-1}],

λ - latent heat of vaporization equals to 2.45 [MJ kg^{-1}],
σ - Stefan – Boltzmann constant equals to 4.903*10^{-9} [MJ m^{-2} K^{-4} d^{-1}],
G_{sc} – solar constant equals to 0.082 [MJ m^{-2} min^{-1}].
Saturation vapour pressure (e_d) [kPa]:

$$e_d = 0.6108 \cdot \exp\left(\frac{17.27 \cdot T}{T + 237.3}\right) \tag{7}$$

Actual vapour pressure (e_a) [kPa]:

$$e_a = \frac{RH}{100} \cdot e_d \tag{8}$$

The slope of the vapour pressure curve (Δ) [kPa °C^{-1}]:

$$\Delta = \frac{4098 \cdot e_d}{(T + 237.3)^2} \tag{9}$$

Wind speed on 10 m level above ground level (v_{10}) [m s^{-1}]:

$$v_{10} = \frac{v_{hi}}{\left(\frac{h_i}{10}\right)^{1/7}} \tag{10}$$

Solar declinations (δ) [rad]:

$$\delta = 0.409 \cdot \sin\left(\frac{2\pi}{365} \cdot J - 1.39\right) \tag{11}$$

Relative distance to the Sun (d_r) [-]:

$$d_r = 1 + 0.033 \cdot \cos\left(\frac{2\pi}{365} \cdot J\right) \tag{12}$$

Time from sunrise to noon (w_s) [rad]:

$$w_s = a\cos(-\tan\varphi \cdot \tan\delta) \tag{13}$$

Possible sunshine (N) [h]:

$$N = \frac{24}{\pi} \cdot w_s \tag{14}$$

Solar radiation at the external atmosphere border (R_a) [W m^{-2}]:

$$R_a = \frac{24 \cdot 60}{\pi} \cdot G_{sc} \cdot d_r \cdot (w_s \cdot \sin\varphi \cdot \sin\delta + \cos\varphi \cdot \cos\delta \cdot \sin w_s) \tag{15}$$

Relation between real radiation to possible radiation – in case when sunshine value is not available there is calculated according to Angstöm criteria:

$$\frac{n}{N} = 1 - \frac{c}{10} \tag{16}$$

The net incoming short wave radiation flux (R_{ns}) [W m-2]:

$$R_{ns} = R_a \cdot (1 - \alpha) \cdot \left(0.209 + 0.565 \cdot \frac{n}{N} \right) \tag{17}$$

The net outgoing long wave radiation flux (R_{nl}) [W m-2]:

$$R_{nl} = \sigma \cdot (T + 273.2)^4 \cdot \left(0.56 - 0.08 \cdot \sqrt{10 \cdot e_a} \right) \cdot \left(0.1 + 0.9 \cdot \frac{n}{N} \right) \tag{18}$$

The net radiation flux (R_n) [W m-2]:

$$R_n = R_{ns} - R_{nl} \tag{19}$$

The aerodynamic factor (E_a) [mm d-1]:

$$E_a = 2.6 \cdot (e_d - e_a) \cdot (1 + 0.4 \cdot v_{10}) \tag{20}$$

Modified Penman reference evapotranspiration (ET_{MP}) [mm d-1]:

$$ET_{MP} = \frac{\Delta}{\gamma + \Delta} \cdot \frac{R_n}{\lambda} + \frac{\gamma}{\gamma + \Delta} \cdot E_a \tag{21}$$

2.2 The reference evapotranspiration computing by the Penman-Monteith method

Among scientists is unanimous the consensus is that the best method of evapotranspiration calculation is a method proposed and developed by John Monteith (1965). Monteith's derivation was built upon that of Penman (1948) in the now well-known combination equation (combination of an energy balance and an aerodynamic formula). The equation describes the evapotranspiration from a dry, extensive, horizontally uniform vegetated surface, which is optimally supplied with water. This equation is known as the Penman-Monteith equation and it is currently recommending by FAO. Potential and even actual evapotranspiration estimates are possible with the Penman-Monteith equation, through the introduction of canopy and air resistance to water vapour diffusion. Nevertheless, since accepted canopy and air resistance may not be available for many crops, a two-step approach is still recommended under field conditions. The first step is the calculation of the reference evapotranspiration as an evapotranspiration of a reference crop for some steady parameters and soil moisture conditions. In the second step the actual evapotranspiration is calculated using the root water uptake reduction due to water stress. The reference crop is defined as "a hypothetical crop which is grass, with a constant, uniform canopy 12 cm tall, constant canopy resistance equals to 70 s m-1, constant albedo equals to 0.23, in conditions of active development and optimally supplied with water" (Łabędzki et al., 1996; Feddes et al., 1997; van Dam et al., 1997; Allen et al., 1998; Howell & Evett, 2004, as cited in: Monteith, 1965). The Penman-Monteith reference evapotranspiration recommended by FAO was calculated by a similar algorithm shown in point 2.1. The difference between the Modified Penman and Penman-Monteith methods bases on solar radiation and an aerodynamic

formula calculation in general. Named factors were calculated according to following formulas shown below (Feddes & Lenselink, 1994).
The following parameters were used:
H- altitude of meteorological station over sea level [m],
T_{Kmin} – daily minimum air temperature [K],
T_{Kmax} – daily maximum air temperature [K],
v – average wind speed on 2 m level [m s^{-1}],
σ - Stefan – Boltzmann constant equals to 5.6745*10^{-8} [W m^{-2} K^{-4}],
Solar radiation at the external atmosphere border (R_a) [W m^{-2}]:

$$R_a = 435 \cdot d_r \cdot \left(w_s \cdot \sin\varphi \cdot \sin\delta + \cos\varphi \cdot \cos\delta \cdot \sin w_s\right) \tag{22}$$

Solar radiation (R_s) [W m^{-2}]:

$$R_s = R_a \cdot \left[0.25 + \left(0.5 \cdot \frac{n}{N}\right)\right] \tag{23}$$

The net incoming short wave radiation flux (R_{ns}) [W m^{-2}]:

$$R_{ns} = (1 - \alpha) \cdot R_s \tag{24}$$

The net outgoing long wave radiation flux (R_{nl}) [W m^{-2}]:

$$R_{nl} = \left(0.9 \cdot \frac{n}{N} + 0.1\right) \cdot \left(0.34 - 0.139 \cdot \sqrt{e_a}\right) \cdot \sigma \cdot \frac{\left(T_{Kmax}^4 + T_{Kmin}^4\right)}{2} \tag{25}$$

The radiation factor (R_n') [mm d^{-1}]:

$$R_n' = 86400 \cdot \frac{(R_{ns} - R_{nl})}{\lambda} \tag{26}$$

The atmospheric pressure [p_a] [kPa]:

$$p_a = 101.3 \cdot \frac{(T + 273.16 - 0.0065 \cdot H)}{T + 273.16} \tag{27}$$

The psychrometric constant (γ) [kPa °C]:

$$\gamma = 1615 \cdot \frac{p_a}{\lambda} \tag{28}$$

Modified psychrometric constant (γ') [kPa °C]:

$$\gamma' = (1 + 0.337 \cdot v) \cdot \gamma \tag{29}$$

The aerodynamic factor (E_a) [mm d^{-1}]:

$$E_a = \frac{900}{(T + 275)} \cdot v \cdot (e_d - e_a) \tag{30}$$

And finally Penman-Monteith reference evapotranspiration (ET_{P-M}) [mm d-1]:

$$ET_{P-M} = \frac{\Delta}{\Delta + \gamma'} \cdot R_n' + \frac{\gamma}{\Delta + \gamma'} \cdot E_a \tag{31}$$

2.3 Crop coefficient

Potential evapotranspiration is calculated by multiplying ET_o by k_c, a coefficient expressing the difference in evapotranspiration between the cropped and reference grass surface. The difference can be combined into a single coefficient, or it can be split into two factors describing separately the differences in evaporation and transpiration between both surfaces. The selection of the approach depends on the purpose of the calculation, the accuracy required, the climatic data available and the time step with which the calculations are executed (Allen et al., 1998). Due to the purpose of this chapter, only the single coefficient approach is taken under consideration. The single crop coefficient combined the effect of crop transpiration and soil evaporation. The crop coefficient expresses crop actual mass and development stage influence on the evapotranspiration value, in sufficient soil moisture content. It is dependant on crop type, development stage and yield. The generalized crop coefficient curve is shown in Figure 1. Shortly after the planting of annuals or shortly after the initiation of new leaves for perennials, the value for k_c is small, often less than 0.4. The k_c begins to increase from the initial k_c value, $k_{c\ ini}$, at the beginning of rapid plant development and reaches a maximum value, $k_{c\ mid}$, at the time of maximum or near maximum plant development. During the late season period, as leaves begin to age and senesce due to natural or cultural practices, the k_c begins to decrease until it reaches a lower value at the end of the growing period equal to $k_{c\ end}$ (Roguski et al., 1988; Allen et al., 1998).

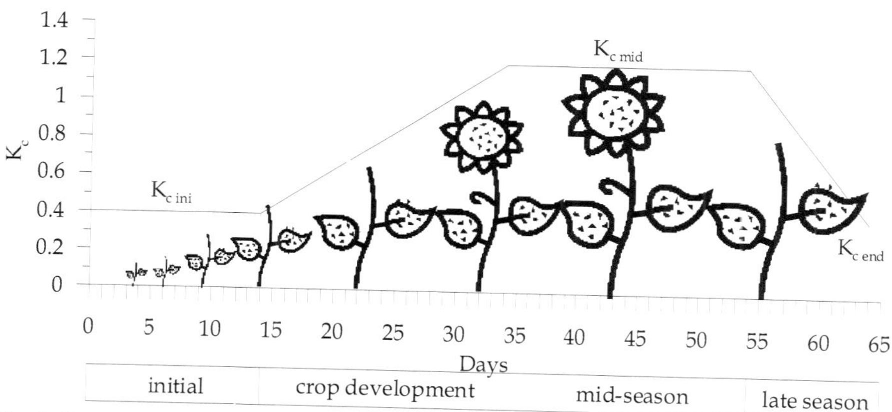

Fig. 1. Crop coefficient due to plant development stage

The objective of this work is to determine the crop coefficient needed when estimating the potential evapotranspiration with the Penman-Monteith method, when the potential evapotranspiration calculated as a product of Modified Penman reference evapotranspiration and appropriate crop coefficient for this method is known. Based on procedures proposed by Feddes et al. (1997), the conversion of the Modified Penman crop coefficient $k_{c\,MP}$ to the Penman-Monteith crop coefficient $k_{c\,P\text{-}M}$ can be write as:

$$ETp = ET_{MP} \cdot k_{cMP} = ET_{P-M} \cdot k_{cP-M} \tag{32}$$

from which:

$$k_{cP-M} = \frac{ET_{MP} \cdot k_{cMP}}{ET_{P-M}} \tag{33}$$

2.4 Potential evapotranspiration estimation by the Thornthwaite method

Both Modified Penman and Penman-Monteith methods required many climatic inputs like: air temperature, relative humidity, wind speed and solar radiation or at least daily sunshine. These are limited or even not available for many regions. Another problem is noncontinuous data series for some periods. Thus using the Modified Penman and Penman-Monteith methods for evapotranspiration calculation is not so easy and problematic in some cases. An alternative commonly used in the United States is the Thornthwaite method, because it requires only air temperature as a input data (Skaags, 1980; Newman, 1981). This method is based on determination of available energy required for the evaporation process. The relationship between average monthly air temperature and potential evapotranspiration is calculated based on a standard 30 days month with 12 hours of daylight each day according to the following equation (Byczkowski, 1979; Newman, 1981; Pereira & Pruitt, 2004):

$$ETp_T = 16.2 \cdot \left(\frac{10 \cdot T_j}{I} \right)^a \tag{34}$$

where:
ETp_T – Thornthwaite monthly potential evapotranspiration (mm),
d_f – correction factor for daylight hours and days in month (-),
T_j – average monthly air temperature (°C),
I – annual heat index as a sum of monthly heat index I_i:

$$I = \sum_{i=1}^{12} I_i = \sum_{i=1}^{12} \left(\frac{T_j}{5} \right)^{1.514} \tag{35}$$

a – coefficient derived from climatological data:

$$a = 6.75 \cdot 10^{-7} \cdot I^3 - 7.71 \cdot 10^{-5} \cdot I^2 + 1.79 \cdot 10^{-2} \cdot I + 0.492 \tag{36}$$

In order to convert the estimates from a standard monthly ETp_T to a decade of evapotranspiration the following correction factor for daylight hours and days in month d_f (-) was used:

$$d_f = \frac{N_{dec}}{360} \tag{37}$$

where:
N_{dec} - possible sunshine for decade (h)
It must to be noted, that the Thornthwaite method is valid for average monthly air temperature from 0 to 26.5 °C.

3. Grasslands and pastures in the north-eastern part of Poland and local condition climate data

As Statistical Yearbook of Agriculture and Rural Areas (2009) presents, grasslands and pastures occupy about 3271.2 thousand hectares which is 20% of the total agricultural land in Poland. According to administrative division, the north-eastern part of Poland are Podlaskie and the eastern part of Warmińsko-Mazurskie voivodships Grasslands and pastures occupy 393.5 thousand hectares (35%) and 290 thousand hectares (28.1%) of these voivodships agricultural land respectively. The valley of the River Biebrza (22° 30′–23° 60′ E and 53° 30′–53° 75′ N) (Fig. 2) is one of the last extensive undrained valley mires in Central Europe. The Biebrza features several types of mires. The dominant types are fens, which account for some 75.9% of the wetland area (Okruszko, 1990). The altitude of the valley ranges from 100 to 130 m above mean sea level and the catchment area of approximately 7000 km^2 has a maximum altitude of 160 m (Byczkowski & Kicinski, 1984). The mean yearly rainfall is 583 mm, of which 244 mm falls in the wet summers. Mean annual temperature is rather low (6.8 °C), and the growing season is quite short (around 200 days) (Kossowska-Cezak, 1984). The part of Warmińsko-Mazurskie voivodship is Warmia region. Main town (former capital of Warmia region) situated on the north part of Warmia region (Fig. 2) is Lidzbark Warmiński (20° 35′ E, 54° 08′ N).

Fig. 2. An approximate location of considered regions in Poland

The altitude of the region ranges from 80 to 100 m above mean sea level on the borders and falls down from 40 to 50 m above mean sea level to the center. Brown Soils and Mollic Gleysols developed from silt and clay dominate in this. These soils are situated on sloping areas with partly well surface water outflow. In the study region average yearly air temperature is equal to 7.1°C and average yearly sum of precipitation equal to 624 mm. The highest amount of rainfall is usually observed in July and August. The vegetation period lasts about 200 days. The snow cover occurs during 60–65 days (Nowicka et al., 1994). The needed meteorological data are available for the 1989-2004 grassland growing seasons derived from the Biebrza meteorological station located in the Middle Biebrza River Basin. The estimation of the pasture evapotranspiration will be based on the meteorological data collected in the Warmia region during the 1999 through 2010 period.

4. Results and discussion

The decade Modified Penman and Penman-Monteith reference evapotranspiration values were calculated both for Warmia Region and Middle Biebrza River Basin. The relationship between reference evapotranspiration values of two kinds of Penman methods was shown on Fig. 3.

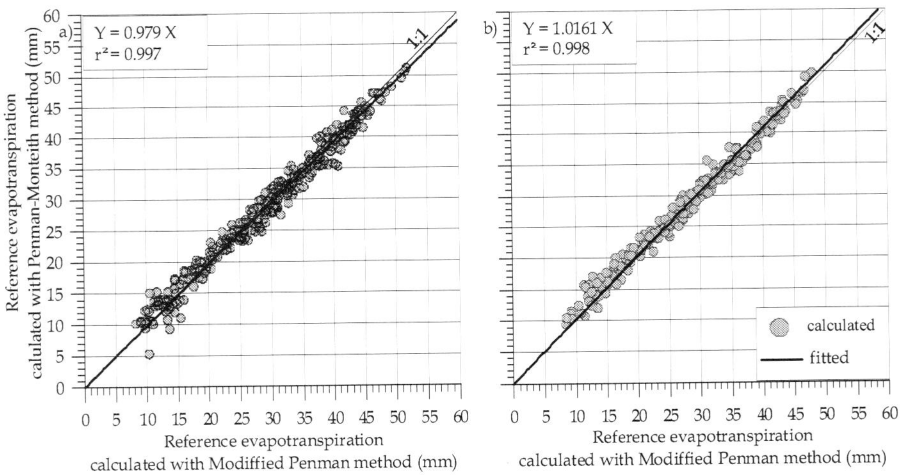

Fig. 3. The relationship between the Modified Penman and the Penman-Monteith reference evapotranspiration for: a) Middle Biebrza River Basin, b) Warmia Region

The relationship was fitted by linear regression through origin. Obtained linear equations indicates there is not significant difference between reference evapotranspiration calculated with Modified Penman and Penman-Monteith methods in both cases. It must to be noted that there is very good correlation between Modified Penman and Penman-Monteith methods. The coefficient of determination r^2 is equall to 99.7% and 99.8% respectively. Due to linear equation, Penman-Monteith reference evapotranspiration values are about 2% lower than values calculated by Modified Penman method for Middle Biebrza River Basin case (Fig. 3a). Whereas, an opposite situation was observed for Warmia Region. Reference evapotranspiration values calculated by the Modified Penman are 1.6% lower than values obtained by the Penman-Monteith method (Fig. 3b).

Consequently, an attempt was made for crop coefficient calculation (Eq. 33) proper for determination of potential evapotranspiration with the Penman-Monteith method. The following croplands were taken under consideration: pasture located in Warmia Region and intensive meadow, extensive meadow and natural wetland plant communities characteristic of Middle Biebrza River Basin. The calculation was conducted for vegetation period decade values of Modified Penman and Penman-Monteith reference evapotranspiration and crop coefficient for the Modified Penman method elaborated by Roguski et al. (1988), Brandyk et al. (1996) and Szuniewicz & Chrzanowski (1996). Considered values of crop coefficient both for Modified Penman ($k_{c\ MP}$) and Penman-Monteith ($k_{c\ P\text{-}M}$) for pasture was presented on Table 1. It can be maintain that $k_{c\ P\text{-}M}$ values for April are about 0.05 lower than $k_{c\ MP}$ values. The values for May, June and July are the same or almost the same – the difference does not exceed 0.02. The most significant differences are present in September, where $k_{c\ P\text{-}M}$ is lower than $k_{c\ MP}$ from 0.09 to 0.21.

Month	Decade	Crop coefficient	
		$k_{c\ MP}$	$k_{c\ P\text{-}M}$
April	1	0.75	0.70
	2	0.80	0.76
	3	0.80	0.76
May	1	0.85	0.84
	2	0.80	0.81
	3	0.95	0.95
June	1	0.70	0.71
	2	0.70	0.71
	3	0.95	0.97
July	1	0.80	0.81
	2	0.85	0.85
	3	0.90	0.89
August	1	0.80	0.79
	2	0.95	0.93
	3	1.05	1.00
September	1	0.95	0.86
	2	1.00	0.87
	3	1.10	0.89

Table 1. Crop coefficient of pasture for Modified Penman and Penman-Monteith methods

Modified Penman crop coefficient for extensive meadows (EM) and natural wetlands plant communities (NWPC) was published by Brandyk et al. (1996) as cited in: Roguski (1985) and Łabędzki & Kasperska (1994). Values of these crop coefficients as well as values of calculated Penman-Monteith crop coefficients was presented on Table 2. It can be maintain that $k_{c\ P\text{-}M}$ values are higher than $k_{c\ MP}$ values from 0.01 to 0.12 for extensive meadow in general. An exception to this rule is the last five decades, when $k_{c\ P\text{-}M}$ values are lower then $k_{c\ MP}$ values from 0.01 to 0.23. A similar tendency can be observed for natural wetland plant communities. But wider differences occur between $k_{c\ P\text{-}M}$ and $k_{c\ MP}$. A value of $k_{c\ P\text{-}M}$ is higher up to 0.08 than $k_{c\ MP}$ value for a few decades and lower until 0.31 for the last decade of September.

Month	Decade	Crop coefficient			
		EM		NWPC	
		$k_{c\ MP}$	$k_{c\ P\text{-}M}$	$k_{c\ MP}$	$k_{c\ P\text{-}M}$
April	1	0.93	1.05	0.62	0.70
	2	0.93	0.97	0.79	0.83
	3	0.85	0.84	0.75	0.74
May	1	0.88	0.90	0.77	0.79
	2	1.04	1.09	1.06	1.10
	3	1.03	1.08	1.21	1.27
June	1	0.76	0.79	1.24	1.30
	2	0.91	0.96	1.28	1.36
	3	0.98	1.04	1.40	1.48
July	1	0.99	1.03	1.32	1.37
	2	1.01	1.06	1.18	1.23
	3	0.98	1.04	1.40	1.48
August	1	0.97	0.98	1.30	1.31
	2	1.07	1.07	1.40	1.39
	3	1.18	1.15	1.40	1.36
September	1	1.34	1.27	1.63	1.55
	2	1.41	1.27	1.85	1.66
	3	1.41	1.18		1.60

Table 2. Crop coefficient of extensive meadow and natural wetland plant communities for Modified Penman and Penman-Monteith methods

The Modified Penman crop coefficient for intensive meadow located in Middle Biebrza River Basin was elaborated by Szuniewicz & Chrzanowski (1996). They based the research on lysimeter experiments conducted on peat –moorsh soil with a ground water level of 35 – 90 cm (optimum soil moisture) during the 1982-1991 period. Researchers had established conditions for 3-cut meadows with different hay yields: 0.10, 0.20, 0.30, 0.40 and 0.50 Mg ha^{-1}. The climate of the considered region is more severe compared to other plain regions in Poland, thus the vegetation period starts about two weeks later. Elaborated by Szuniewicz & Chrzanowski crop coefficients for the Modified Penman method as well as calculated crop coefficients for Penman-Monteith was presented on Table 2. There are not significant differences between $k_{c\ P\text{-}M}$ and $k_{c\ MP}$ values for the first two decades of the vegetation period.

The differences increase during successive decades of May and June from 0.02 up to 0.07. Next, they decrease from 0.04 to 0.02 in July. There are not significant differences again for first and second decades of July. The difference begins it's increase from the third decade of July up to the second decade of September. The values of $k_{c\ P\text{-}M}$ are even 0.12 - 0.18 lower than $k_{c\ MP}$ for the second decade of September. There is also a clear tendency towards an increase of differences between crop coefficients $k_{c\ P\text{-}M}$ and $k_{c\ MP}$ values due to an increase of potential hay yield. The $k_{c\ P\text{-}M}$ values get higher from 0.02 to 0.07 in May and June. However, the opposite tendency can be observed in September, when $k_{c\ P\text{-}M}$ get lower from 0.06 to even 0.18.

Month	Decade	Cut	Crop coefficient at hay yields Mg ha^{-1}									
			0.10		0.20		0.30		0.40		0.50	
			$k_{c\ MP}$	$k_{c\ P\text{-}M}$	$k_{c\ MP}$	$k_{c\ P\text{-}M}$	$k_{c\ MP}$	$k_{c\ P\text{-}M}$	$k_{c\ MP}$	$k_{c\ P\text{-}M}$	$k_{c\ MP}$	$k_{c\ P\text{-}M}$
April	2	I	0.93	0.96	0.93	0.96	0.93	0.96	0.93	0.96	0.93	0.96
	3		0.78	0.77	0.85	0.84	0.9	0.89	0.95	0.94	0.99	0.98
May	1		0.77	0.79	0.88	0.90	0.97	0.99	1.06	1.08	1.13	1.15
	2		0.89	0.93	1.04	1.09	1.17	1.22	1.28	1.34	1.39	1.45
	3		0.86	0.90	1.03	1.08	1.18	1.24	1.31	1.38	1.43	1.50
June	1	II	0.76	0.80	0.76	0.80	0.76	0.80	0.76	0.80	0.76	0.80
	2		0.86	0.91	0.91	0.96	0.95	1.01	0.99	1.05	1.02	1.08
	3		0.87	0.92	0.98	1.04	1.08	1.14	1.17	1.24	1.25	1.32
July	1		0.85	0.89	0.99	1.03	1.11	1.16	1.21	1.26	1.30	1.36
	2		0.86	0.90	1.01	1.06	1.15	1.20	1.27	1.33	1.38	1.44
	3	III	0.78	0.80	0.78	0.80	0.78	0.80	0.78	0.80		
August	1		0.89	0.90	0.97	0.98	1.04	1.05	1.09	1.10		
	2		0.95	0.94	1.07	1.06	1.17	1.16	1.26	1.25		
	3		0.96	0.94	1.18	1.15	1.36	1.33	1.52	1.48		
September	1		1.12	1.06	1.34	1.27	1.52	1.44	1.68	1.59		
	2		1.16	1.04	1.41	1.27	1.63	1.47	1.82	1.64		

Table 3. Crop coefficient of 3-cut meadow for Modified Penman and Penman-Monteith methods

The next step of this work use to be an comparison potential evapotranspiration calculated as a product of Penman-Monteith reference evapotranspiration and determined crop coefficient ($k_{c\ P\text{-}M}$) with alternative potential evapotranspiration by Thornthwaite. In order to solve the problem, decade values of Thornthwaite potential evapotranspiration was calculated (Eq. 34-37) and Penman-Monteith potential evapotranpiration applying crop coefficient for proper land use. The relationship between Thornthwaite potential evapotranspiration and Penman-Monteith potential evapotranspiration was presented on Fig. 4. The relationship was fitted by linear regression through origin. Analyzing obtained results, it can be maintain that Penman-Monteith evapotranspiration values are lower by about 25% for pasture (Fig. 4a) and 8% for extensive meadow than the Thornthwaite method

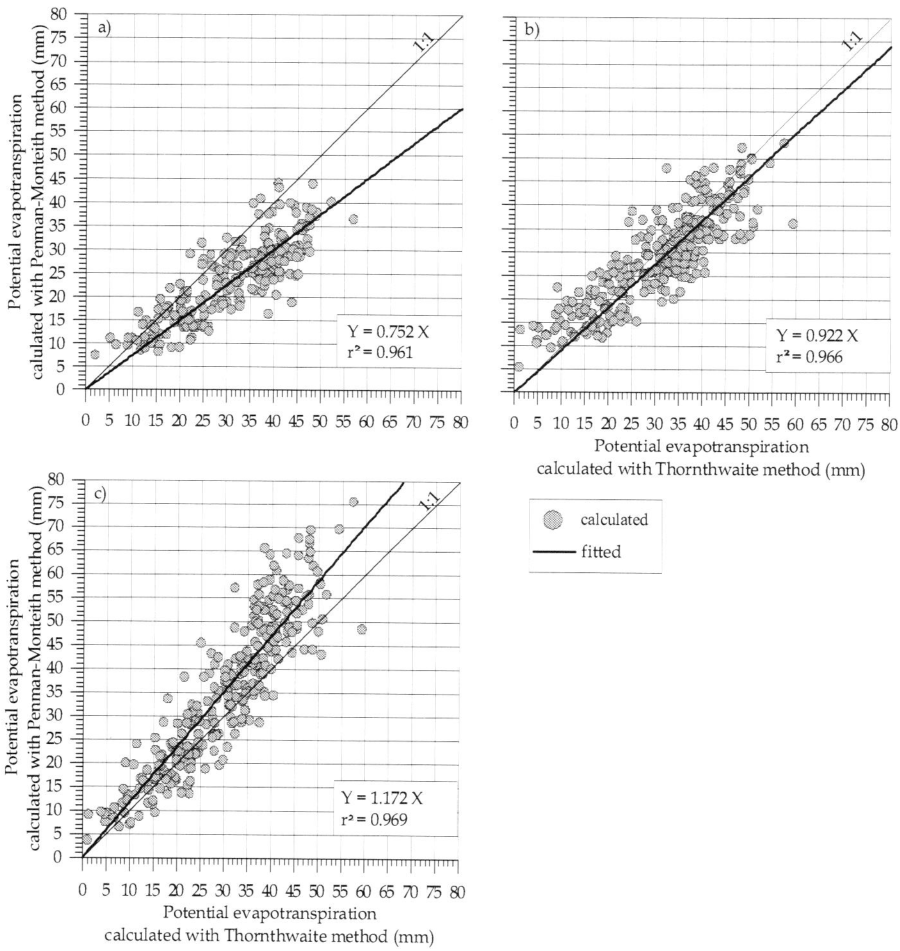

Fig. 4. The relationship between Thornthwaite potential evapotranspiration and Penman potential evapotranspiration for: pasture (a), extensive meadow (b) and natural wetland plant communities (c)

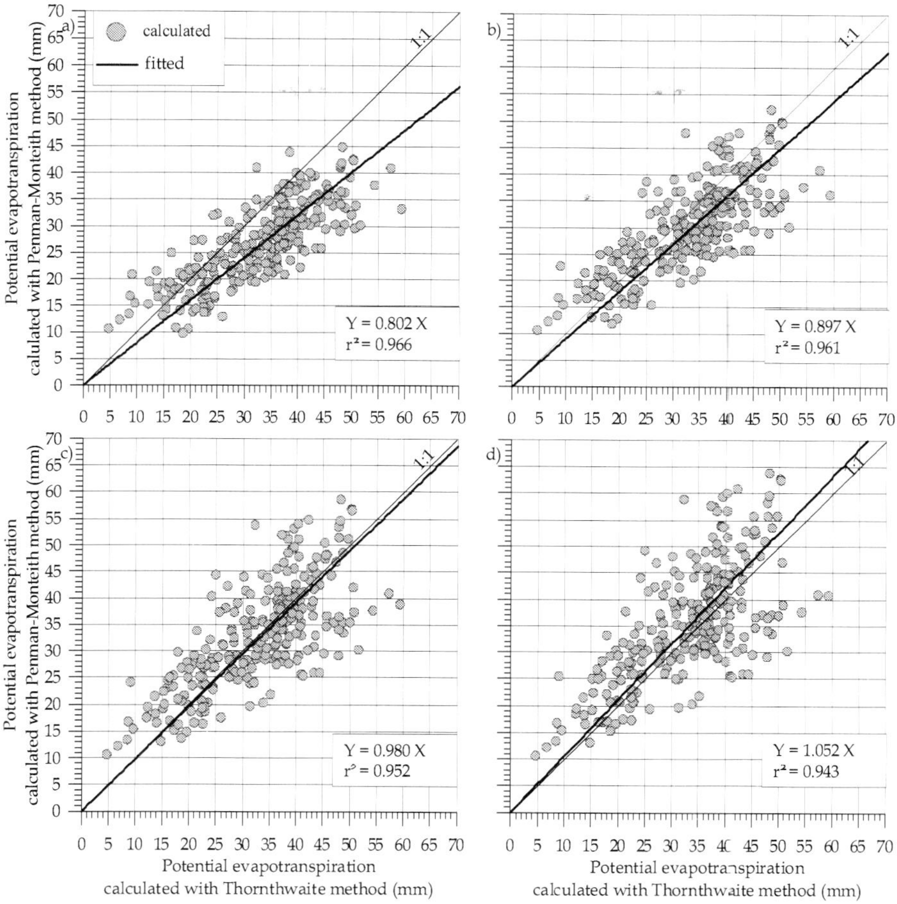

Fig. 5. The relationship between Thornthwaite potential evapotranspiration and Penman potential evapotranspiration of 3-cut meadow for hay yield Mg ha^{-1}: 0.10 (a), 0.20 (b), 0.30 (c) and 0.40 (d)

(Fig. 4b). Whereas in case of natural wetland plan community evapotranspiration, values calculated with Penman-Monteith method are of about 17% higher then values calculated with Thornthwaite method. It must to be noted, that coefficient of determination is almost equal ($r^2 \approx 97\%$) for all three cases. The relationship between Thornthwaite potential evapotranspiration and Penman-Monteith potential evapotranspiration for 3-cut meadow was presented on Fig. 5. Analyzing obtained results, it can be maintained that Penman-Monteith evapotranspiration values are very close to Thornthwaite evapotranspiration values for 0.30 Mg ha^{-1} hay yield. An evapotranspiration calculated with the Thornthwaite method is just about 2% higher than Penman-Monteith evapotranspiration. The highest overestimation (20%) of the Thornthwaite method is observed for the lowest hay yield

(0.10 Mg ha^{-1}). The case of 0.20 Mg ha^{-1} hay yield characterizes about a 10% overestimation of the Thornthwaite method. An opposite case is the case of 0.40 Mg ha^{-1} hay yield, where the Thornthwaite method underestimates evapotranspiration by about 5%. Coefficients of determination vary between 94.3% (0.40 Mg ha^{-1} hay yield) and 96.6% (0.10 Mg ha^{-1} hay yield).

5. Conclusion

Based on the performed research the following conclusions can be formulated:
There are not significant differences between reference evapotranspiration calculated with the Modified Penman and Penman-Monteith methods of the Warmia Region as well as Middle Biebrza River Basin for entire vegetation period (April - September). Due to linear equation, Penman-Monteith reference evapotranspiration values are about 1.6 % higher than values calculated by the Modified Penman method for the Warmia Region case. Whereas, values of Modified Penman reference evapotranspiration are about 2.0% lower than values obtained with the Penman-Monteith method. From a practical point of view, the difference of total vegetation period reference evapotranspiration equals about 8 mm for the Warmia Region and 10 mm for Middle Biebrza River Basin due to 513 mm (Warmia Region) and 486 mm (Middle Biebrza River Basin) of average vegetation period reference evapotranspiration assumption.
Crop coefficients calculated for the Penman-Monteith evapotranspiration method are comparable or lower than crop coefficients for the Modified Penman method in case of pasture. Taking under consideration crop coefficient differences for extensive meadow and natural wetland plant communities it can be found that $k_{c\ P\text{-}M}$ values are higher than $k_{c\ MP}$ values from 0.01 to 0.12 for most of the vegetation period in general. An exception to this rule is the last five decades, when $k_{c\ P\text{-}M}$ values were lower then $k_{c\ MP}$ values from 0.01 even to 0.31. There are not significant differences between $k_{c\ P\text{-}M}$ and $k_{c\ MP}$ values for the first and second decades of vegetation period as well as for the first and second decades of July in the case of 3-cut meadow. The difference begins to from the third decade of July up to the second decade of September. The values of $k_{c\ P\text{-}M}$ are even 0.12 - 0.18 lower than $k_{c\ MP}$ for the second decade of September. Summarizing, crop coefficients calculated for Penman-Monteith method are almost equal or slightly higher compare to Modified Penman crop coefficients for most of a vegetation period in all considered land use. An exception are last three to four decades of vegetation period when values of $k_{c\ P\text{-}M}$ are clearly lower compared to $k_{c\ MP}$ values. These differences are equal during the entire vegetation period. But they can have essential meaning in certain parts (decades) of vegetation period when a crop water requirement is determined.
Potential evapotranspiration values calculated with the Thornthwaite method are overestimated in ratio to values calculated with the Penman-Monteith method in the following cases by about: 25% for pasture, 20% for 3-cut meadow (0.10 Mg ha^{-1} hay yield), 10% for 3-cut meadow (0.20 Mg ha^{-1} hay yield) and 8% for extensive meadow. Whereas, one time Thornthwaite potential evapotranspiration values were lower by about 5% for 3-cut meadow (0.40 Mg ha^{-1} hay yield). The best convergence of the considered methods is observed for 3-cut meadow in case of 0.30 Mg ha^{-1}. It has to be said, that coefficient of determination r^2 exceeds 94% of the value for all cases. Summarized, the Thornthwaite potential evapotranspiration method is comparable with the Penman-Monteith method for 3-cut meadow with a high value of hay yield and extensive meadow.

Future research should be focused on trials to find correlations between Thornthwaite and Penman-Monteith potential evapotranspiration for individual months of vegetation period. Another aim could be crop coefficient calculation for the Penman-Monteith method for field crops like grains, potatoes or sugar beets.

6. Acknowledgment

A part of this work considered to evapotranspiration calculation of Warmia Region was supported by the grant of Polish Ministry of Science and Higher Education No N N305 039234.

Special thanks to friend of mine Dr Jan Szatyłowicz for help with Penman's methods evapotranspiration calculation for Middle Biebrza River Basin.

7. References

Allen R.G., Pereira L.S., Raes D. & Smith M. (1998). Crop evapotranspiration - Guidelines for computing crop water requirements. *FAO Irrigation and Drainage Paper*, No. 56, pp. 290, ISBN 92-5-104219-5, FAO, Rome, Retrieved from: http://www.fao.org/docrep/x0490e/x0490e00.htm#Contents

Brandyk T., Szuniewicz J., Szatyłowicz J. & Chrzanowski S. (1996). Potrzeby wodne roślinności obszarów hydrogenicznych. *Zesz. Probl. Post. Nauk Rol.* 432, pp. 91-104, ISSN 0084-5477

Byczkowski A. (1979). *Hydrologiczne podstawy projektów wodnomelioracyjnych. Przepływy charakterystyczne*, ISBN 83-09-00035-9, Wyd. PWRiL, Warszawa

Byczkowski, A. & Kicinski, T. (1984). Surface waters in the Biebrza drainage basin. Pol. Ecol.Stud. 10, pp. 271–299, ISSN 0324-8763

Doorenbos J. & Pruitt W.O. (1977). Guidelines for predicting crop water requirements. *Irrigation and Drainage Paper*, No. 24, pp. 290, ISBN 92-5-100279-7, FAO, Rome

Feddes R.A. & Lenselink K.J. (1994). Evapotranpiration, In: *Drainage Principles and Application*. Ritzema H.P., (Ed.), ILRI, Publication 16, Second Edition, ISBN 90 70754 3 39, Wageningen, The Netherlands

Feddes R.A., Koopmans R.W.R. & Van Dam J.C. (1997). *Agrohydrology*, Wageningen University, Department of Environmental Sciences, Sub-department Water Resources

Howell T.A. & Evett S.R., (2004). *The Penman-Monteith Method*, USDA-Agricultural Research Service Conservation & Production Research Laboratory, Bushland, Texas, USA, Retrieved from: www.cprl.ars.usda.gov

Kędziora A. (1999). *Podstawy Agrometeorologii*, ISBN: 8309016417, Wyd. PWRiL, Poznań

Kossowska-Cezak U. (1984). Climate of the Biebrza ice-marginal valley. Pol. Ecol. Stud.10, 253–270, ISSN 0324-8763

Kowalik P. (1995). *Obieg wody w ekosystemach lądowych*. ISSN 0867-7816, Monografia PAN. Zeszyt 3., Warszawa

Łabędzki L., Szajda J. & Szuniewicz J. (1996). Ewapotranspiracja upraw rolniczych – terminologia, definicje, metody obliczania. *Materiały Informacyjne IMUZ*, pp. 1 – 7, Wyd. IMUZ, Falenty

Łabędzki L. (1997). Potrzeby nawadniania użytków zielonych – uwarunkowania przyrodnicze i prognozowanie. Wydawnictwo IMUZ Falenty

Newman J.E. (1981). Weekly Water Use Estimates by Crops and Natural Vegetation in Indiana. *Station Bulletin* No. 344, pp. 1-2, Department of Agronomy, Agricultural Experimental Station Purdue University. West Lafayette, Indiana

Nowicka A., Banaszkiewicz B. & Grabowska K. (1994). The selected meteorological elements for Olsztyn region in 1951–1990 years with comparison to averages for 1881–1930 period. *Mat. Konf. XXV zjazd agrometeorologow,* Olsztyn-Mierki, Poland, 27-29.09.1994

Okruszko H. (1990). *Wetlands of the Biebrza Valley, their Value and Future Management,* ISBN 83-00-03461, Polish Academy of Science, Warszawa

Pereira A.R. & Pruitt W.O. (2004). Adaptation of the Thornthwaite scheme for estimating daily reference evapotranspiration. *Agric. Water Manag.* 66, pp. 251-257, ISSN: 0378-3774

Roguski W., Sarnacka S. & Drupka S. (1988). Instrukcja wyznaczania potrzeb i niedoborów wodnych roślin uprawnych i użytków zielonych. *Materiały Instruktażowe* 66, pp. 90., ISSN 0860-0813, Wyd. IMUZ, Falenty

Skaggs R.W., (1980). Drainmod Reference Report. U.S. Department of Agriculture, Soil Conservation Service, North Carolina State University. Raleigh, North Carolina, pp. 19–23

Statistical Yearbook of Agriculture and Rural Areas (2009). ISSN 1895-121X, Zakład Wydawnictw Statystycznych, Warszawa

Szuniewicz J. & Chrzanowski S. (1996). Współczynniki roślinne do obliczania ewapotranspiracji łąki trzykośnej na glebie torfowo-murszowej w Polsce północno-wschodniej. *Wiad. IMUZ XVIII(4),* pp. 109-118, ISBN 83-85735-28-3

Van Dam J.C., Huygen J., Wesseling J.G., Feddes R.A., Kabat P., Van Walsum P.E.V., Groenendijk P. & Van Diepen C.A. (1997). *Theory of SWAP version 2.0,* ISSN 0928-0944, Technical Document 45 DLO Winand Staring Centre, Wageningen

Woś A. (1995). *ABC meteorologii.* ISBN 8323207097, U.A.M. Poznań

10

The Role of Evapotranspiration in the Framework of Water Resource Management and Planning Under Shortage Conditions

Giuseppe Mendicino and Alfonso Senatore
Department of Soil Conservation, University of Calabria, Arcavacata di Rende (CS) Italy

1. Introduction

The increased availability of observed data and of advanced techniques for the analysis of meteo-hydrological information allows an even more detailed description of the evolution of global climate. The results showed by the Fourth Assessment Report (FAR) of the International Panel on Climate Change (IPCC, 2007) about the changes that, starting from 1950, are affecting the atmosphere, the cryosphere and the oceans, confirm global warming. The global average surface temperature has increased in the last 100 years by 0.74°C ± 0.18°C, accelerating in the last 50 years (0.13°C ± 0.03°C per decade), especially over land (about 0.27 °C per decade) and at higher northern latitudes. As a consequence, the higher available energy on the surface has speeded up the hydrological cycle. The concentration of the water vapor in the troposphere has increased (1.2 ± 0.3% per decade from 1988 to 2004), while long-period precipitation trends (both positive and negative) in many regions have been observed by analyzing time series from the year 1900 to the year 2005. Changes in temperature and precipitation regimes strongly affect the hydrological cycle. As an example, the increase in temperature has produced a substantial reduction in snow cover in several regions, mainly in spring, and a reduction in the areas covered by seasonal frozen ground (reduction of about 7% in the northern hemisphere over the latter half of the 20th century). Direct long-term measurements of all the main components of the hydrological cycle are not widely available: in order to assess soil moisture long-term changes, due to the lack of direct measurements the primary approach is to calculate Palmer Drought Severity Index, while long-term stream flow gauge records do not cover entirely and uniformly the world, and they present gaps and different record lengths. However, generally stream flow trends are positively correlated to precipitation, while a common effect of climate change is arising independently on precipitation trends: starting from the '70s a considerable increase of the frequency of extreme hydrological events (floods and droughts) has been observed. Also concerning actual evapotranspiration, direct measurements over global land areas are still very limited, but already the Third Assessment Report (TAR) reported that actual evapotranspiration increased during the second half of the 20th century over most dry regions of the USA and Russia, and, by means of observed precipitation, temperature, cloudiness-based surface solar radiation and a land surface model, Qian et al. (2006) found that global land evapotranspiration closely follows variations in land precipitation.

Following the FAR, it is extremely unlikely (<5% probability) that the global warming trend observed in the last half century, whose remarkable characteristics in the history of the Earth seem to be confirmed even by paleo-climatic studies, could be explained without considering external forcings, and is very likely (>90%) that the production of greenhouse gases is the main cause of the observed increase in temperature.

Human activities negatively impact on water resource availability, not only contributing to the water cycle changes on a global scale, but also in a more direct way, through the pollution of water courses and aquifers. This pollution is specifically generated by the over-exploitation of the soil and chemical contaminants due to agriculture and forestry, by urban waste, transportation and building, and by the over-exploitation of the coastal aquifers, which generates saline water intrusion.

Many of the problems connected to water shortage and to bad water quality are due to not efficient or even inexistent water resources planning and management. Recently, most advanced planning studies have adopted tools for integrated water resources management. Specifically, by now among planners the idea is diffused that a reactive approach, based on the implementation of actions after a drought event has occurred and is perceived, is not adequate and a proactive approach is needed (Yevjevich et al., 1983; Rossi, 2003), based on the development of plans allowing the identification of long- and short-term actions to face drought, and the implementation of such plans, on the basis of timely information provided by a drought monitoring system.

Different measures can be used to cope with water resource crises due to drought. Rossi et al. (2007) show several classifications of these measures: first, the one suggested by Yevjevich et al. (1978) that distinguishes among measures aimed at increasing water supply, reducing demand and minimizing impacts; next, considering the one differentiating reactive and proactive measures (Yevjevich et al., 1983); and finally, the one between long- and short-term measures. The Water Scarcity Drafting Group (2006) disseminated a document specifying a series of mitigation measures that can be adopted in the EU countries. Pereira (2007), starting from a conceptual distinction between water conservation (referred to the measures for the conservation and preservation of water resource) and water saving (referred to the measures aimed at limiting and/or controlling water demand), points out a set of actions that can be adopted in agriculture to reduce the impacts of drought resulting economically, socially and environmentally more competitive than the "classical" proposal of realizing artificial reservoirs, the latter being an alternative preferred in even fewer cases in the countries where water resource planning is more advanced (e.g. Cowie et al., 2002). Finally, the European Commission in the Communication "Addressing the challenge of water scarcity and droughts in the European Union", adopted on July 18, 2007 (COM, 2007), while stating the necessity of progressing towards the full implementation of the Water Framework Directive 2000/60/EC (WFD), underlines the huge potential for water saving across Europe, where people continue to waste at least 20% of water due to inefficiency, indeed leakages greater than 50% have been recorded in the irrigation networks. A report connected to the EU Communication (Dworak et al., 2007) estimates a potential water saving in the EU of about 40%. Regarding the strategic paths for future interventions, the enhancement of drought risk management should be achieved also through: developing drought risk management plans; developing an observatory (an European Drought

Observatory is now available at http://edo.jrc.ec.europa.eu) and an early drought warning system; further optimizing the use of the EU Solidarity Fund and European Mechanism for Civil Protection; fostering water efficient technologies and practices; fostering the emergence of a water-saving culture in Europe.

In this framework, evapotranspiration assessment is of outstanding importance both for planning and monitoring purposes. Its magnitude (mainly referring to potential evapotranspiration) is comparable to the main forcing of the water balance, i.e. precipitation, and for this reason several climatic classifications are based on comparisons between these two quantities, with the aim of determining specific climate conditions for different areas (e.g. Rivas-Martinez, 1995). Furthermore, evapotranspiration is the only component of the water balance with a central role also in the energy and carbon balance, since it directly accounts for hydrological, agricultural and ecological effects of drought events. Specifically, in agriculture evapotranspiration can be closely related to water demand. This means that the role of evapotranspiration, and losses due to evapotranspiration in agriculture (which are foreseeable to a certain extent) can be handled in a way allowing to assure the best conditions for agricultural needs, if water resources management is correctly planned and implemented. Hence, in this chapter evapotranspiration assessment/water demand fulfillment will be considered within the wider framework of water resources management and planning, both for a correct evaluation of the water balance (considering both the hydrological balance and the differences between water requirements and availability), and for determining incoming drought events through appropriate indices (drought monitoring). The issue of reducing water requirements, meaning loss reductions and/or evapotranspiration reductions (mainly in agriculture) will only be touched on, while dealing with methods and tools for water resource management under shortage conditions.

In the next sections, after an analysis of the available water resource and water demand in a southern Italian region (Calabria), the chapter highlights some weaknesses of the regional water system in rainfall deficit conditions, drafting the main strategies of intervention to be adopted to face the different aspects of drought. Then, some guidelines for the proactive management of drought in agriculture are proposed and specifically, by means of a case-study related to one of the most important agricultural areas in southern Italy (the Sibari Plain), the development of the three most important operational management tools is shown, i.e. the Strategic Plan for long-term interventions, the Management Plan for short-term interventions and the Contingency Plan for emergency conditions. Drought indices are important tools for correctly drafting these plans: a specific section will provide some insight about them. Finally, some climatologic and hydrologic scenarios over a specific basin are hypothesized, with the aim of assessing water resource availability in the second half of the present century and of verifying whether the intense and prolonged drought periods currently affecting the Calabria region will become ordinary situations in the near future.

2. Natural water resource

Since no useful information is available for an estimate of the direct runoff volume on the whole region, natural water resource was determined using a distributed monthly water balance model described by Mendicino & Versace (2007) and Mendicino et al. (2008a), which extends the approach proposed by Thornthwaite & Mather (1955) and simulates soil moisture variations, evapotranspiration, and runoff on a 5 km regular grid (Fig. 1) using data sets that include climatic drivers, vegetation, and soil properties. This model does not consider the horizontal motion of water on the land surface, or in the soil (hence no flow routing algorithms are required), and it is based on a simplified mass balance:

$$\Delta W = P + SM - SA - ET - Q \tag{1}$$

where ΔW is the change in soil moisture storage, P the precipitation, SM the snow melt, SA the snow accumulation, ET the actual evapotranspiration, and Q is the runoff (all the quantities are evaluated in mm month^{-1}). In the model, potential evapotranspiration PET is estimated through the Priestley-Taylor method (Priestley & Taylor, 1972), requiring only temperature, air pressure and net radiation data, overcoming the lack of observed wind speed and air humidity data in the analyzed area before the year 2000. In the case of net radiation, monthly values were obtained starting from a modified version of the model originally suggested by Moore et al. (1993). Actual evapotranspiration ET is calculated starting from PET and considering the Accumulated Potential Water Loss (APWL), such as suggested by Thornthwaite & Mather (1955), which represents the total amount of unsatisfied potential evapotranspiration to which the soil has been subjected.

Because of the significant reforestation campaigns carried out in Calabria after the Second World War, whose results were evident already at the end of 1950s, the starting period for the analysis was assumed to be 1957. The assumption of constant soil use (derived by the Corine Land Cover 2000 project) is justified by the coarse resolution of the model (5 km grid). The model schematized in figure 1 was improved also considering: i) that a portion of the rainfall is directly transformed into "instantaneous" runoff (depending on the ratio between actual soil moisture and soil water holding capacity WHC, in its turn derived by combining soil use with a detailed soil texture map of Calabria); ii) an additional very simple snow module, which partitions snow and rain precipitation and regulates snow melt just referring to the current monthly temperature in the cell; iii) that the hydraulic subsoil characteristics are simulated with reservoirs whose rates of depletion vary with the predominating geo-lithological characteristics in the single cells of the model (Mendicino et

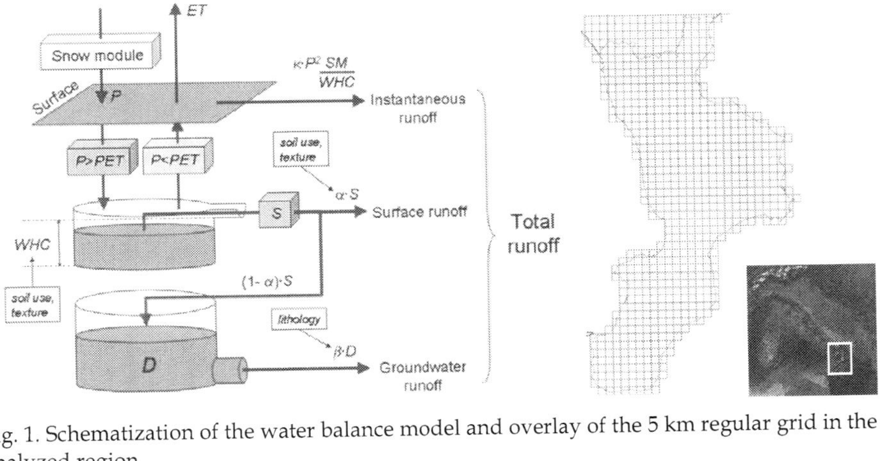

Fig. 1. Schematization of the water balance model and overlay of the 5 km regular grid in the analyzed region.

al., 2005). The different characteristics of subsoil leaded to the subdivision of the region into three categories: I) areas with a high capability of producing perennial flow (rocks with high permeability not in the plain); II) areas with mean capability of producing perennial flow (rocks with mean permeability); III) areas with low capability of producing perennial flow (rocks with low permeability or with high permeability in the plain).

The monthly water balance model was validated considering about 2900 monthly runoff values observed in 14 Calabrian catchments during the period 1955-2006 (Fig. 2). Figure 2 also shows the quite satisfactory performance of the model that, besides reproducing the monthly average behaviour of each considered catchment, provided values of the slopes of the regression curves obtained comparing observed and simulated runoff values varying from a minimum of 0.791 (Alli Orso) to a maximum of 1.135 (Esaro La Musica), while the correlation coefficients *r* varied from 0.447 (Coscile Camerata) to 0.939 (Corace Grascio).

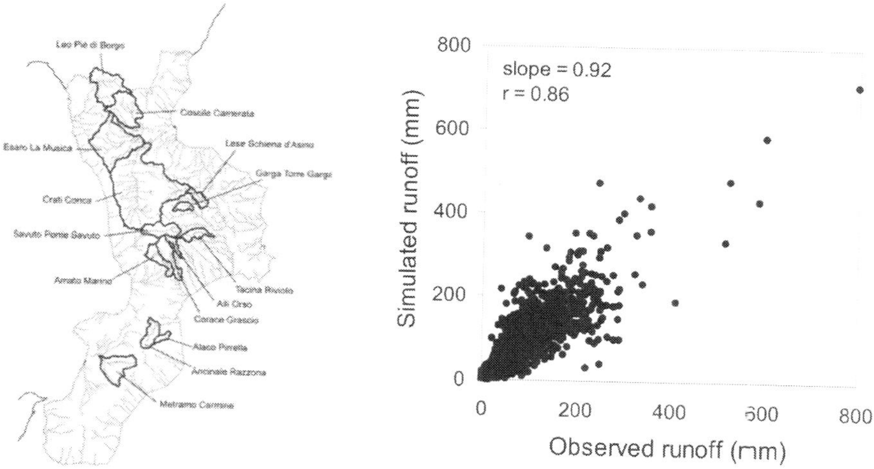

Fig. 2. Spatial distribution of the gauged catchments and comparison between all observed and simulated runoff during the period 1960-2006.

The monthly water balance model was applied on the whole territory of Calabria for the period 1960-2006 on a 5 km regular grid, where each cell was independent from the others, determining the main components of the hydrological balance in the whole region: precipitation, actual evapotranspiration, soil moisture storage, groundwater volume and instantaneous, surface and subsurface runoff. In several areas of the region a negative trend was observed for many of these variables. Specifically, while the potential evapotranspiration trend was strongly related to increasing temperature, actual evapotranspiration was affected also by changes (reduction) in precipitation. Considering the whole region, the average annual actual evapotranspiration estimated in the analyzed period is 581 mm, equal to about 57.8% of the average cumulated annual rainfall (potential evapotranspiration is about 110%). Figure 3 (left side) shows the average monthly values in the whole region for actual and potential evapotranspiration. The months where a significant difference can be observed are the months from May to September. In these months (the less rainy and warmest ones), evaporation of soil moisture accumulated in wintertime exceeds rainfall, requiring irrigation in most of the agricultural areas. Figure 3 (right side) also shows the trend of cumulated annual actual evapotranspiration. The decrease in time of this quantity due to rainfall reduction is partly balanced by the increasing temperatures, hence the negative trend is not significant. It is noteworthy that peaks and troughs are generally dependent on rainy (e.g. 2005) or not rainy (e.g. 2001) years, even though rainfall distribution during the single year also affects the evapotranspirative phenomenon. The correlation coefficient between cumulated annual actual evapotranspiration and precipitation in the period 1960-2006 was 0.638.

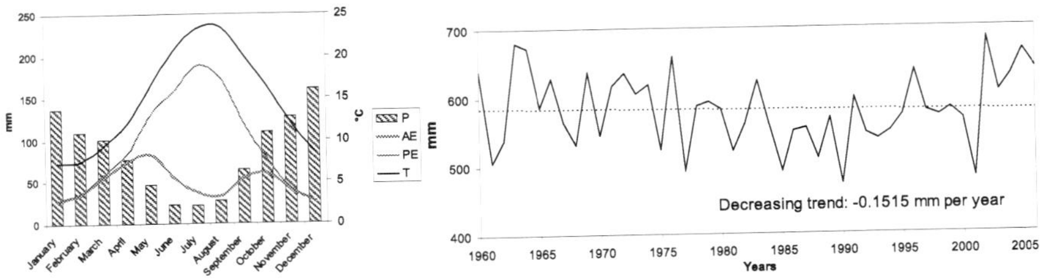

Fig. 3. Left: average monthly values in Calabria of actual (AE) and potential evapotranspiration (PE), precipitation (P) and temperature (T) during the period 1960-2006. Right: trend of cumulated annual actual evapotranspiration.

3. Water demand and availability

The water balance between available water resource and water demand is the starting point for a correct water management. One of the main problems occurring in this phase is the general lack of observed data, obliging to synthetic estimates of water availability and several levels of approximation in the assessment of water needs, mainly for irrigation and for determining the management rules of the reservoirs.

In this context, the water balance on the Calabrian region was carried out considering also withdrawals from springs, streams, reservoirs and wells for irrigation and for potable uses, adopting two sequential simulation models. The former is a modified version of the distributed hydrological model, where the natural water balance is integrated with the withdrawal for irrigation and potable uses, producing (output variable) a residual availability. This water availability is used in a second GIS-based model considering the effects of diversions and reservoirs.

In the first model, inside a single 5 km squared cell can co-exist both wells and springs used to feed small irrigation systems or few users, located in the same cell, and wells and springs used for water mains collecting water outside the cell. If both the points where the water is withdrawn and used are inside the same cell (this happens only for wells for irrigation purposes), the schematization shown in figure 4a is adopted, hypothesizing that inside the cell a known volume is transferred monthly from the subsoil reservoir to the surface as an "added" precipitation (owing to the irrigation). This volume has to be summed to the meteorological precipitation and is subjected to the cycle simulated by the water balance, increasing the soil moisture and actual evapotranspiration and eventually feeding the aquifer from which it has been withdrawn. Instead, if the cell where the water is withdrawn does not coincide with the cell where it is used (that is only the case of regional water mains) then the schematization shown in figure 4b is adopted. The source cell is subjected to a reduction of the volume of the subsoil reservoir, while the water is hypothesized to reach directly the water stream in the destination cell, feeding the surface runoff with a restitution coefficient equal to 0.7.

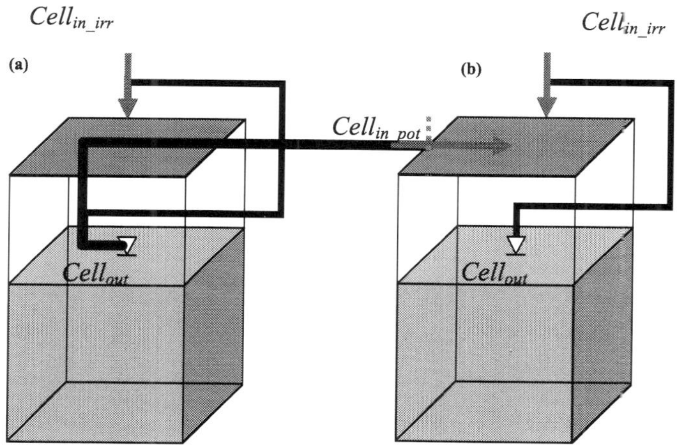

Fig. 4. Schematization of the modified water balance model considering withdrawals for irrigation and potable uses.

Summarizing, the proposed model allows that every month for each cell a volume $cell_{out}$ can be extracted from the subsoil reservoir, which is equal to the withdrawals for irrigation and potable purposes, that a volume $cell_{in\_irr}$ can be added like a supplementary precipitation representing the water derived from the same cell and used for irrigation, and finally, that a volume $cell_{in\_pot}$ can be added like a supplementary surface runoff accounting for the water come in the cell to satisfy the potable uses. All the data related to potable and irrigation withdrawals were derived from several official sources, even if sometimes incomplete, and were aggregated at the resolution of the water balance model. Figure 5 shows the distribution of the regional water mains and of the local water distribution systems.

The modified natural water balance is the input of the commercial GIS-based model Mike Basin (DHI Software), accounting for the effects of diversions and reservoirs aimed at satisfying irrigation, hydro-power, civil and industrial requirements (Fig. 6). The lack of actual information about the management rules of reservoirs led to hypothesize several working schemes for the definition of the optimal water balance. Finally, for all the analyzed reservoirs the minimum flow requirements were considered following two different approaches: the former proposed by the Regional Basin Authority (very conservative, especially for some typical Calabrian rivers, called *fiumare*, characterized by no flow conditions for a relevant part of the year) and; the latter based on the $Q_{7,10}$ flow, i.e. the lowest 7 consecutive-day average flow characterized by a 10 years time period.

In the case of the irrigation demand (i.e. water requirements for balancing evapotranspiration losses), a detailed analysis was carried out on each irrigation district

(Fig. 7) during the irrigation season April - September. Specifically, the assessment of the effective water consumption was determined by considering different seasonal (spring, summer, autumn) soil use spatial distributions (e.g. in Table 1). For each soil use the seasonal irrigation demand (m^3/ha) of the crops (Table 2) was achieved. The same was split monthly taking into account that the highest request is obtained during the trimester June - August (Table 3). An adequately detailed knowledge of the irrigation network allowed the correct estimate of the possible uptake of volumes to/from other cells. It is noteworthy to highlight that all the information related to soil use and water requirements were aggregated at the resolution of the model, i.e. 5×5 km^2, for the whole region.
In the proposed analysis the quite small volumes related to industrial areas were neglected.

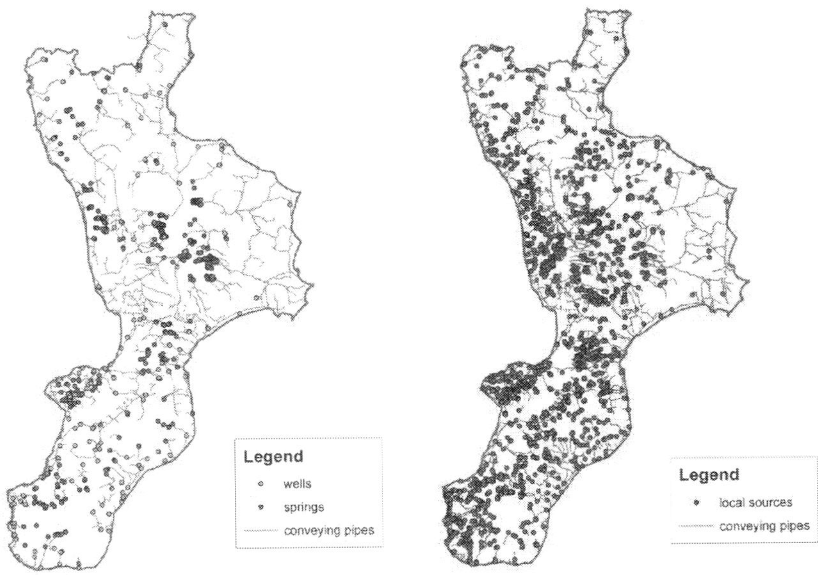

Fig. 5. Left: regional water mains (479 springs, 281 wells and about 2000 conveying pipes). Right: local water distribution systems (over 1200 springs and wells).

Water balance results showed that, for average conditions, the residual annual water availability is great, even if some weaknesses arise. Among these, the strong differences in the seasonal precipitation, which is mainly concentrated in the wet winter period (80-90%), require an accurate management of the volumes stored in natural and artificial reservoirs for facing the hot and dry Mediterranean summer. Furthermore, the decrepitude of several conveying pipes has to be considered with remarkable water losses, and the negative precipitation trend due to climate change that seems to be relevant in Calabria (a preliminary analysis about future climate scenarios in Calabria is shown in the 6th section).
The weaknesses pointed out in normal conditions suggested water resources availability should be analyzed when drought conditions occur. Specifically, through the use of the Standardized Precipitation Index (SPI, McKee et al., 1993) intensity and duration of droughts were determined on the whole Calabrian region.

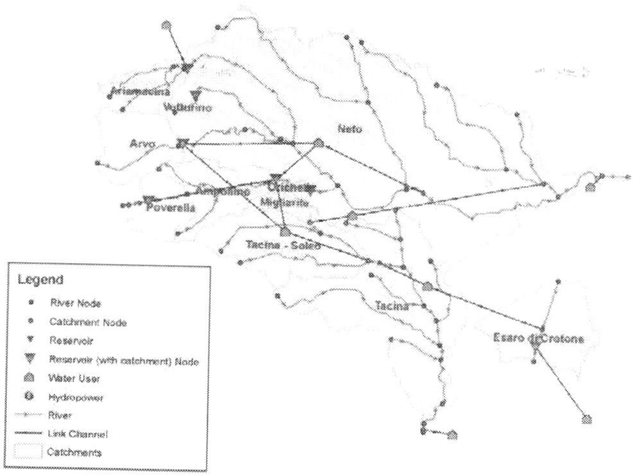

Fig. 6. Example of water system schematization realized within the GIS-based model Mike Basin.

Crops		Spring	Summer	Autumn
Code	Description	Soil use (ha)		
2121	Spring-summer herbaceous crops		14.115	
2122	Summer-autumn/spring horticultural crops			14.115
2123	Spring-summer horticultural crops			
2211	Irrigated vineyards	4.536	4.536	4.536
2221	Irrigated orchards	328.791	328.791	328.791
2231	Irrigated olive groves	95.788	95.788	95.788

Table 1. Seasonal soil use for a generic irrigation district.

For each of the most significant Calabrian basins, and for each month of the period 1960-2006, a mean SPI areal value was calculated for different time scales (1-, 3-, 6-, 12-, 24- and 48-months), with the aim of highlighting the longest and most intense drought periods (Fig. 8). Drought indices are essential at all levels of the planning process. The reader is referred to section 5 for a brief review of the most diffused ones.

Usually, the beginning of a drought period can be defined when SPI values are lower than -1.0, and its end when the values come back positive. Nevertheless, based on a historical analysis of the official declarations of "natural disaster" in Calabria due to drought, even a 12-month SPI value equal to -0.7 was observed to be adequate as a drought threshold. Hence, when a generic month presented a 12-month SPI value lower than -0.7, it was considered a drought month, and the correspondent total runoff simulated with the water balance model was taken into account. The aggregation, from January to December, of the average runoff estimated during the drought months leaded to the definition of a so-called "scarce year" whose runoff values, even if statistically less probable than the ones of the single months, pointed out the possibility of extremely critical situations in Calabria, with a reduction of total runoff up to 43%. This analysis introduces issues related both to the management of water shortage and to the mitigation of drought through the use of restrictive measures. The development and implementation of strategic and emergency plans are primary tools to face the different aspects of drought phenomenon, as it is shown in the next paragraph.

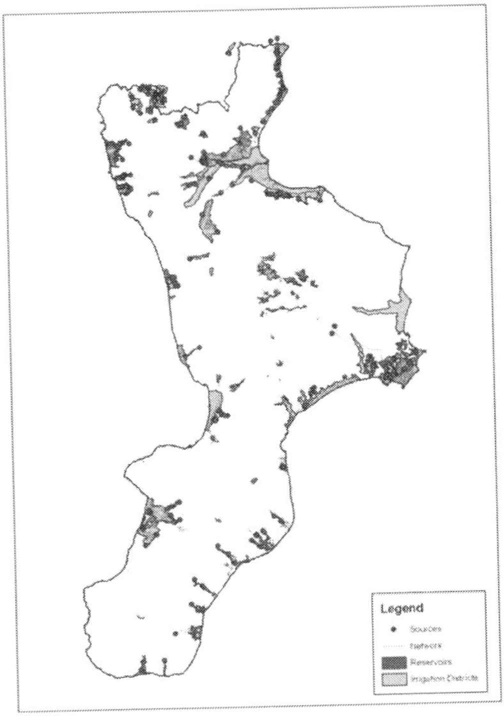

Fig. 7. Calabrian irrigation districts and network systems.

Code	Description	Irrigation demand (m^3/ha)
2121	Spring-summer herbaceous crops	7000
2122	Summer-autumn/spring horticultural crops	7600
2123	Spring-summer horticultural crops	5000
2125	Greenhouse crops	9000
213	Rice fields	15000
2211	Irrigated vineyards	3500
2221	Irrigated orchards	5000
2231	Irrigated olive groves	3000

Table 2. Seasonal irrigation demand (m^3/ha) of the crops.

Code	Description	A	M	J	J	A	S	TOT
2121	Spring-summer herbaceous crops	0	858	1497	2337	1445	863	7000
2122	Summer-autumn/spring horticultural crops	163	745	1719	2450	1663	861	7600
2123	Spring-summer horticultural crops	164	751	1733	2352	0	0	5000
2125	Greenhouse crops	193	882	2035	2901	1969	1020	9000
213	Rice fields	3780	3240	3240	3240	1500	0	15000
2211	Irrigated vineyards	0	0	1063	1411	1026	0	3500
2221	Irrigated orchards	0	0	1349	2125	1168	458	5000
2231	Irrigated olive groves	0	0	894	1249	857	0	3000

Table 3. Monthly irrigation demand (m^3/ha) of the crops.

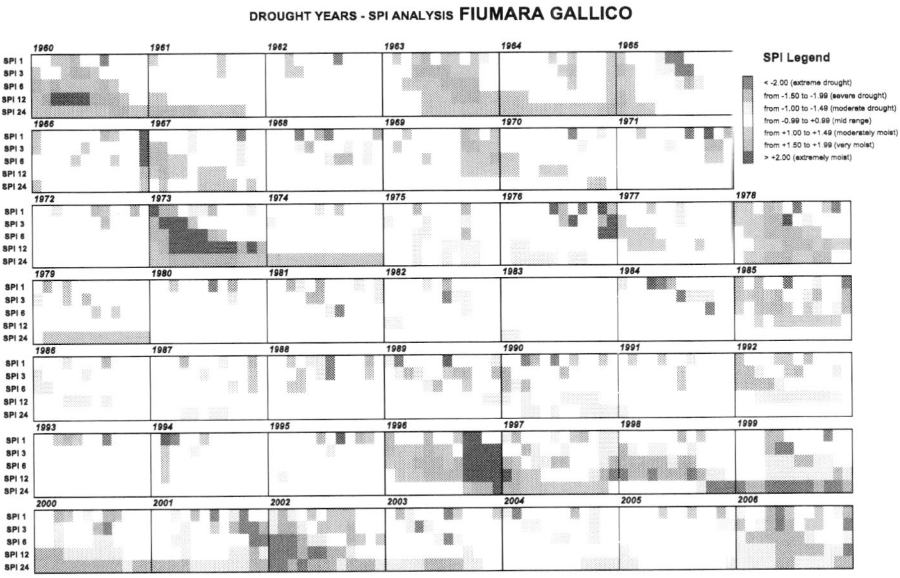

Fig. 8. Temporal evolution of SPI values in a generic Calabrian river basin. Red squares correspond to drier periods.

4. Water resource management under shortage conditions

In its 2007 Communication (COM, 2007) the European Commission stated that the challenge of water scarcity and droughts needs to be addressed both as an essential environmental issue and as a precondition for sustainable economic growth in Europe, and highlighted the necessity of progressing towards full implementation of the EU Water Framework Directive (WFD) 2000/60. The WFD is the EU's flagship Directive on water policy, explicitly defining

long-term planning as the main tool for ensuring good status of water resources. Nevertheless, it does not indicate criteria and actions to face risk of drought, delegating National Legislations to concretely realize its framework (after a series of yearly follow-up reports, a policy review is foreseen for 2012 at the EU level).
In Italy the EU WFD was taken into account with the Legislative Decree 152/2006 on environmental protection. Though this act is quite recent, it seems to be far from being adequate to actually cope with drought, mainly because it does not stress the necessity of passing from a reactive to a proactive approach, based on preparedness and mitigation actions planned in advance with the contribution of all the involved stakeholders, ready to be implemented when drought phenomena occur.
Within a comprehensive drought management planning process, Rossi et al. (2007) proposed the identification of three main tools: Strategic Water Shortage Preparedness Plan, Water Supply System Management Plan and Drought Contingency Plan. Following, an example of application of the proposed guidelines is shown for the planning of the best mix of measures needed for coping with drought phenomena on one of the most important agricultural areas in southern Italy, the Low Esaro and Sibari Plain (Mendicino et al., 2008b). It is noteworthy that in the proposed example (water shortage planning in the agricultural sector) water demand is strictly correlated to the amount of water needed from crops for facing lack of precipitation and high potential evapotranspiration during summer (see Table 3). Hence, in this case the planning process is triggered by the need of coping with the high water loss due to evapotranspiration in a particularly dry period of the year. As it is explained in the next sections, this objective can be reached by means of demand reduction, water supply increase or impacts minimization measures, and considering long-, medium- and short-term actions.

4.1 Methods and tools

The Agricultural Strategic Water Shortage Preparedness Plan (ASP) is aimed at obtaining the reduction of drought vulnerability in the analyzed area through the implementation in normal conditions of long term mitigation measures, consisting in a series of structural and non-structural actions applied in the water supply system. Usually, structural measures are economically expensive and require the use of many human resources. However, their effects are easier to be foreseen than the effects produced by the non-structural mitigation actions, in their turn usually more accepted by all the stakeholders. The long term mitigation measures are specifically indicated in the systems characterized by a low level of reliability and are oriented at improving the water balance in the analyzed system. These actions not only enhance the reliability of the system through fulfilling water requirements, but also reduce its vulnerability with respect to future drought events, fulfilling three main objectives: i) water demand reduction; ii) water supply increase and improvement of the efficiency of the system; iii) minimization of the impacts. Within the actions reducing water demand, some are directly aimed at reducing evapotranspiration by adopting appropriate agronomic techniques, such as e.g. irrigating during non windy periods for minimizing wind drift losses, or early defoliation to reduce crop transpiration surface (for a deeper description, the reader is referred to Pereira, 2007). In table 4 the long term measures that can be potentially adopted in agriculture are listed, subdivided considering their main objectives.

Category	Long-term actions
Demand reduction	Economic incentives for water saving and sanctions for wastes
	Agronomic techniques and irrigation systems for reducing water consumption (e.g. Pereira, 2007)
	Dry crops in place of irrigated crops
Water supply increase and improvement of the efficiency of the system	Conveyance networks for bi-directional exchanges
	Reuse of treated wastewater
	Inter-basin and within-basin water transfers
	Construction of new reservoirs or increase of storage volume of existing reservoirs
	Use of aquifers as groundwater reservoirs
	Non conventional sources (particularly desalination of brackish or saline waters)
	Control of seepage and evaporation losses
	Elimination of the possible risks of pollution of the sources
	Modernization and restructuring of the irrigation network
Impacts minimization	Reallocation of water resources based on water quality requirements
	Development of early warning systems
	Implementation of Agricultural Management Plans and Contingency Plans
	Insurance programs
	Education activities for improving drought preparedness and/or permanent water saving

Table 4. Main long term drought mitigation measures in agriculture (adapted from Rossi et al., 2007, and Georgia Dept. Of Natural Resources, 2003).

Since the ASP has to be drawn up choosing among several combinations of long-term mitigation measures, a suitable evaluation procedure has to be adopted. A multi-criteria technique could provide an as objective as possible comparison among different alternatives, according to a series of economic, environmental and social criteria, and taking into account the point of view of all the stakeholders. The tool adopted in this study for multi-criteria analysis is the software NAIADE (Munda, 1995).

The ASP should be prepared by the Basin or Hydrographic District Authorities, which are the bodies responsible for planning, and corresponds to the Drought Management Plan included into the River Basin Management Plan provided in the WFD.

Once the long-term mitigation measures are defined, an Agricultural Water Supply System Management Plan (AMP) has to be developed with the aim of: defining the best mix of long and short-term measures to avoid the beginning of a real water emergency; estimating the costs and the financing sources for the chosen mitigation measures, and; fostering the stakeholder participation and exchanges. It is prepared by the authority responsible for agricultural water management (i.e. the Land Reclamation Consortium), and the operative measures defined have to be adopted according to the values of early warning indicators, showing Normal, Pre-Alert or Alert conditions. The threshold values of the indicators can be chosen through an objective function or, if several aspects have to be accounted for, through a multi-criteria analysis.

In table 5 the short term measures that can be potentially adopted in agriculture are shown, subdivided on the basis of their principal objectives. With respect to the long-term mitigation measures, in this case the actions in the "demand reduction" category implicitly accept a certain percentage of water stress for the crops, because they are only aimed to reduce water consumption, without taking into account crop conditions. On the contrary, former long-term mitigation measures suggested some structural actions (i.e. actions to be adopted always) aimed at limiting some additional evapotranspiration due, e.g., to not correct irrigation practices, and that could be avoided without consequences for the crop. In brief, adopting the AMP evapotranspiration losses could be not completely compensated, and the farmers should be supported in assessing how to minimize water stress effects adopting even more specific agronomic techniques.

Category	Short-term actions
Demand reduction	Public information campaign for water saving
	Restriction of irrigation of annual crops
	Pricing (discourage excessive water use)
	Mandatory rationing
Water supply increase	Improvement of existing water systems efficiency (leak detection programs, new operating rules, etc.)
	Use of emergency sources (additional sources of low quality and/or high exploitation cost)
	Over exploitation of aquifers (use of strategic reserves)
	Increased diversion by relaxing ecological or recreational use constraints
Impacts minimization	Temporary reallocation of water resources
	Public aids to compensate income losses
	Tax reduction or delay of payment deadline
	Public aids for crops insurance

Table 5. Main short term mitigation measures in agriculture (adapted from Rossi et al., 2007).

If a particularly severe drought occurs, and the indicators signal Alarm conditions, the Agricultural Drought Contingency Plan (ACP) has to be adopted, defining the most appropriate short-term measures to reduce the impact of emergency situations. In this case the efforts are turned to protect the essential activities of the agricultural system, and the threshold values of the indicators have to be chosen taking into account this objective, preferably using a probabilistic approach, that allows the decision-makers to evaluate the effective risk of having water deficit for different scenarios. The ACP should be prepared by the Basin or Hydrographic District Authorities, with the collaboration of the Civil Protection.

Such as in the AMP, also in the ACP the assessment of crop losses can be made through production functions. In the case of extreme and particularly prolonged drought also the damage to perennial crops, the excessive decrease of the water tables of the aquifers, sea water intrusion, ecological damages to aquatic flora and fauna have to be considered. Some of this damage can be irreversible and can also influence crop production in the following years.

4.2 Case study

The core of the analyzed water supply system is the Farneto Dam (Fig. 9), closing the Esaro Catchment (about 245.4 km^2) in southern Italy. The dam is aimed at: (i) containing the ordinary floods and mitigating the extraordinary ones, according to the condition that the reservoir level is maintained almost empty from October to March; (ii) supplying water (about 30 hm^3 from April to September) to the downstream agricultural area (about 85 km^2), sited in the Low Esaro and Sibari Plain. At present about 63% of the irrigable area is based on open channel irrigation systems.

Fig. 9. Study area for the development of the planning process.

4.3 Applying the Agricultural Strategic Plan

Table 6 shows 13 selected alternatives (from A to M), obtained combining the following six long-term mitigation measures: 0) System in current configuration; 1) Modernization of the irrigation network for reducing water losses and evaporation (it has been calculated that the efficiency of the actual scenario is equal to 67%, while the efficiency of the "modernized" scenario will be 80%; Mendicino et al., 2008b); 2) Construction of farm ponds; 3) Construction of a new upstream dam; 4) Economic incentives and educational activities for water saving; 5) Allowing the dam to store a little volume during the winter (i.e. dam not empty in March).

	Alternatives												
Measure	A	B	C	D	E	F	G	H	I	J	K	L	M
0	X												
1		X				X	X			X			X
2			X								X		
3								X				X	X
4				X		X			X	X	X	X	X
5					X		X	X	X	X	X	X	X

Table 6. Long-term mitigation measures and alternatives.

The alternatives were compared within the DSS tool NAIADE according to 4 economic criteria (construction costs of infrastructures, operation and maintenance costs, crop yield losses and amount of public aids needed), 2 environmental criteria (failures to meet ecological requirements and reversibility of the alternatives) and 4 social criteria (system vulnerability, temporal reliability, realization time of the infrastructures and employment increase). Since the observed period is short in order to evaluate the criteria and is characterized by few drought events, two monthly synthetic temperature and precipitation series of 1000 years were generated as input of the water balance model providing the corresponding runoff values.

Within the analysis carried out with NAIADE the final ranking of the alternatives comes from the intersection of two separate rankings. The former Φ^+ is based on the "better" and "much better" preference relations, hence it points out how an alternative is "better" than the others. The latter Φ^- is based on the "worse" and "much worse" preference relations, and indicates how an alternative is "worse" than the others.

The two rankings are different, since one alternative could result slightly better than the others with respect to few criteria and at the same time could result worse with respect to many criteria, or vice versa. In figure 10 the partial rankings and the final ranking are shown. The most efficient alternative is the "J", where measures 1, 4 and 5 are considered together. The alternative "M", mainly characterized by the construction of a new upstream dam, is the best only in the Φ^+ ranking. A sensitivity analysis, carried out to assess the robustness of the achieved solution, showed a substantial stability of the ranking, constantly confirming alternative J as the optimal one. It is pointed out that alternative J is made up also by measure 1), allowing a reduction of evaporation losses.

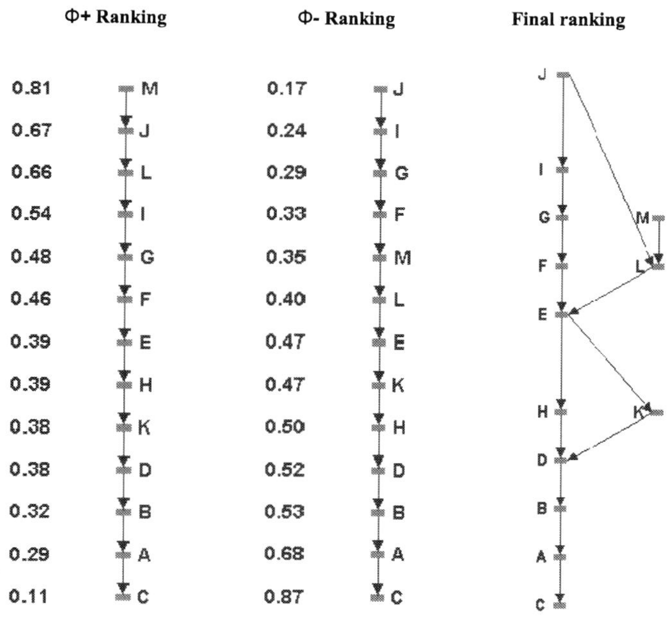

Fig. 10. Partial and final ranking of the drought mitigation alternatives in the Esaro River Basin.

4.4 Applying the Agricultural Management Plan

The AMP is aimed at defining the indicators and the triggers for establishing the Normal, Pre-Alert and Alert conditions for the agricultural areas of the system. It has to take into account the guidelines provided by the ASP. In fact, it has to select the best combination among the optimal long-term mitigation measure previously determined (J) and the several short-term measures that can be adopted to manage water deficits on the analyzed area. Whereas the long-term measure J is adopted continuously, the short-term measures vary following the status of the system. Specifically, for this case study:

- in Normal condition no short-term actions are taken;
- when Pre-Alert condition occurs, then exploitation of the groundwater resources in the irrigated area till 1/3 of maximum estimated volume is considered;
- when Alert condition occurs, then exploitation of the groundwater resources like in the Pre-Alert condition, the reduction of the release for minimum instream flow till 50% and the reduction of the release for irrigation (till 80% of the requirements) are taken into account. When alert condition occurs, the farmer is aware that the evapotranspiration losses cannot be completely compensated.

With the aim of determining the threshold values of the indices indicating the passage from one status to another, for every month from April to September a multicriteria analysis of the effects through NAIADE was carried out. The conflicting objectives to minimize are:

- the vulnerability of the system (including the assessment of crop losses due to reduced irrigation, made through specific production functions);
- groundwater withdrawals;
- the failures to meet the minimum instream flow.

For each month, starting from April, an impact matrix was achieved where, on the basis of the criteria selected for the fulfillment of the objectives, the optimal combination of the thresholds triggering the Pre-Alert and Alert status was selected (Fig. 11). The selected index for the definition of the drought thresholds is the volume stored in the dam from May to

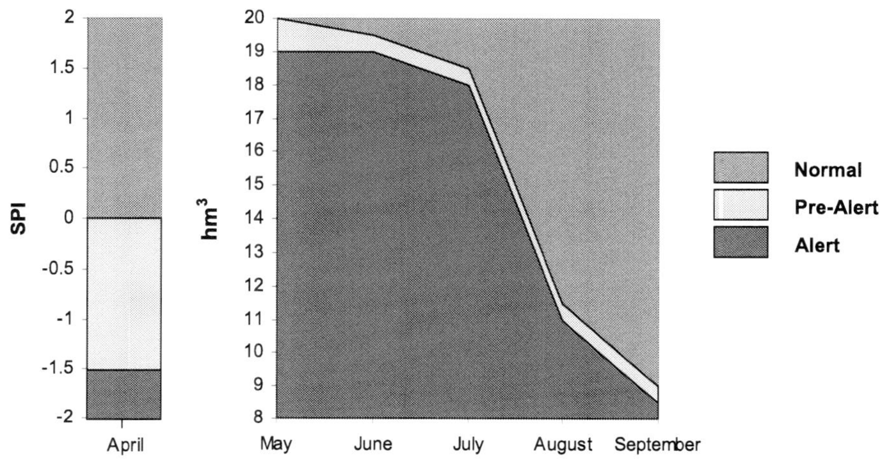

Fig. 11. Pre-Alert and Alert thresholds defined in the AMP.

September, while for the month of April a meteorological index was chosen, since owing to the rules adopted for dam management, at the end of March the dam level is not a significant index. For the month of April an analysis was carried out relating the yearly irrigation deficit to the 6 month-SPI calculated in March (considering in this way the first six months of the hydrologic year, from October to March). In the selection of the threshold values a rule was followed considering that, if the multicriteria analysis provides more optimal solutions, the one with the lowest irrigation deficit is selected.

4.5 Applying the Agricultural Contingency Plan

The first objective of the ACP is the definition of indices and their thresholds for univocally establishing the beginning of an emergency situation. Since the hydrologic analysis in April shows that the water demand is always less than the water availability in the Farneto del Principe Dam, and that every year the volume stored increases during this month, the thresholds are selected starting from May, choosing as an index, such as in the AMP, the volume stored in the dam. Furthermore, since using the 1000-year series of generated meteorological data the application of the two previous Plans determined a very high temporal reliability of the system (98.7%), it is not useful to evaluate the emergency thresholds considering the few residual years. Hence, the adopted approach was based on a probabilistic analysis of the system failures and deficit percentage of the demand.

Specifically, hypothesizing that all the short-term measures were already adopted, the 1000-year series of generated meteorological data, for every month and for different fixed initial volumes stored, were used to assess the probability of having failures in fulfilling demand either in the same month or in the subsequent irrigation period, and the deficit percentage with respect to demand. The results, allowing the decision-makers to evaluate the effective risk of having water deficit for a specific storage in a specific month, are shown (from May to August) in figure 12.

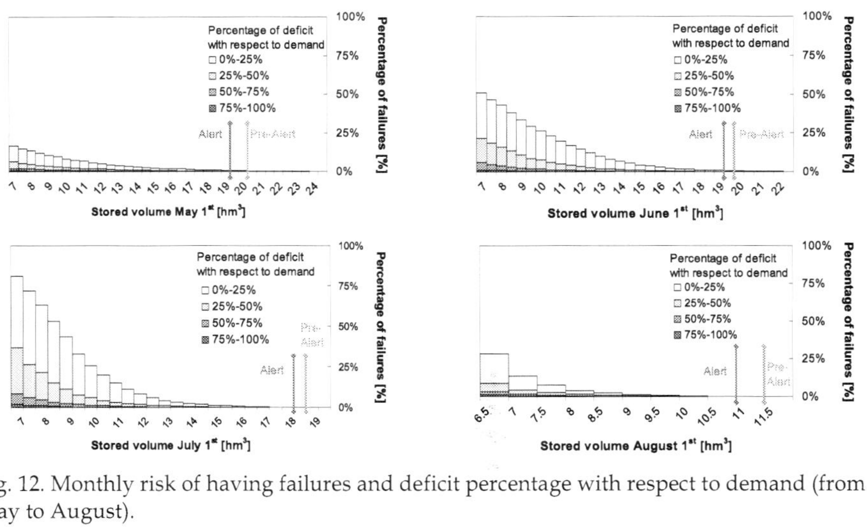

Fig. 12. Monthly risk of having failures and deficit percentage with respect to demand (from May to August).

5. Drought indices

Drought indices are tools necessary at all levels of the planning process: as it was shown in the previous sections, in the Strategic Plan they are used to identify the zones most exposed to drought risk in the analyzed areas, whereas in the Management Plan and in the Contingency Plan they are used to define trigger values for the activation of the measures for impact prevention or mitigation.

Most of the proposed methodologies for the characterization and the monitoring of drought phenomena are based on drought indices with the capability of synthetically summarizing drought conditions in a specific moment for a particular area. Nevertheless, drought is difficult to represent through a single index, hence frequently more indices or aggregate indices are used.

In rainfed agriculture meteorological indices are particularly suitable, because they give the opportunity of establishing a direct spatial correlation between the drought event and the agricultural production, allowing drought risk maps to be drawn.

Many authors provide lists describing the characteristics of the main drought indices (e.g. Ntale & Gan, 2003; Tsakiris et al., 2007a). Among them, the most widely used are the Palmer Drought Severity Index (PDSI; Palmer, 1965), the most "classical" drought index formulated to evaluate prolonged periods of both abnormally wet and abnormally dry weather conditions, and the Standardized Precipitation Index (SPI; McKee et al., 1993), a meteorological drought index based on the precipitation amount in a period of n months. Since SPI just needs precipitation data to be calculated, it has found widespread application. Guttman (1998) shows that the PDSI has a complex structure with an exceptionally long memory, while the SPI is an easily interpreted, simple moving average process. Hayes et al. (1999) describe the three main advantages in using SPI: the first and primary is its simplicity, the second is its variable time scale, and the third is its standardization. Nevertheless, the SPI is a meteorological index unable to take into account the effects of aquifers, soil, land use characteristics, crop growth and temperature anomalies, which influence agricultural and hydrological droughts.

Besides SPI, in the process of drought identification the MEDROPLAN Guidelines (Tsakiris et al., 2007a) suggest using also: the Reconnaissance Drought Index (RDI, Tsakiris et al., 2007b), also accounting for temperature anomalies (therefore for an eventual excessive evapotranspiration); deciles (Gibbs & Maher, 1967), used by the Australian Drought Watch System, which compare monthly observed precipitation values with the quantiles corresponding to the not exceeded frequencies of 10%, 20%,... 100% achieved from a long enough monthly precipitation series; the Surface Water Supply Index (SWSI, Shafer & Dezman, 1982), aggregating information about precipitation, runoff, volumes stored in the reservoirs and snowpack, and expressing drought conditions in a standardized way. Furthermore, owing to their diffusion, other two indices are recalled: the run method (Yevjevich, 1967), based on the comparison between the time series of the analyzed hydrological index and a representative threshold of "normal" conditions, and the Palmer Hydrological Drought Index (Karl, 1986), a modified version of the PDSI for real-time monitoring.

An interesting way to account for soil and land use effects (in some respects, the way followed by Palmer to calculate PDSI) is to derive the drought indices starting from hydrological modeling. These indices can be called "comprehensive" drought indices, because they allow a more comprehensive picture of the water cycle and its elements

(Niemeyer, 2008). A typical example of comprehensive drought index is the Groundwater Resource Index (GRI) derived by Mendicino et al. (2008a) using the monthly water balance model shown in figure 1. For each single element where the model was applied (5 km regular cell), the monthly values of groundwater detention (i.e. the storage D) were standardized (for almost all the cells and months the skewness test of normality showed that the series were normally distributed) through the following equation:

$$GRI_{y,m} = \frac{D_{y,m} - \mu_{y,m}}{\sigma_{y,m}} \tag{2}$$

where $GRI_{y,m}$ and $D_{y,m}$ are respectively the values of the index and of the groundwater detention for the year y and the month m, while $\mu_{D,m}$ and $\sigma_{D,m}$ are respectively the mean and the standard deviation of groundwater detention values D simulated for the month m in a defined number of years (at least 30). This simple index, but based on several pieces of information provided by the water balance model, allows assessment of the deviation from the mean values of the available groundwater in a spatially-distributed way for the whole territory where the model is applied. Figure 13 shows the maps of the GRI distribution in northern Calabria for the months of April from 1979 to 2006. Examining the maps immediately the years with lower GRI values (the driest years, with brighter colors) are recognizable, as are the wettest years (darkest colors).

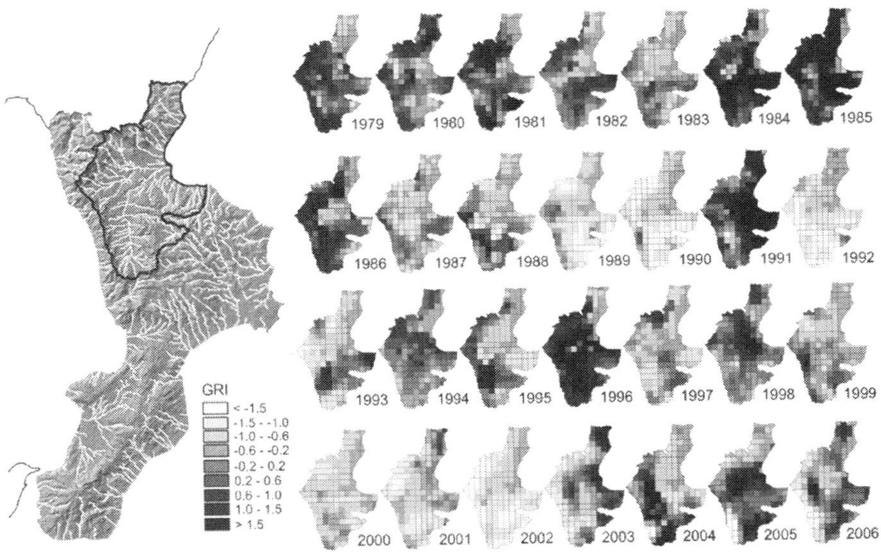

Fig. 13. Boundaries of the selected study area in Calabria and GRI distribution in north-eastern Calabria for the months of April from 1979 to 2006 (from Mendicino et al., 2008a).

Other comprehensive indices were developed by Narasimhan & Srinivasan (2005), who using the Soil and Water Assessment Tool (SWAT) model, derived two drought indices for agricultural drought monitoring, the Soil Moisture Deficit Index (SMDI) and the Evapotranspiration Deficit Index (ETDI), based respectively on weekly soil moisture and evapotranspiration (ET) deficit. Also Matera et al. (2007) derived a new agricultural drought index, called DTx, based on the daily transpiration deficit calculated by a water balance model.

In the few last years the possibility of using long data series coming from remote sensing has opened new and promising perspectives to satellite-derived drought indices, which have the advantage of being intrinsically spatially distributed. Anderson et al. (2007) provide a brief presentation of TIR-based drought indices, while a list of many NOAA-AVHRR images-derived drought indices is presented by Bayarjargal et al. (2006). Zhang et al. (2005) exploit the capabilities of the MODerate resolution Imaging Spectroradiometer (MODIS) for monitoring and forecasting crop production using a satellite-based Climate-Variability Impact Index.

Several remote sensing-derived drought indices depend on the ratio ET/PET, where ET is actual evapotranspiration and PET potential evapotranspiration (e.g. Crop Water Stress Index (CWSI), Jackson et al., 1981; Drought Severity Index (DSI), Su et al., 2003; Evaporative Drought Index (EDI), Anderson et al., 2007; Yao et al., 2010). While PET is generally calculated by means of ground based measurements, ET is easily estimated through "residual" methods (e.g. SEBAL, Bastiaanssen et al., 1998; and Bastiaanssen, 2000; SEBI, Menenti & Choudhury, 1993; S-SEBI, Roerink et al., 2000; SEBS, Su, 2002; TSEB, Norman et al., 1995; DisAlexi, Anderson et al., 1997; METRIC, Allen et al., 2007), where the evapotranspirative term is the residual term of the energy balance equation:

$$\lambda E = R_n - G - H \tag{3}$$

with R_n net radiation, G soil heat flux, H sensible heat flux and λE latent heat flux, from which ET is derived.

Even though at this stage very seldom they are used as operational tools, remote sensing-derived indices are potentially very useful because they intrinsically provide space-time variation of drought phenomena, and the ratio ET/PET can be reasonably related to soil water content. For instance, the relative evaporation Λ_r can be directly linked to the soil degree of saturation θ / θ_s (Su et al., 2003). As an example, figure 14 shows the space-time evolution of the DSI, derived from SEBS and MODIS images, during summer 2006 in Northern Calabria. DSI is equal to 1 - λE / λE_{wet} (where λE_{wet} is the latent heat flux estimated for the so-called "wet" pixel), hence higher DSI values indicate low actual evapotranspiration. A graph shown at the top of the figure provides information about precipitation in a micrometeorological station placed almost in the middle of the area (these data are only roughly representative, owing to the extension of the whole area). Figure 14 shows that the maps with the highest DSI values (e.g. July 20, but also September 4 and October 31), indicating drought stress conditions, are related to some of the most distant days from antecedent significant precipitation events.

To complete this brief review, a much-discussed issue is mentioned, i.e. the possibility of using the drought indices (especially SPI) to forecast stochastically the possible evolution of an ongoing drought (Cancelliere et al., 1996; Lohani et al., 1998; Bordi et al., 2005; Cancelliere et al., 2007). Several studies are also aimed at explaining and predicting possible drought conditions through the analysis of sea surface temperature (SST) and atmospheric circulation patterns (e.g. Wilby et al., 2004; Kim et al., 2006; Cook et al., 2007).

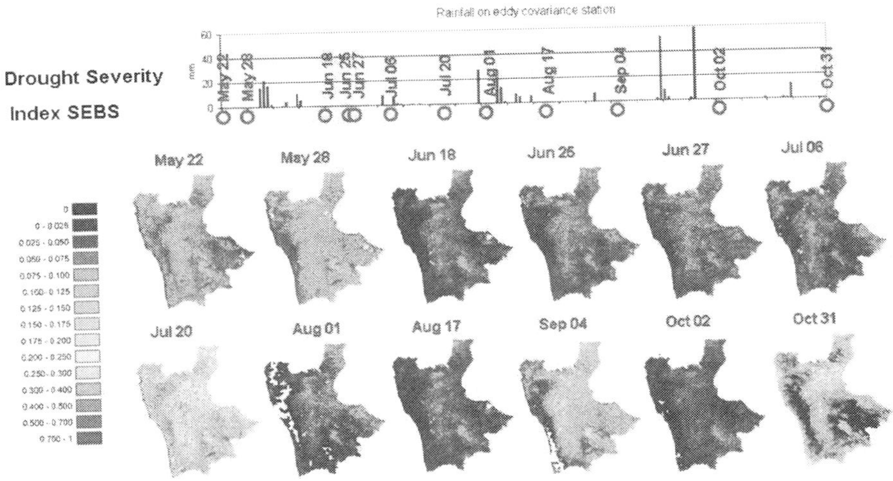

Fig. 14. Evolution of the DSI derived from SEBS in northern Calabria, from May 22 to October 31 2006. Top graph shows precipitation events on the representative micrometeorological station, placed approximately in the middle of the analyzed area.

However, when dealing with complex systems, where irrigated agriculture assumes a greater importance, one single index is often not able to capture the different features of drought and to take in account the effects of human activities (use of irrigation, water from reservoirs, wells, etc.) on the hydrological cycle. On the other hand, it is more practical to declare drought condition considering only one indicator. Thus, there is a growing interest in aggregating more indices. Keyantash & Dracup (2004) use an Aggregate Drought Index that considers all relevant variables of the hydrological cycle through Principal Component Analysis (but they do not include groundwater in the suite of variables); instead Steinemann & Cavalcanti (2006) use the probabilities of different indicators of drought and shortage, selecting the trigger levels on the basis of the most severe level of the indicator or the level of the majority of the indicators.

6. Future scenarios

The most critical scenarios discussed in the previous paragraphs could become "normal" circumstances if global climate change increases the prolonged and intense drought periods. At the end of the proposed analysis, it is useful to hypothesize some future climatic scenarios, with the aim of steering decision makers towards suitable water management policies, as it is suggested by the European Commission (COM, 2009).

The methodology usually followed to assess the hydrological consequences of climate change basically consists of a three-step process (Xu et al., 2005): (1) the development and use of general circulation models (GCMs) to provide future global climate scenarios under the effect of increasing greenhouse gases, (2) the development and use of downscaling techniques (both statistical methods and nested regional climate models, RCMs, which are being continuously improved) for "downscaling" the GCM output to the scales compatible with hydrological models, and (3) the development and use of hydrological models to simulate the effects of climate change on hydrological regimes at various scales. However, uncertainties within this framework have to be taken into account such as the internal variability of the climate system, model structure and parameterizations at different spatial and temporal scales, the downscaling techniques and bias correction methods and the choice of future climate scenarios. Several different approaches were chosen for providing operational solutions to these drawbacks (Xu et al., 2005). However, numerous GCM simulations show almost univocal trends for global climate evolution. Giorgi & Lionello (2008) highlight a robust and consistent description specifically for the Mediterranean area, with a significant reduction in precipitation, mainly in summertime. In the same area, according to Giorgi (2006), a major increase in climatic variability is also expected.

Below, some results obtained by Senatore et al. (2011) are shown related to future water availability in the main basin of northern Calabria (Crati River Basin, 1332 km^2, Fig. 15) at the end of the XXI century. Future scenarios were made by applying the outputs of three Regional Climate Models (RCMs) RegCM, HIRHAM and COSMO-CLM to the newly developed Intermediate Space Time Resolution Hydrological Model (In-STRHyM). The analysis was performed using two time slices (1961–1990 and 2070–2099) with the SRES A2 (GCM HAD3AM) and A1B (GCM ECHAM5/MPI-OM) scenarios. Observed biases in simulated precipitation and temperature fields during the control period (1961-1990) were corrected before using meteorological outputs from each RCM as input for In-STRHyM.

In-STRHyM is a fully distributed hydrological model detailed enough to describe the hydrological processes of several small-medium sized Mediterranean basins. It has a relatively simple structure and is suitable for long period simulations to be undertaken within acceptable time frames. Specifically, In-STRHyM calculates separately transpiration and evaporation, depending on a remote sensing-derived vegetation fraction. Both transpiration and bare soil evaporation are estimated through the crop coefficient approach suggested by Allen et al. (1998), considering a water stress coefficient of the canopy depending on soil moisture conditions, and the reference values calculated through the Priestley & Taylor (1972) equation.

The RCMs predict an increase in mean annual temperature from 3.5 °C to 3.9 °C, and a decrease in mean annual precipitation from 9% to 21%. The effects of the changes in the forcing meteorological variables are relevant for all the hydrological output variables. Here we highlight results achieved for actual evapotranspiration (*ET*). This variable tends to decrease with reduced precipitation, but it increases with higher temperatures. Lower decrease in precipitation predicted by HIRHAM, together with the higher temperatures, leads to an average year *ET* increase of +2.5%, while for RegCM and CLM the annual mean reduction is equal to -5.1% (Fig. 15) and -8.3%, respectively. However, in the summer period, that is the irrigation period, in all cases an *ET* reduction is achieved (from -1.0% with HIRHAM to -9.1% with RegCM, Fig. 15), indicating a decrease in water availability for plants and soil. This water stress is better highlighted when considering simulated root zone soil moisture. For this variable a reduction is predicted, differently from *ET*, during the whole year (-20.7%±1.9%, -12.8%±1.9% and -17.6%±1.8% with RegCM, HIRHAM and CLM, respectively). Figure 16 shows as an example the daily changes computed using RegCM (the behavior considering the other RCMs is similar): they are less relevant in winter and spring, but the reduction is dramatic in summer and early autumn, due to the increased evaporative demand (up to -40% with RegCM).

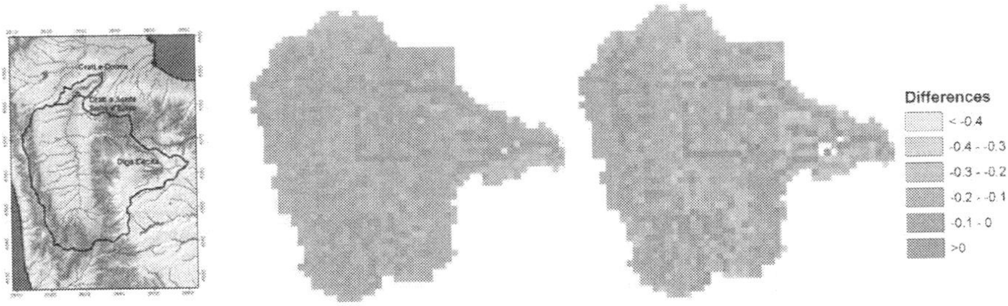

Fig. 15. Location of the Crati River Basin (left) and spatially distributed percentage changes in annual actual evapotranspiration (middle) and in actual evapotranspiration during the April–September irrigation period (right) simulated using RegCM (2070-2099 vs 1961-1990) (adapted from Senatore et al., 2011).

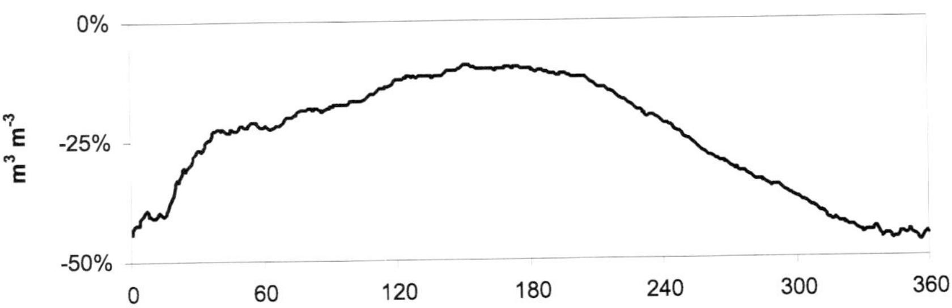

Fig. 16. Daily changes in root zone soil moisture computed using RegCM. RCM values are rescaled over 360 days, with the first day being October the 1st (readapted from Senatore et al., 2011).

7. Summary and conclusions

Evapotranspiration deeply affects the water resources availability in Calabria (average annual actual evapotranspiration estimated equal to almost 60% of the average cumulated annual rainfall). Highest water requirements come from agriculture, where losses due to evapotranspiration demand have to be re-equilibrated by huge amounts of water, mainly in the summer hot and dry period. The analysis of the comparison between the available water resource and the water demand was carried out considering both the "normal" conditions due to meteorological forcing, and the most critical derived by intense and prolonged drought periods. In the first case, neglecting the very conservative constraints proposed by the Regional Basin Authority for the minimum flow requirements, specific issues are not observed, the residual water availability being sufficient. Several problems arise instead when drought conditions occur: in these cases the development of guidelines is essential to define operative aspects about the individuation of the water use priorities, to characterize different drought levels, to individuate the main objectives of water management related to these levels, and to determine and apply the mitigation measures.

The proposed example of water resource management under shortage conditions in the agricultural area of the Sibari Plain shows the benefits that a proactive approach may provide with respect to the classical approaches based on emergency measures, which are usually expensive and not efficient. Within a proactive approach, specific care should be taken into account for reducing evapotranspiration losses through appropriate agronomic techniques. This action has to be considered as a strategic measure, with an impact on water scarcity reduction comparable to the effect of structural measures.

The review of drought indices showed that evapotranspiration could provide useful insights: i) when adopted within comprehensive indices, considering the effects of the whole water balance, and not only of some components, on water resources availability; ii) and mainly, when dealing with optical remote sensing techniques, because these allow to estimate in a relatively easy way the spatially distributed actual evapotranspiration over a specific area, and then they can relate this quantity to soil moisture and to the incoming of drought events.

Finally, applying some future scenarios with different GCMs and RCMs, it was observed that in Calabria the issues related to water resource management under shortage conditions in the next few years will be more frequent and intense, affecting wider areas. Evapotranspiration will be "tied down" by reduced precipitation (reducing its magnitude) and by higher temperatures (providing an opposite effect). It will not clearly increase or decrease on an annual basis, but in any case it will contribute to reduce useable water from the soil, needed for agricultural purposes. The hypothesized scenarios of climate change, though subject to uncertainty, have to be intended as an important part of knowledge for the planning of future interventions on the water resource by the Public Authorities, and for defining the optimal criteria to evaluate the amount of public investments.

8. References

Allen, R., Pereira, L.S., Raes, D. & Smith, M. (1998). Crop Evapotranspiration - Guidelines for Computing Crop Water Requirements. *FAO Irrigation and Drainage Paper* 56, FAO, Rome.

Allen, R.G., Tasumi, M. & Trezza, R. (2007). Satellite-based Energy Balance for MApping Evapotranspiration with Internalized Calibration (METRIC) – *Model. J. Irrig. and Drain. Engrg*, ASCE 133(4), 380-394.

Anderson, M., Norman, J., Diak, G., Kustas, W. & Mecikalski, J. (1997). A two-source time-integrated model for estimating surface fluxes from thermal infrared satellite observations. *Remote Sensing of Environment* 60: 195–216.

Anderson, M.C., Norman, J.M., Mecikalski, J.R., Otkin, J.A. & Kustas, W.P. (2007). A climatological study of evapotranspiration and moisture stress across the continental United States based on thermal remote sensing: 2. Surface moisture climatology. *J. Geophysical Research*; 112: art. No. D11112.

Bastiaanssen, W.G.M., Menenti, M., Feddes, R.A. & Holtslag, A.A.M. (1998) A remote sensing surface energy balance algorithm for land (SEBAL). 1. Formulation. *Journal of Hydrology* 212-213: 198–212.

Bastiaanssen, W.G.M. (2000). SEBAL-based sensible and latent heat fluxes in the irrigated Gediz Basin, Turkey. *Journal of Hydrology* 229: 87-100.

Bayarjargal, Y., Karnieli, A., Bayasgalan, M., Khudulmur, S., Gandush, C. & Tucker, C.J. (2006). A comparative study of NOAA–AVHRR derived drought indices using change vector analysis. *Remote Sensing of Environment*; 105: 9-22.

Bordi, I., Fraedrich, K., Petitta, M. & Sutera, A. (2005). Methods for predicting drought occurrences. *Proc. of the 6th EWRA International Conference*, Menton, France, 7-10 September 2005.

Cancelliere, A., Rossi, G. & Ancarani, A. (1996). Use of Palmer Index as drought indicator in Mediterranean regions. *Proc. IAHR Congress "From flood to drought"*, Sun City, South Africa, August 5-7, 1996, pp. S4.12. 1-25.

Cancelliere, A., Di Mauro, G., Bonaccorso, B. & Rossi, G. (2007). Stochastic forecasting of drought indices. In: Rossi, G., Vega, T., Bonaccorso, B. (Eds.), Methods and tools for drought analysis and management, *Water Science and Technology Library*, vol. 62, Springer, Dordrecht, 83-100.

COM - Commission of the European Communities, (2007). Addressing the challenge of water scarcity and droughts in the European Union, Communication from the Commission to the European Parliament and the Council, 414, 18.7.2007, Brussels, 14 pp.

COM - Commission of the European Communities, (2009). Towards a comprehensive climate change agreement in Copenhagen, Communication from the Commission to the European Parliament, the Council, the European Economic and Social Committee and the Committee of the Regions, 39, 28.1.2009, Brussels, 14 pp.

Cook, E.R., Seager, R., Cane, M.A. & Stahle, D.W. (2007). North American drought: Reconstructions, causes, and consequences. *Earth-Science Review*; 81: 93-134.

Cowie, G., Davis, M., Holmbeck-Pelham, S., Freeman, B., Freeman, M., Hatcher, K., Jackson, R., Miller Keyes, A., Merrill, M., Meyer, J., Sutherland, E. & Wenger, S. (2002). Reservoirs in Georgia: *Meeting Water Supply Needs While Minimizing Impacts*, University of Georgia's River Basin Science and Policy Center, Athens, Georgia,

Dworak, T., Berglund, M., Laaser, C., Strosser, Roussard, P. J., Grandmougin, B., Kossida, M., Kyriazopoulou, I., Berbel, J., Kolberg, S., Rodrigues-Diaz, J.A. & Montesinos, P. (2007). EU water saving potential, *Ecologic-Institute for International and European Environmental Policy*, ENV.D.2/ETU/2007/0001r.

Georgia Department of Natural Resources: Georgia Drought Management Plan, Atlanta, GA, (2003).

Gibbs, W.J. & Maher, J.V. (1967). Rainfall deciles as drought indicators. *Bureau of Meteorology Bulletin No. 48*, Commonwealth of Australia, Melbourne.

Giorgi, F. (2006). Climate change hot-spots. *Geophys. Res. Lett.* 33 L08707.

Giorgi, F. & Lionello, P. (2008). Climate change projections for the Mediterranean region. *Glob. Planet. Change* 63 90–104.

Guttman, N.B. (1998). Comparing the Palmer Drought Index and the standardized precipitation index. *J. Am.Water Res. Assoc.*; 34 (1): 113–121.

Hayes, M.J., Svoboda, M.D., Wilhite, D.A. & Vanyarkho, O.V. (1999). Monitoring the 1996 drought using the standardized precipitation index. *Bulletin of the American Meteorological Society*; 80: 429–438.

IPCC (2007). *Climate Change 2007. The Physical Science Basis.* Contribution of Working Group I to the Fourth Assessment Report of the Intergovernmental Panel on Climate Change [Solomon, S., D. Qin, M. Manning, Z. Chen, M. Marquis, K.B. Averyt, M. Tignor and H.L. Miller (eds.)]. Cambridge University Press, Cambridge, United Kingdom and New York, NY, USA, 996 pp., 2007.

Jackson, R.D., Idso, S.B., Reginato, R.J. & Printer, P.J. (1981). Canopy temperature as a crop water stress indicator. *Water Res. 17*, 1133-1138.

Karl, T.R. (1986). The Sensitivity of the Palmer Drought Severity Index (PDSI) and Palmer's Z-index to their calibration coefficients including potential evapotranspiration. *Journal of Climate and Applied Meteorology*; 25: 77-86.

Keyantash, J.A. & Dracup, J.A. (2004). An aggregate drought index: Assessing drought severity based on fluctuations in the hydrologic cycle and surface water storage. *Water Resources Research*; 40(9): Art. No. W09304.

Kim, T.W., Valdés, J.B., Nijssen, B. & Roncayolo, D. (2006). Quantification of linkages between large-scale climatic patterns and precipitation in the Colorado River Basin. *J. Hydrology*; 321: 173-186.

Kunstmann, H., Schneider, K., Forkel, R. & Knoche, R. (2004). Impact analysis of climate change for an alpine catchment using high resolution dynamic downscaling of ECHAM4 time slices, *Hydrology and Earth System Sciences*, Vol. 8, 1031-1045.

Lohani, V.K., Loganathan, G.V. & Mostaghimi, S. (1998). Long-term analysis and short-term forecasting of dry spells by Palmer Drought Severity Index. *Nord. Hydrol.*; 29(1): 21-40.

Matera, A., Fontana, G., Marletto, V., Zinoni, F., Botarelli, L. & Tomei, F. (2007). Use of a new agricultural drought index within a regional drought observatory. In: Rossi, G., Vega, T., Bonaccorso, B. (Eds.), Methods and tools for drought analysis and management, *Water Science and Technology Library*, vol. 62, Springer, Dordrecht, 103-124.

McKee, T.B., Doesken, N.J. & Kleist, J. (1993). The relationship of drought frequency and duration to time scales, *Proceedings of the 8th Conference on Applied Climatology*, American Meteorological Society, Anaheim, CA, Boston, MA, pp. 179–184.

Mendicino, G., Senatore, A. & Versace, P. (2005). I deflussi minimi annuali, stagionali e di magra nei corsi d'acqua calabresi, *Acts of the 25th Corso di Aggiornamento in Tecniche per la Difesa dall'Inquinamento*, ed. G. Frega, BIOS, Cosenza, pp. 89-117.

Mendicino, G. & Versace, P. (2007). Integrated Drought Watch System: A Case Study in Southern Italy, *Water Resour. Manage.*, Vol. 21, 1409–1428.

Mendicino, G., Senatore, A. & Versace, P. (2008a). A Groundwater Resource Index (GRI) for drought monitoring and forecasting in a Mediterranean climate. *J. Hydrol.*, Vol. 357(3-4), pp. 282-302.

Mendicino, G., Senatore, A. & Versace, P. (2008b). "Water resources management in agriculture under drought and water shortage conditions: a case study in southern Italy", *European Water 23/24*, pp. 41-56.

Menenti, M. & Choudhury, B.J. (1993). Parameterization of land surface evapotranspiration using a location dependent potential evapotranspiration and surface temperature range. In: Bolle H.J. et al. (Eds.), *Exchange Processes at the Land Surface for a Range of Space and Time Scales*, IAHS Publication 212, 561-568.

Moore, I.D., Norton, T.W & Williams, J.E. (1993). Modelling environmental heterogeneity in forested landscapes, *J. Hydrol.*, Vol. 150, 717-747.

Munda, G. (1995). Multicriteria Evaluation in a Fuzzy Environment, Series: *Contributions to Economics*, Physica-Verlag, Heidelberg.

Nakícenovíc, N., Alcamo, J., Davis, G., de Vries, B., Fenhann, J., Gaffin, S., Gregory, K., Grübler, A., Jung, T.Y., Kram, T., Emilio la Rovere, E., Michaelis, L., Mori, S., Morita, T., Pepper, W., Pitcher, H., Price, L., Riahi, K., Roehrl, A., Rogner, H.-H., Sankovski, A., Schlesinger, M.E., Shukla, P.R., Smith, S., Swart, R.J., van Rooyen, S., Victor, N. & Dadi, Z. (2000). *Special Report on Emissions Scenarios*, Cambridge University Press, Cambridge, available on-line: http://www.grida.no/climate/ipcc/emission/.

Narasimhan, B. & Srinivasan, R. (2005). Development and evaluation of Soil Moisture Deficit Index (SMDI) and Evapotranspiration Deficit Index (ETDI) for agricultural drought monitoring. *Agric. For. Meteorol.*; 133: 69-88.

Niemeyer, S. (2008). New drought indices. Options méditerranéennes, SERIE A: *Séminaiers Méditerranéens*, N. 80, "Drought Management: Scientific and Technological Innovations", pp. 267-274.

Norman, J., Kustas, W. & Humes, K. (1995). A two-source approach for estimating soil and vegetation energy fluxes from observations of directional radiometric surface temperature. *Agricultural and Forest Meteorology* 77: 263-293.

Ntale, H.K. & Gan, T.H. (2003). Drought indices and their application to East Africa. *Int. J. Climatol.*; 23: 1335-1357.

Palmer, W.C. (1965). Meteorological drought, Research Paper 45. U.S. Department of Commerce, Weather Bureau, Washington, DC.

Pereira ,L.S. (2007). Drought impacts in agriculture: water conservation and water saving practices and management, in: G. Rossi et al. (Eds.), *Water Science and Technology Library*, vol. 62: Methods and Tools for Drought Analysis and Management, Springer Netherlands, pp.349-373.

Priestley, C.H.B. & Taylor, R.J. (1972). On the assessment of the surface heat flux and evaporation using large-scale parameters, *Monthly Weather Review*, Vol. 100, 81-92.

Qian, T., Dai, A., Trenberth, K.E. & Oleson, K.W. (2006). Simulation of global land surface conditions from 1948-2004. Pt I: Forcing data and evaluations. *J. Hydrometeorol.*, 7, 953-975.

Rivas-Martinez, S. (1995). Bases para una nueva clasificacion bioclimatica de la Tierra. *Folia Botanica Matritensis 16.*

Roerink, G.J., Su, B. & Menenti, M. (2000). S-SEBI A simple remote sensing algorithm to estimate the surface energy balance. *Physics and Chemistry of the Earth (B)* 25(2): 147-157.

Rossi, G. (2003). An integrated approach to drought mitigation in Mediterranean regions, in: G. Rossi et al. (Eds.), *Tools for drought mitigation in Mediterranean regions*, Kluwer Academic Publishing, Dordrecht, pp. 3-18.

Rossi, G., Castiglione, L. & Bonaccorso, B. (2007). Guidelines for planning and implementing drought mitigation measures, in: G. Rossi et al. (Eds.), *Water Science and Technology Library*, Vol. 62: Methods and Tools for Drought Analysis and Management, Springer Netherlands, pp.325-347.

Senatore, A., Mendicino, G., Smiatek, G. & Kunstmann, H. (2011). Regional climate change projections and hydrological impact analysis for a Mediterranean basin in southern Italy. *Journal of Hydrology*, 399(1-2), 70-92.

Shafer, B.A. & Dezman, L.E. (1982). Development of a Surface Water Supply Index (SWSI) to assess the severity of drought conditions in snowpack runoff areas. *Proceedings of the Western Snow Conference*, Colorado State University, Fort Collins, Colorado, 164-175.

Steinemann, A.C. & Cavalcanti, L.F.N. (2006). Developing multiple indicators and triggers for drought plans. J. *water resources planning and management, ASCE* 132(3): 164-174.

Su, Z. (2002). The surface energy balance system (SEBS) for estimation of turbulent heat fluxes. *Hydrology and Earth System Sciences* 6(1): 85–99.

Su, Z., Yacob, A., Wen, J., Roerink, G., He, Y., Gao, B., Boogaard, H. & van Diepen, C. (2003). Assessing relative soil moisture with remote sensing data: Theory, experimental validation, and application to drought monitoring over the North China Plain. *Physics and Chemistry of the Earth (B)* 28(1-3): 89-101.

Thornthwaite, C.W. & Mather, J.R. (1955). The water balance, *Climatology*, Drexel Inst. Of Technology, Centeron, New Jersey.

Tsakiris, G., Loukas, A., Pangalou, D., Vangelis, H., Tigkas, D., Rossi, G. & Cancelliere, A. (2007a). Drought characterization, in: *Drought Management Guidelines Technical Annex*, Options méditerranéennes, Série B: Etudes et Recherches, Numéro 58.

Tsakiris, G., Pangalou, D. & Vangelis, H. (2007b). Regional drought assessment based on the Reconnaissance Drought Index (RDI). *Water Resources Management*; 21(5): 821-833.

Water Scarcity Drafting Group (2006). Water scarcity management in the context of WFD, Salzburg, June 1-2.

Wilby, R.L., Wedgbrow, C.S. & Fox, H.R. (2004). Seasonal predictability of the summer hydrometeorology of the River Thames, UK. *J. Hydrology*; 295: 1-16.

WHO/Unicef (2006). Protecting and Promoting Human Health, in: *The 2nd UN World Water Development Report*: "Water, a shared responsibility", available on line: http://www.unesco.org/water/wwap/wwdr/wwdr2/.

Xu, C., Widén, E. & Halldin, S. (2005). Modelling Hydrological Consequences of Climate Change-Progress and Challenges. *Advances in Atmospheric Sciences*, vol.22, no.6, 789-797.

Yao, Y., Liang, S., Qin, Q. & Wang, K., (2010). Monitoring Drought over the Conterminous United States Using MODIS and NCEP Reanalysis-2 Data. *Journal of Applied Meteorology and Climatology*, 49, 1665-1680.

Yevjevich, V. (1967). An objective approach to definitions and investigations of continental hydrologic droughts. *Hydrology Papers*, Colorado State University, Fort Collins, pp. 23.

Yevjevich, V., Hall, W.A. & Salas, J.D. (1978). Drought research needs, *Water Resources Publications*, Fort Collins, Colorado.

Yevjevich, V., Da Cunha, L. & Vlachos, E. (1983). Coping with droughts, *Water Resources Publications*, Littleton, Colorado.

Zhang, P., Anderson, B., Tan, B., Huang, D. & Myneni, R. (2005). Potential monitoring of crop production using a satellite-based Climate-Variability Impact Index. *Agric. For. Meteorol.*; 132: 344-358.

11

Guidelines for Remote Sensing of Evapotranspiration

Christiaan van der Tol and Gabriel Norberto Parodi
University of Twente. Faculty of ITC
The Netherlands

1. Introduction

This chapter describes the possibilities, the limitations and the future of remote sensing of evapotranspiration (ET). The principles behind the techniques of remote sensing of ET are presented systematically. The mathematical formulations of the key equations are used to highlight the critical parts and the variables that remote ET is most sensitive to. The focus will be on the input data. Which input data do we definitively need, and with what accuracy? How can we select the best methodology to estimate ET spatially? A number of new developments will be introduced, and priorities for the near future formulated.

There is no global, validated ET product available today. We can find products of other components of the terrestrial water cycle, like rainfall and soil moisture, but not of ET. This means that remote estimation of ET is custom made, and that it requires specific skills. At first glance, this is surprising, because the idea of remote sensing of evapotranspiration is more than three decades old (Jackson et al., 1977; Jackson et al., 1987; Seguin, 1988). In this chapter we hope to clarify the reasons why the operational dissemination of remote sensing evapotranspiration products lags behind.

The one fundamental problem with estimating ET is that it cannot be measured directly. This is well illustrated by borrowing an allegory from the evangelist Billy Graham: "I've seen the effects of the wind but I've never seen the wind". This quote is certainly true, in a literal sense, for evapotranspiration. Evapotranspiration affects the water and energy balances, and it is these effects of evapotranspiration which are observed. For example, evapotranspiration reduces soil moisture content and its cools the land surface. By studying changes in soil moisture with lysimeters, or by studying patterns in land surface temperature with remote sensing, ET is estimated. The problem is that ET is not the only factor affecting the water and energy budgets of the surface. Other processes and physical properties play important roles as well. For this reason, quite a large number of input variables are needed to achieve reasonably accurate estimates of ET; estimates that are at least better than rule-of-thumb a priori estimates. This is the case for field techniques, and even more so for remote techniques. In remote sensing an additional issue is that the input data are not from just one source. Often data of different satellite platforms are needed in conjunction with meteorological data from ground stations.

In the process of collecting the data, the user faces practical and scientific problems. The practical problems of the user include the data collection, merging the data in a GIS database, and programming the algorithm for the calculation of evapotranspiration. Each of these steps requires expertise and access to dedicated GIS software packages, because most algorithms are not available 'on the shelf'. The scientific problem is that the procedure requires merging of data measured at different time and spatial scales. The merging of data for the calculation of evapotranspiration is a typical example of a data assimilation problem. Ideally, both the accuracy and the representativeness of each of the input data are taken into account in the calculation of the final product: the spatial map of evapotranspiration. Although algorithms for the calculation of evapotranspiration are available in the literature and on the internet, there is no consistent data assimilation procedure attached that calculates the accuracy and reliability of the final product.

In recent years there have been a number of initiatives to build global products of evapotranspiration (Vinukullu et al., 2011; www.wacmos.org), addressing the above mentioned issues. The priorities for the near future are to establish a consistent way of merging the input data, improve the data assimilation techniques and to validate algorithms. In addition, new sources of data may be introduced. A promising tool is the use of satellite based laser altimetry for surface roughness estimation (Rosette et al., 2008).

It is the aim of this chapter to focus on principles that the algorithms have in common, and on the input data. Reviews of the history of remote sensing of ET can be found in the scientific literature (Courault et al., 2005; Glenn et al., 2007; Gowda et al., 2007; Kalma et al., 2008). In addition to these reviews, we would like to provide some anchor points and guidelines for the selection of a methodology for estimating ET in basin hydrology. We will quantify and evaluate the error of each of the input data, and show how this error propagates into the final result. For this analysis we will use theoretical considerations, a remote sensing model, and a selection of field data.

2. Principles of remote sensing algorithms for evapotranspiration

Although remote sensing of evapotranspiration has evolved since the first initiatives in the 1970's, the fundamental principle has remained the same. All remote sensing based evapotranspiration estimates make use of the thermal and visible bands and the formulation of the energy balance of the surface. The instantaneous latent heat flux of evaporation is calculated as a residual of the energy balance, and this latent heat flux is in turn converted into an evapotranspiration rate after time integration. An inherent problem of this approach is that the errors in the various terms of the energy balance are affecting the latent heat flux in a manner that is difficult to predict. For this reason it is necessary to evaluate the different terms of the energy balance individually.

In the evaluation of the remote sensing algorithms presented in this section, we will discuss the terms, and indicate at what spatial and temporal resolution the data can be collected. It will become clear that the land surface temperature is the most important state variable. It plays a crucial role in sensible heat flux, ground heat flux and the balance of long wave radiation. Apart from the collection of accurate land surface temperature data, important selection criteria for a methodology are the heterogeneity of the land cover, the topography and the spatial resolution (sampling) of the remote data.

Neglecting the energy used in the process of photosynthesis, the instantaneous energy balance equation (EBE) over crops reads:

$$R_n = G + H + \lambda E \tag{1}$$

R_n is the net radiation remaining in the system, G the ground heat flux, H the sensible heat flux and λE is the latent heat flux that is the energy consumed in evapotranspiration (all in

W m^{-2}). Radiation fluxes are positive when directed towards the land surface, the other fluxes are positive when pointed away from the surface. The partition of energy between the terms is largely controlled by the availability of water or moisture in the system. When moisture is not restricted, λE reaches a maximum and H is small.

In order to estimate ET, Eq. 1 is solved for λE. When applied to remote sensor retrievals, R_n is solved entirely from a combination of radiation counting at sensor level and few ground information. Ground or soil heat flux is a minor component in densely vegetated areas, but a large term in non-vegetated or sparsely vegetated areas (Heusinkveld et al., 2004). The importance of a better evaluation of the soil heat flux is gaining attention, mainly to ensure the EBE closure in such areas. The evaluation of H is the major difficulty. There are several models and approaches to solve for H (SEB models) and a number of parameters and assumptions are still under debate. The remote sensing models for ET mainly differ in the way H is treated.

In the following sections, the individual terms of the EBE (Eq. 1) will be discussed in further detail. A theoretical description is presented for each term in the EBE, in combination with a discussion on the feasibility of data acquisition from remote and ground sources.

2.1 Net radiation

Net radiation R_n is the dominant term in the EBE, since it represents the source of energy that must be balanced by the thermodynamic equilibrium of the other terms. The net radiation can also be expressed as an electromagnetic balance of all incoming and outgoing radiation reaching and leaving a flat horizontal and homogeneous surface as:

$$R_n = S\downarrow - S\uparrow + L\downarrow - L\uparrow \tag{2}$$

Where S is the shortwave radiation, nominally between 0.25 to 3 μm and L is the long wave radiation, nominally between 3 to 100 μm. The arrows show the direction of the flux entering '$\downarrow$' or leaving '$\uparrow$' the system.

Equation 2 is very convenient from the data acquisition point of view since each term can either be obtained from available models, or directly from instruments at ground stations or remote platforms. As remote sensors are positioned looking to Earth, they measure outgoing radiation only. The incoming fluxes must be either modelled or derived through alternative methodologies.

The instantaneous incoming shortwave radiation (also called global radiation), $S\downarrow$, is commonly measured at ground stations by means of pyranometers or solarimeters. These instruments usually work in the shortwave broadband range (usually 0.305 - 2.4 μm). This range comprises almost 96% of the spectral interval of the solar irradiance. Recently there are remote sensing products and clearinghouses that account for the incoming and outgoing shortwave and long wave radiation. The use of them may reduce the need of permanently operational ground radiometers.

The outgoing shortwave radiation is the portion of the shortwave reflected back to the atmosphere. It is characterized by the albedo. The reflectance is the ratio between the reflected and the incoming radiation in a certain wavelength over an arbitrary horizontal plane. The integrated value over all visible bands defines the albedo, r_0. Since albedo is a reflective property of the material, it can be evaluated from remote sensors multi-spectral bands, and the integration to full shortwave range is approached by a linear model that might include the atmospheric correction. The shortwave radiation balance reads:

$$\Delta S = S\downarrow - S\uparrow = (1 - r_0)\cdot S\downarrow \tag{3}$$

For all bodies, the total incident radiation is either reflected by the body, absorbed by it or transmitted through. This is expressed by the Kirchoff's law:

$$1 = \rho_\lambda + \tau_\lambda + \alpha_\lambda \tag{4}$$

where ρ_λ is the reflectivity, τ_λ the transmissivity and α_λ the absorptivity. A blackbody is defined as a body that absorbs all the radiation that receives. A blackbody is a physical abstraction that does not exist in nature. To keep a body temperature constant, it should emit the same radiation that absorbs. As a consequence a property of blackbodies, the absorptivity, is equal to the emissivity, and both are equal to 1, while reflectivity and transmissivity are equal to zero.

Terrestrial materials behave more as grey bodies, meaning that part of the received radiation is reflected back to the atmosphere, or in other words, not all the energy that receives is absorbed. In order to keep the temperature constant, the absorbed radiation should equal the emission, so again emissivity is equal to absorptivity. Because the reflectivity is not zero, emissivity of real bodies is smaller than 1.

The longwave radiation terms are calculated with Planck's equation extended to real bodies. A blackbody having a kinetic temperature T_0 [K] emits in a single wavelength a radiation that corresponds to:

$$L_\lambda^{bb} = \frac{3.74\cdot 10^8}{\lambda^5}\cdot\frac{1}{\left[\exp\left(\frac{1.44\cdot 10^4}{\lambda\cdot T_0}\right)-1\right]} \tag{5}$$

Where $L_\lambda{}^{bb}$ is the blackbody energy emission [W m^{-2} μm^{-1}] and λ is the wavelength [μm]. The kinetic temperature is the temperature as it would be measured by a standard thermometer in contact with the surface of the body. Emissivity ε_λ at a chosen wavelength is the ratio of the radiation emitted by a real body at temperature T_0 to the radiation emitted by a blackbody at the same temperature. By definition, a blackbody has a constant emissivity equal to one for all wavelengths, whereas the real emissivity varies with wavelength. For natural bodies, the thermal emission can then be written as:

$$L_\lambda(T_0) = \varepsilon_\lambda \cdot L_\lambda^{bb}(T_0) \tag{6}$$

Integration of L over all wavelengths leads to:

$$L(T_0)\uparrow = \int_0^\infty \varepsilon_\lambda \cdot L_\lambda^{bb}(T_0)\cdot d\lambda = \sigma\cdot\varepsilon_0\cdot T_0^4 \tag{7}$$

where σ = 5.67 x10^{-8} W m^{-2} K^{-4} is the Stefan-Boltzmann constant and ε_0 is a broadband surface emissivity. A remote sensor working within a spectral range of the thermal channels measures only a portion of $L(T_0)\uparrow$. The outgoing longwave radiation at any sensor channel is calculated by integration over the spectral range of the sensor:

$$L_i^{sat}(T_0)\uparrow = \int_i \varepsilon_\lambda \cdot L_\lambda^{bb}(T_0)\cdot d\lambda \tag{8}$$

The surface temperature T_0 is retrieved from Eq (8), once the surface emissivity ε_λ in the considered thermal channel is estimated. Once 'T_0' is obtained and ε_0 estimated, $L\uparrow$ is retrieved from Eq 7. Before the application of Eq (8), an atmospheric correction process is needed to derive $L_i^{sur}(T_0)\uparrow$ at the surface, because $L_i^{sat}(T_0)\uparrow$ as measured at the satellite sensor is affected by atmospheric interference. Atmospheric correction in the thermal range and in the shortwave is out of the scope of this chapter. We only mention that $L_i^{sur}(T_0)\uparrow$ can be obtained from $L_i^{sat}(T_0)\uparrow$ and using atmospheric correction model, in which water vapour and aerosol concentrations are the main input variables.

The incoming long wave radiation cannot be derived directly from remote sensors. It can either be determined from ground data or derived after atmospheric modelling. It varies with cloudiness (water vapour), air temperature and atmospheric constituents. For clear skies, the notion of effective thermal infrared emissivity of the atmosphere or apparent emissivity of the atmosphere ($\varepsilon_a^{'}$) introduces an overall emission value for all constituents. If the air temperature T_a at screen level is available, $L\downarrow$ is estimated as:

$$L\downarrow = \sigma \cdot \varepsilon_a^{'} \cdot T_a^4 \tag{9}$$

There are several models simple to evaluate $\varepsilon_a^{'}$. The apparent emissivity of the atmosphere is usually estimated with equations based on vapour pressure and temperature at standard meteorological stations. For clear skies a common formulation, among others, is (Brutsaert, 1975):

$$\varepsilon_a^{'} = 1.24\left(\frac{e_a}{T_a}\right)^{\frac{1}{7}} \tag{10}$$

Where T_a is the air temperature [K] and e_a is the vapour pressure [mbar], everything measured at screen level. A portion $L\downarrow$ reaching the Earth surface is reflected back to the atmosphere. Since the surface is opaque the transmissivity is zero, the reflection of $L\downarrow$ can be evaluated with Kirchoff law. As ε_0 describes the emissivity of a body in the thermal range, (1- ε_0) accounts for the reflection. The final expression for R_n becomes:

$$R_n = (1 - r_o)\cdot S\downarrow + \varepsilon_a^{'} \cdot \sigma \cdot T_a^4 - (1-\varepsilon_0)\cdot \varepsilon_a^{'} \cdot \sigma \cdot T_a^4 - \varepsilon_0 \cdot \sigma \cdot T_0^4 \tag{11}$$

Eq. 11 is valid for instantaneous observations. The conversion to a daily value is briefly discussed in Sect 2.4.

It is not always necessary to carry out the calculations of Eqs 3-11 manually. Some organizations provide atmospherically corrected components of the radiation platforms directly. For example LandSaf (landsaf.meteo.pt) provides MeteoSat Second Generation (MSG) products of atmospherically corrected $S\downarrow$ and $L\downarrow$ with a 15 minute resolution and daily albedo for South America, Africa and Europe. An emissivity product will be released soon. Validation over ground based measurements for a site in Spain over sparse vegetation shows that these products are rather reliable (Fig 1.).

As an alternative to the use of satellite data, a computation of the radiation terms from synoptic weather stations is also possible. The recommendations by the FAO (Doorenbos and Pruitt, 1977; Allen et al., 1998) could be followed. The daily short wave radiation $S\downarrow_{day}$ [MJm^{-2}day^{-1}], is measured at agrometeorological stations with pyranometers and integrated to daytime hours. In most areas in the world, only sunshine hours are measured with periheliometers. In that case, the daily incoming shortwave radiation $S{\downarrow}_{day}$ can be obtained from the following empirical relationship:

$$S_{day}\downarrow = (a_s + b_s \cdot \frac{n}{N}) \cdot S_{0,day}\downarrow \quad (12)$$

where a_s is the fraction of the extraterrestrial radiation reaching the ground in a complete overcast day (when n=0), a_s + b_s the fraction of the extraterrestrial radiation reaching the ground in a complete clear day (n=N), n the duration of bright sunshine per day [hours], N the total daytime length [hours], $S_{0,\ day}\downarrow$ is the terrestrial radiation [MJm^{-2} day^{-1}]. Local instrumentation can be used to calibrate a_s and b_s for local conditions.

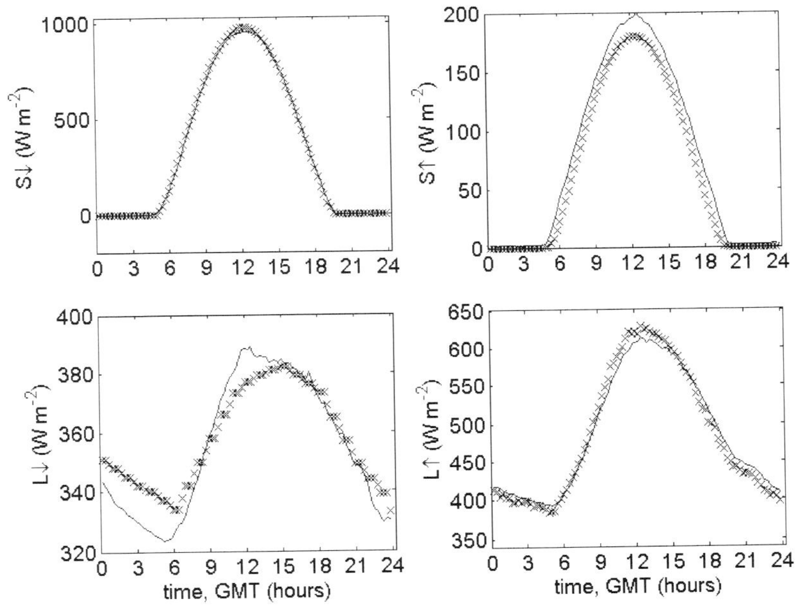

Fig. 1. Comparison between MeteoSat Second Generation radiation products (symbols) with 5-minute interval ground based measurements (lines) for a pixel with sparse vegetation in central Spain, for 5 July 2010.

The net daily shortwave radiation ΔS_{day} is estimated as in Eq. (3), assuming an average daily (sun hours only) albedo r_{0day}. The daily longwave radiation exchange between the surface and the atmosphere is very significant. Since on average the surface is warmer than the atmosphere and $\varepsilon_0 > \varepsilon_a'$, there is usually a net loss of energy as thermal radiation from the ground. The daily net shortwave radiation ΔL_{day} [W m^{-2}] between vegetation and soil on the one hand, and atmosphere and clouds on the other, can be represented by the following radiation law:

$$\Delta L_{day} = -f \cdot \varepsilon_{a,day}^{'} \cdot \sigma \cdot (T_{a,mean} + 273.15)^4 \quad (13)$$

Where $\varepsilon'_{a,\text{day}}$ [-] is the daily net emissivity between the atmosphere and the ground, f a cloudiness factor and $T_{a,\text{mean}}$ is the mean daily air temperature at screen level [°C]. Parameter $\varepsilon'_{a,\text{day}}$ can be estimated from data from meteorological stations as:

$$\varepsilon'_{a,\text{day}} = a_e + b_e \cdot \sqrt{e_{d,\text{mean}}/10} \tag{14}$$

Where a_e is a correlation coefficient (ranging from 0.34 to 0.44, with a default of 0.34), b_e a correlation coefficient (ranging from -0.14 to –0.25 with a default of -0.14), $e_{d\,\text{mean}}$ the average vapour pressure at temperature [mbar]. If true $e_{d,\text{mean}}$ is not available, then it can be calculated from daily average relative humidity RH_{mean} and mean air temperature $T_{a,\text{mean}}$ [°C]:

$$e_{d,\text{mean}} = \frac{RH_{\text{mean}}}{100} \cdot e_{s,\text{mean}} \quad \text{and} \quad e_{s,\text{mean}} = 6.108 \cdot \exp\left[\frac{17.27 \cdot T_{a,\text{mean}}}{T_{a,\text{mean}} + 237.15}\right] \tag{15}$$

The cloudiness factor f is equal to 1 in case of a perfect clear day and 0 in a complete overcast day. In case the station has solar radiation data from pyranometers f can be calculated as:

$$f = a_c \cdot \frac{S\downarrow_{\text{day}}}{S\downarrow_{\text{clear,day}}} + b_c \quad \text{or} \quad f = a_c \cdot \frac{S\downarrow_{\text{day}}}{(a_s + b_s \cdot) \cdot S_0\downarrow_{\text{day}}} + b_c \tag{16}$$

Where a_s, b_s, a_c and i_c are calibration values to be estimated through specialized local studies which involve measuring longwave radiation values. Average values for a_c and b_c in arid and humid environments can be found in Table 1:

Climate	a_c	b_c	a_s	b_s
Arid	1.35	-0.35	0.25	0.50
Humid	1.00	0.00	0.25	0.50

Table 1. Typical values the coefficients a_c, b_c, a_s and b_s for arid and humid climates (Maidment, 1992).

If only data on sunshine hours data are available, then:

$$f = \left(a_c \cdot \frac{b_s}{a_s + b_s}\right) \cdot \frac{n}{N} + \left(b_c + \frac{a_s}{a_s + b_s} \cdot a_c\right) \tag{17}$$

2.2 Sensible heat flux

The sensible heat flux (H) is the exchange of heat through air as a result of a temperature gradient between the surface the atmosphere. Since the surface temperature during the day is usually higher than the air temperature, the sensible heat flux is normally directed upwards. During the night the situation may be reversed. Close to the surface, the sensible heat transport takes place mostly by diffusive processes, whereas at some distance away from the surface turbulent transport becomes more important.

The mathematical formulation of the sensible heat flux is based on the theory of mass transport of heat and momentum between the surface and the near-surface atmospheric environment (surface boundary layer). All existing remote sensing algorithms for turbulent sensible heat flux use the analogy of Ohm law of resistance driven by a gradient of temperature:

$$H = \rho_a \cdot c_p \cdot \frac{T_s - T_a}{r_{\mathrm{ah}}} \tag{18}$$

where ρ_a is the density of moist air [kg m^{-3}], c_p is the air specific heat at constant pressure [J kg^{-1} K^{-1}], r_{ah} is the aerodynamic resistance to heat transport between the surface and the reference level [s m^{-1}] and T_s - T_a is the driving temperature gradient between the surface (with temperature T_s) and the reference height (with temperature T_a).

Equation 18 shows that the estimation of sensible heat flux has two main elements: a temperature difference between two heights and the corresponding resistance. As a first approximation we can conclude that the error in the sensible heat is linearly proportional to the error in the temperature gradient, and linearly proportional to the error in the inverse of the resistance. Equation 1 shows that this error (W m^{-2}) is directly transferred to the latent heat flux. We will now show that this is only approximately true, because the equation is not linear and the aerodynamic resistance itself depends on the temperature gradient. We will show that because of this, r_{ah} can only be solved iteratively.

Understanding the physical concepts involved in the calculation of sensible heat flux, and in particular the aerodynamic resistance, is essential for an evaluation of remote sensing techniques. The evaluation of r_{ah} is the most complicated issue of all in the whole EBE procedure for AET estimates. It is our experience that lack of or incomplete knowledge of the entire formulation, image pre-processing and atmospheric correction processes leads to severe flaws in the intermediate and final outputs. Many researchers are still seeking for alternatives, procedures and methods to improve the accuracy of ET estimates form the EBE - RS approach. The actual parameterization is not optimal in the sense that some sensitive information can only be strictly evaluated under controlled experimental research, and not in a routine fashion.

Near the ground two phenomena take place simultaneously in the transfer of heat between the surface and the atmosphere: free convection produced by temperature gradient T_s-T_a and forced convection by the dragging forces of the wind. Then, the estimation of the turbulent heat fluxes requires a description of the turbulent wind profile near the surface. The starting point of the analysis is the wind profile in a neutral atmosphere (no convection, and T_s=T_a). In this situation and for an open site, the horizontal wind speed u [m s^{-1}] varies logarithmically with height above the ground z [m]:

$$u(z) = A \cdot \ln(z) + B \tag{19}$$

B is usually replaced by A .ln (z_{om}) where z_{0m} is the aerodynamic roughness length of the surface for momentum transport and represents the value of z for which Eq 19 predicts $u(z)$ = 0 (see also Fig 2):

$$u(z) = A \cdot \ln\left(\frac{z}{z_{0\mathrm{m}}}\right) \tag{20}$$

In Eq 20, A must have the dimension of velocity and it should be independent of z since the profile description is given by the logarithmic term. Over plant communities of uniform height h, the turbulent boundary layer behaves as if the vertically distributed elements of the community were located at a certain distance d from the ground. Parameter d is called the zero plane displacement or displacement height level of the flow. It acts as a correction to the level where z=0, and thus z in Eqs 19 and 20 should be directly replaced by z-d in vegetated areas:

$$z_{\text{shift}} = z - d \tag{21}$$

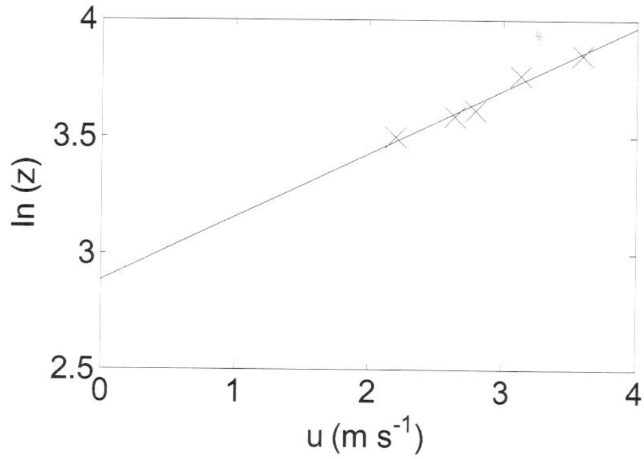

Fig. 2. Example plot of mean wind speed u against ln(z), measured above a 31 m tall forest in The Netherlands in August 2009. The intercept with the vertical axis leads to $z_{0m}+d$ = 17.9 m.

The displacement height d is usually rather large: it ranges between 60 to 80 percent of the plant community height. Common values for well developed wheat are found in Verma and Bartfield (1979). A relation d= 0.67h is usually adopted for other vegetation types (Allen et al., 1998). An exact estimation can be carried out when a wind speed measurements at three or more heights are available. A plot of ln(z-d) versus wind speed should then give a straight line for the correctly calibrated value of d. Strictly speaking, d also depends on plant density; for sparse vegetation, d is often neglected. In many occasions, insufficient data are available to accurately predict its values for discrete crop canopies (Verma and Bartfield, 1979). If no valid field data are present, then it is suggested to leave d out of the equations altogether.

The logarithmic wind profile in neutral conditions forms the basis for calculating the aerodynamic resistance for heat transport (and transport of water vapour) by mechanical turbulence. To include thermal turbulence (or convection or buoyancy), the Monin-Obukhov theory is needed in addition (Obukhov, 1946). We will now summarize the equations leading to the aerodynamic resistance of mechanical transport, followed by the modification for non-neutral conditions, when convection plays a role.

The transfer of momentum in the direction of the flux takes place through molecular and turbulent eddy activity. Random vertical movements of the air cause air with different horizontal wind speeds to mix. This causes a momentum sink at the surface in a form of shear stress, τ. For convenience the shear stress is expressed as a function of a scalar u_*, usually called eddy velocity or friction velocity [m s^{-1}]:

$$\tau = \rho_a u_*^2 \tag{22}$$

Since wind is produced by turbulent eddy motion, it is postulated that A in Eq 20 is proportional to the speed of the internal eddies. Then it can be demonstrated that:

$$A = \frac{u_*}{k} \tag{23}$$

Where k is Von Kármán's constant, experimentally found to be 0.41. The final expression of the wind profile under neutral atmosphere is:

$$u(z) = \frac{u_*}{k} \cdot \ln\left(\frac{z}{z_{0m}}\right) \tag{24}$$

Then the gradient of wind speed with height can be expressed as:

$$\frac{du}{dz} = \frac{u_*}{k \cdot z} \tag{25}$$

In the theory of estimating the aerodynamic resistance from the wind profile, only vertical transport is considered. In the atmosphere the steepest gradients of heat, wind speed and humidity are found in the vertical direction. Horizontal variation is present in the order of tens of kilometres (Brutsaert, 2005), but these are considered negligible. This implies that horizontal advection effects (for example between pixels) are not considered in the remote sensing approach, a serious restrictions in patchy (wet and dry) environments.

In analogy to horizontal wind speed, heat and water vapour also have vertical profiles near the surface. Vertical mixing then causes a transport of heat and vapour too, resulting in vertical fluxes of sensible and latent heat. The three fluxes, of momentum (F_u), heat (F_h) and vapour (F_v), can be expressed as the covariance of vertical wind speed (w') and concentration of the admixture (u', T' and q'):

$$\begin{aligned} F_u &\equiv -\tau = \rho \cdot \overline{w' \cdot u'} \\ F_h &\equiv -H = \rho \cdot c_p \cdot \overline{w' \cdot T'} \\ F_v &\equiv -\lambda E = \rho \cdot \overline{w' \cdot q'} \end{aligned} \tag{26}$$

These equations can be linked with the approach of electrical analogy (Eq 18) by approximating the covariance to simply the product of the vertical gradient of the quantities at two different heights (Brutsaert, 2005). A dimensionless parameter C is needed to fit the equality. This can be done for all three quantities. For example, for heat:

$$\overline{w' \cdot T'} = C_h \cdot \left(\overline{u_2} - \overline{u_1}\right) \cdot \left(\overline{T_4} - \overline{T_3}\right) \tag{27}$$

In this case, C_h will depend on the heights 1, 2, 3 and 4. It is convenient to choose the heights 4 and 3 equal to 2 and 1:

$$\overline{w' \cdot T'} = C_h \cdot \left(\overline{u_2} - \overline{u_1}\right) \cdot \left(\overline{T_2} - \overline{T_1}\right) \tag{28}$$

Similarly, for momentum transport (shear stress):

$$\overline{w' u'} = C_d \cdot \left(\overline{u_2} - \overline{u_1}\right)^2 \tag{29}$$

The coefficient C_d can be calculated by combining Eqs 22, 24, 26 and 29. Using $z_1 = z_{0m}$, and considering that at $z = z_{0m}$, the wind speed is zero:

$$C_d = \left[k \Big/ \ln\left(\frac{z_2}{z_{0m}}\right) \right]^2 \tag{30}$$

In neutral conditions it can be assumed that $C_h=C_v=C_d$, and thus:

$$H = -\rho \cdot c_p C_d \cdot u_2 \cdot \left(\overline{T_2} - \overline{T(z_{0m})}\right) \tag{31}$$

The appearance of the average temperature at height z_{0m} in Eq 31 is inconvenient. It can be eliminated by assuming a logarithmic wind profile for temperature too, by defining a scalar roughness height for heat transfer z_{0h} at which the extrapolated temperature profile fitted through $\overline{T_2}$ and $\overline{T(z_{0m})}$ becomes T_0, i.e. the kinematic surface temperature. Using this we finally express the aerodynamic resistance r_{ah} in neutral conditions as:

$$r_{ah} = \frac{\ln\left(\frac{z_2}{z_{0m}}\right)\ln\left(\frac{z_2}{z_{0h}}\right)}{k^2 \cdot u} = \frac{\ln\left(\frac{z_2}{z_{0h}}\right)}{ku_*} \tag{32}$$

The roughness height, z_{0h}, changes with surface characteristics, atmospheric flow and thermal dynamic state of the surface (Blümel, 1999; Massman, 1999). It can be shown that:

$$z_{0h} = z_{0m} / \exp\left(kB^{-1}\right) \tag{33}$$

where B^{-1} is the inverse Stanton number, a dimensionless heat transfer coefficient.
Free convection might alter the forced convective eddies generated by wind turbulence.
During daytime or when temperature decreases with height, convection amplifies the vertical eddy motions (unstable condition). During the night or when inversion conditions occur, and temperature increases with height, the horizontal eddy motions are enhanced (stable conditions).
Mechanical turbulence and buoyancy coexists in a form of a hybrid regime known as mix-convection. Monin and Obukhov showed that these conditions eventually lead to an alteration of the wind and temperature profiles (Brutsaert, 1982). The Monin-Obukhov similarity theory uses dimensional analysis to correct the wind profile produced by buoyancy effects in such conditions. A non-dimensional correction factor for momentum transfer $\varphi_m(\xi)$ is introduced to correct the wind profile gradient for conditions different from neutral, in which ξ is the ratio of thermal to mechanical turbulence:

$$\frac{du}{dz} = \frac{u_*}{kz} \cdot \varphi_m(\xi) \tag{34}$$

They introduced semi-empirical functions to correct the wind profile depending on the stability, based on dimensional analysis, of the form:

$$u_* = \frac{\overline{u(z)}}{k} \cdot \left[\ln\left(\frac{z-d_0}{z_{0m}}\right) - \Psi_m\left(\frac{z-d_0}{L}\right) + \Psi_m\left(\frac{z_{0m}}{L}\right) \right] \tag{35}$$

$$H = \frac{T_s - \overline{T(z)}}{k \cdot u_* \cdot \rho \cdot c_p} \cdot \left[\ln\left(\frac{z-d_0}{z_{0h}}\right) - \Psi_h\left(\frac{z-d_0}{L}\right) + \Psi_h\left(\frac{z_{0h}}{L}\right) \right] \tag{36}$$

Where L is defined as the Monin-Obukhov length ($L = z \cdot \xi$) [m], calculated as:

$$L = -\frac{\rho_a \cdot C_p \cdot u_*^3 \cdot \overline{T(z)}}{k \cdot g \cdot H} \tag{37}$$

Where g is the gravity constant (9.81 m s^{-2}). Semi-empirical expressions for the stability corrections Ψ_h and Ψ_m can be found in the literature, for example Paulson (1970) and Brutsaert (1982). It is important to realize that L depends on air temperature and sensible heat flux, while sensible heat flux and air temperature in turn depend on L. For this reason, an iterative procedure is needed to calculate L, u_* and H using Eqs 35-37.

2.3 Ground heat flux

The ground heat flux has received relatively little attention compared to the other terms. This is often justified, because ground heat flux is usually the smallest of all terms. Moreover, the 24-hour sum of ground heat flux is close to zero, because the heat absorbed during the day is released during the night.

At the moment of a satellite overpass, ground heat flux is not necessarily negligible. At midday it usually varies from 10% of net radiation for dense vegetation to 45% of net radiation for bare soil (Clothier et al., 1986). Often a vegetation cover dependent ratio between G and R_n is assumed at satellite overpass (Kustas et al., 1990).

If more accurate estimates of ground heat flux are required, for example in areas with sparse vegetation, then remote estimates of ground heat flux are possible with the method of Van Wijk and De Vries (1963). For this method, diurnal cycles of land surface temperature and net radiation are needed (Verhoef, 2004; Murray and Verhoef, 2007); this means that time series of data of a geostationary satellite are required.

An equation for ground heat flux can be derived the thermal diffusion equation, assuming a periodic land surface temperature:

$$G(t) = \Gamma \sum_{k=0}^{n} \sqrt{k\omega/2} \cdot \left(A_k \cdot \sin(\omega k t) + B_k \cdot \cos(\omega k t) \right) \tag{38}$$

where Γ is the thermal inertia of the soil (J m^{-2} K^{-1} s$^{-1/2}$), which depends on texture and soil moisture, t is time (s), $\omega = (2\pi/N)$ is the radial frequency (s^{-1}), N the length of the time series [s], A and B integration coefficients [°C], and n the number of harmonics. The coefficients A and B are fitted against the observed land surface temperature time series, for a chosen number of harmonics. The thermal inertia Γ can be estimated from soil texture and soil moisture, or calibrated against night time radiation, by assuming that night time radiation equals the night-time ground heat flux.

2.4 Latent heat flux

Latent heat flux is finally calculated as a residual of the energy balance (Eq. 1). Because H, G and R_n are instantaneous measurements, it is necessary to find a procedure to integrate to daily totals. A common way to carry out this integration, is by making use of the evaporative fraction, Λ. The evaporative fraction (Brutsaert and Sugita, 1992) is the energy used for the evaporation process divided by the total amount of energy available for the evaporation process:

$$\Lambda = \frac{\lambda E}{\lambda E + H} = \frac{\lambda E}{R_n - G} \tag{39}$$

It is assumed that the evaporative fraction remains constant throughout the day.

$$\Lambda_{inst} = \Lambda_{24hrs} \tag{40}$$

Assuming that the ground heat flux integrated over 24-hours is negligible, the evapotranspiration rate over 24 hours can be calculated as:

$$E_{24hrs} = \frac{8.64 \cdot 10^7 \, \Lambda_{inst}}{\lambda \rho_w} R_{n,24hrs} \tag{41}$$

Where λ= 2.0501-0.00236 T_{water} MJ kg^{-1} (T in °C), ρ_w= 1000 kg m^{-3} and R_{n24} is the average net radiation over 24 hours [W m^{-2}].

The assumption of a constant evaporative fraction may lead to underestimates of daily evaporation, because the evaporative fraction in reality has a diurnal cycle with a concave shape (Gentine, et al., 2007). The concave shape is caused by changes in weather conditions (wind, advection, humidity), a phase difference between ground heat flux and net radiation, and stomatal regulation. There is an alternative to the assumption of constant evaporative fraction if hourly weather data are available. It may then be assumed that the ratio of actual to reference evaporation is constant over the day; hourly values of reference evaporation can be calculated (Allen et al., 2007). The ratio of actual to reference evaporation is more stable, because it eliminates the effects of diurnal variations in weather conditions.

3. Data requirements and sensitivity

Every remote sensing based SEB model requires a sequence of dedicated ground and remote sensing data to properly operate. Efforts increasingly focus on the remote estimation of the necessary variables, but ground data are still needed in addition.

All models require net radiation and land surface temperature retrieved from remote sensing. The additional required information varies among algorithms. As an example we list the input needed for the remote sensing model SEBS (Su, 2002). This model explicitly solves Eqs 35-37. It also includes an algorithm to estimate kB^{-1} from vegetation cover fraction.

SEBS requires the following data, most of which cannot be retrieved from remote sensing, but is obtained from ground-based meteorological data instead:

1. Reference height z_{ref} [m]: height from the ground where measurements of temperature, wind, pressure and specific humidity are made [m].
2. Air Temperature at reference height (T_a) [°C].
3. Specific humidity [kg.kg^{-1}] or relative humidy [%], for calculation of emissivity of the sky.
4. Wind speed at the reference height (u_{ref}) [m.s^{-1}].
5. Air Pressure at reference height [Pa].
6. Air Pressure at land surface and reference height [Pa].
7. The planetary boundary (PBL) height h_i [m], required for the calculation of stability. It can be estimated by radiosounding or using atmospheric model outputs. By default h_i=1000 m.
8. A map of vegetation heights, or alternatively, classes associated with vegetation height values, a map of Leaf Area Index (LAI) from which vegetation height is estimated.

All meteorological input must be instantaneous information collected at the time of satellite overpass, interpolated and re-sampled to the pixel size. Other models require similar input. Two source models do not use the concept of kB^{-1}, but require separate resistances for soil and vegetation. Although the exact input data varies per algorithm, the most important are those related to the calculation of sensible heat flux, in particular the surface-air gradient and the corresponding aerodynamic resistance r_{ah}. The success or failure of a SEB relies on the skills of the research team to extract realistic values for these two variables. For this reason, we will discuss these in more detail in the following sections.

3.1 The temperature gradient

For the temperature gradient we need to estimate both the air and the land surface temperature. The issue is that sensible heat flux is proportional to a difference between two temperatures which are obtained from two different sources in the same vertical. For this reason great care should be taken to retrieve both temperatures accurately.

For the air temperature at reference height, interpolated data of meteorological stations are commonly used. We need the air temperature well above the canopy, in the atmospheric surface layer, for which the aerodynamic resistance is defined. The standard measurement height in meteorological stations of 2 m cannot be used for vegetation taller than this height. Thus a conversion of temperatures from the meteorological stations to a higher reference height is needed. Another option is to use temperature profiles disseminated by organizations like EUMetsaf (www.eumetsat.int).

For the surface temperature it is necessary to take a closer look at the concepts first. As discussed before, the radiometric temperature is the temperature as it is retrieved from a remote radiometer by inverting Stefan-Boltzmann's law, assuming a bulk emissivity for the thermal spectrum range of the radiometer. The kinematic temperature is the real, contact temperature. A third definition is needed here: the aerodynamic temperature, which is hypothetic temperature obtained when extrapolating the vertical profile of air temperature to the depth z_{0h}. The aerodynamic temperature is a conceptual model parameter that is close to the kinetic temperature, but they are not equal. The reason is that kinematic temperature varies between the elements of the surface within a remote sensing pixel. For example, sunlit and shaded parts of the soil and canopy may have rather different temperatures. This is particularly the case in a heterogeneous landscape, where bare soil, vegetated and paved areas are mixed. It is even the case in a homogeneous land cover, where leaf temperatures may differ depending on their vertical position in the crown. This is illustrated in Fig 3, showing the diurnal variations of contact temperatures of a needle forest in the Netherlands, the Speulderbos site, measured during a field campaign in The Netherlands on 16 June 2006 (Su et al., 2009). The lines are ensembles of 8 soil surface and 9 needle temperature sensors, mounted at different heights of the canopy of just a few trees.

The heterogeneity of the soil and canopy temperatures will affect the radiometric surface temperature. The radiometric temperature is predominantly affected by the upper, visible, part of the canopy. Lower canopy layers also contribute to the outgoing upward radiation, but their contribution will be relatively low due to re-absorption of radiation. The radiometric temperature also depends on the solar angle and the observation angle of the satellite.

A new model to analyse these effects is the model SCOPE (**S**oil **C**anopy **O**bvservation of **P**hotosynthesis and the **E**nergy balance). This model is a radiative transfer model combined with an energy balance model for homogeneous vegetation (Van der Tol et al., 2009). With SCOPE one can analyse the relation between the sensible heat flux, the kinematic

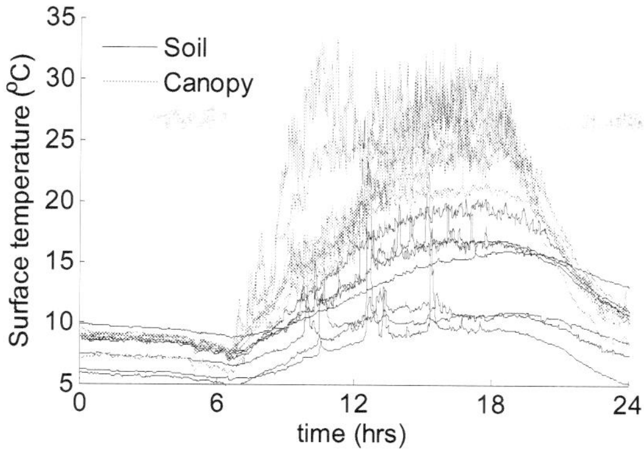

Fig. 3. Diurnal cycle of 8 soil and 9 needle contact temperature measurements (with NTC sensors) at the Speulderbos needle forest site in the Netherlands, on 16 June 2006.

temperatures of different elements in the canopy and the radiometric temperature. An example of an analysis carried out with SCOPE is shown in Fig 4. This figure shows simulated radiometric temperature of sparse but homogeneous crop as a function of the satellite observation azimuth (the counter-clockwise rotation angle from the top in the graph) and zenith angle (distance from the centre of the graph). We can see pronounced differences in the observed radiometric temperatures. Cleary visible is the hotspot, the situation where the solar zenith and azimuth angles equal those of the sensor. In the hotspot the radiometric temperature is higher than outside the hotspot. Radiometric temperatures are also higher at lower zenith angles compared to NADIR observations (vertically downward, in the centre of the graph). The differences in temperature are up to 2 °C, indicating that care should be taken of the observation angle relative to the solar angle. It is also possible to exploit the differences in radiometric surface temperature observed at different angles in order to separate soil and canopy kinetic temperatures (Timmermans et al, 2009).

How severe is an error of 2 °C for the estimation of sensible heat flux? Equation 18 shows that the error in sensible heat flux is proportional to the ratio of the temperature gradient to the resistance. This means that for the same error in the temperature gradient, the error in the sensible heat flux will be larger if the aerodynamic resistance is low than if the aerodynamic resistance is high.

In order to consider the sensitivity more precisely, we take the example of a situation where the aerodynamic resistance is low: the Speulderbos forest site in The Netherlands. This site is equipped with a 46-m tall eddy covariance measurement tower. Because of the low aerodynamic resistance, the sensitivity to temperature is expected to be relatively high. For this site, we calculated the friction velocity and the sensible heat flux with Eqs 35-37, using a canopy height of 30 m, the measured wind speed and temperature at 45 m height, radiometric temperature measured with a long-wave radiometer, and assuming that z_{0m} = 0.12 h and d = 0.67 h.

Figure 5 shows the results for 15-18 July 2009. The day-time friction velocity matches well with the measurements, showing that the calculation of aerodynamic resistance was

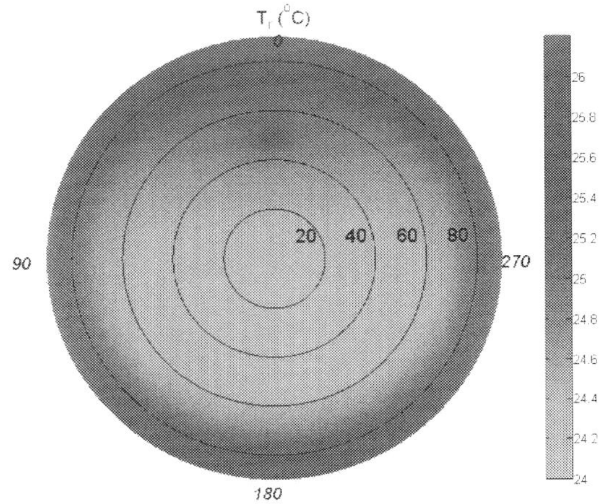

Fig. 4. Hemispherical graph of simulated radiometric surface temperature of a thinned maize crop with a LAI of 0.25, as a function of viewing zenith angle and viewing azimuth angle (relative to the solar azimuth). Zenith angle varies with the radius, the azimuth angle (in italic) increases while rotating anticlockwise from north. The solar zenith angle was 48° (after Van der Tol et al., 2009).

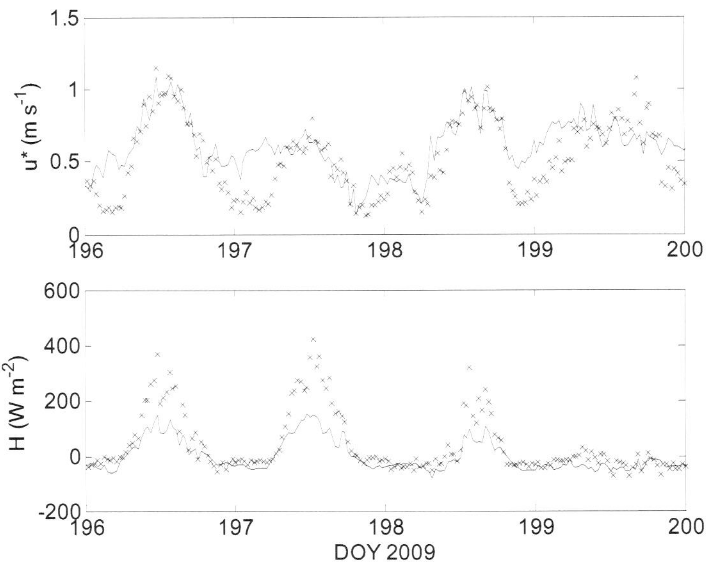

Fig. 5. Measured (symbols) and modeled (line) friction velocity u^* and sensible heat H flux versus Julian day number (14-19 July 2009) for an eddy covariance tower in the Speulderbos forest site, The Netherlands.

accurate. During the night (stable conditions), the performance is worse, but this is not a large problem because the night-time sensible heat flux proved very small. However, there is a 50% error in afternoon sensible heat flux. Can this be related to an error in the surface temperature? Figure 6 shows the result of a sensitivity analysis to surface temperature. A consistent bias was added to the measured time series (x-axis), and the resulting root mean square error (RMSE) of friction velocity and sensible heat flux was calculated (y-axis). The RMSE reaches a minimum when surface temperature is 0.5 °C above the measured value, but it rises to unacceptably high values of the absolute temperature bias is greater than 2 °C. In this example, field data of radiometric surface temperature were used. What if remote sensing data are available? Satellite products are available at either high temporal (geostationary satellites) or at high spatial resolution (polar orbiting satellites). Data are available at a spatial resolution of 3-5 km and a temporal resolution of 15 minutes (MeteoSat or GOES) to 1 km resolution at a daily time scale (AVHRR, MODIS or MERIS), or 60 m with a repetition time of weeks to months (LANDSAT, ASTER). The low temporal resolutions are not really useful, because of the dynamic nature of the turbulent heat fluxes. The daily revisits are useful provided that reasonable assumptions are made about the diurnal cycles of the fluxes (see Sect 2.4). The orbits are designed to overpass at the same solar time every day. The 15-minute intervals are ideal, but the spatial resolution makes the estimation of an effective aerodynamic resistance difficult, as we will see later.

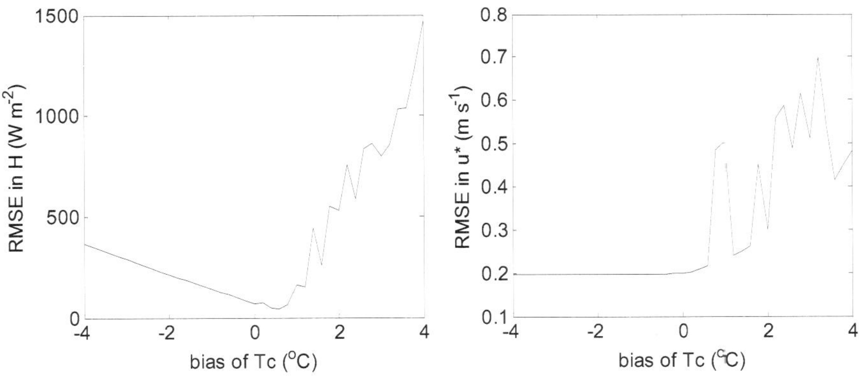

Fig. 6. Root mean square error of modelled sensible heat flux and friction velocity versus a forced bias in the observed radiometric surface temperatures for the Speulderbos forest site, for 14-19 July 2009.

A final issue that needs to be considered is the topography. In areas with large elevation differences, the interpolation technique for air temperature data is crucial. Errors of several degrees in the air temperature are easily introduced if an incorrect adiabatic lapse rate is used.

It is possible to circumvent the problem of estimating the temperature gradient by using an image based calibration (Bastiaanssen et al., 1998), in which assumptions are made for the energy balance state at the hottest and the coolest pixel in the image. In the first versions of this approach the calculated fluxes depended on the size of the image that was selected and the on assumption that the hottest pixel is dry, but more recent developments do not suffer from this drawback. The Mapping Evapotranspiration with Internalized Calibration (METRIC) model (Allen et al., 2007) uses reference evapotranspiration of alfalfa to calibrate the relation between the temperature gradient and the measured surface temperature. In the METRIC model it is assumed that the evaporation in the wettest pixel is 5% above the reference evapotranspiration, and the evaporation of the driest pixel is estimated with a soil-vegetation-atmosphere model. This has the additional advantage that the evaporation values are bound to a realistic minimum and a realistic maximum rate. The METRIC model also accounts for topography by correcting radiation for slope and aspect and temperature for elevation using a local lapse rate.

3.2 Sensitivity to the aerodynamic resistance

The roughness length z_{0m} (and often displacement height is linked to it) is recognized as the main source of error in the remote estimate of ET. Currently, there are several methods that can be used to approach a good z_{0m} (see Table 2).

When near surface wind speed and vegetation parameters (height and leaf area index) are available, the within-canopy turbulence model proposed by Massman (1999) can be used to estimate aerodynamic parameters, *d*, the displacement height, and, z_{0m}, the roughness height for momentum. This model has been shown by Su et al. (2001) to produce reliable estimates of the aerodynamic parameters. If only the height of the vegetation is available, the relationships proposed by Brutsaert (1982) can be used. If a detailed land use classification is available, for example based on LandSat images, the tabulated values of Wieringa (1993) can be used. By using literature values, errors in the canopy height of the order of decimetres to several metres are likely to occur, and errors in the roughness length in the order of decimetres.

Method	Input needed	Remark
z_{0m} = 0.136 h	Vegetation height map (h)	
from Lookup table (LUT)	Vegetation map & z_{0m} LUT	
From vegetation index	Vegetation index maps	
z_{0m} from modelling	Landuse & veg. structure	
LIDAR	Experimental. Costly.	Costly method
Retrievals from wind profiles	Wind speed profiles	Point values only

Table 2. Methods for the estimation of z_{0m} (After: Su, 2002).

When all of the above information is not available, then the aerodynamic parameters can be related to vegetation indices derived from satellite data. However in this case, care must be taken, because the vegetation indices saturate at higher vegetation densities and the relationships are vegetation type dependent. For example, characteristic of the land surface are sometimes calculated from indices like the Normalized Difference Vegetation Index (NDVI), but there is no reason why NDVI should have a universal relation with surface roughness. A grass field may have a similar NDVI to that of a forest, but a roughness length that is an order of magnitude smaller. For this reason, literature data or ground-truth data are indispensible for an accurate estimate of the surface roughness. A relation between NDVI and surface roughness can only be made for low vegetation, normally irrigated. In that case a non-linear relation with vegetation structure is first established by assigning a maximum value and a minimum value of height corresponding to values of NDVI.

We illustrate the sensitivity of the aerodynamic resistance model with the data set of the Dutch forest site introduced in the previous section. Now, the height of the forest was varied between 5 and 45 m, and the RMSE of sensible heat flux and friction velocity evaluated (Fig 7). Note that the vertical scale in Fig 7 is much smaller than in Fig 6, which indicates that for forest, the model is less sensitive to errors in the canopy height than errors in surface temperature.

Sparse canopies require special attention. In sparse canopies the temperature differences between canopy and soil may be over 20 °C. In addition, no canopy height can be defined, which makes it difficult to estimate the roughness length z_{0m}, and normally d is neglected. A solution to these problems is to program a two-source model (e.g. Norman et al., 1995). An alternative solution is to modify the parameter kB^{-1} to incorporate the differences in surface temperature implicitly in the value of z_{0h} (Verhoef et al., 1997). We will illustrate the latter solution with a simple example.

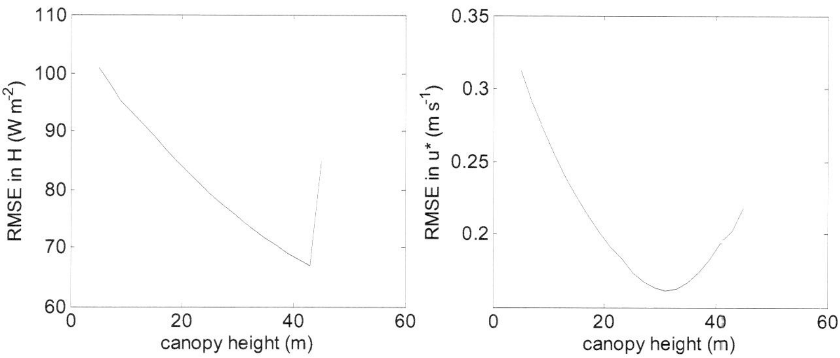

Fig. 7. Root mean square error of modelled sensible heat flux H and friction velocity u^* velocity versus assumed canopy height of the Speulderbos forest site, for 14-19 July 2009.

The sparse canopy of our example is a study site in the province of León, Spain. The vegetation cover fraction is 11%, consisting of patches of 6-m tall *Quercus ilex* and *Quercus pyrenaica*. Data of an eddy covariance flux tower are used for validation of the satellite product. For this site, a roughness length of z_{0m}=0.2 m was assumed, and a displacement height of d=0. The friction velocity and sensible heat flux were again calculated from Eqs 35-37. For z_{0h}, a value of 0.02 m was initially assumed (kB^{-1} = 2.3), and for wind speed, the field measurements at the flux tower were used. For net radiation and surface temperature, 15-minute interval MeteoSat Second Generation (MSG) satellite data were used. The top panels in Fig 8 show the results of the satellite based algorithm. The friction velocity observations are accurately reproduced, but the modelled sensible heat flux is extremely high, even double the net radiation. The overestimate is solved when we reduce z_{0h} by four orders of magnitude (kB^{-1} = 11.5). The reduction in z_{0h} needed to match the model with the observations is large. This problem was discussed earlier after the Hapex-Sahel measurement campaign (Verhoef et al., 1997). It was then concluded that the whole concept of kB^{-1} is questionable. It is indeed recommended to avoid the use of kB^{-1}, and this can be done in two ways: (1) by using more complicated two-source models for sparse vegetation, or (2) to use image-based calibration to relate surface temperature to a temperature gradient between two heights well above z_{0h}. The second approach is used in models such as SEBAL (Bastiaanssen et al., 1998) or METRIC (Allen et al., 2007).

A model exists to estimate the kB^{-1} from vegetation density (Su et al., 2001). This model is used in the remote sensing algorithm SEBS (Su, 2002). However, care should be taken with any kB^{-1} model for areas where no detailed information on cover or other field data are available for calibration.

In the future, global maps of surface roughness may become timely available. Through synthesis of LiDAR with high resolution optical remote sensing, the roughness parameters have been successfully estimated spatially (Tian et al., 2011). Surface maps produced with laser satellites (NASA's ICESat and the future ICESat2) are also promising tools for estimating roughness (Roxette et al., 2008).

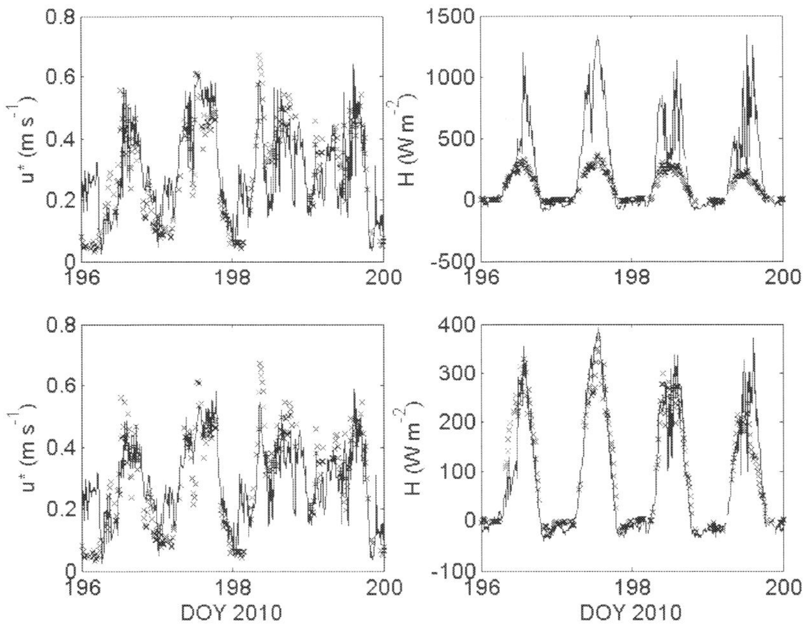

Fig. 8. Measured (symbols) and modeled (line) friction velocity u^* and sensible heat H flux versus Julian day number (14-19 July 2010) for an eddy covariance tower in the sparsely vegetated area of Sardon, Spain. Top graphs: using z_{0m} = 0.2 and kB^{-1} = 2.3. Bottom graphs: using z_{0m} = 0.2 and kB^{-1} = 11.5.

4. Conclusions

All remote sensing algorithms for ET make use of the energy balance equation (EBE). In this equation, latent heat flux is calculated as a residual of the energy balance. Net radiation can

be estimated from remote sensing products relatively easily. Ground heat flux can only be retrieved with geostationary satellites for sparsely vegetated areas or bare land. It is usually a minor term in vegetated areas that causes relatively small errors in the final ET product. The most critical component of the energy balance is the sensible heat flux. In the calculation of the sensible heat flux, both the temperature difference (land surface temperature minus the air temperature) and the aerodynamic resistance need careful attention.
In areas with high elevation differences, the errors in temperature are usually so high, and temperature correction using local lapse rates is necessary. In flat areas, a local sensitivity analysis is recommended. For forest, the accuracy of the temperature gradient should be better than 2 °C in order to achieve reasonable results. In sparse vegetation two source models are preferred over single-source models, because in the latter, parameterization of z_{0h} on operational basis is no better than a wild guess. If a two-source model is not an option, then image based calibration using reference evaporation is a good alternative in these areas. Accurate roughness information (z_{0m}) is required; the information is preferably verified and monitored on the ground. Satellite laser altimetry provides a promising tool for better roughness estimates in the near future.

5. References

Allen, R.G., Pereira, L.S., Raes, D. & Smith M. (1998). Crop Evapotranspiration. Guidelines for computing crop water requirements. *FAO Publication* 56, Rome, Italy, pp. 300

Allen, R.G., Tasumi, M. & Trezza, R. (2007). Satellite-based energy balance for mapping evapotranspiration with internalized calibration (METRIC) – model. *J. Irr. Drainage Engineering ASCE* 133 (4), pp. 380-394

Bastiaanssen, W.G.M., Menenti, M., Feddes, R.A. & Holtslag, A.A.M. (1998). A remote sensing surface energy balance algorithm for land (SEBAL). 1. Formulation. *J. Hydrol.* 212-213, pp. 198–212

Blümel, K. (1999). A simple formula for estimation of the roughness length for heat transfer over partly vegetated surfaces, *J. Appl. Meteorol.*, 38, pp. 814–829

Brutsaert, W. (1975). On a derivable formula for long-wave radiation from clear skies. Water Resour Research 11 (5), pp. 742-744, doi:10.1029/WR011i005p00742

Brutsaert, W. (1982). Evaporation into the atmosphere. *D. Reidel Publishing Co.*, Dordrecht, The Netherlands, pp. 300

Burtsaert, W., (2005). *Hydrology - an introduction.* Cambridge University Press: Cambridge, UK, pp. 605

Brutsaert, W. & Sugita, M. (1992). Application of self-preservation in diurnal evolution of the surface energy budget to determine daily evaporation. *J. Geophysical Res* 97, pp. 18377-18382

Courault, D., Seguin, B. & Olioso, A. (2005). Review on estimation of evapotranspiration from remote sensing data: From empirical to numerical modeling approaches. *Irrigation Drain Syst* 19, pp. 223–249

Clothier, B.E., Clawson, K.L., Pinter, P.J. Jr., Moran, M.S., Reginato, R.J. & Jackson, R.D. (1986). Estimation of Soil Heat Flux from Net Radiation During Growth of Alfalfa. Agric. *For. Meteorol.* 37, pp. 319-329

Doorenbos, J. & Pruitt, W.O. (1977). Guidelines for predicting crop water requirements, *Irrigation and Drainage Paper 24*, FAO, Rome, Italy, pp. 179.

Gentine P., Entekhabi D., Chehbouni A., Boulet G. & Duchemin B. (2007). Analysis of evaporative fraction diurnal behaviour. *Agric. Forest Meteorol.*, 143 (1-2), pp. 13-29

Gieske, A.S.M. (2003). Operational solutions of actual evapotranspiration. In: Understanding water in a dry environment. Hydrological processes in arid and semi-arid zones. I. Simmers (Ed.), *International Association of Hydogeologists*, Balkema Publishers, The Netherlands

Glenn, E.P., Huete, A.R., Nagler, P.L., Hirschboeck, K.K. & Brown, P. (2007). Integrating remote sensing and ground methods to estimate evapotranspiration. *Crit Rev Plant Sci* 26(3), pp. 139–168, doi:10.1080/07352680 701402503

Gowda, P.H., Chavez, J.L., Colaizzi, P.D., Evett, S.R., Howell, T.A. & Tolk, J.A. (2007). Remote sensing based energy balance algorithms for mapping ET: current status and future challenges. *Trans Am Soc Agric Biol Engineers* 50(5), pp. 1639–1644

Heusinkveld, B.G. Jacobs, A.F.G., Holtslag, A.A.M. & Berkowicz, S. M. (2004). Surface energy balance closure in an arid region: role of soil heat flux. *Agric. Forest Meteorol.* 122(1-2), 21-37, doi:10.1016/j.agrformet.2003.09.005

Heusinkveld, B.G., Jacobs, A.F.G., Holtslag, A.A.M. & Berkovicz, S.M. (2004). Surface energy balance closure in an arid region: role of soil heat flux. *Agric. Forest Meteorol.* 122 (1-2), pp. 21-37

Jackson, R.D., Reginato, R.J. & Idso, S.B. (1977). Wheat canopy temperatures: A practical tool for evaluating water requirements. *Water Res. Res* 13, pp. 651–656

Jackson R.D., Moran, M.S., Gay, L.W. & Raymond, L.H. (1987). Evaluating Evaporation from Field Crops Using Airborne Radiometry and Ground-based Meteorological Data. *Irrig. Sci.* 8, pp. 81-90.

Kalma, J.D., McVicar, T.R. & McCabe, M.F. (2008). Estimating land surface evaporation: A review of methods using remote sensed surface temperature data. *Surveys in Geophysics*, 29, pp. 421-469

Kustas, W.P. & Daughtry, C.S.T. (1990). Estimation of the Soil Heat Flux/Net Radiation Ratio from Spectral Data. *Agric. For. Meteorol.* 49, pp. 205-223

Maidment, D. (1992). *Handbook of Hydrology*. Ed: D.R. Maidment, McGraw-Hill, Inc.

Massman, W.J. & Weil, J.C. (1999). An analytical one-dimensional second-order closure model of turbulence statistics and the Lagrangian time scale within and above plant canopies of arbitrary structure. *Boundary-Layer Meteorol.* 91, pp. 81–107

Murray, T. & Verhoef, A. (2007). Moving towards a more mechanistic approach in the determination of soil heat flux from remote measurements I. A universal approach to calculate thermal inertia. *Agric Forest Meteorol.* 147, pp. 80–87

Norman, J.M., Kustas, W.P. & Humes, K.S. (1995). Source approach for estimating soil and vegetation energy fluxes in observations of directional radiometric surface temperature. *Agric. Forest Meteorol.* 77 (3-4), pp. 263-293

Obukov, A. (1946). Turbulence in an atmosphere with a non-uniform temperature. *Bound. Layer Meteorol.* 2, pp. 7-29

Paulson, C.A. (1970). The mathematical representation of wind speed and temperature profiles in the unstable atmospheric surface layer. *J. Applied Meteorol.* 9 (6), pp. 857-861

Reginato R., Jackson, R. & Pinter, J.Jr. (1985). Evapotranspiration Calculated from Remote Multispectral and Ground Station Meteorological Data. *Remote Sens. Environ.* 18, pp. 75-89

Rosette, J.A.B., North, P.R.J. & Suárez, J.C. (2008). Vegetation height estimates for a mixed temperate forest using satellite laser altimetry. *Int. J. of Rem. Sens.*, 29(5), pp. 1475-1493.

Seguin, B. (1988). Use of Surface Temperature in Agrometeorology. In: *Applications of Remote Sensing to Agrometeorology*. Ed F. Toselli, pp. 221-240

Su, Z. (2002). The Surface Energy Balance System (SEBS) for estimation of turbulent heat fluxes. *Hydrol. Earth Syst. Sci.* 6, pp. 85-99

Su, Z., Schmugge, T., Kustas, W.P. & Massman, W.J., (2001). An evaluation of two models for estimation of the roughness height for heat transfer between the land surface and the atmosphere. *J. Appl. Meteorol.* 40, pp. 1933–1951

Su, Z., Timmermans, W. J., van der Tol, C., Dost, R., Bianchi, R.,Gómez, J A., House, A., Hajnsek, I., Menenti, M., Magliulo, V., Esposito, M., Haarbrink, R., Bosveld, F., Rothe, R., Baltink, H. K., Vekerdy, Z., Sobrino, J. A., Timmermans, J., van Laake, P., Salama, S., van der Kwast, H., Claassen, E., Stolk, A., Jia, L., Moors, E., Hartogensis, O. & Gillespie, A. (2009). EAGLE 2006 – Multi-purpose, multi-angle and multi-sensor in-situ and airborne campaigns over grassland and forest, *Hydrol. Earth Syst. Sci.*, 13, pp. 833–845

Tian, X. Li, Z.Y., Van der Tol, C., Su, Z., Li, X., He, Q.S., Bao, Y.F., Chen, E.X. & Li, L.H. (2011). Estimating Zero-Plane Displacement Height and Aerodynamic Roughness Length using Synthesis of LiDAR and SPOT-5 data, *Remote Sens. Environ.* 115, 2330-2341.

Timmermans, J., Verhoef, W., Su, Z. & Van der Tol, C. (2009). Retrieval of canopy component temperatures through Bayesian inversion of directional thermal measurements. *Hydrol. Earth Syst. Sci.*, 13, pp. 1249–1260

Van der Tol, C., Verhoef, W., Timmermans, J., Verhoef, A. & Su, Z. (2009). An integrated model of soil-canopy spectral radiances, photosynthesis, fluorescence, temperature and energy balance, *Biogeosciences* 6, pp. 3109–3129

Van Wijk, W.R., De Vries, D.A. (1963). Periodic temperature variations in a homogeneous soil. In: W.R. Van Wijk (Ed.), Physics of Plant Environment, North-Holland, Amsterdam, pp. 102–143

Verhoef, A., 2004. Remote estimation of thermal inertia and soil heat flux for bare soil. Agric. Forest Meteorol. 123, pp. 221–236

Verhoef, A. (2004). Remote estimation of thermal inertia and soil heat flux for bare soil. *Agric. Forest Meteorol.* 123, pp. 221–236

Verhoef, A., De Bruin, H.A.R. & Van Den Hurk, B.J.J.M. (1997). Some Practical Notes on the Parameter kB−1 for Sparse Vegetation. *J. Appl. Meteor.* 36, pp. 560–572, doi: 10.1175/1520-0450(1997)036<0560:SPNOTP>2.0.CO;2

Verma, S.B. & Bartfield, B.J. (1979). Aerial and Crop Resistances Affecting Energy Transport. In: *Modification of Aerial Environment of Crops*. Ed. B.J. Bartfield & J.F. Gerber, SAE, pp. 230-248

Vinukollo, R.K., Wood, E.F., Ferguson, C.R., & Fisher, J.B. (2011). Global estimates of evapotranspiration for climate studies using multi-sensor remote sensing data:

Evaluation of three process-based approaches. *Remote Sens. Environ.* 115, pp. 801-823

Wieringa, J. (1993). Representative Roughness Parameters for Homogeneous Terrain, *Boundary-Layer Meteorol.* 63, pp. 323–363

12

Estimation of the Annual and Interannual Variation of Potential Evapotranspiration

Georgeta Bandoc
University of Bucharest, Department of Meteorology and Hydrology
Center for Coastal Research and Environmental Protection
Romania

1. Introduction

Knowledge of ecological factors for all natural systems, including human-modified natural systems, is essential for determining the nature of changes in these systems and to establish interventions that must be achieved to ensure optimal functioning of these systems.

The purpose of this chapter is to identify annual and interannual variations of potential evapotranspiration, in conjunction with climate changes in recent years, on the coastal region of Sfântu Gheorghe - Danube Delta. Under natural conditions, evapotranspiration flows continuously throughout the year, representing a main link in the water cycle and an important heat exchange factor affecting ecosystems. Potential evapotranspiration is the maximum amount of water likely to be produced by a soil evaporation and perspiration of plants in a climate.

Real balance between the amount of precipitation fallen named P and the amount of water taken from the atmosphere as vapour, called potential evapotranspiration PET is of particular importance in characterizing climate, representing an expression of power absorption by the atmosphere and expressing quantity water on soil and vegetation that request (Henning & Henning, 1981).

The difference between precipitation (P) and potential evapotranspiration (PET), i.e. $P-PET$ known as ΔP is denoted by excess precipitation to PET (E) or deficit of precipitation to PET (D) if the difference is positive or, respectively, negative. The intensity of water loss through evaporation from the soil or by transpiration from the leaf surface is largely determined by vapour pressure gradient, i.e. the vapour pressure difference between leaf and soil surface and atmospheric vapour pressure (Berbecel et al, 1970).

The vapour pressure gradient is determined, in turn, by the characteristics of air and soil factors, such as: radiant energy, air temperature, vertical and horizontal movements of the air saturation deficit, the degree of surface water supply evaporation, plant biology and soil characteristics.

Heat factor also has a significant influence on evapotranspiration as temperature, on one hand, intensify of water vapour increases and, on the other hand, increases air capacity to maintain water vapour saturation state, reducing atmosphere's evaporated power (Eagleman, 1967).

2. General issues related to estimate

Potential evapotranspiration evidence and interannual variations of *PET* potential evapotranspiration and water balance, climate charts are used based on measurements from weather stations hydrothermal (Walter & Lieth, 1960; Walter, 1955, 1999; Köppen, 1900, 1936 etc.).

In 2005, Oudin et al. compile lists 25 methods for estimating potential evapotranspiration based on a series of meteorological parameters (Douglas et al, 2009).

Estimation of potential evapotranspiration can be done using the indirect method based on air temperature readings and diagrams and on Thornthwaite's tables (Thorntwaite, 1948; Donciu, 1958; Walter & Lieth, 1960).

Recent studies use Penman's equation for this purpose, Penman (Penman, 1946), Penman - Monteith (Thomas, 2000a, 200b; Choudhury, 1997; Allen et al., 1998; Chen et al, 2005). Also, in determining the potential evapotranspiration, other formulas have been used with results almost similar with the ones of direct measurements, such as formulas of Bouchet (Bouchet, 1964), Turc (Turc, 1954), Hargreaves (Hargreaves & Samani, 1982), Papadakis (1966), Hamon (1963), Priestley - Taylor (1972), Makkink (1957) (Lu et al, 2005) and Blaney-Criddle (1950) (Ponce, 1989).

Potential evapotranspiration *PET* is of great temporal variability and thus an estimation can be done based on heat and water vapour from the atmosphere (Dugas et al, 1991; Celliar and Brunet, 1992; Rana & Katerji, 1996; Droogers at al, 1996; Frangi et al, 1996; Linda et al, 2002 as cited in Chuanyan et al, 2004).

Another model of estimation for *PET* is based on soil moisture and rainfall (model Century) (Metherell et al 1992; Zhou et al, 2008 as cited in Liang et al, 2010). On the interaction of global precipitation and air temperature estimates can be done for potential evapotranspiration (Raich & Schlesinger, 1992; Buchmann, 2000; Andréassian et al, 2004; Li et al, 2008a, 2008b; Casals et al, 2009).

Estimation of potential evapotranspiration can be achieved also based on satellite measurements related to air humidity and wind characteristics, but only in case of high-resolution satellite images (Irmak, 2009).

In the estimation of *PET* remote sensing methods are applied (Chaudhury, 1997; Granger, 1997; Stefano & Ferro, 1997; Caselles et al, 1998; Stewarta et al, 1999). These methods are based using geographic information system using GIS spatial modeling (Baxter et al, 1996; Srinivasan et al 1996; Moore, 1996; Cleugh et al, 2007, Tang et al, 2010).

Other studies use numerical modeling to simulate various weather variables in a particular location, variables used to calculate potential evapotranspiration (Kumar et al, 2002; Smith et al, 2006; Torres et al 2011).

3. Research on characteristics of coastal area of potential evapotranspiration

The location where this study has been made is the south-east of Salt and marine field is bordered by the Black Sea coast in the east, marine low deltaic plain in the west and north-west and the arm of Sfântu Gheorghe –Danube Delta (fig. 1). To the south of arm of Sfântu Gheorghe is the marine plain Dranov , Sfântu Gheorghe secondary delta and Sacalin Island.

In the context of global climate change, interannual evolution analysis, annual and multiannual magnitudes that characterize the climate of a region are of particular interest (Palutikov et al, 1994; Chattopadhyay et al, 1997; Kouzmov, 2002; Oguz et al, 2006). This interest increases when it is a coastal region where sea atmosphere - interactions induce very specific issues.

Fig. 1. Study area location

The Danube Delta combines the temperate semi-arid climate space typical for the Pontic steppes. The aquatic very wide plane spaces, differently covered by vegetation and intrerrupted by the sandy islands of the marine fields, make up an active area specific to the delta and to the adjacent lagoons but totally different from that belonging to the Pontic stepps. This active area reacts upon the total radiation intercepted by the general circulation of atmosphere, resulting in a mosaic of microclimates (Vespremeanu, 2000, 2004).

For determining how climate changes affect the interannual potential evapotranspiration in the Sfântu Gheorghe costal area it was started, primarily from the fact that *PET* potential evapotranspiration has strong fluctuations in time and space as a direct consequence of the variation factors leads. Thus, in order to achieve the intended purpose of this chapter, interannual and annual potential evapotranspiration values were determined according to Thornthwaite's method, both for the period 1961 - 1990, taken as a reference period and analyzed for the studied period 2000 - 2009. Interannual differences $P - PET$ as well as annual amounts of the differences of the same sign, $\sum(P - PET)^+$ and $\sum(P - PET)^-$ as well as the annual review, are important climatic indicators. The determination of the efficiency of precipitation was done by calculating the difference $P - PET$ taken as reference period 1961 - 1990 and for the period under study from 2000 to 2009. Positive differences indicate excess water from rainfall, water shortages $\sum \Delta P^+$ and the negative ones indicate deficit of precipitation, water requirements from the atmosphere $\sum \Delta P^-$.

It was also determined the precipitation deficit offset by previously accumulated surpluses and deficits of precipitation uncompensated by previous surpluses.

To identify climate changes in coastal Sfântu Gheorghe area and deviations from the average annual values of air temperature and precipitation, diagrams were drawn, type Walter and Leith, to identify dry periods and also different indices and specific factors were calculated such as: Martonne arid index (I_{ar}), retention index offset (I_{hc}), the amount of rainfall in the period with t ≥ 10 ° C temp ($P_{t\geq 10^0 C}$) rainfall amount of soil loading in the months from November to March (P_{XI-III}), the amount of summer rainfall in July and August ($P_{VII-VIII}$), Lang precipitation index for the period with t ≥ 10 ° C ($L_{t\geq 10^0 C}$), precipitation index for summer Lang ($L_{VI-VIII}$) and Lang precipitation index for spring season (L_{III-V}) and annual and interannual precipitation deficits (*D*) and excess (*E*) respectively , comparing to potential evapotranspiration of 10 mm, 20 mm, 30 mm etc. These indices and ratios were calculated based on meteorological measurements for the period 1961 - 1990, taken as a reference period for the 2000 - 2009 period under study. In this chapter, climate charts are playing an important role in the knowledge of the climate changes in the studied area and also helps in determining the precipitation – evapotranspiration, and hence the temperature deficit or surplus in the form of precipitation from evapotranspiration. Dryness site layout is determined in this study. Curve surplus or deficit of precipitation from evapotranspiration is crucial in environmental hydrothermal annual and interannual knowledge of an area.

Climate chart includes curved surfaces and values of temperature, precipitation at the scale 1/5 and 1/3 and potential evapotranspiration after Thornthwaite, interannual, annual and for certain periods (the amount of rainfall during the period from November - March, yet soil load, and the summer period July - August). The diagram also contains interannual surpluses and deficits and total rainfall to *PET*, the deficits in compensated and uncompensated previous surpluses, Walter - Lieth dry period, the annual aridity index, retention index offset, Lang rainfall index, calculated for the period temperature $t \geq 10^0 C$, for summer and spring time.

At the bottom of the chart months of the year and intra-annual values ΔP are indicated to express the character of moisture or dryness of the climate in different months, the monthly differences in classification categories *E* and *D* for each 10 mm, 20 mm, 30 mm etc. On the diagram, for 1 degree of temperature correspond 5 mm, 3 mm respectively of precipitation. Scale 1/5 was chosen in order to maximally achieve principle of the rainfall curve to be above the temperature when precipitation *PET* outperforms, and below it, when *PET* exceeds precipitation. Scale 1/3 was chosen to determine the dry period after the Walter - Lieth, which lasts as long as the rainfall curve is well below that of temperature.

It is important to know to what extent and interannual deficit of precipitation to *PET* during the growing season is offset by the surplus of precipitation to *PET* during the loading of the soil with water from precipitation (late autumn - winter). In this way deficit or surplus annual and interannual of effective precipitation is obtained comparing to *PET*.

In case of no loss of water through surface runoff and water infiltration or gains, the excess water is retained during loading or accumulated in the soil, and it is called full hydrologic soil (Chiriță et al, 1977).

The entire accumulated surplus of precipitation is the main reserve of soil water in the vegetation is gradually consumed and evapotranspiration together with new fallen rains (Donciu, 1983).
For the studied area, where the climate is characterized by periods of dryness, the water reserve accumulated in the soil is gradually depleted by evapotranspiration and biomass formation. This last amount of water should be considered as an element of water balance, important in the quantitative ratio (Chiriță et al, 1977).
Until finishing the accumulated precipitation of the soil in each month, water loss through evapotranspiration and precipitation is compensated by the previous reserve accumulation. Once this reserve is ended, precipitation deficit starts for the area studied. Evapotranspiration consumes current rainfall, leaving an additional demand of the atmosphere, dissatisfied with the precipitation.
The deficit of precipitation in this period presents the quantitative nature of *PET*'s dry climate and soil and thus the existence of a period of severe water available to vegetation.

4. Results and discussion

The analysis of average monthly *PET* value as obtained for Sfântu Gheorghe, was a functional correlation of these values with the mean monthly air temperature T (Bandoc & Golumbeanu, 2010). For both analyzed periods, the correlations are straightforward.
From the calculation of correlation coefficient r and determination coefficient r^2 between potential evapotranspiration air temperature values of this coefficient $r = 0,98$, $r = 0,97$ and $r^2 = 96,04\%$, $r^2 = 94,04\%$ resulted, for the reference period 1961 - 1990 and for 2000 - 2009 period under study (fig. 2).
From the climate charts made for coastal Sfântu Gheorghe area (fig. 2, fig. 3, fig. 4, fig. 5, fig. 6, fig. 7, fig. 8 and fig. 9) for the analyzed periods, the result is a series of changes comparing to the duration of dryness reference interval 1961 - 1990.
Analizing the data the drought period for 2000-2009 was found to be 7 months which compared to the reference 1961-1990 (average drought) period of 6 months shows a increase of one month of drought per year.
Arid annual index Martonne calculation to determine the ratio between the amount of rainfall and temperatures $\left(I_{ar} = \dfrac{P}{T+10} \right)$ showed that for the period 2000 - 2009, there was a decrease in the value of the index with 17,71 % which leads to increased awareness of dryness for the studied area (fig. 10).
Rain index called Lang index or Lang factor of the period with temperatures ≥ 10 ^{0}C ($L_{t \geq 10^0 C}$), spring (L_{III-V}) and summer ($L_{VI-VIII}$) determined as a ratio of the average monthly precipitation values and P values of monthly average air temperature T $\left(L = \dfrac{P}{T} \right)$.
The results obtained for these intervals revealed that the index $L_{t \geq 10^0 C}$ values decreased by 20,22 % for the period with t ≥ 10 ^{0}C, for spring period L_{III-V} rainfall index fell 26,05 %, while during the summer $L_{VI-VIII}$ value of this index was 37,20 % compared to the reference period 1961 - 1990 (fig. 10).

Offset fluid index $\left(I_{hc} = \frac{\sum \Delta P^{+}}{\sum \Delta P^{-}}\right)$ expresses the extent of precipitation deficits are compensated by the surpluses. Values lower than the 1 $\left(I_{hc} \le 1\right)$ expressed precipitation deficits unabated. Following determination of the index for the two periods analyzed, that index values are 0,24 for the reference period 1961 - 1990 and 0,15 for the period 2000 - 2009. From the two values determined using the formula (0,24 and 0,15), for the past 10 years interval, results that the fluid compensation index decreased by 37,5 % compared to the reference period 1961 - 1990.

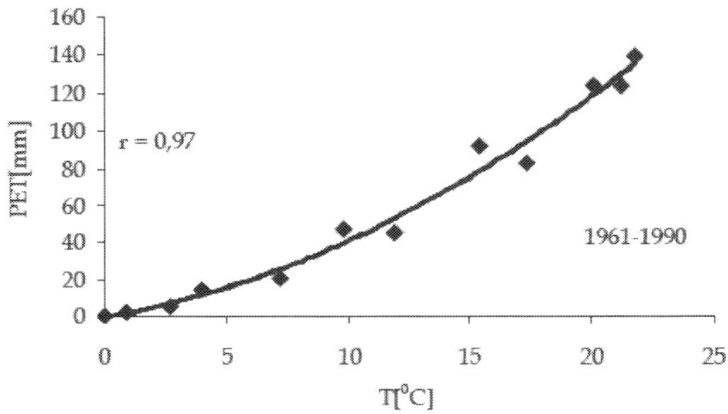

Fig. 2. Correlation between the potential evapotranspiration *PET* and air temperature *T* in the coastal region Sfântu Gheorghe for reference period 1961 - 1990

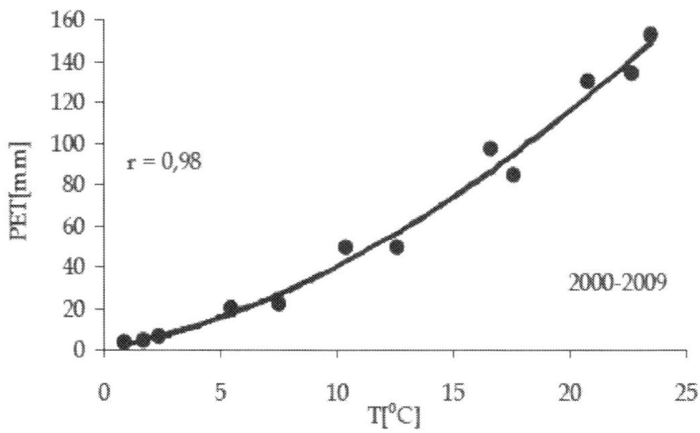

Fig. 3. Correlation between the potential evapotranspiration *PET* and air temperature *T* in the coastal region Sfântu Gheorghe for the period 2000 – 2009

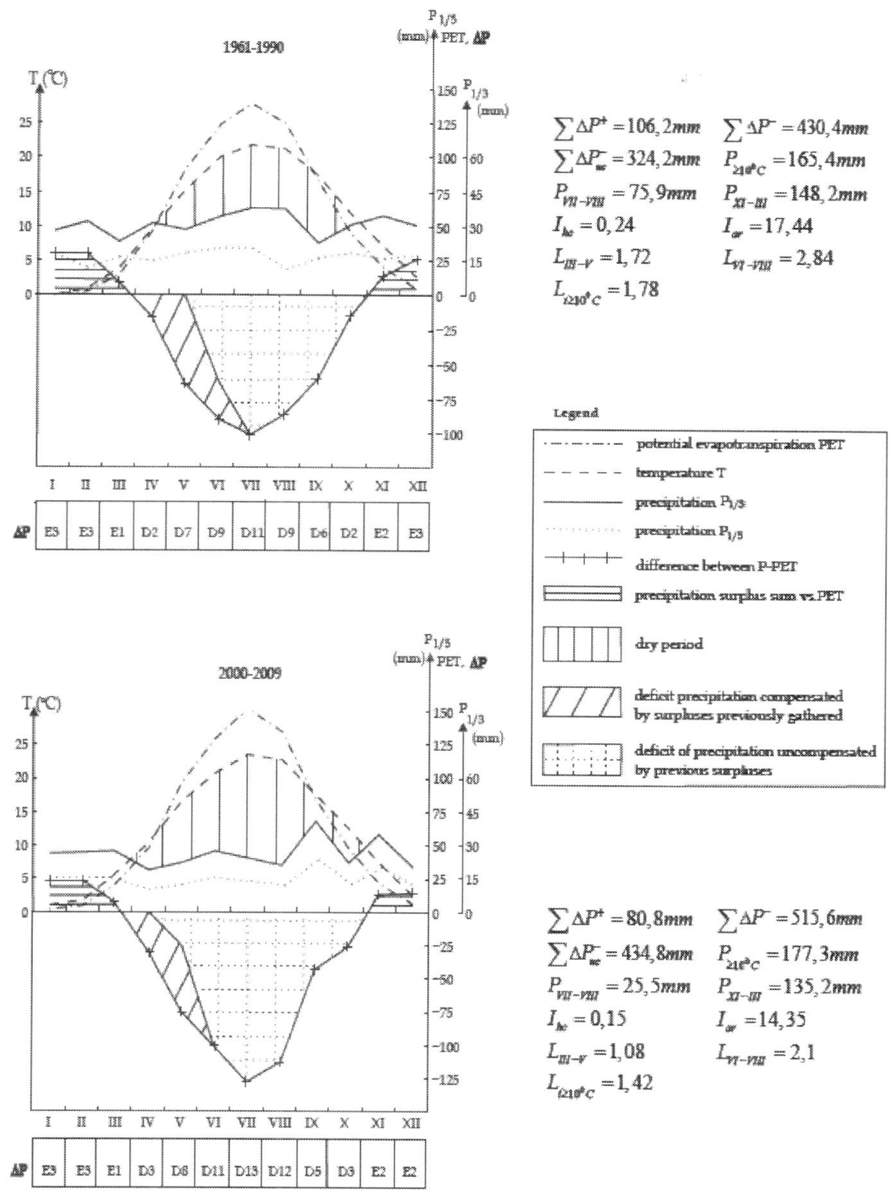

Fig. 4. Climate diagrams for reference interval 1960 - 1990 and interval 2000-2009 with characteristics sizes determined for reviewed site

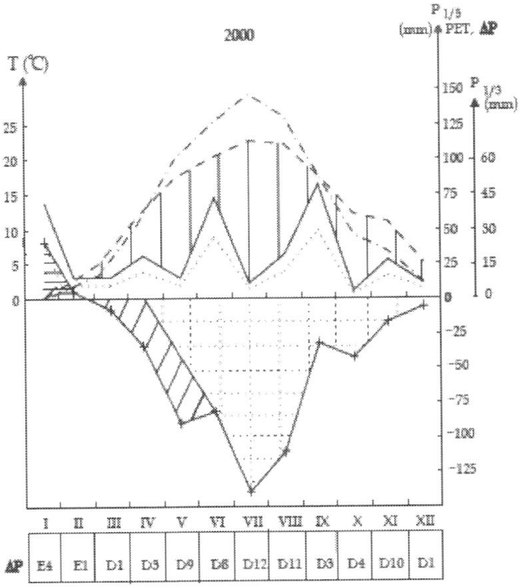

$\sum \Delta P^{+} = 45,7mm$ $\sum \Delta P^{-} = 561,9mm$

$\sum \Delta P^{-}_{vc} = 470,5mm$ $P_{\geq 10^{\circ}C} = 165,4mm$

$P_{VII-VIII} = 25,5mm$ $P_{XI-III} = 83,2mm$

$I_{hc} = 0,088$ $I_{ar} = 10,46$

$L_{III-V} = 1,05$ $L_{VI-VIII} = 1,04$

$L_{(\geq 10^{\circ}C)} = 1,23$

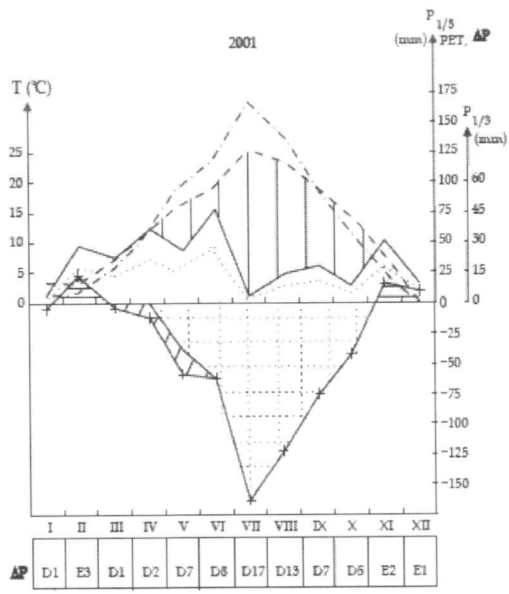

$\sum \Delta P^{+} = 48mm$ $\sum \Delta P^{-} = 557,4mm$

$\sum \Delta P^{-}_{vc} = 509,4mm$ $P_{\geq 10^{\circ}C} = 153,4mm$

$P_{VII-VIII} = 17,8mm$ $P_{XI-III} = 93,7mm$

$I_{hc} = 0,086$ $I_{ar} = 11,28$

$L_{III-V} = 0,94$ $L_{VI-VIII} = 2,51$

$L_{(\geq 10^{\circ}C)} = 1,53$

Fig. 5. Climate charts for years 2000 and 2001 and characteristics sizes determined for reviewed site

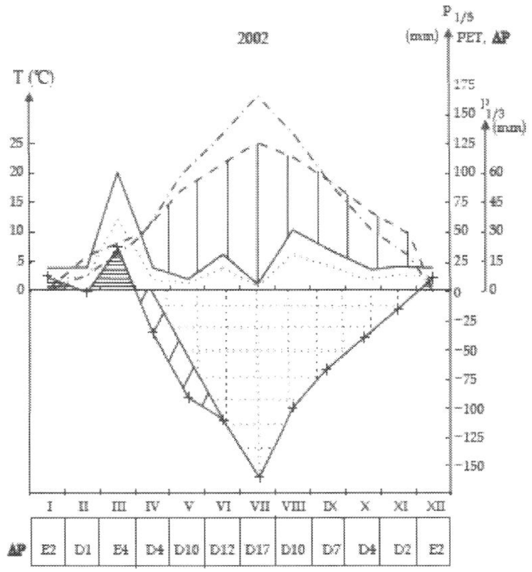

$\sum \Delta P^{+} = 59,7mm$ $\sum \Delta P^{-} = 626,9mm$

$\sum \Delta P^{-}_{w} = 509,4mm$ $P_{\geq 10^{\circ}C} = 88,4mm$

$P_{VII-VIII} = 34mm$ $P_{XI-III} = 19,1mm$

$I_{hc} = 0,095$ $I_{ar} = 9,33$

$L_{III-V} = 0,76$ $L_{VI-VIII} = 2,26$

$L_{t\geq 10^{\circ}C} = 0,75$

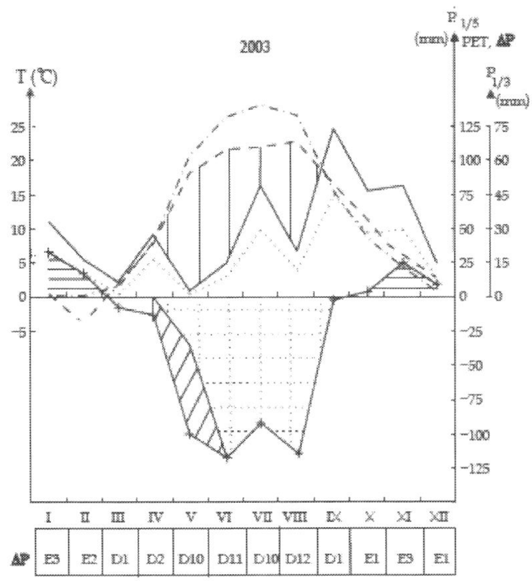

$\sum \Delta P^{+} = 86,9mm$ $\sum \Delta P^{-} = 446,6mm$

$\sum \Delta P^{-}_{w} = 359,7mm$ $P_{\geq 10^{\circ}C} = 207mm$

$P_{VII-VIII} = 69mm$ $P_{XI-III} = 114,1mm$

$I_{hc} = 0,241$ $I_{ar} = 16,75$

$L_{III-V} = 1,26$ $L_{VI-VIII} = 1,09$

$L_{t\geq 10^{\circ}C} = 1,84$

Fig. 6. Climate charts for years 2002 and 2003 and characteristics sizes determined for reviewed site

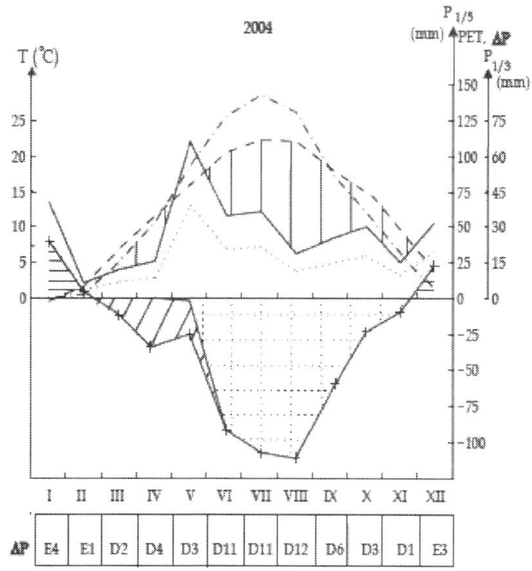

$\sum \Delta P^{+} = 68mm$ $\quad\sum \Delta P^{-} = 472,6mm$

$\sum \Delta P^{-}_{we} = 404,6mm$ $\quad P_{\geq 10^{\circ}C} = 225,4mm$

$P_{VII-VIII} = 55,1mm$ $\quad P_{XI-III} = 105,2mm$

$I_{he} = 0,143$ $\quad I_{ar} = 15,16$

$L_{III-V} = 1,38$ $\quad L_{VI-VIII} = 2,82$

$L_{t\geq 10^{\circ}C} = 1,83$

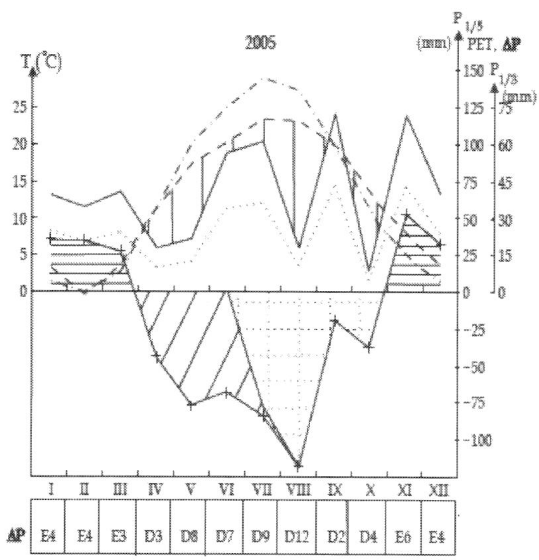

$\sum \Delta P^{+} = 183,8mm$ $\quad\sum \Delta P^{-} = 435,1mm$

$\sum \Delta P^{-}_{we} = 251,3mm$ $\quad P_{\geq 10^{\circ}C} = 236,4mm$

$P_{VII-VIII} = 78,6mm$ $\quad P_{XI-III} = 229,5mm$

$I_{he} = 0,42$ $\quad I_{ar} = 22,23$

$L_{III-V} = 1,02$ $\quad L_{VI-VIII} = 1,29$

$L_{t\geq 10^{\circ}C} = 2,06$

Fig. 7. Climate charts for years 2004 and 2005 and characteristic sizes determined for reviewed site

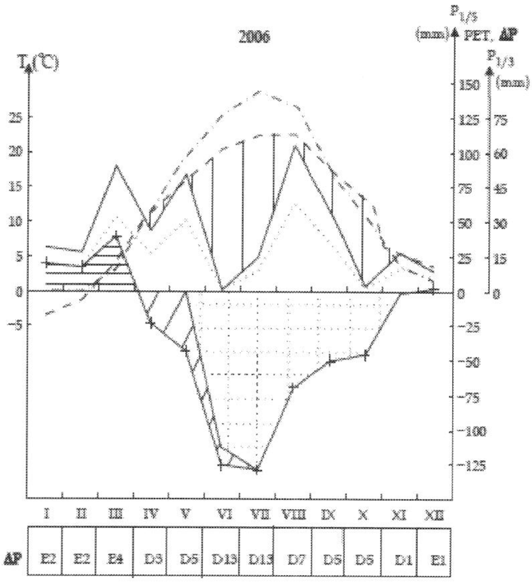

$\sum \Delta P^{+} = 76,5mm$ $\sum \Delta P^{-} = 486mm$

$\sum \Delta P^{-}_{sc} = 409,5mm$ $P_{\geq 10^{\circ}C} = 187,2mm$

$P_{VII-VIII} = 77,6mm$ $P_{XI-III} = 114,1mm$

$I_{hc} = 0,15$ $I_{ar} = 14,49$

$L_{III-V} = 1,2$ $L_{VI-VIII} = 4,35$

$L_{t\geq 10^{\circ}C} = 1,55$

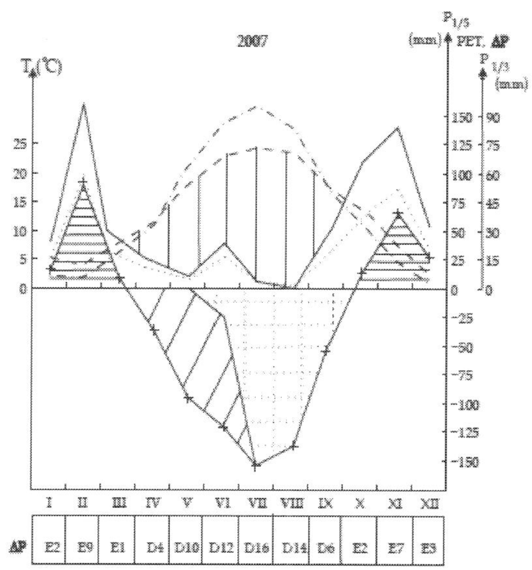

$\sum \Delta P^{+} = 218mm$ $\sum \Delta P^{-} = 588,2mm$

$\sum \Delta P^{-}_{sc} = 370,2mm$ $P_{\geq 10^{\circ}C} = 191,6mm$

$P_{VII-VIII} = 3,6mm$ $P_{XI-III} = 114,1mm$

$I_{hc} = 0,37$ $I_{ar} = 18,01$

$L_{III-V} = 0,38$ $L_{VI-VIII} = 1,49$

$L_{t\geq 10^{\circ}C} = 1,54$

Fig. 8. Climate charts for years 2006 and 2007 and characteristic sizes determined for reviewed site

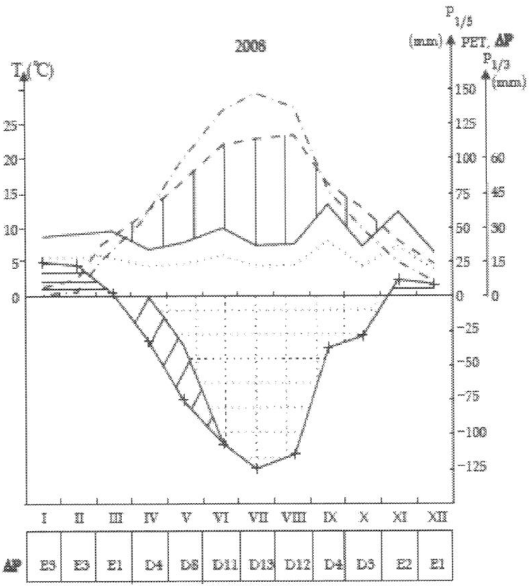

$\sum \Delta P^{+} = 70,8mm$ $\sum \Delta P^{-} = 511,1mm$

$\sum \Delta P^{-}_{ve} = 440,3mm$ $P_{\geq 10^{\circ}C} = 180,4mm$

$P_{VII-VIII} = 45,1mm$ $P_{XI-III} = 138,8mm$

$I_{hc} = 0,13$ $I_{ar} = 23,7$

$L_{III-V} = 1,1$ $L_{VI-VIII} = 1,35$

$L_{t \geq 10^{\circ}C} = 1,44$

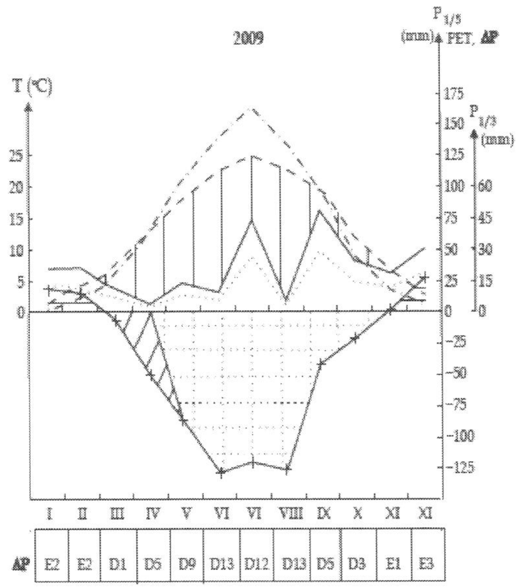

$\sum \Delta P^{+} = 59,7mm$ $\sum \Delta P^{-} = 580mm$

$\sum \Delta P^{-}_{ve} = 520,3mm$ $P_{\geq 10^{\circ}C} = 158,6mm$

$P_{VII-VIII} = 49,4mm$ $P_{XI-III} = 103,3mm$

$I_{hc} = 0,10$ $I_{ar} = 19,01$

$L_{III-V} = 0,84$ $L_{VI-VIII} = 0,9$

$L_{t \geq 10^{\circ}C} = 1,22$

Fig. 9. Climate charts for years 2008 and 2009 and characteristic sizes determined for reviewed site

All the obtained values places the deltaic coast Sfântu Gheorghe in area with a dry climate (Bandoc, 2009).

Regarding the average annual values of the variation of potential evapotranspiration, we can say that, for the period 2000 - 2009 is an increase *PET* value to the annual average of the reference period 1961 - 1990 at a rate of 7 %. Highest increases were registered in 2002, 2007 and 2009, years in which temperatures were recorded over annual average values of the reference period.

The observed values of *PET* in these years are on average 11 % higher than the reference period 1961 - 1990, while during other years the annual increases are in the range 0,07 ... 1 6 % for the period 2000 - 2009 (fig. 11).

Concluding, it can be stated that for Sfântu Gheorghe coastal region there is a significant increase in the potential evapotranspiration *PET* for the last 10 years compared to the reference 1961-1990.

The method used to calculate potential evapotranspiration is Thorntwaite's method, using average monthly air temperature values. Based on the values obtained for *PET* using the method of Thornthwaite (Thornthwaite diagram), one can say that there are significant variations in *PET* for the period under study from 2000 to 2009 compared with the reference period 1961 - 1990, both as annual values and mean interannual values (fig. 12).

The interannual distribution of *PET* in the period 2000 - 2009 shows that these values were, in most months in each year of the analyzed interval over the average interannual values of the reference period 1961 - 1990. It appears that for the months of July and August all *PET* values are over the annual average calculated for the same month of the reference period 1961 - 1990. For instance, for the months of July in 2000-2009 period compared to the the reference values in 1961-1990, PET values are above the multiannual July average (fig.12). Notable years for July values are 2001, 2007 and 2009 where the increase above the multiannual monthly average were 20.14%, 13.66% and 17.98% respectively.

In the same time the following indices were calculated: monthly differences $P-PET$, annual amounts of differences with the same sign $\sum(P-PET)^{+}$ and $\sum(P-PET)^{-}$, as well as the yearly balance $\sum(P-PET)_{A}$, all these being important climatic indices. Calculations for the two analyzed periods led to the following results regarding water deficit and excess from precipitation presented below:

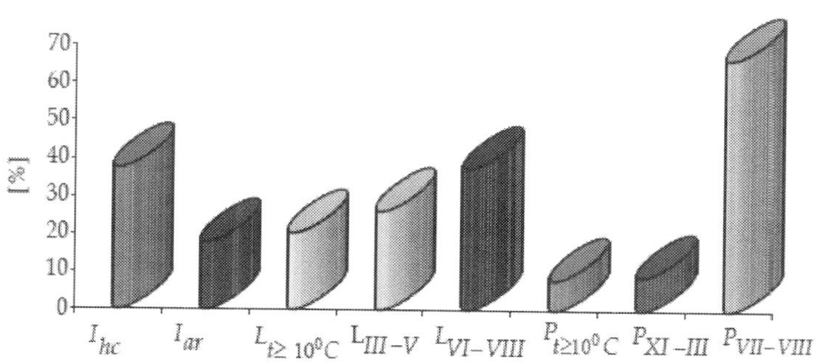

Fig. 10. Increases of the average annual percentage values of main indices for the period 2000 - 2009 for the studied site comparing to the specific values of the reference period 1961 – 1990.

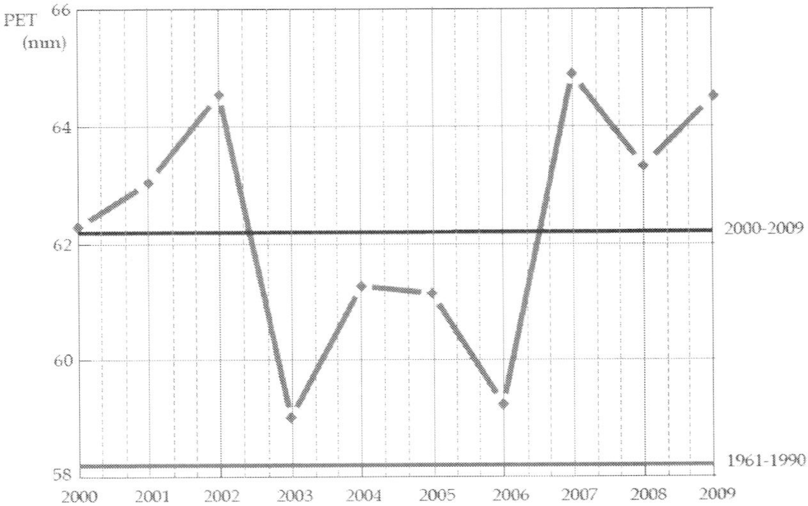

Fig. 11. Changes in annual and multiannual average values of *PET* for the period 2000 - 2009. Comparison with the 1961 - 1990 annual average for the chosen location.

$$\Sigma(P-PET)^{-}_{1961-1990} = 430,4mm\ ;\ \Sigma(P-PET)^{-}_{2000-2009} = 515,2mm\ ;$$

$$\Sigma(P-PET)^{+}_{1961-1990} = 106,2mm\ ;\ \Sigma(P-PET)^{+}_{2000-2009} = 80,8mm$$

The annual balance sheet $\Sigma(P-PET)_{A:2000-2009}$ shows a significant increase, with 31,6 % of the water deficit comparing to the period 1961 - 1990 for which the balance reference value is $\Sigma(P-PET)_{A:1961-1990} = -330,2mm$.

The obtained values show that there is an increase in the deficit for the last 10 years by 19,7 % compared to the reference period and a decrease of 23,9 % in terms of excess rainfall for the period 2000 - 2009 (fig. 13).

For emphasizing very clear each month's character, at the bottom of the chart climate values ΔP were given indicating each month's category in terms of surplus *E* or deficit *D* of precipitation versus potential evapotranspiration. Thus, there are determined the interannual values for the period 2000 - 2009 as well as average multiannual values for the two periods under study.

Based on measurements one could build a mosaic of surpluses *E* and deficits *D* of precipitation variation comparing to potential evapotranspirationfor in the period 2000-2009, comparison with average multianual of *E* and *D* of the periods 2000-2009 and 1961-1990 intervals (fig. 14).

Values for excess precipitation comparing to potential evapotranspiration reached a maximum of *E9* (>80 mm) and *E7* (>60 mm) in February and November 2007 respectively, values much higher than multiannual average of the reference period when the values were *E3* and *E2* (see fig. 14).

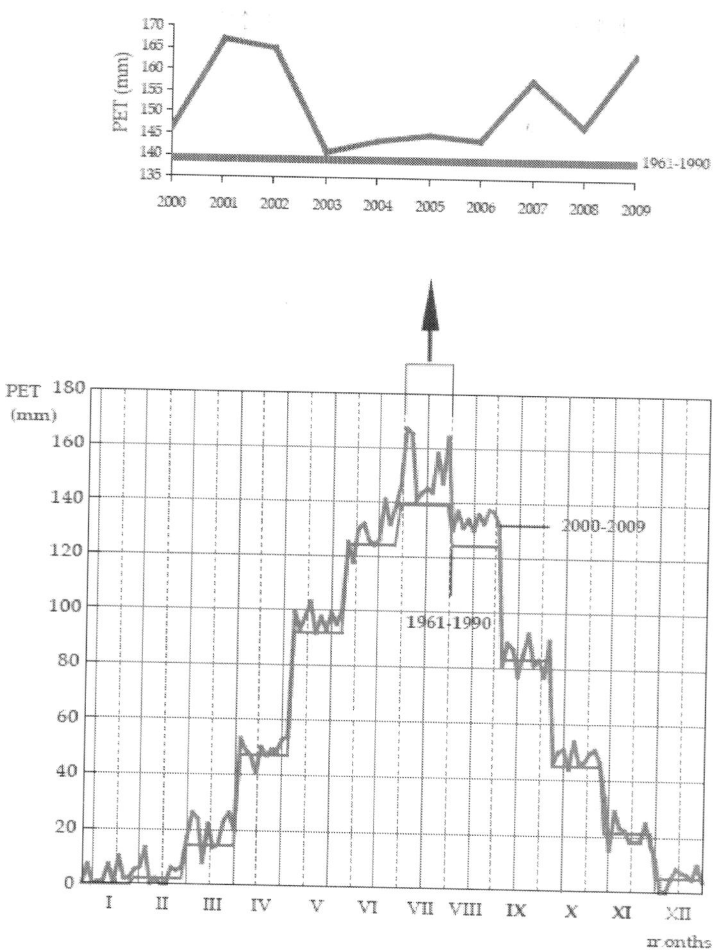

Fig. 12. Interannual distribution of *PET* in the period 2000 - 2009 comparing to the annual average of the reference period 1961 - 1990 for the studied area.

In addition, a reduction of the months with surplus between 2000 - 2009 for the years 2000, 2001, 2003 and 2004 can be seen. Also, there is a reduction in the number of months with a precipitation surplus for 2000, 2001, 2003 and 2004. In these years the precipitation excedent over *PET* period narrowed to 2 months in 2000 and 3 months in 2001, 2002, 2003 compared to 5 months in the reference 1961-1990 period (fig. 14).

As for the precipitation - potential evapotranspiration deficit it can be stated that the deficits suffered a significant increase compared to the reference period. Thus, there can be noticed maximum values of deficits *D17* (>160 mm) to be recorded in 2001 and 2002.

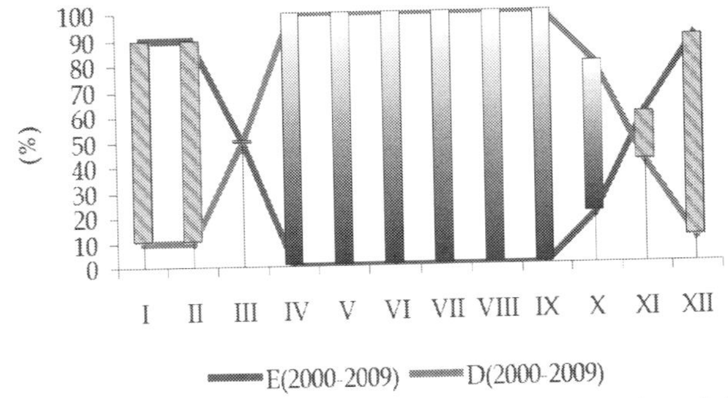

Fig. 13. Percent interannual variations of deficits *D* and surpluses *E* of precipitation to potential evapotranspiration for the period 2000 - 2009.

It appears that while the deficit intervals of the average multiannual values is seven months, the interannual period with deficit intervals is a few months longer between 2000 - 2009. Thus, in 2000, 2001 and 2004 this period has increased by three months and two months respectively compared to that of reference period (fig. 14).

D1 D2 D3 D4 D5 D6 D7 D8 D9
D10 D11 D12 D13 D14 D15 D16 D17
E1 E2 E3 E4 E5 E6 E7 E8 E9

	I	II	III	IV	V	VI	VII	VIII	IX	X	XI	XII
2000	E5	E1	D1	D4	D9	D9	D14	D12	D4	D5	D2	D1
2001	D1	E3	D1	D2	D7	D8	D17	D13	D7	D5	E2	E1
2002	E2	D2	E4	D4	D10	D12	D17	D10	D7	D4	D2	E2
2003	E3	E2	D1	D2	D10	D11	D10	D12	D1	E1	E3	E1
2004	E4	E1	D2	D3	D3	D11	D11	D12	D6	D3	D1	E3
2005	E4	E4	E3	D3	D8	D7	D9	D12	D2	D4	E6	E4
2006	E2	E2	E4	D3	D5	D13	D13	D7	D5	D5	D1	E1
2007	E2	E9	E1	D4	D10	D12	D16	D14	D6	E2	E7	E3
2008	E3	E3	E1	D4	D8	D11	D13	D12	D4	D3	E2	E1
2009	E2	E2	D1	D5	D9	D13	D12	D13	D5	D3	E1	E3
1961-1990	E3	E3	E1	D5	D7	D9	D11	D9	D6	D2	E2	E3
2000-2009	E3	E3	E1	D3	D8	D11	D13	D12	D5	D3	E2	E2

Fig. 14. Distribution of surpluses *E* and deficits *D* of precipitation comparing to potential evapotranspiration in the period 2000 - 2009; comparison with average multiannual of *E* and *D* of the periods 2000 - 2009 and 1961 – 1990.

Analysis of reference period in terms of deficit and surplus, highlights that the studied area is characterized by a lack of *D3* compared to the same period last years when the average value increased to a deficit of *D4*, which means a 17,06 % increase in the deficit.

5. Conclusions

The research results concerning yearly and monthly potential evapotranspiration in the Sfantu Gheorghe coastal area, synthetized in this chapter revealed for years 2001 to 2009 changes in the humidity periods, an increase in air temperature (Busuioc et al, 2010), a diminished atmospheric precipitation amount and also an increase of precipitation to potential evapotranspiration deficit compared to 1961-1990 reference period.

All these changes lead to high vulnerability and low adaptive capacity to adverse impacts from climate change of this area (Liubimtseva & Henebry, 2009).

Thus, by drawing Walter and Leith diagrams, significant increase of dryness periods and decrease of moisture periods were observed with implications upon potential evapotranspiration and upon the shore phytocoenoses.

There are also changes in the length of the periods with precipitation surplus and deficit compared to potential evapotranspiration that means increasing periods of deficit and decreasing periods of surplus.

The following calculated characteristic measurements include the delta coast in Sfântu Gheorghe in arid climate and climatic changes show that the period 2000 - 2009 led to a trend towards increasing aridity: *Martonne arid index* (I_{ar}), *retention index offset* (I_{hc}), *the amount of rainfall in the period with temperature* $T \geq 10\ °\ C$ ($P_{t\geq10^0C}$), *the amount of rainfall the soil load in the months from November to March* (P_{XI-III}), *the amount of summer rainfall July and August* ($P_{VII-VIII}$), *Lang precipitation index for the period with* $t \geq 10$ °C ($L_{t\geq10^0C}$), *Lang precipitation index for the summer season* ($L_{VI-VIII}$) *and Lang precipitation index for the spring season* (L_{III-V}).

From the differences in monthly $P-PET$ calculation of amounts $\Sigma(P-PET)^+$, $\Sigma(P-PET)^-$ of the precipitation deficit offset by previously accumulated $\Sigma\Delta P^+$, surpluses and deficits of precipitation uncompensated by previous surpluses $\Sigma\Delta P_{uc}^-$ and the annual balance $\Sigma(P-PET)_A$ for the period under study year 2000 - 2009 and for the reference period 1961 - 1990, there was a deficit increase and a decrease of excess water from precipitation, an extension of periods of water shortage against period with excess of water and a significant increase by about 23,9 % for deficit of water that gathers negative differences uncompensated during periods of surplus.

Therefore, the research presented in this article have highlighted significant changes in potential evapotranspiration in relation to climate changes for the 2000 - 2009 studied period, in Sfântu Gheorghe area - Danube Delta, showing an increase of precipitation deficit and an increase of climate aridity .

Indirect method used in this paper work to determine the potential evapotranspiration was based on the values of air temperature and Thornthwaite's diagrams and tables. In this way a general view of a time variation of *PET* for Sfântu Gheorghe area - Danube Delta, has been created.

The advantages of this indirect method results from the fact that it doesn't require a large number of measured meteorological parameters and that it can be easily applied obtaining good estimates.
In the future it is intended that research should continue in order to see whether the growth trend of a interannual and annual potential evaporation is kept over the period 2000 - 2009.
No doubt that climate change is underway affecting Earth's biodiversity.
Biggest challenge in this respect is related to the marine area, but it is unclear to what extent these changes in climate will affect ecosystems.
What is known is that the temperatures that rise steadily and increasingly frequent extreme weather events are those that have influence on migrating wildlife and also causes invasive species.
Coastal areas offer considerable benefits to society while human activities are exerting considerable pressure on coastal ecosystems. Therefore, these benefits to society are in danger (Nobre, 2009).

6. Acknowledgment

Research carried out were conducted at the Center for Coastal Research and Environmental Protection, Department of Meteorology and Hydrology at the University of Bucharest, Romania.

7. References

Allen, R.G.; Pereira, L.S.; Raes, D. & Smith, M. (1998). Crop Evapotranspiration—Guidelines for Computing Crop Water Requirements. Food and Agriculture Organization of the United Nations. FAO Irrigation and drainage, Rome, ISBN 92-5-104219-4

Andréassian, V.; Perrin, Ch. & Michel, C. (2004). Impact of imperfect potential evapotranspiration knowledge on the efficiency and parameters of watershed models. *Journal of Hydrology*, Vol. 286, pp.19–35, ISSN 0022-1694

Bandoc, G. (2009). Costal phenologic cycles for Sfantu Gheorghe station (Danube Delta). *Journal of Environ. Protection and Ecology*, Vol. 9, No. 4, pp 953-960, ISSN 1311 - 5065

Bandoc, G. & Golumbeanu, M (2010). Climate variability influence to the potential evapotranspiration regime of Sfantu Gheorghe Delta Shore. *Journal of Environmental Protection and Ecology*, Vol. 10, No. 1, pp.172 -181, ISSN 1311 - 5065

Baxter, E.V.; Nadim, S.; Farajalla & Nalneesh, G. (1996). Integrated GIS and distributed storm water runoff modeling. In: Goodchild, et al. (Eds.), *GIS and Environmental Modeling Progress and Research Issues.* Donald F. Hemenway Jr., Fort Collins, pp. 199–204, ISBN 0470-236-779

Berbecel, L.; Socor,O. & Roşca, V. (1970). Current concepts in studying the phenomenon evapotranspiration (in romanian). *Rev. Hidrotehnica*, Vol. 15, No. 5, pp. 265-274

Bouchet, R. J. (1964). Évaporation réelle, évaporation - transpiration potentielle et production agricole, în l'eau et la production végétale, Inst. Nat. *De la Rech. Agr.*, Paris, pp. 151 - 232

Buchmann, N. (2000). Biotic and abiotic factors controlling soil respiration rates in Picea abies stands. *Soil Biol. Biochem*, Vol. 32, pp. 1625–1635, ISSN 0038-0717

Busuioc,A; Caian, M.; Cheval, S.; Bojariu, R.; Boroneant, C.; Baciu, M.; Dumitrescu, A. (2010). *Climate variability and change in Romania*, Ed. ProUniversitaria, pp. 59-72, ISBN 978-973-129-549-7, Bucureşti, România

Casals, P.; Gimeno, C.; Carrara, A.; Lopez-Sangil, L. & Sanz, M. (2009). Soil CO_2 efflux and extractable organic carbon fractions under simulated precipitation events in a Mediterranean Dehesa. *Soil Biol. Biochem*, Vol. 41, pp. 1915–1922, ISSN 0038-0717

Caselles, V.;Artigao, M.M.; Hurtado; E.; Coll, C. & Brasa, A. (1998) Mapping actual evapotranspiration by combining landsat TM and NOAA-AVHRR images: application to the Barrax Area, Albacete, Spain. *Remote Sensing of Environment*, Vol. No. 63, pp. 1–10, ISSN 0034-4257

Chattopadhyay, N. & Hulme, M. (1997). Evaporation and potential evapotranspiration in India under conditions of recent and future climatic change. *Agricultural and Forest Meteorology* , Vol. 87, No. 1, pp. 55-75. ISSN 0168-1923

Chiriţă, C.; Vlad, I.; Păunescu, C.; Pătrăşcoiu, N.; Roşu, C. & Iancu, I. (1977). *Forest sites* (in romanian). Ed. Academiei RSR, Bucureşti, România

Choudhury, B.J. (1997). Global pattern of potential evaporation calculated from the Penman–Monteith equation using satellite and assimilated data. *Remote Sens. Environ*, Vol. 61, pp. 64–81, ISSN 0034-4257

Chen, D.; Gao, G.; Xu, C.-Y. & Ren, G. (2005). Comparison of the Thornthwaite method and pan data with the standard Penman-Monteith estimates of reference evapotranspiration in China. *Climate research*, Vol. 28, pp. 123-132 ISSN 1616-1572

Chuanyana, Z.; Zhongrena, N. & Zhaodonga, F. (2004). GIS-assisted spatially distributed modeling of the potential evapotranspiration in semi-arid climate of the Chinese Loess Plateau. *Journal of Arid Environments*, Vol. 58, pp. 387–403, ISSN 0140-1963

Cleugh, H.A.; Leuning, R.; Mu, Q. & Running, S.W. (2007). Regional evaporation estimates from flux tower and MODIS satellite data. *Remote Sens. Environ*, Vol. 106, pp. 285–304, ISSN 0034-4257

Donciu, C. (1958). Evapotranspiration in the RPR (in romanian). *Rev. Hidrotehnica*, Vol. 3, No. 1, pp. 129-135

Donciu, C. (1983). Evapotranspiration and soil water balance (in romanian), *Memoriile Secţiilor Ştiinţifice*, Seria IV, tom VI, nr. 2, pp. 347-366, Edit. Acad. R.S.R., Bucureşti

Douglas, E. M.; Jacobs, J. M.; Sumner, D, M. & Ray, R. L. (2009). A comparison of models for estimating potential evapotranspiration for Florida land cover types. *Journal of Hydrology*, Vol. 373, pp. 366–376, ISSN 0022-1694

Eagleman, J. R. (1967). Pan evaporation, potential and actual evaporation, *Journal of Applied Meteorology*, Vo. 6, No 3, pp. 482-488, ISSN 1520-0450

Granger, R.J. (1997). Comparison of surface and satellite derived estimates of evapotranspiration using a feedback algorithm. In: Kite, G.W., Pietroniro, A., Pultz, T. (Eds.), Applications of Remote Sensing in Hydrology. Proceedings of the Symposium No. 17 NHRI, Saskatoon, Canada. National Hydrology Research Institute (NHRI), pp. 21–81.

Hargreaves, G.H. & Samani, Z.A. (1982). Estimating potential evapotranspiration (Tech. Note). *Journal of Irrigation and Drainage Engineering*, Vol. 108, No 3, pp. 225–230, ISSN 0733-9437

Henning, I. & Henning, D. (1981). Potential evapotranspiration in mountain geo – ecosystems of different altitudines and latitudes. *Mountain Research and Development* Vol. 1, pp. 267-274, ISSN 0276-4741

Irmak, A. & Kamble, B. (2009). Evapotranspiration data assimilation with genetic algorithms and SWAP model for on-demand irrigation. Irrigation Science, Vol. 28, No.1, pp.101-112 , ISSN 1432-1319

Köppen, W. (1900). Versuch einer Klassifikation der Klimate, vorzugsweise nach ihren Beziehungen zur Pflanzenwelt. *Geogr. Zeitschr*

Kouzmov, K. (2002). Climatic changes in the region of Vidin and their efect on the agroclimatic resources. *Journal of Environmental Protection and Ecology*, Vol. 3, No.3, pp. 126-131, ISSN 1311 - 5065

Kumar, M.; Raghuwanshi, N.S.; Singh, R.; Wallender, W.W. & Pruitt, W.O. (2002). Estimating evapotranspiration using Artificial Neural Network. *Journal of Irrigation and Drainage Engineering*, ASCE 128, pp. 224–233, ISSN 0733-9437

Li, H.; Yan, J.; Yue, X. & Wang, M. (2008a). Significance of soil temperature and moisture for soil respiration in a Chinese mountain area. *Agric. Forest. Meteorol*, Vol. 148, pp. 490-503, ISSN 0168-1923

Li, Z.; Wang, Y.; Zhou, Q.; Wu, J.; Peng, J. & Chang, H. (2008b). Spatiotemporal variability of land surface moisture based on vegetation and temperature characteristics in Northern Shaanxi Loess plateau, China. *Journal of Arid Environments*, Vol. 72, pp. 974–985, ISSN 0140-1963

Lioubimtseva, E. & Henebry, G.M. (2009). Climate and environmental change in arid Central Asia: Impacts, vulnerability and adaptations. *Journal of Arid Environments*, Vol. 73. pp. 963–977, ISSN 0140-1963

Lu, J.; Sun, G.; McNulty,S.; & Amatya, D. M. (2005). A comparison of six potential evapotranspiration methods for regional use in the Southeastern United States. *Journal of the American Water Resources Association* (*JAWRA*) Vol. 41 (3), pp. 621-633, ISSN 1752-1688

Monteith, J.L. (1965). Evaporation and environment. *Symposium of the Society for Experimental Biology*, Vol. 19, pp. 205–224

Moore, I.D. (1996). Hydrologic modeling and GIS. In: Goodchild, et al. (Eds.), *GIS and Environmental Modeling Progress and Research Issues*. Donald F. Hemenway Jr., Fort Collins, pp. 143–149, ISBN 0470-236-779

Nobre, A.M.; Ferreira, J.G; Nunes, J.P; Yan, X; Bricker, S.; Corner, R.; Groom,S.; Gu, H.; Hawkins, A.J.S.; Hutson, R.; Dongzhao Lan, D.; Lencart e Silva,J.D.; Pascoe,P.; Telfer, T.; Zhang, X. & Zhu, M. (2010). Assessment of coastal management options by means of multilayered ecosystem models. Estuarine, *Coastal and Shelf Science*, Vol. 87, No. 1, pp. 43-62, ISSN 0272-7714

Oguz, T.; Dippner, J. & Kaymaz, Z. (2006). Climatic regulation of the Black Sea hydro-meteorological and ecological properties at interannual-to-decadal time scales. *Journal of Marine Systems* Vol. 60, No. 3-4, pp. 235–254, ISSN 0924-7963

Oudin, L.; Michel, C. & Anctil, F. (2005a). Which potential evapotranspiration input for a lumped rainfall-runoff model? Part 1 - Can rainfall-runoff models effectively handle detailed potential evapotranspiration inputs?. *Journal of Hydrology*, Vol. 303, pp. 275–289, ISSN 0022- 1694

Oudin, L.; Hervieu, F.; Michel, C.; Perrin, C.; Andreassian, V.; Anctil, F. & Loumagne, C. (2005b). Which potential evapotranspiration input for a lumped rainfall-runoff model? Part 2 - Towards a simple and efficient potential evapotranspiration model for rainfall-runoff modeling. *Journ. of Hydr.*, Vol. 303, pp. 290–306, ISSN 0022-1694

Palutikof, J.P.; Goddes, S.C.M. & Guo, X. (1994). Climate change, potential evapotranspiration and moisture availability in the Mediterranean Basin. *International Journal of Climatology*, Vol. 14 , No. 8, pp. 853-869, ISSN 0899-8418

Penman, H. L. (1946). Natural evaporation from open water, bare soil and grass. *Proceedings of the Royal Society of London.* Series A, Mathematical and Physical Sciences, Vol. 193, No. 1032 (Apr. 22, 1948), pp. 120-145

Ponce, V.M. (1989). Engineering hydrology: principles *and practices*. John Wiley and Sons, New York, pp. 48-51, ISBN: 0471147354

Raich, J.W. & Schlesinger, W.H. (1992). The global carbon dioxide flux in soil respiration and its relationship to vegetation and climate. *Tellus* Vol. 44B, pp. 81–89, ISSN 1600-0899

Smith, B.A.; McClendon, R.W. & Hoogenboom, G. (2006). Improving air temperature prediction with Artificial Neural Networks. *International Journal of Computational Intelligence,* Vol. 3, No. 3, pp. 179–186, ISSN 0883-9514

Srinivasan, R.; Arnold, J.; Rosenthal, W. & Muttiah, R.S. (1996). Hydrologic Modeling of Texas Gulf Basin using GIS. In: Goodchild, et al. (Eds.), *GIS and Environmental Modeling Progress and Research Issues*, Donald F. Hemenway Jr., Fort Collins, pp. 213–219, ISBN 0470-236-779

Stefano, C.D. & Ferro, V. (1997). Estimation of evapotranspiration by Hargreaves formula and remotely sensed data in semi-arid Mediterranean areas. *Journal of Agricultural Engineering Research*, Vol. 68, pp. 189–199, ISSN 0021-8634

Stewarta, J.B.; Watts, C.J.; Rodriguez, J.C.; De Bruin, H.A.R.; van den Berg, A.R. & Garatuza-Payan, J. (1999). Use of satellite data to estimate radiation and evaporation for Northwest Mexico. *Agricultural Water Manag.*, Vol. 38, pp. 181–193, ISSN 0378-3774

Tang, R.L.; Li, Z.L. & Tang, B.H. (2010). An application of the Ts–VI triangle method with enhanced edges determination for evapotranspiration estimation from MODIS data in and semi-arid regions: implementation and validation. *Remote Sensing of Environment*, Vol. 114, No. 3,pp. 540–551, ISSN 0034-4257

Thornthwaite, C.W. (1948). An approach towards a rational classification of climate. *Geographical Revue*,Vol. 38, pp. 55-94

Thomas, A. (2000a). Spatial land temporal characteristics of potential evapotranspiration trends over China. *Inter. Journal of Climatology*, Vol. 20, pp. 381-395, ISSN 0899-8418

Thomas, A. (2000b). Climatic changes in yield index and soil water deficit trends in China. *Agricultural and Forests Meteorology*, Vol. 102, pp. 71-81, ISSN 0168-1923

Torres, A.F., Walker, W.R. & McKee, M. (2011). Forecasting daily potential evapotranspiration using machine learning and limited climatic data. *Agricultural Water Management*, Vol. 98 , pp.553–562, ISSN 0378-3774

Turc, L. (1954). Calcul du bilan de l'eau evaluation en function des precipitations et des temperatures. In *Association International d'Hydrology*, Assemblée Génrale de Rome, Tome III , No.3, pp. 188–202

Vespremeanu, E. (2000). *The Danube Delta tourist map* 1:200 000. Ed Amco Press, Bucureşti, România

Vespremeanu, E. (2004). *Geography of the Black Sea*, Ed. Univ. din Bucureşti, ISBN 973-575-925-X, Bucureşti, România

Walter, H. (1955). Die Klimadiagramme als Mittel zur Beurteilung der Klimaverhältnisse für ökologische, vegetationskundliche und landwirtschaftliche Zwecke. *Berichte der Deutschen Botanischen Gesellschaft* Vol. 68, pp. 331-344

Walter, H., & Lieth, H. (1960). Kimadiagramm-Weltatlas, Fischer-Verlag, Jena

Walter, H. (1999). Vegetation und Klimazonen. Grundriß der globalen Ökologie. Ulmer, ISBN 3-8252-0014-0, Stuttgart, Germania

*****Climate of Romania* (2008). National Meteorological Administration. Ed. Academiei Române, Bucureşti, România

13

Evapotranspiration of Partially Vegetated Surfaces

L.O. Lagos[1,2], G. Merino[1], D. Martin[2], S. Verma[2] and A. Suyker[2]
[1]*Universidad de Concepción Chile*
[2]*University of Nebraska-Lincoln*
[1]*Chile*
[2]*USA*

1. Introduction

Latent heat flux equivalent to Evapotranspiration (ET) is the total amount of water lost via transpiration and evaporation from plant surfaces and the soil in an area where a crop is growing. Since 80-90% of precipitation received in semiarid and subhumid climates is commonly used in evapotranspiration, accurate estimations of ET are very important for hydrologic studies and crop water requirements. ET determination and modelling is not straightforward due to the natural heterogeneity and complexity of agricultural and natural land surfaces. In evapotranspiration modelling it is very common to represent vegetation assuming a single source of energy flux at an effective height within the canopy. However, when crops are sparse, the single source/sink of energy assumption in such models is not entirely satisfied. Improvements using multiple source models have been developed to estimate ET from crop transpiration and soil evaporation. Soil evaporation on partially vegetated surfaces over natural vegetation and orchards includes not only the soil under the canopy but also areas of bare soil between vegetation that contribute to ET. Soil evaporation can account for 25-45% of annual ET in agricultural systems. In irrigated agriculture, partially vegetated surfaces include fruit orchards (i.e. apples, oranges, vineyards, avocados, blueberries, and lemons among others), which cover a significant portion of the total area under irrigation.

In semiarid regions, direct soil evaporation from sparse barley or millet crops can account for 30% to 60% of rainfall (Wallace et al., 1999). On a seasonal basis, sparse canopy soil evaporation can account for half of total rainfall (Lund & Soegaard, 2003). Allen (1990) estimated the soil evaporation under a sparse barley crop in northern Syria and found that about 70% of the total evaporation originated from the soil. Lagos (2008) estimated that under irrigated maize conditions soil evaporation accounted for around 26-36% of annual evapotranspiration. Under rain-fed maize conditions annual evaporation accounted for 36-39% of total ET. Under irrigated soybean the percentage was 41%, and under rainfed soybean conditions annual evaporation accounted for 45-47% of annual ET. Massman (1992) estimated that the soil contribution to total ET was about 30% for a short grass steppe measurement site in northeast Colorado. In a sparse canopy at the middle of the growing season, and after a rain event, more than 50% of the daily ET corresponds to directly soil evaporation (Lund & Soegaard, 2003). Soil evaporation can be maximized under frequent rainfall or irrigation events, common conditions in agricultural systems for orchard with drip or micro sprinklers systems. If some of this unproductive loss of water could be retained in the soil and used as transpiration, yields could be increased without increased rainfall or the use of supplemental irrigation (Wallace et al., 1999). The measurement and modelling of soil evaporation on partially vegetated surfaces is crucial to estimate how much water is lost to the atmosphere via soil evaporation. Consequently, better water management can be proposed for water savings.

Partially vegetated surface accounts for a significant portion of land surface. It occurs seasonally in all agricultural areas and throughout the year in or chard and natural land covers. Predictions of ET for these conditions have not been thoroughly researched. In Chile, agricultural orchards with partially vegetated surfaces include apples, oranges, avocados, cherries, vineyards, blueberries, and berries, among others. According to the agricultural census (INE, 2007) the national orchard surface covers more than 324,000 ha, representing 30% of the total surface under irrigation.

Similar to the Shuttleworth and Wallace (1985), Choudhury and Monteith (1988) and Lagos (2008) models, the modelling of evapotranspiration for partially vegetated surfaces can be accomplished using explicit solutions of the equations that define the conservation of heat and water vapor fluxes for partially vegetated surfaces and soil. Multiple-layer models offer the possibility to represent these conditions to solve the surface energy balance and consequently, estimate evapotranspiration. Modelling is essential to predict long-term trends and to quantify expected outcomes. Since ET is such a large component of the hydrologic cycle in areas with partially vegetated surfaces, small changes in the calculation of ET can result in significant changes in simulated water budgets. Thus, good data and accurate modelling of ET is essential for predicting not only water requirements for agricultural crops but also to predict the significance of irrigation management decisions and land use changes to the entire hydrologic cycle.

Currently, several methods and models exist to predict natural environments under different conditions. More complex models have been developed to account for more variables affecting model performance. However, the applicability of these models has been limited by the difficulties and tedious algorithms needed to complete estimations. Mathematical algorithms used by multiple-layer models can be programmed in a software package to facilitate and optimize ET estimation by any user. User-friendly software facilitates the use of these improved methods; users (i.e. students) can use the computer model to study the behaviour of the system from a set of parameters and initial conditions.

Accordingly, in this chapter, a review of models that estimate ET for partially covered surfaces that occur normally in agricultural systems (i.e. orchards or vineyards) is presented, and the needs for further research are assessed.

2. ET modelling review

Evapotranspiration (ET) is the total amount of water lost via transpiration and evaporation from plant surfaces and the soil in an area where a crop is growing. Traditionally, ET from agricultural fields has been estimated using the two-step approach by multiplying the weather-based reference ET (Jensen et al., 1971; Allen et al., 1998 and ASCE, 2002) by crop coefficients (Kc) to make an approximate allowance for crop differences. Crop coefficients are determined according to the crop type and the crop growth stage (Allen et al., 1998). However, there is typically some question regarding whether the crops grown compare with the conditions represented by the idealized Kc values (Parkes et al., 2005; Rana et al.,

2005; Katerji & Rana, 2006; Flores, 2007). In addition, it is difficult to predict the correct crop growth stage dates for large populations of crops and fields (Allen et al., 2007).

A second method is to make a one-step estimate of ET based on the Penman-Monteith (P-M) equation (Monteith, 1965), with crop-to-crop differences represented by the use of crop-specific values of surface and aerodynamic resistances (Shuttleworth, 2006) ET estimations using the one-step approach with the P-M model have been studied by several authors (Stannard, 1993; Farahani & Bausch, 1995; Rana et al., 1997; Alves & Pereira, 2000; Kjelgaard & Stockle, 2001; Ortega-Farias et al., 2004; Shuttleworth, 2006; Katerji & Rana, 2006; Flores, 2007; Irmak et al., 2008). Although different degrees of success have been achieved, the model has generally performed more satisfactorily when the leaf area index (LAI) is large (LAI>2). Results shows that the "big leaf" assumption used by the P-M model is not satisfied for sparse vegetation and crops with partial canopy cover.

A third approach consists of extending the P-M single-layer model to a multiple-layer model (i.e. two layers in the Shuttleworth-Wallace (S-W) model (Shuttleworth-Wallace, 1985) and four layers in the Choudhury-Monteith (C-M) model (Choudhury & Monteith, 1988). Shuttleworth and Wallace (1985) combined a one-dimensional model of crop transpiration and a one-dimensional model of soil evaporation. Surface resistances regulate the heat and mass transfer in plant and soil surfaces, and aerodynamic resistances regulate fluxes between the surface and the atmospheric boundary layer. Several studies have evaluated the performance of the S-W model to estimate evapotranspiration (Farahani & Baush,1995. Stannard, 1993; Lafleur & Rouse, 1990; Farahani & Ahuja, 1996; Iritz et al. 2001; Tourula & Heikinheimo, 1998; Anadranistakis et al., 2000; Ortega-Farias et al., 2007). Field tests of the model have shown promising results for a wide range of both agricultural and non-agricultural vegetation.

Farahani and Baush (1995) evaluated the performance of the P-M model and the S-W model for irrigated maize. Their main conclusion was that the Penman-Monteith model performed poorly when the leaf area index was less than 2 because soil evaporation was neglected in calculating surface resistance. Results of the S-W model were encouraging as it performed satisfactorily for the entire range of canopy cover. Stannard (1993) compared the P-M, S-W and Priestley-Taylor ET models for sparsely vegetated, semiarid rangeland. The P-M model was not sufficiently accurate (hourly r^2 =0.56, daily r^2=0.60); however, the S-W model performs significantly better for hourly (r^2=0.78) and daily data (r^2=0.85). Lafleur and Rouse (1990) compared the S-W model with evapotranspiration calculated from the Bowen Ratio Energy Balance technique over a range of LAI from non-vegetated to fully vegetated conditions. The results showed that the S-W model was in excellent agreement with the measured evapotranspiration for hourly and day-time totals for all values of LAI. Using the potential of the S-W model to partition transpiration and evaporation, Farahani and Ahuja (1996) extended the model to include the effects of crop residues on soil evaporation by the inclusion of a partially covered soil area and partitioning evaporation between the bare and residue-covered areas. Iritz et al. (2001) applied a modified version of the S-W model to estimate evapotranspiration for a forest. The main modification consisted of a two-layer soil module, which enabled soil surface resistance to be calculated as a function of the wetness of the top soil. They found that the general seasonal dynamics of evaporation were fairly well simulated with the model. Tourula and Heikinheimo (1998) evaluated a modified version of the S-W model in a barley field. A modification of soil surface resistance and aerodynamic resistance, over two growing seasons, produced daily and hourly ET estimates in good agreement with the measured evapotranspiration. The performance of the S-W model was evaluated against two eddy covariance systems by Ortega-Farias et al. (2007) over a Cabernet

Sauvignon vineyard. Model performance was good under arid atmospheric conditions with a correlation coefficient (r^2) of 0.77 and a root mean square error (RMSE) of 29 Wm^{-2}.

Although good results have been found using the Shuttleworth-Wallace approach, the model still needs an estimation or measurement of soil heat flux (G) to estimate ET. Commonly, G is calculated as a fixed percentage of net radiation (Rn). Shuttleworth and Wallace (1985) estimated G as 20% of the net radiation reaching the soil surface. In the FAO56 method, Allen et al. (1998) estimated daily reference ET (ETr and ETo), assuming that the soil heat flux beneath a fully vegetated grass or alfalfa reference surface is small in comparison with Rn (i.e. G=0). For hourly estimations, soil heat flux was estimated as one tenth of the Rn during the daytime and as half of the Rn for the night time when grass was used as the reference surface. Similarly, G was assumed to be 0.04xRn for the daytime and 0.2xRn during the night time for an alfalfa reference surface. A more complete surface energy balance was presented by Choudhury and Monteith (1988). The proposed method developed a four-layer model for the heat budget of homogeneous land surfaces. The model is an explicit solution of the equations which define the conservation of heat and water vapor in a system consisting of uniform vegetation and soil. An important feature was the interaction of evaporation from the soil and transpiration from the canopy expressed by changes in the vapor pressure deficit of the air in the canopy. A second feature was the ability of the model to partition the available energy into sensible heat, latent heat, and soil heat flux for the canopy/soil system.

Similar to Shuttleworth-Wallace (1985), the Choudhury-Monteith model included a soil surface resistance to regulate the heat and mass transfer at the soil surface. However, residue effects on the surface energy balance are not included in the model. Crop residue generally increases infiltration and reduces soil evaporation. Surface residue affects many of the variables that determine the evaporation rate. These variables include Rn, G, aerodynamic resistance and surface resistances to transport of heat and water vapor fluxes (Steiner, 1994; Horton et al., 1996; Steiner et al., 2000).

Caprio et al. (1985) compared evaporation from three mini-lysimeters installed in bare soil and in a 14 and 28 cm tall standing wheat stubble. After nine days of measurements, evaporation from the lysimeter with stubble was 60% of the evaporation measured from bare soil. Enz et al. (1988) evaluated daily evaporation for bare soil and stubble-covered soil surfaces. Evaporation was always greater from the bare soil surface until it was dry, then evaporation was greater from the stubble covered-surface because more water was available. Evaporation from a bare soil surface has been described in three stages. An initial, energy-limited stage occurs when enough soil water is available to satisfy the potential evaporation rates. A second, falling rate stage is limited by water flow to the soil surface, while the third stage has a very low, nearly constant evaporative rate from very dry soil (Jalota & Prihar, 1998). Steiner (1989) evaluated the effect of residue (from cotton, sorghum and wheat) on the initial, energy-limited rate of evaporation. The evaporation rate relative to bare soil evaporation was described by a logarithmic relationship. Increasing the amount of residue on the soil surface reduced the relative evaporation rate during the initial stage. Bristow et al. (1986) developed a model to predict soil heat and water budgets in a soil-residue-atmosphere system. Results from application of the model indicate that surface residues decreased evaporation by roughly 36% compared with simulations from bare soil.

With the recognition of the potential of multiple-layer models to estimate ET, a modified surface energy balance model (SEB) was developed by Lagos (2008) and Lagos et al. (2009) to include the effect of crop residue on evapotranspiration. The model relies mainly on the Schuttleworth-Wallace (1985) and Choudhury and Monteith (1988) approaches and has the potential to predict evapotranspiration for varying soil cover ranging from partially residue-covered soil to closed canopy surfaces. Improvements to aerodynamic resistance, surface canopy resistance and soil resistances for the transport of heat and water vapor were also suggested.

2.1 The SEB model

The modified surface energy balance (SEB) model has four layers (Figure 1), the first extended from the reference height above the vegetation and the sink for momentum within the canopy, a second layer between the canopy level and the soil surface, a third layer corresponding to the top soil layer and a lower soil layer where the soil atmosphere is saturated with water vapor. The soil temperature at the bottom of the lower level was held constant for at least a 24h period.

The SEB model distributes net radiation (Rn), sensible heat (H), latent heat ($\lambda \Xi$), and soil heat fluxes (G) through the soil/residue/canopy system. Horizontal gradients of the potentials are assumed to be small enough for lateral fluxes to be ignored, and physical and biochemical energy storage terms in the canopy/residue/soil system are assumed to be negligible. The evaporation of water on plant leaves due to rain, irrigation or dew is also ignored.

The SEB model distributes net radiation (Rn) into sensible heat (H), latent heat (λE), and soil heat fluxes (G) through the soil-canopy system (Figure 2). Total latent heat (λE) is the sum of latent heat from the canopy (λEc), latent heat from the soil (λEs) and latent heat from the residue-covered soil (λEr). Similarly, sensible heat is calculated as the sum of sensible heat from the canopy (Hc), sensible heat from the soil (Hs) and sensible heat from the residue covered soil (Hr).

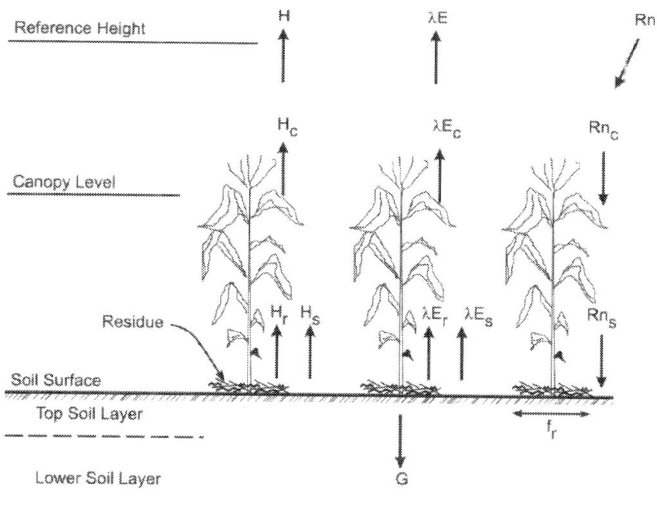

Fig. 1. Fluxes of the surface energy balance model (SEB).

The total net radiation is divided into that absorbed by the canopy (Rnc) and the soil (Rns) and is given by Rn = Rnc + Rns. The net radiation absorbed by the canopy is divided into latent heat and sensible heat fluxes as Rnc = λEc +Hc. Similarly, for the soil Rns = Gos + Hs, where Gos is a conduction term downwards from the soil surface and is expressed as Gos = λEs + Gs, where Gs is the soil heat flux for bare soil. Similarly, for the residue-covered soil Rns = Gor + Hr where Gor is the conduction downwards from the soil covered by residue. The conduction is given by Gor = λEr + Gr where Gr is the soil heat flux for residue-covered soil. Total latent heat flux from the canopy/residue/soil system is the sum of the latent heat from the canopy (transpiration), latent heat from the soil and latent heat from the residue-covered soil (evaporation), calculated as:

$$\lambda E = \lambda E_c + (1 - fr) \cdot \lambda E_s + fr \cdot \lambda E_r \tag{1}$$

where fr is the fraction of the soil affected by residue. Similarly, the total sensible heat is given by:

$$H = H_c + (1 - fr) \cdot H_s + fr \cdot H_r \tag{2}$$

The differences in vapor pressure and temperature between levels can be expressed with an Ohm's law analogy using appropriate resistance and flux terms (Figure 2). The sensible and latent heat fluxes from the canopy, from bare soil and soil covered by residue are expressed by (Shuttleworth & Wallace, 1985):

$$H_c = \frac{\rho \cdot c_p \cdot (T_1 - T_b)}{r_1} \quad \text{and} \quad \lambda E_c = \frac{\rho \cdot C_p \cdot (e_1^* - e_b)}{\gamma \cdot (r_1 + r_c)} \tag{3}$$

$$H_s = \frac{\rho \cdot C_p \cdot (T_2 - T_b)}{r_2} \quad \text{and} \quad \lambda E_s = \frac{\rho \cdot C_p \cdot (e_L^* - e_b)}{\gamma \cdot (r_2 + r_s)} \tag{4}$$

$$H_r = \frac{\rho \cdot C_p \cdot (T_{2r} - T_b)}{r_2 + r_{rh}} \quad \text{and} \quad \lambda E_r = \frac{\rho \cdot C_p \cdot (e_{Lr}^* - e_b)}{\gamma \cdot (r_2 + r_s + r_r)} \tag{5}$$

where, ρ is the density of moist air, C_p is the specific heat of air, γ is the psychrometric constant, T_1 is the mean canopy temperature, T_2 is the temperature at the soil surface, T_b is the air temperature within the canopy, T_{2r} is the temperature of the soil covered by residue, r_1 is an aerodynamic resistance between the canopy and the air, r_c is the surface canopy resistance, r_2 is the aerodynamic resistance between the soil and the canopy, r_s is the resistance to the diffusion of water vapor at the top soil layer, r_{rh} is the residue resistance to transfer of heat, r_r is the residue resistance to the transfer of vapor acting in series with the soil resistance r_s, e_b is the vapor pressure of the atmosphere at the canopy level, e_1^* is the saturation vapor pressure in the canopy, e_L^* is the saturation vapor pressure at the top of the wet layer, and e_{Lr}^* is the saturation vapor pressure at the top of the wet layer for the soil covered by residue.
Conduction of heat for the bare-soil and residue-covered surfaces are given by:

$$G_{os} = \frac{\rho \cdot C_p \cdot (T_2 - T_L)}{r_u} \quad \text{and} \quad G_s = \frac{\rho \cdot C_p \cdot (T_L - T_m)}{r_L} \tag{6}$$

$$G_{or} = \frac{\rho \cdot C_p \cdot (T_{2r} - T_{Lr})}{r_u} \quad \text{and} \quad G_r = \frac{\rho \cdot C_p \cdot (T_{Lr} - T_m)}{r_L} \tag{7}$$

where; r_u and r_L are resistance to the transport of heat for the upper and lower soil layers, respectively, T_L and T_{Lr} are the temperatures at the interface between the upper and lower layers for the bare soil and the residue-covered soil, and T_m is the temperature at the bottom of the lower layer which was assumed to be constant on a daily basis.

Choudhury and Monteith (1988) expressed differences in saturation vapor pressure between points in the system as linear functions of the corresponding temperature differences. They found that a single value of the slope of the saturation vapor pressure, Δ, when evaluated at the air temperature, T_a, gave acceptable results for the components of the heat balance. The vapor pressure differences were given by:

$$e_1^* - e_b^* = \Delta \cdot (T_1 - T_b) e_L^* - e_b^* = \Delta \cdot (T_L - T_b) e_b^* - e_a^* = \Delta \cdot (T_b - T_a) \quad \text{and} \quad e_{Lr}^* - e_b^* = \Delta \cdot (T_{Lr} - T_b) \tag{8}$$

The above equations were combined and solved to estimate fluxes. Details are provided by Lagos (2008). The solution gives the latent and sensible heat fluxes from the canopy as:

$$\lambda E_c = \frac{\Delta \cdot r_1 \cdot Rn_c + \rho \cdot C_p \cdot (e_b^* - e_b)}{\Delta \cdot r_1 + \gamma \cdot (r_1 + r_c)} \quad \text{and} \quad H_c = \frac{\gamma \cdot (r_1 - r_c) \cdot Rn_c - \rho \cdot C_p \cdot (e_b^* - e_b)}{\Delta \cdot r_1 + \gamma \cdot (r_1 + r_c)} \tag{9}$$

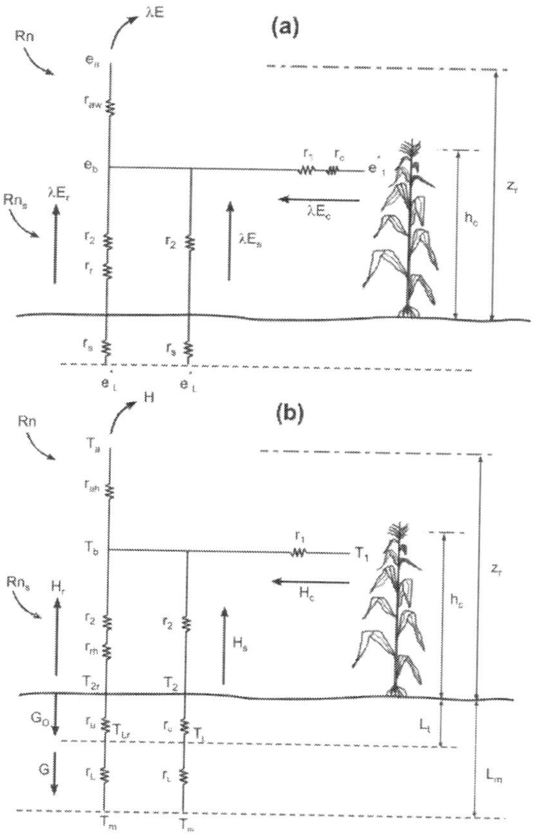

Fig. 2. Schematic resistance network of the Surface Energy Balance (SEB) model a) Latent heat flux and b) Sensible heat flux.

Similarly, latent and sensible heat fluxes from bare soil surfaces are estimated by:

$$\lambda E_s = \frac{Rn_s \cdot \Delta \cdot r_2 \cdot r_L + \rho \cdot C_p \cdot [(e_b^* - e_b) \cdot (r_u + r_L + r_2) + (T_m - T_b) \cdot \Delta \cdot (r_u + r_2)]}{\gamma \cdot (r_2 + r_s) \cdot (r_u + r_L + r_2) + \Delta \cdot r_L \cdot (r_u + r_2)} \quad (10)$$

$$H_s = \frac{Rn_s \cdot r_L \cdot \Delta - \lambda E_s \cdot [r_L \cdot \Delta + \gamma \cdot (r_2 + r_s)] + \rho \cdot C_p \cdot (e_b^* - e_b) - \rho \cdot C_p \cdot \Delta \cdot (T_b - T_m)}{r_L \cdot \Delta} \quad (11)$$

The latent and sensible heat fluxes from the residue-covered soil are simulated with:

$$\lambda E_r = \frac{Rn_s \cdot \Delta \cdot (r_2 + r_{rh}) \cdot r_L + \rho \cdot C_p \cdot [(e_b^* - e_b) \cdot (r_u + r_L + r_2 + r_{rh}) + (T_m - T_b) \cdot \Delta \cdot (r_u + r_2 + r_r)]}{\gamma \cdot (r_2 + r_s + r_r) \cdot (r_u + r_L + r_2 + r_{rh}) + \Delta \cdot r_L \cdot (r_u + r_2 + r_{rh})} \quad (12)$$

$$H_r = \frac{Rn_s \cdot r_L \cdot \Delta - \lambda E_r \cdot [r_L \cdot \Delta + \gamma \cdot (r_2 + r_s + r_r)] + \rho \cdot C_p \cdot (e_b^* - e_b) - \rho \cdot C_p \cdot \Delta \cdot (T_b - T_m)}{r_L \cdot \Delta} \quad (13)$$

Values for T_b and e_b are necessary to estimate latent heat and sensible heat fluxes. The values of the parameters can be expressed as:

$$e_b = \left(T_b \cdot (\Delta \cdot A_2 - A_3) + \frac{A_1}{\rho \cdot C_p} - \Delta \cdot A_2 \cdot T_a + A_2 \cdot e_a^* + T_m \cdot A_3 + \frac{e_a}{\gamma \cdot r_{aw}}\right) \cdot \left(\frac{\gamma \cdot r_{aw}}{1 + A_2 \cdot \gamma \cdot r_{aw}}\right) \quad (14)$$

$$T_b = \left[\frac{B_1}{\rho \cdot C_p} + T_a \cdot \left(\frac{1}{r_{ah}} - \Delta \cdot B_2\right) + (e_a^* - e_b) \cdot B_2 + T_m \cdot B_3\right] \cdot \left(\frac{r_{ah}}{1 - \Delta \cdot B_2 \cdot r_{ah} + B_3 \cdot r_{ah}}\right) \quad (15)$$

where, r_{ah} is the aerodynamic resistance for heat transport, r_{aw} is the aerodynamic resistance for water vapor transport, e_a is the vapor pressure at the reference height, and e_a^* is the saturated vapor pressure at the reference height. Six coefficients (A_1, A_2, A_3 and B_1, B_2 and B_3) are involved in these expressions. These coefficients depend on environmental conditions and other parameters. The expressions to compute the coefficients are given by (Lagos, 2008):

$$A_1 = \frac{\Delta \cdot r_1 \cdot Rn_c}{\Delta \cdot r_1 + \gamma \cdot (r_1 + r_c)} + (1 - f_r) \cdot \frac{Rn_s \cdot \Delta \cdot r_2 \cdot r_L}{\gamma \cdot (r_2 + r_s) \cdot (r_u + r_L + r_2) + \Delta \cdot r_L \cdot (r_u + r_2)} + f_r \cdot \frac{Rn_s \cdot \Delta \cdot (r_2 + r_{rh}) \cdot r_L}{\gamma \cdot (r_2 + r_s + r_r) \cdot (r_u + r_L + r_2 + r_{rh}) + \Delta \cdot r_L \cdot (r_u + r_2 + r_{rh})} \quad (16)$$

$$A_2 = \frac{1}{\Delta \cdot r_1 + \gamma \cdot (r_1 + r_c)} + (1 - f_r) \cdot \frac{(r_u + r_L + r_2)}{\gamma \cdot (r_2 + r_s) \cdot (r_u + r_L + r_2) + \Delta \cdot r_L \cdot (r_u + r_2)} + f_r \cdot \frac{(r_u + r_L + r_2 + r_{rh})}{\gamma \cdot (r_2 + r_s + r_r) \cdot (r_u + r_L + r_2 + r_{rh}) + \Delta \cdot r_L \cdot (r_u + r_2 + r_{rh})} \quad (17)$$

$$A_3 = \left[(1 - f_r) \cdot \frac{\Delta \cdot (r_u + r_2)}{\gamma \cdot (r_2 + r_s) \cdot (r_u + r_L + r_2) + \Delta \cdot r_L \cdot (r_u + r_2)} + f_r \frac{\Delta \cdot (r_u + r_2 + r_{rh})}{\gamma \cdot (r_2 + r_s + r_r) \cdot (r_u + r_L + r_2 + r_{rh}) + \Delta \cdot r_L \cdot (r_u + r_2 + r_{rh})}\right] \quad (18)$$

$$B_1 = \left[Rn_c \cdot \frac{\gamma \cdot (r_1 + r_c)}{\Delta \cdot r_1 + \gamma \cdot (r_1 + r_c)} + Rn_s \cdot \begin{pmatrix} (1 - f_r) \cdot (1 - \Delta \cdot r_2 \cdot r_L \cdot X_s) + \\ f_r \cdot (1 - \Delta \cdot (r_2 + r_{rh}) \cdot r_L \cdot X_r) \end{pmatrix} \right] \tag{19}$$

$$B_2 = \frac{-1}{\Delta \cdot r_1 + \gamma \cdot (r_1 + r_c)} + (1 - f_r) \cdot \left(\frac{1}{r_L \Delta} - (r_u + r_L + r_2) \cdot X_s \right)$$
$$+ f_r \cdot \left(\frac{1}{r_L \Delta} - (r_u + r_L + r_2 + r_{rh}) \cdot X_r \right) \tag{20}$$

$$B_3 = \left[(1 - f_r) \cdot \left(\frac{1}{r_L} - \Delta \cdot (r_u + r_2) \cdot X_s \right) + f_r \cdot \left(\frac{1}{r_L} - \Delta \cdot (r_u + r_2 + r_{rh}) \cdot X_r \right) \right] \tag{21}$$

$$X_s = \left(\frac{1}{\gamma \cdot (r_2 + r_s) \cdot (r_u + r_L + r_2) + \Delta \cdot r_L \cdot (r_u + r_2)} \right) \left(\frac{(r_L \cdot \Delta + \gamma \cdot (r_2 + r_s))}{r_L \cdot \Delta} \right) \text{ and}$$

$$X_r = \left(\frac{1}{\gamma \cdot (r_2 + r_s + r_r) \cdot (r_u + r_L + r_2 + r_{rh}) + \Delta \cdot r_L \cdot (r_u + r_2 + r_{rh})} \right) \left(\frac{(r_L \cdot \Delta + \gamma \cdot (r_2 + r_s + r_r))}{r_L \cdot \Delta} \right) \tag{22}$$

These relationships define the surface energy balance model which is applicable to conditions ranging from closed canopies to surfaces with bare soil or those partially covered with residue. Without residue, the model is similar to that by Choudhury and Monteith (1988).

2.1.1 Determination of the SEB model parameters

In the following sections, the procedures to compute parameter values for the model are detailed. The parameters are as important as the formulation of the energy balance equations.

2.1.1.1 Aerodynamic resistances

Thom (1972) stated that heat and mass transfer encounter greater aerodynamic resistance than the transfer of momentum. Accordingly, aerodynamic resistances to heat (r_{ah}) and water vapor transfer (r_{aw}) can be estimated as:

$$r_{ah} = r_{am} + r_{bh} \qquad \text{and} \qquad r_{aw} = r_{am} + r_{bw} \tag{23}$$

where r_{am} is the aerodynamic resistance to momentum transfer, and r_{bh} and r_{bw} are excess resistance terms for heat and water vapor transfer.

Shuttleworth and Gurney (1990) built on the work of Choudhury and Monteith (1988) to estimate r_{am} by integrating the eddy diffusion coefficient over the sink of momentum in the canopy to a reference height z_r above the canopy, giving the following relationship for r_{am}:

$$r_{am} = \frac{1}{k \cdot u^*} \cdot \mathrm{Ln}\left(\frac{z_r - d}{h - d} \right) + \frac{h}{\alpha \cdot K_h} \cdot \left[\exp\left(\alpha \cdot \left(1 - \frac{z_o + d}{h} \right) \right) - 1 \right] \tag{24}$$

where k is the von Karman constant, u* is the friction velocity, z_o is the surface roughness, d is the zero-plane displacement height, K_h is the value of eddy diffusion coefficient at the top of the canopy, h is the height of vegetation, and α is the attenuation coefficient. A value of α = 2.5, which is typical for agricultural crops, was recommended by Shuttleworth and Wallace (1985) and Shuttleworth and Gurney (1990).

Verma (1989) expressed the excess resistance for heat transfer as:

$$r_{bh} = \frac{k \cdot B^{-1}}{k \cdot u^*} \tag{25}$$

where B^{-1} represents a dimensionless bulk parameter. Thom (1972) suggests that the product kB^{-1} equal approximately 2 for most arable crops.
Excess resistance was derived primarily from heat transfer observations (Weseley & Hicks 1977). Aerodynamic resistance to water vapor was modified by the ratio of thermal and water vapor diffusivity:

$$r_{bw} = \frac{k \cdot B^{-1}}{k \cdot u^*}\left(\frac{k_1}{D_v}\right)^{2/3} \tag{26}$$

where, k_1 is the thermal diffusivity and D_v is the molecular diffusivity of water vapor in air. Similarly, Shuttleworth and Gurney (1990) expressed the aerodynamic resistance (r_2) by integrating the eddy diffusion coefficient between the soil surface and the sink of momentum in the canopy to yield:

$$r_2 = \frac{h \cdot \exp(\alpha)}{\alpha \cdot K_h} \cdot \left[\exp\left(\frac{-\alpha \cdot z_o'}{h}\right) - \exp\left(\frac{-\alpha \cdot (d + z_o)}{h}\right)\right] \tag{27}$$

where z_o' is the roughness length of the soil surface. Values of surface roughness (z_o) and displacement height (d) are functions of leaf area index (LAI) and can be estimated using the expressions given by Shaw and Pereira (1982).
The diffusion coefficients between the soil surface and the canopy, and therefore the resistance for momentum, heat, and vapor transport are assumed equal although it is recognized that this is a weakness in the use of the K theory to describe through-canopy transfer (Shuttleworth & Gurney, 1990). Stability is not considered.

2.1.1.2 Canopy resistances

The mean boundary layer resistance of the canopy r_1, for latent and sensible heat flux, is influenced by the surface area of vegetation (Shuttleworth & Wallace, 1985):

$$r_1 = \frac{r_b}{2 \cdot LAI} \tag{28}$$

where r_b is the resistance of the leaf boundary layer, which is proportional to the temperature difference between the leaf and surrounding air divided by the associated flux (Choudhury & Monteith, 1988). Shuttleworth and Wallace (1985) noted that resistance r_b exhibits some dependence on in-canopy wind speed, with typical values of 25 s m^{-1}. Shuttleworth and Gurney (1990) represented r_b as:

$$r_b = \frac{100}{\alpha} \cdot \left(\frac{w}{u_h}\right)^{1/2} \cdot \left(1 - \exp\left(\frac{-\alpha}{2}\right)\right)^{-1} \tag{29}$$

where w is the representative leaf width and u_h is the wind speed at the top of the canopy. This resistance is only significant when acting in combination with a much larger canopy surface resistance, and Shuttleworth and Gurney (1990) suggest that r_1 could be neglected for foliage completely covering the ground. Using r_b = 25 s m^{-1} with an LAI = 4, the corresponding canopy boundary layer resistance is r_1 = 3 s m^{-1}.

Canopy surface resistance, r_c, can be calculated by dividing the minimum surface resistance for a single leaf (r_l) by the effective canopy leaf area index (LAI). Five environmental factors have been found to affect stomata resistance: solar radiation, air temperature, humidity, CO_2 concentration and soil water potential (Yu et al., 2004). Several models have been developed to estimate stomata conductance and canopy resistance. Stannard (1993) estimated r_c as a function of vapor pressure deficit, leaf area index, and solar radiation as:

$$r_c = \left[C_1 \cdot \frac{LAI}{LAI_{max}} \cdot \frac{C_2}{C_2 + VPD_a} \cdot \frac{Rad \cdot (Rad_{max} + C_3)}{Rad_{max} \cdot (Rad + C_3)}\right]^{-1} \tag{30}$$

where LAI_{max} is the maximum value of leaf area index, VPD_a is vapor pressure deficit, Rad is solar radiation, Rad_{max} is the maximum value of solar radiation (estimated at 1000 W m^{-2}) and C_1, C_2 and C_3 are regression coefficients. Canopy resistance does not account for soil water stress effects.

2.1.1.3 Soil resistances

Farahani and Bausch (1995), Anadranistakis et al. (2000) and Lindburg (2002) found that soil resistance (r_s) can be related to volumetric soil water content in the top soil layer. Farahani and Ahuja (1996) found that the ratio of soil resistance when the surface layer is wet relative to its upper limit depends on the degree of saturation (θ/θ_s) and can be described by an exponential function as:

$$r_s = r_{so} \cdot \exp\left(-\beta \cdot \frac{\theta}{\theta_s}\right) \quad \text{and} \quad r_{so} = \frac{L_t \cdot \tau_s}{D_v \cdot \emptyset} \tag{31}$$

where L_t is the thickness of the surface soil layer, τ_s is a soil tortuosity factor, D_v is the water vapor diffusion coefficient and $\emptyset$ is soil porosity, θ is the average volumetric water content in the surface layer, θ_s is the saturation water content, and β is a fitting parameter. Measurements of θ from the top 0.05 m soil layer were more effective in modeling r_s than θ for thinner layers.

Choudhury and Monteith (1988) expressed the soil resistance for heat flux (r_L) in the soil layer extending from depth L_t to L_m as:

$$r_L = \frac{\rho \cdot C_p \cdot (L_m - L_t)}{K} \tag{32}$$

where K is the thermal conductivity of the soil. Similarly, the corresponding resistance for the upper layer (r_u) of depth L_t and conductivity K' as:

$$r_u = \frac{\rho \cdot C_p \cdot L_t}{K'} \tag{33}$$

2.1.1.4 Residue resistances

Surface residue is an integral part of many cropping systems. Bristow and Horton (1996) showed that partial surface mulch cover can have dramatic effects on the soil physical environment. The vapor conductance through residue has been described as a linear function of wind speed. Farahani and Ahuja (1996) used results from Tanner and Shen (1990) to develop the resistance of surface residue (r_r) as:

$$r_r = \frac{L_r \cdot \tau_r}{D_v \cdot \emptyset_r}(1 + 0.7 \cdot u_2)^{-1} \tag{34}$$

where L_r is residue thickness, τ_r is residue tortuosity, D_v is vapor diffusivity in still air, $\emptyset_r$ is residue porosity and u_2 is wind speed measured two meters above the surface. Due to the porous nature of field crop residue layers, the ratio $\tau_r/\emptyset_r$ is about one (Farahani & Ahuja, 1996).

Similar to the soil resistance, Bristow and Horton (1996) and Horton et al. (1996) expressed the resistance of residue for heat transfer, r_{rh}, as:

$$r_{rh} = \frac{\rho \cdot C_p \cdot L_r}{K_r} \tag{35}$$

where K_r is the residue thermal conductivity.

The fraction of the soil covered by residue (f_r) can be estimated using the amount and type of residue (Steiner et al., 2000). The soil covered by residue and the residue thickness are estimated using the expressions developed by Gregory (1982).

2.1.2 SEB model inputs

Inputs required to solve multiple layer models (i.e. Shuttleworth and Wallace (1985), Choudhury and Monteith (1988) and Lagos (2008) models) are net radiation, solar radiation, air temperature, relative humidity, wind speed, LAI, crop height, soil texture, soil temperature, soil water content, residue type, and residue amount. In particular, net radiation, leaf area index, soil temperatures and residue amount are variables rarely measured in the field, other than at research sites. Net radiation and soil temperature models can be incorporated into surface energy balance models to predict evapotranspiration from environmental variables typically measured by automatic weather stations.

Similar to the Shuttleworth and Wallace (1985) and Choudhury and Monteith (1988) models, measurements of net radiation and estimations of net radiation absorbed by the canopy are necessary for the SEB model. Beer's law is used to estimate the penetration of radiation through the canopy and estimates the net radiation reaching the surface (Rn_s) as:

$$Rn_s = Rn \cdot \exp(-C_{ext} \cdot LAI) \tag{36}$$

where Cext is the extinction coefficient of the crop for net radiation. Consequently, net radiation absorbed by the canopy (Rnc) can be estimated as Rnc = Rn – Rn_s.

2.1.3 SEB model evaluation

An irrigated maize field site located at the University of Nebraska Agricultural Research and Development Center near Mead, NE (41°09′53.5″N, 96°28′12.3″W, elevation 362 m) was used for model evaluation. This site is a 49 ha production field that provides sufficient upwind fetch of uniform cover required for adequately measuring mass and energy fluxes using eddy covariance systems. The area has a humid continental climate and the soil corresponds to a deep silty clay loam (Suyker & Verma, 2009). The field has not been tilled since 2001. Detailed information about planting densities and crop management is provided by Verma et al. (2005) and Suyker and Verma (2009).

Soil water content was measured continuously at four depths (0.10, 0.25, 0.5 and 1.0 m) with Theta probes (Delta-T Device, Cambridge, UK). Destructive green leaf area index and biomass measurements were taken bi-monthly during the growing season. The eddy covariance measurements of latent heat, sensible heat and momentum fluxes were made using an omnidirectional three dimensional sonic anemometer (Model R3, Gill Instruments Ltd., Lymington, UK) and an open-path infrared CO_2/H_2O gas analyzer system (Model LI7500, Li-Cor Inc, Lincoln, NE). Fluxes were corrected for sensor frequency response and variations in air density. More details of measurements and calculations are given in Verma et al. (2005). Air temperature and humidity were measured at 3 and 6 meters (Humitter 50Y, Vaisala, Helsinki, Finland), net radiation at 5.5 m (CNR1, Kipp and Zonen, Delft, NLD) and soil heat flux at 0.06 m (Radiation and Energy Balance Systems Inc, Seattle, WA). Soil temperature was measured at 0.06, 0.1, 0.2 and 0.5 m depths (Platinum RTD, Omega Engineering, Stamford, CT). More details are given in Verma et al. (2005) and Suyker and Verma (2009).

Evapotranspiration predictions from the SEB model were compared with eddy covariance flux measurements during 2003 for an irrigated maize field. To evaluate the energy balance closure of eddy covariance measurements, net radiation was compared against the sum of latent heat, sensible heat, soil heat flux and storage terms. Storage terms include soil heat storage, canopy heat storage, and energy used in photosynthesis. Storage terms were calculated by Suyker and Verma (2009) following Meyers and Hollinger (2004). During these days, the regression slope for energy balance closure was 0.89 with a correlation coefficient of $r^2 = 0.98$.

For model evaluation, 15 days under different LAI conditions were selected to initially test the model, however further work is needed to test the model for entire growing seasons and during longer periods. Hourly data for three 5-day periods with varying LAI conditions (LAI = 0, 1.5 and 5.4) were used to compare measured ET to model predictions. Input data of the model included hourly values for: net radiation, air temperature, relative humidity, soil temperature at 50 cm, wind speed, solar radiation and soil water content. During the first 5-day period, which was prior to germination, the maximum net radiation ranged from 240 to 720 W m^{-2}, air temperature ranged from 10 to 30°C, soil temperature was fairly constant at 16°C and wind speed ranged from 1 to 9 m s^{-1} but was generally less than 6 m s^{-1} (Figure 3). Soil water content in the evaporation zone averaged 0.34 m^3 $m^{-3}$and the residue density was 12.5 ton/ha on June 6, 2003. Precipitation occurred on the second and fifth days, totaling 17 mm.

Evapotranspiration estimated with the SEB model and measured using the eddy covariance system is given in Figure 4. ET fluxes were the highest at midday on June 6, reaching approximately 350 W m^{-2}. The lowest ET rates occurred on the second day. Estimated ET tracked measured latent heat fluxes reasonably well. Estimates were better for days without precipitation than for days when rainfall occurred. The effect of crop residue on evaporation from the soil is shown in Figure 4 for this period. Residue reduced cumulative evaporation by approximately 17% during this five-day period. Evaporation estimated with the SEB model on June 6 and 9 was approximately 3.5 mm/day, totaling approximately half of the total evaporation for the five days.

During the second five-day period, when plants partially shaded the soil surface (LAI = 1.5), the maximum net radiation ranged from 350 to 720 W m^{-2} and air temperature ranged from 10 to 33°C (Figure 5). The soil temperature was nearly constant at 20°C. Wind speed ranged from 0.3 to 8 m s^{-1} but was generally less than 6 m s^{-1}. The soil water content was about 0.31

m³ m⁻³ and the residue density was 12.2 ton/ha on June 24, 2003. Precipitation totaling 3mm occurred on the fifth day. The predicted rate of ET estimated with the SEB model was close to the observed data (Figure 6). Estimates were smaller than measured values for June 24, which was the hottest and windiest day of the period. The ability of the model to partition ET into evaporation and transpiration for partial canopy conditions is also illustrated in Figure 6. Evaporation from the soil represented the majority of the water used during the night, and early or late in the day. During the middle of the day transpiration represented approximately half of the hourly ET flux.

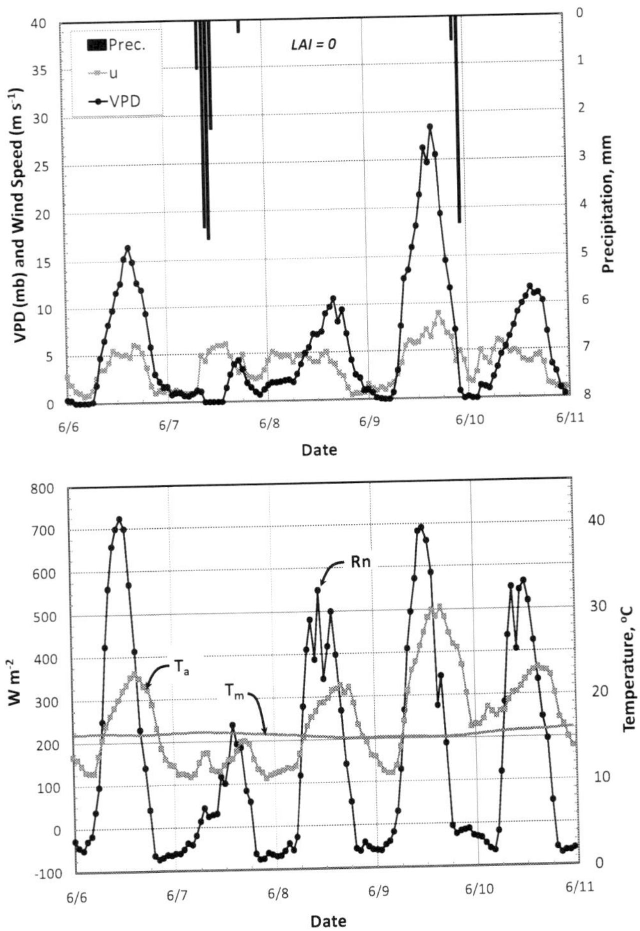

Fig. 3. Environmental conditions during a five-day period without canopy cover for net radiation (Rn), air temperature (T_a), soil temperature (T_m), precipitation (Prec.), vapor pressure deficit (VPD), and wind speed (u).

The last period represents a fully developed maize canopy that completely shaded the soil surface. The crop height was 2.3 m and the LAI was 5.4. Environmental conditions for the period are given in Figure 7. The maximum net radiation ranged from 700 to 740 W m^{-2} and air temperature ranged from 15 to 36 °C during the period. Soil temperature was fairly constant during the five days at 21.5°C and wind speed ranged from 0.3 to 4 m s^{-1}. The soil water content was about 0.25 m^3 m^{-3} and the residue density was 11.8 ton/ha on July 16, 2003. Precipitation totaling 29 mm occurred on the third day. Observed and predicted ET fluxes agreed for most days with some differences early in the morning during the first day and during the middle of several days (Figure 8). Transpiration simulated with the SEB model was nearly equal to the simulated ET for the period as evaporation rates from the soil was very small.

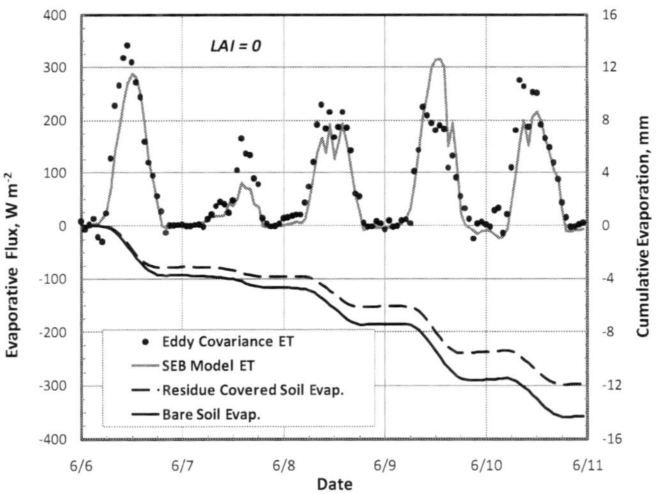

Fig. 4. Evapotranspiration estimated by the Surface Energy Balance (SEB) model and measured by an eddy covariance system and simulated cumulative evaporation from bare and residue-covered soil for a period without plant canopy cover.

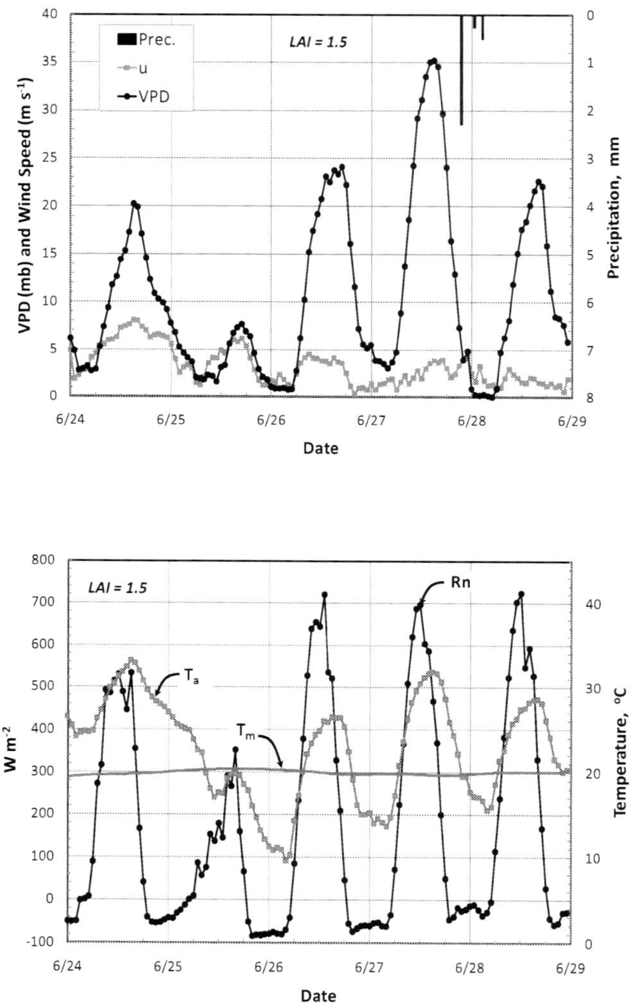

Fig. 5. Environmental conditions for a five-day period with partial crop cover for net radiation (Rn), air temperature (Ta), soil temperature (Tm), precipitation (Prec), vapor pressure deficit (VPD), and wind speed (u).

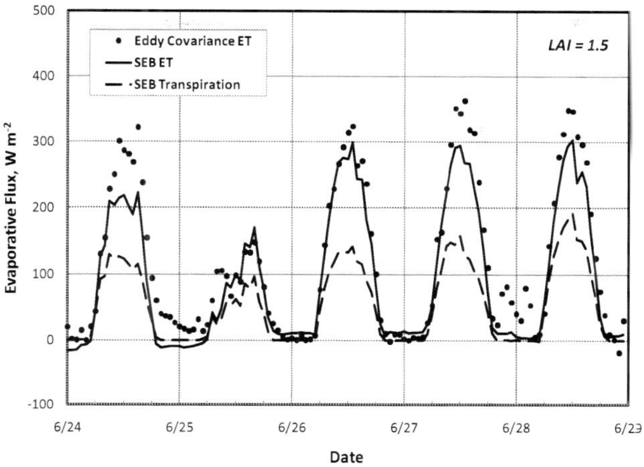

Fig. 6. Evapotranspiration and transpiration estimated by the Surface Energy Balance (SEB) model and ET measured by an eddy covariance system for a 5-day period with partial canopy cover.

Hourly measurements and SEB predictions for the three five-day periods were combined to evaluate the overall performance of the model (Figure 9). Results show variation about the 1:1 line; however, there is a strong correlation and the data are reasonably well distributed about the line. Modeled ET is less than measured for latent heat fluxes above 450 W m^{-2}. The model underestimates ET during hours with high values of vapor pressure deficit (Figure 6 and 8), this suggests that the linear effect of vapor pressure deficit in canopy resistance estimated with equation (30) produce a reduction on ET estimations. Further work is required to evaluate and explore if different canopy resistance models improve the performance of ET predictions under these conditions. Various statistical techniques were used to evaluate the performance of the model. The coefficient of determination, Nash-Sutcliffe coefficient, index of agreement, root mean square error and the mean absolute error were used for model evaluation (Legates & McCabe 1999; Krause et al., 2005; Moriasi et al., 2007; Coffey et al. 2004). The coefficient of determination was 0.92 with a slope of 0.90 over the range of hourly ET values. The root mean square error was 41.4 W m^{-2}, the mean absolute error was 29.9 W m^{-2}, the Nash-Sutcliffe coefficient was 0.92 and the index of agreement was 0.97. The statistical parameters show that the model represents field measurements reasonably well. Similar performance was obtained for daily ET estimations (Table 1). Analysis is underway to evaluate the model for more conditions and longer periods. Simulations reported here relied on literature-reported parameter values. We are also exploring calibration methods to improve model performance.

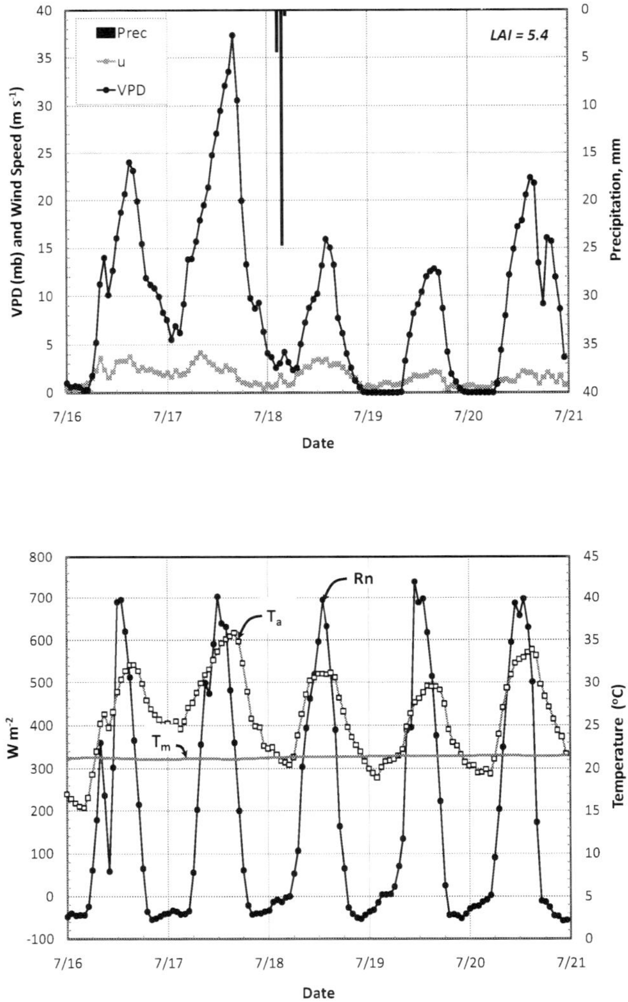

Fig. 7. Environmental conditions for 5-day period with full canopy cover for net radiation (Rn), air temperature (Ta), soil temperature (Tm), precipitation (Prec), vapor pressure deficit (VPD) and wind speed (u).

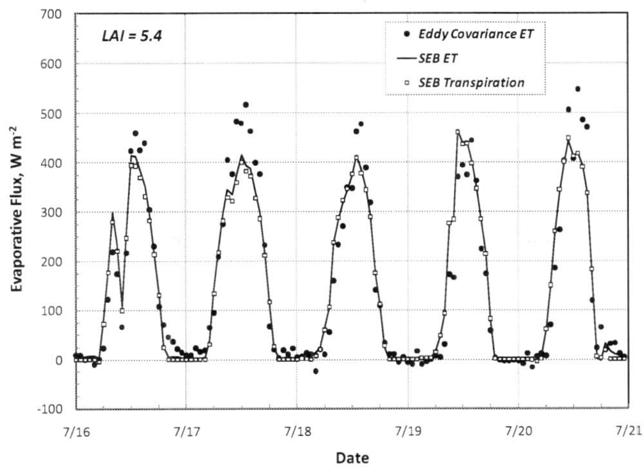

Fig. 8. Evapotranspiration and transpiration estimated by the Surface Energy Balance (SEB) model and ET measured by an eddy covariance system during a period with full canopy cover.

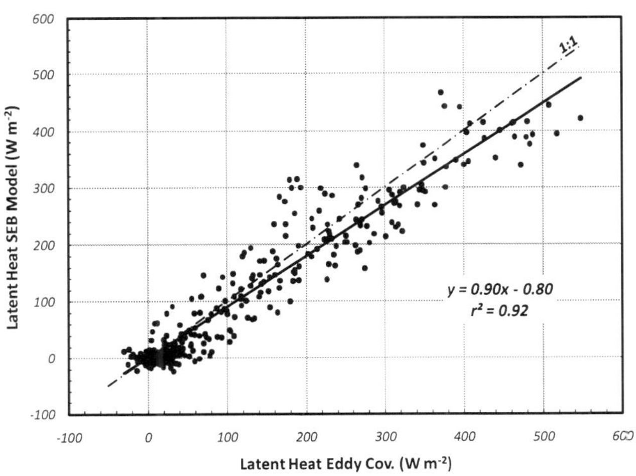

Fig. 9. Measured versus modeled hourly latent heat fluxes.

	LAI	Evapotranspiration (mm day^{-1})	
Date	m^2 m^{-2}	SEB	EC
6-Jun	0	3.2	3.7
7-Jun	0	0.7	1.4
8-Jun	0	2.3	3.2
9-Jun	0	3.5	2.7
10-Jun	0	2.4	3.5
24-Jun	1.5	2.9	4.4
25-Jun	1.5	1.7	2.1
26-Jun	1.5	4.1	4.3
27-Jun	1.5	4.0	5.0
28-Jun	1.5	3.8	4.7
16-Jul	5.4	5.1	5.1
17-Jul	5.4	5.8	6.8
18-Jul	5.4	5.2	5.0
19-Jul	5.4	5.0	4.1
20-Jul	5.4	5.1	5.4

Table 1. Daily evapotranspiration estimated with the Surface Energy Balance (SEB) model and measured from the Eddy Covariance (EC) system.

2.2 The modified SEB model for Partially Vegetated surfaces (SEB-PV)

Although good performance of multiple-layer models has been recognized, multiple-layer models estimate more accurate ET values under high LAI conditions. Lagos (2008) evaluated the SEB model for maize and soybean under rainfed and irrigated conditions; results indicate that during the growing season, the model more accurately predicted ET after canopy closure (after LAI=4) than for low LAI conditions. The SEB model, similar to S-W and C-M models, is based on homogeneous land surfaces. Under low LAI conditions, the land surface is partially covered by the canopy and soil evaporation takes place from soil below the canopy and areas of bare soil directly exposed to net radiation. However, in multiple-layer models, evaporation from the soil has been only considered below the canopy and hourly variations in the partitioning of net radiation between the canopy and the soil is often disregarded. Soil evaporation on partially vegetated surfaces & inorchards and natural vegetation include not only soil evaporation beneath the canopy but also evaporation from areas of bare soil that contribute directly to total ET.

Recognizing the need to separate vegetation from soil and considering the effect of residue on evaporation, we extended the SEB model to represent those common conditions. The modified model, hereafter the SEB-PV model, distributes net radiation (Rn), sensible heat (H), latent heat (λE), and soil heat fluxes (G) through the soil/residue/canopy system. Similar to the SEB model, horizontal gradients of the potentials are assumed to be small enough for lateral fluxes to be ignored, and physical and biochemical energy storage terms in the canopy/residue/soil system are assumed to be negligible. The evaporation of water on plant leaves due to rain, irrigation or dew is also ignored.

The SEB-PV model has the same four layers described previously for SEB (Figure 10):the first extended from the reference height above the vegetation and the sink for momentum within the canopy, a second layer between the canopy level and the soil surface, a third layer corresponding to the top soil layer and a lower soil layer where the soil atmosphere is saturated with water vapor.

Total latent heat (λE) is the sum of latent heat from the canopy (λEc), latent heat from the soil (λEs) beneath the canopy, latent heat from the residue-covered soil (λEr) beneath the canopy, latent heat from the soil (λEbs) directly exposed to net radiation and latent heat from the residue-covered soil (λEbr) directly exposed to net radiation.

$$\lambda E = [\lambda E_c + \lambda E_s(1 - f_r) + \lambda E_r f_r]F_V + [\lambda E_{bs}(1 - f_r)](1 - F_V) \quad (37)$$

Where fr is the fraction of the soil affected by residue and Fv is the fraction of the soil covered by vegetation. Similarly, sensible heat is calculated as the sum of sensible heat from the canopy (Hc), sensible heat from the soil (Hs) and sensible heat from the residue covered soil (Hr), sensible heat from the soil (Hbs) directly exposed to net radiation and latent heat from the residue-covered soil (Hbr) directly exposed to net radiation.

$$H = [Hc + Hs\,(1 - fr) + Hr\,fr\,]\,Fv + [\,Hbs\,(1 - fr) + Hbr\,fr]\,(1 - Fv) \quad (38)$$

For the fraction of the soil covered by vegetation, the total net radiation is divided into that absorbed by the canopy (Rnc) and the soil beneath the canopy (Rns) and is given by Rn = Rnc + Rns. The net radiation absorbed by the canopy is divided into latent heat and sensible heat fluxes as Rnc = λEc + Hc. Similarly, for the soil Rns = Gos + Hs, where Gos is a conduction term downwards from the soil surface and is expressed as Gos = λEs + Gs, where Gs is the soil heat flux for bare soil. Similarly, for the residue covered soil Rns = Gor + Hr where Gor is the conduction downwards from the soil covered by residue. The conduction is given by Gor = λEr + Gr where Gr is the soil heat flux for residue-covered soil. For the area without vegetation, total net radiation is divided into latent and sensible heat fluxes as Rn = λEbs +λEbr + Hbs + Hbr.

The differences in vapor pressure and temperature between levels can be expressed with an Ohm's law analogy using appropriate resistance and flux terms (Figure 10) Latent and sensible flux terms with in the resistance network were combined and solved to estimate total fluxes. The solution gives the latent and sensible heat fluxes from the canopy, the soil beneath the canopy and the soil covered by residue beneath the canopy similar to equations (9), (10), (11), (12) and (13).

The new expressions for latent heat flux of bare soil and soil covered by residue, both directly exposed to net radiation are:

For bare soil:

$$\lambda E_{bs} = \frac{(R_n \cdot \Delta \cdot (r_{2b}) \cdot r_L + \rho \cdot C_p \cdot ((e_b^* - e_b) \cdot r_u + r_L + r_{2b}) + (T_m - T_b) \cdot \Delta \cdot (r_u + r_{2b}))}{\gamma \cdot (r_{2b} + r_s) \cdot (r_u + r_L + r_{2b}) + \Delta \cdot r_L \cdot (r_u + r_{2b})} \quad (39)$$

For residue covered soil:

$$\lambda E_{br} = \frac{R_n \cdot \Delta \cdot (r_{2b} + r_{rh}) \cdot r_L + \rho \cdot C_p \cdot ((e_b^* - e_b) \cdot (r_u + r_L + r_{2b} + r_{rh}) + (T_m - T_b) \cdot \Delta \cdot (r_u + r_{2b} + r_r))}{\gamma \cdot (r_{2b} + r_s + r_r) \cdot (r_u + r_L + r_{2b} + r_{rh}) + \Delta \cdot r_L \cdot (r_u + r_{2b} + r_{rh})} \quad (40)$$

These relationships define the surface energy balance model, which is applicable to conditions ranging from closed canopies to surfaces partially covered by vegetation. If Fv = 1 the model SEB-PV is similar to the original SEB model and with Fv=1 without residue, the model is similar to that by Choudhury and Monteith (1988).

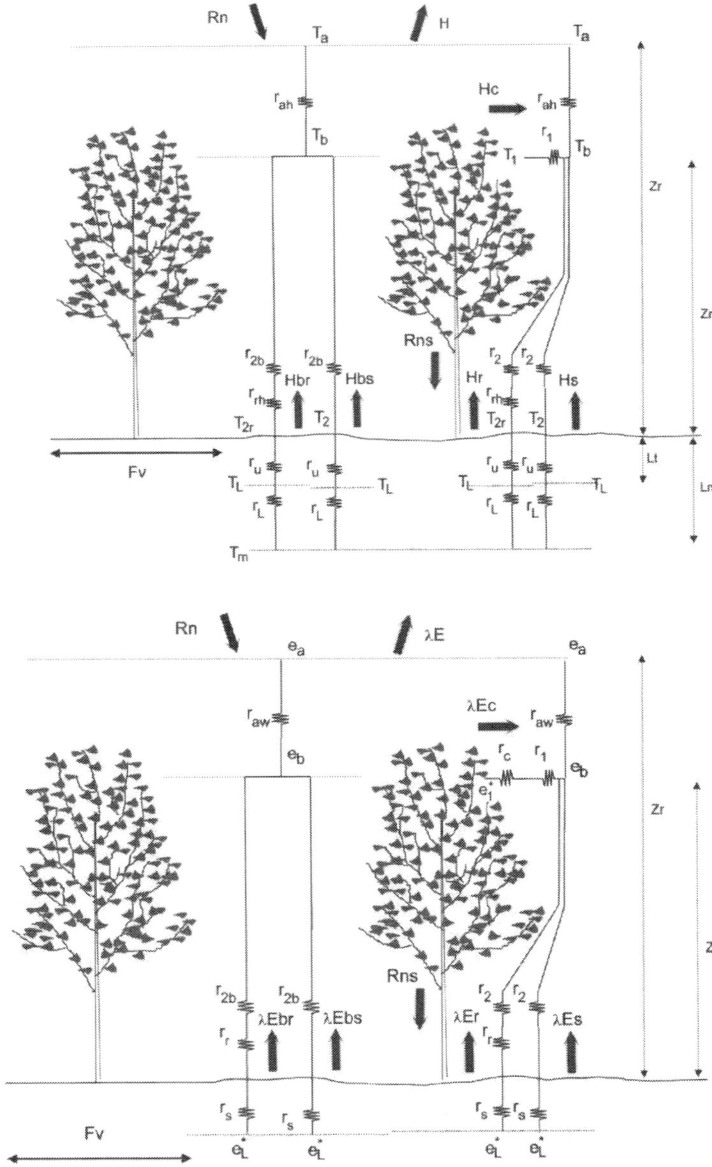

Fig. 10. Schematic resistance network of the modified Surface Energy Balance (SEB - PV) model for partially vegetated surfaces a) Sensible heat flux and b) Latent heat flux.

2.2.1 Model resistances

Model resistances are similar to those described by the SEB model; however, a new aerodynamic resistance (r_{2b}) for the transfer of heat and water flux is required for the surface without vegetation.

The aerodynamic resistance between the soil surface and Zm (r_{2b}) could be calculated by assuming that the soil directly exposed to net radiation is totally unaffected by adjacent vegetation as:

$$r_{as} = \frac{\ln\left(\frac{z_m}{z_o'}\right)^2}{k^2 u} \tag{41}$$

According to Brenner and Incoll (1997), actual aerodynamic resistance (r_{2b}) will vary between r_{as} for Fv=0 and r_2 when the fractional vegetative cover Fv=1. The form of the functional relationship of this change is not known, r_{2b} was varied linearly between r_{as} and r_2 as:

$$r_{2b} = FV(r_2) + (1 - FV)(r_{as}) \tag{42}$$

2.2.2 Model inputs

The proposed SEB-PV model requires the same inputs of the SEB model plus the fraction of the surface covered by vegetation (Fv).

2.3 Sensitivity analysis

A sensitivity analysis was performed to evaluate the response of the SEB model to changes in resistances and model parameters. Meteorological conditions, crop characteristics and soil/residue characteristics used in these calculations are given in Table 2. Such conditions are typical for midday during the growing season of maize in southeastern Nebraska. The sensitivity of total latent heat from the system was explored when model resistances and model parameters were changed under different LAI conditions. The effect of the changes in model parameters and resistances were expressed as changes in total ET (λE) and changes in the crop transpiration ratio. The transpiration ratio is the ratio between crop transpiration (λEc) over total ET (transpiration ratio= λEc / λE).

The response of the SEB model was evaluated for three values of the extinction coefficient (Cext = 0.4, 0.6 and 0.8), three conditions of vapor pressure deficit (VPDa = 0.5 kPa, 0.1 kPa and 0.25 kPa) three soil temperatures (T_m=21°C, 0.8xT_m=16.8 °C and 1.2xT_m=25.2 °C) (Figure 11), changes in the parameterization of aerodynamic resistances (the attenuation coefficient, α= 1, 2.5 and 3.5), the mean boundary layer resistance, r_b (±40%) the crop height, h (±30%)), selected conditions for the soil surface resistance, r_s (0, 227, and 1500 s m^{-1}) (Figure 12), four values for residue resistance, r_r (0, 400, 1000, and 2500 s m^{-1}), and changes of ±30% in surface canopy resistance, r_c (Figure 13).

In general, the sensitivity analysis of model resistances showed that simulated ET was most sensitive to changes in surface canopy resistance for LAI > 0.5 values, and soil surface resistance and residue surface resistance for small LAI values (LAI < ~3). The model was less sensitive to changes in the other parameters evaluated.

Variable	Symbol	Value	Unit
Net Radiation	Rn	500	W m^{-2}
Air temperature	Ta	25	^oC
Relative humidity	RH	68	%
Wind speed	U	2	m s^{-1}
Soil Temperature at 0.5 m	Tm	21	^oC
Solar radiation	Rad	700	W m^{-2}
Canopy resistance coeff.	C1, C2, C3	5, 0.005, 300	
Maximum leaf area index	LAImax	6	m^2 m^{-2}
Soil water content	Θ	0.25	m^3 m^{-3}
Saturation soil water content	Θs	0.5	m^3 m^{-3}
Soil porosity	ϕ	0.5	m^3 m^{-3}
Soil tortuosity	τs	1.5	
Residue fraction	Fr	0.5	
Thickness of the residue layer	Lr	0.02	m
Residue tortuosity	τr	1	
Residue porosity	ϕ r	1	
Upper layer thickness	Lt	0.05	m
Lower layer depth	Lm	0.5	m
Soil roughness length	Zo′	0.01	m
Drag coefficient	Cd	0.07	
Reference height	Z	3	m
Attenuation coefficient	α	2.5	
Maximum solar radiation	Radmax	1000	W m^{-2}
Extinction coefficient	Cext	0.6	
Mean leaf width	W	0.08	m
Water vapor diffusion coefficient	Dv	2.56×10^{-5}	m^2 s^{-1}
Fitting parameter	β	6.5	
Soil thermal conductivity, upper layer	K	2.8	W m^{-1} $^oC^{-1}$
Soil thermal conductivity, lower layer	K′	3.8	W m^{-1} $^oC^{-1}$

Table 2. Predefined conditions for the sensitivity analysis.

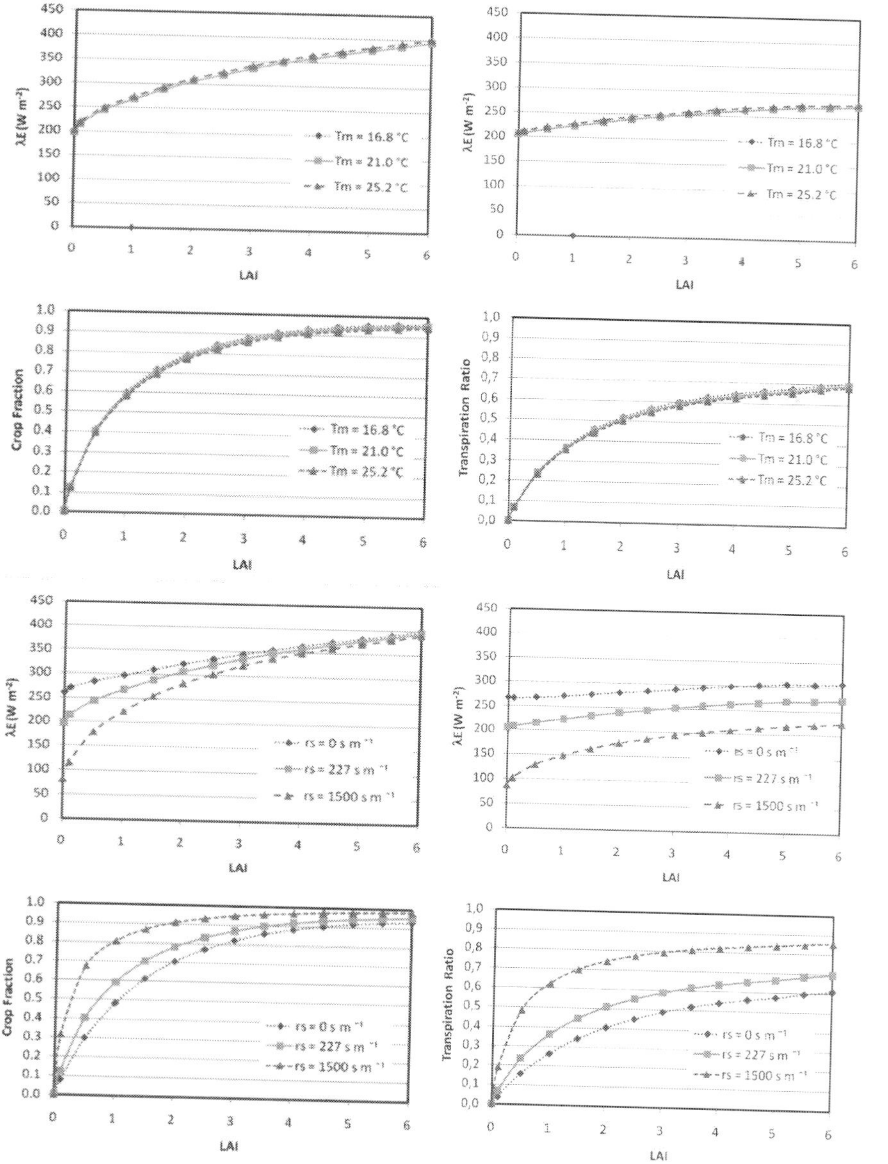

Fig. 11. Sensitivity analysis of the SEB-PV model for Fv=1 (left) and Fv=0,5 (right) under different soil temperatures Tm, and soil resistance conditions.

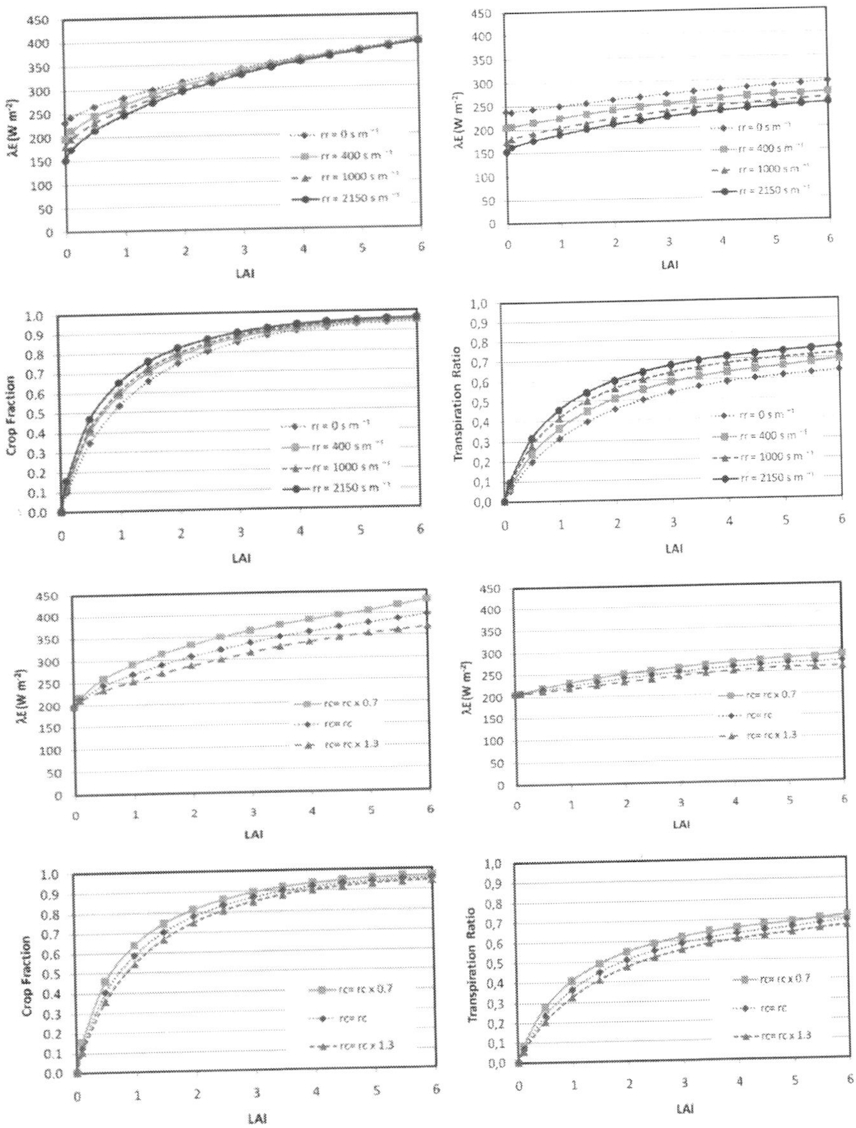

Fig. 12. Sensitivity analysis of the SEB-PV model for Fv=1 (left) and Fv=0,5 (right) under different residue and canopy conditions.

3. Conclusions

A surface energy balance model (SEB) based on the Shuttleworth-Wallace and Choudhury-Monteith models was developed to account for the effect of residue, soil evaporation and canopy transpiration on ET. The model describes the energy balance of vegetated and residue-covered surfaces in terms of driving potential and resistances to flux. Improvements in the SEB model were the incorporation of residue into the energy balance and modification of aerodynamic resistances for heat and water transfer, canopy resistance for water flux, residue resistance for heat and water flux, and soil resistance for water transfer. The model requires hourly data for net radiation, solar radiation, air temperature, relative humidity, and wind speed. Leaf area index and crop height plus soil texture, temperature and water content as well as the type and amount of crop residue are also required. An important feature of the model is the ability to estimate latent, sensible and soil heat fluxes. The model provides a method for partitioning ET into soil/residue evaporation and plant transpiration, and a tool to estimate the effect of residue ET on water balance studies. Comparison between estimated ET and measurements from an irrigated maize field provide support for the validity of the surface energy balance model. Further evaluation of the model is underway for agricultural and natural ecosystems during growing seasons and dormant periods. We are developing calibration procedures to refine parameters and improve model results.

The SEB model was modified for modeling evapotranspiration of partially vegetated surfaces given place to the SEB-PV model. The SEB-PV model can be used for partitioning total ET on canopy transpiration and soil evaporation beneath the canopy and soil directly exposed to net radiation. The model can be used for partitioning net radiation into not only latent heat fluxes but also sensible heat fluxes from each surface. A preliminary sensitivity analysis shows that similar to the SEB model, the proposed modification was sensitive to soil surface resistance, residue resistance, canopy resistance and vapor pressure deficit. Further model evaluation is needed to test this approach. A model to estimate Rn and a model to estimate soil temperature T_m from air temperature and soil conditions are also required to reduce the required inputs of the model.

4. List of variables

Rn Net Radiation (W m^{-2}).
Rn_c Net Radiation absorbed by the canopy (W m^{-2}).
Rn_s Net Radiation absorbed by the soil (W m^{-2}).
λE Total latent heat flux (W m^{-2}).
λE_c Latent heat flux from the canopy (W m^{-2}).
λE_s Latent heat flux from the soil (W m^{-2}).
λE_r Latent heat flux from the residue-covered soil (W m^{-2}).
λE_{bs} Latent heat from the soil directly exposed to net radiation (W m^{-2}).
λE_{br} Latent heat from the residue-covered soil directly exposed to net radiation (W m^{-2}).
H Total Sensible heat flux (W m^{-2}).
H_c Sensible heat flux from the canopy (W m^{-2}).
H_s Sensible heat flux from the soil (W m^{-2}).
H_r Sensible heat flux from the residue-covered soil (W m^{-2}).
G_{os} Conduction flux from the soil surface (W m^{-2}).
G_{or} Conduction flux from the residue-covered soil surface (W m^{-2}).
G_s Soil heat flux for bare soil (W m^{-2}).

G_r Soil heat flux for residue-covered soil (W m^{-2}).
f_r Fraction of the soil covered by residue (0-1).
ρ Density of moist air (Kg m^{-3}).
C_p Specific heat of air (J Kg^{-1} $^oC^{-1}$).
γ Psychrometric constant (Kpa $°C^{-1}$).
T_a Air temperature (^oC).
T_b Air temperature at canopy height (^oC).
T_1 Canopy temperature (^oC).
T_2 Soil surface temperature (^oC).
T_{2r} Soil surface temperature below the residue (^oC).
T_L Soil temperature at the interface between the upper and lower layers for bare soil (^oC).
T_{Lr} Soil temperature at the interface between the upper and lower layers for residue-covered soil (^oC).
T_m Soil temperature at the bottom of the lower layer (^oC).
e_a Vapor pressure of the air (mb).
e_b Vapor pressure of the air at the canopy level (mb).
e_1^* Saturated vapor pressure at the canopy (mb).
e_L^* Saturated vapor pressure at the top of the wet layer (mb).
e_b^* Saturated vapor pressure at the canopy level (mb).
e_a^* Saturated vapor pressure of the air (mb).
e_{Lr}^* Saturated vapor pressure at the top of the wet layer for the residue-covered soil (mb).
r_{am} Aerodynamic resistance for momentum transfer (s m^{-1}).
r_{ah} Aerodynamic resistance for heat transfer (s m^{-1}).
r_{aw} Aerodynamic resistance for water vapor (s m^{-1}).
r_{bh} Excess resistance term for heat transfer (s m^{-1}).
r_{bw} Excess resistance term for water vapor (s m^{-1}).
r_1 Aerodynamic resistance between the canopy and the air at the canopy level (s m^{-1}).
r_b Boundary layer resistance (s m^{-1}).
r_2 Aerodynamic resistance between the soil and the air at the canopy level (s m^{-1}).
r_{2b} Actual aerodynamic resistance between the soil surface and Zm (s m^{-1}).
r_{as} Aerodynamic resistance between the soil surface and Zm totally unaffected by adjacent vegetation (s m^{-1}).
r_c Surface canopy resistance (s m^{-1}).
r_r Residue resistance for water vapor flux (s m^{-1}).
r_s Soil surface resistance for water vapor flux (s m^{-1}).
r_{rh} Residue resistance to transfer of heat (s m^{-1}).
r_r Residue resistance for heat flux (s m^{-1}).
r_u Soil heat flux resistance for the upper layer (s m^{-1}).
r_L Soil heat flux resistance for the lower layer (s m^{-1}).
Δ Slope of the saturation vapor pressure (mb $^oC^{-1}$).
h Vegetation height (m).
LAI Leaf area index (m^2 m^{-2}).
LAI_{max} Maximum value of leaf area index (m^2 m^{-2}).
d Zero plane displacement (m).
z_r Reference height above the canopy (m).
Z_m Reference height (m).
z_o Surface roughness length (m).
z_o' Roughness length of the soil surface (m).

k Von-Karman Constant.
k_h Diffusion coefficient at the top of the canopy ($m^2 s^{-1}$).
u^* Friction velocity ($m s^{-1}$).
α Attenuation coefficient for eddy diffusion coefficient within the canopy.
B^{-1} Dimensionless bulk parameter.
VPD_a Vapor pressure deficit (mb).
Rad Solar radiation ($W m^{-2}$).
Rad_{max} Maximum value of solar radiation ($W m^{-2}$).
w Mean leaf width (m).
u_h Wind speed at the top of the canopy ($m s^{-1}$).
L_t Thickness of the surface soil layer (m).
L_m Thickness of the surface and bottom soil layers (m)
r_{so} Soil surface resistance to the vapor flux for a dry layer ($m s^{-1}$).
τ_s Soil tortuosity.
D_v Water vapor diffusion coefficient ($m^2 s^{-1}$).
k_1 Thermal diffusivity ($m^2 s^{-1}$).
ϕ Soil porosity.
β Fitting parameter.
θ Volumetric soil water content ($m^3 m^{-3}$).
θ_s Saturation water content of the soil ($m^3 m^{-3}$).
L_r Residue thickness (m).
τ_r Residue tortuosity.
ϕ_r Residue porosity.
u_2 Wind speed at two meters above the surface ($m s^{-1}$).
K Thermal conductivity of the soil, upper layer ($W m^{-1} {}^oC^{-1}$).
K′ Thermal conductivity of the soil, lower layer ($W m^{-1} {}^oC^{-1}$).
K_r Thermal conductivity of the residue layer ($W m^{-1} {}^oC^{-1}$).
C_{ext} Extinction coefficient.
Fv Fraction of the soil covered by vegetation.
H_{bs} Sensible heat from the soil ($W m^{-2}$).
H_{br} Latent heat from the residue-covered soil ($W m^{-2}$).

5. Acknowledgments

We thank the University of Nebraska Program of Excellence, the University of Nebraska-Lincoln Institute of Agriculture and Natural Resources, Fondo Nacional de Desarrollo Cientifico y Tecnologico (FONDECYT 11100083) and Fondo de Fomento al Desarrollo Cientifico y Tecnologico (FONDEF D09I1146) Their support is gratefully recognized.

6. References

Allen, S.J., (1990). Measurement and estimation of evaporation from soil under sparse barley crops in northern Syria. Agric. For. Meteorology, 49: 291-309.

Allen R G, Pereira LS, Raes D and Smith M (1998) Crop Evapotranspiration: Guidelines for computing crop requirement. (Irrigation and Drainage Paper No 56) FAO, Rome, Italy.

Allen, R.G., Tasumi M., and Trezza, R., (2007). Satellite-based energy balance for mapping evapotranspiration with internalized calibration (METRIC)-model. Journal of Irrigation and Drainage Engineering, 133 (4): 380-394.

Alves I., and Pereira, L. S., (2000). Modeling surface resistance from climatic variables? Agricultural Water Management, 42, 371-385.

Anadranistakis M, Liakatas A, Kerkides P, Rizos S, Gavanosis J, and Poulovassilis A (2000) Crop water requirements model tested for crops grown in Greece. Agric Water Manage 45:297-316.

ASCE (2002) The ASCE Standardized equation for calculating reference evapotranspiration, Task Committee Report.Environment and Water Resources Institute of ASCE, New York.

Brenner AJ and Incoll LD (1996) The effect of clumping and stomatal response on evaporation from sparsely vegetated shrublands. Agric For Meteorol 84:187-205.

Bristow KL, Campbell GS, Papendick RI and Elliot LF (1986) Simulation of heat and moisture transfer through a surface residue-soil system. Agric For Meteorol 36:193-214.

Bristow KL and Horton R (1996) Modeling the impact of partial surface mulch on soil heat and water flow. Theor Appl Clim 56(1-2):85-98.

Caprio J, Grunwald G and Snyder R (1985) Effect of standing stubble on soil water loss by evaporation. Agric For Meteorol 34:129-144.

Choudhury BJ and Monteith JL (1988) A four layer model for the heat budget of homogeneous land surfaces. Quarterly J Royal Meteorol Soc 114:373-398.

Coffey ME, Workman SR, Taraba JL and Fogle AW (2004) Statistical procedures for evaluating daily and monthly hydrologic model predictions. Trans ASAE 47:59-68.

Enz J, Brun L and Larsen J (1988) Evaporation and energy balance for bare soil and stubble covered soil. Agric For Meteorol 43:59-70.

Farahani HJ and Ahuja L R (1996) Evapotranspiration modeling of partial canopy/residue covered fields. Trans ASAE 39:2051-2064.

Farahani HJ and Bausch W (1995) Performance of evapotranspiration models for maize - bare soil to closed canopy. Trans ASAE 38:1049-1059.

Flores H (2007) Penman-Monteith formulation for direct estimation of maize evapotranspiration in well watered conditions with full canopy. PhD Dissertation, University of Nebraska-Lincoln, Lincoln, Nebraska.

Gregory JM (1982) Soil cover prediction with various amounts and types of crop residue. Trans ASAE 25:1333-1337.

Horton R, Bristow KL, Kluitenberg GJ and Sauer TJ (1996) Crop residue effects on surface radiation and energy balance-review.Theor Appl Clim 54:27-37.

INE, 2007. Instituto Nacional de Estadisticas.Censo agropecuario. www.censoagropecuario.cl.

Irmak, S., Mutiibwa D., Irmak A., Arkebauer T., Weiss A., Martin D., and Eisenhauer D., (2008). On the scaling up leaf stomatal resistance to canopy resistance using photosynthetic photon flux density. Agricultural and Forest Meteorology, 148 1034-1044.

Iritz Z, Tourula T, Lindroth A and Heikinheimo M (2001) Simulation of willow short-rotation forest evaporation using a modified Shuttleworth-Wallace approach. Hydrolog Processes 15:97-113.

Jalota SK and Prihar SS (1998) Reducing soil water evaporation with tillage and straw mulching. Iowa State University, Ames, Iowa.

Jensen ME, Burman RD and Allen RG (1990) Evapotranspiration and irrigation water requirements. ASCE Manuals and Reports on Engineering Practice No. 70 332 pp.

Jensen, J.M., Wright, J.L., and Pratt , B.J., 1971. Estimating soil moisture depletion from climate, crop and soil data. Transactions of the ASAE, 14 (6): 954-959.

Katerji, N. and Rana, G. (2006). Modeling evapotranspiration of six irrigated crops under Mediterranean climate conditions. Agricultural and Forest Meteorology, 138 142-155.

Kjelgaard JF and CO Stockle (2001) Evaluating surface resistance for estimating corn and potato evapotranspiration with the Penman-Monteith model. Trans ASAE 44:797-805.

Krause P, Boyle DP and Base F (2005) Comparison of different efficiency criteria for hydrological model assessment. Adv Geosciences 5:89-97.

Lafleur P and Rouse W (1990) Application of an energy combination model for evaporation from sparse canopies. Agric For Meteorol 49:135-153.

Lagos LO (2008) A modified surface energy balance to model evapotranspiration and surface canopy resistance. PhD Dissertation University of Nebraska-Lincoln Lincoln, Nebraska.

Lagos L.O. Martin D.L., Verma S. Suyker A. and Irmak S. (2009). Surface Energy Balance Model of Transpiration from Variable Canopy Cover and Evaporation from Residue Covered or Bare Soil Systems. Irrigation Science. (28)1:51-64.

Legates DR and McCabe GJ (1999) Evaluating the use of goodness of fit measures in hydrologic and hydroclimatic model validation. Water Res Res 35(1): 233-241.

Lindburg M (2002) A soil surface resistance equation for estimating soil water evaporation with a crop coefficient based model. M.Sc. Thesis University of Nebraska-Lincoln Lincoln, Nebraska.

Lund M.R. and Soegaard H. (2003). Modelling of evaporation in a sparse millet crop using a two-source model including sensible heat advection within the canopy. Journal of Hydrology, 280: 124-144.

Massman, W.J.(1992). A Surface energy balance method for partitioning evapotranspiration data into plant and soil components for a surface with partial canopy cover. Water Resources Research. 28:1723-1732.

Meyers TP and Hollinger SE (2004) An assessment of storage terms in the surface energy balance of maize and soybean. Agric For Meteorol 125:105-115.

Monteith JL (1965) Evaporation and the environment. Proc Symposium Soc Expl Biol 19:205-234.

Monteith J.L., and Unsworth M.H. (2008). Principles of environmental physics.Academic Press, Burlington, MA USA.

Moriasi DN, Arnold J G, Van Liew MW, Bingner RL, Harmel RD and Veith TL (2007) Model evaluation guidelines for systematic quantification of accuracy in watershed simulations. Trans ASAE 50:885-900.

Ortega-Farias S, Carrasco M and Olioso A (2007) Latent heat flux over Cabernet Sauvignon vineyard using the Shuttleworth and Wallace model. Irrig Sci 25:161-170.

Ortega-Farias S, Olioso A and Antonioletti R (2004) Evaluation of the Penman-Monteith model for estimating soybean evapotranspiration. Irrig Sci 23:1-9.

Parkes M., Jiang W., and Knowles R. (2005). Peak crop coefficient values for shaanxi north-west China. Agricultural Water Management, 73 149-168.

Penman H L (1948) Natural evaporation from open water, bare soil and grass. Proc Royal Soc London , Series A, 193:120-146.

Rana G, Katerji N, Mastrorilli M, El Moujabber M and Brisson N (1997) Validation of a model of actual evapotranspiration for water stressed soybeans. Agric For Meteorol 86:215-224.

Rana, G., Katerji, N., and De Lorenzi F. (2005). Measurement and modeling of evapotranspiration of irrigated citrus orchard under Mediterranean conditions. Agricultural and Forest Meteorology, 128 199-209.

Shaw RH and Pereira AR (1982) Aerodynamic roughness of a plant canopy: A numerical experiment. Agric Meteorol 26(1):51-65.

Shuttleworth WJ (2006) Towards one-step estimation of crop water requirements. Trans ASAE 49: 925-935.

Shuttleworth WJ and Gurney R (1990) The theoretical relationship between foliage temperature and canopy resistance in sparse crops. Quarterly J Royal Meteorol Soc 116:497-519.

Shuttleworth WJ and Wallace JS (1985) Evaporation from sparse crops-an energy combination theory. Quarterly J Royal Meteorol Soc 111:839-855.

Stannard DI (1993) Comparison of Penman-Monteith, Shuttleworth-Wallace, and modified Priestley-Taylor evapotranspiration models for wildland vegetation in semiarid rangeland. Water Res Res 29(5):1379-1392.

Steiner J (1989) Tillage and surface residue effects on evaporation from soils. Soil Sci Soc Am J 53:911-916.

Steiner J (1994) Crop residue effects on water conservation. Managing agricultural residues, Unger P ed, Lewis Publisher, Boca Raton, Florida, 41-76.

Steiner J, Schomberg H, Unger P and Cresap J (2000) Biomass and residue cover relationships of fresh and decomposing small grain residue. Soil Sci Soc Am J 64:2109-2114.

Suyker A and Verma S (2009) Evapotranspiration of irrigated and rainfed maize-soybean cropping systems. Agric For Meteorol 149:443-452.

Tanner B and Shen Y (1990) Water vapor transport through a flail-chopped corn residue. Soil Sci Soc Am J 54(4):945-951.

Thom AS (1972) Momentum, mass and heat exchange of vegetation. Quarterly J Royal Meteorol Soc 98:124-134.

Todd RW, Klocke NL, Hergert GW and Parkhurst AM (1991) Evaporation from soil influenced by crop shading, crop residue, and wetting regime. Trans ASAE 34:461-466.

Tourula T and Heikinheimo M (1998) Modeling evapotranspiration from a barley field over the growing season.Agric For Meteorol 91:237-250.

Verma S (1989) Aerodynamic resistances to transfer of heat, mass and momentum. Proc (Estimation of Areal Evapotranspiration) Vancouver BC Canada IAHS Publ #177:13-20.

Verma SB, Dobermann A, Cassman KG, Walters DT, Knops JM, Arkebauer TJ, Suyker AE, Burba GG, Amos B, Yang H, Ginting D, Hubbard KG, Gitelson AA, and Walter-Shea EA (2005) Annual carbon dioxide exchange in irrigated and rainfed maize-based agroecosystems. Agric For Meteorol 131:77-96.

Wallace J.S., Jackson N.A. and C.K. Ong. (1999). Modelling soil evaporation in an agroforestry system in Kenya. Agricultural and Forest Meteorology 94: 189-202.

Weseley ML and Hicks BB (1977) Some factors that affect the deposition rates of sulfur dioxide and similar gases on vegetation. J Air Pollution Control Assoc 27(11):1110-1116.

Yu Q, Zhnag Y, Liu Y and Shi P (2004) Simulation of the stomatal conductance of winter wheat in response to light, temperature and CO2 changes. Annals of Bot 93:435-441.

14

Evapotranspiration – A Driving Force in Landscape Sustainability

Martina Eiseltová[1,2], Jan Pokorný[3], Petra Hesslerová[3,4] and Wilhelm Ripl[5]

[1]*Crop Research Institute*
[2]*Environment and Wetland Centre*
[3]*Enki, o.p.s.*
[4]*Faculty of Environmental Sciences, Czech University of Life Sciences, Prague*
[5]*Aquaterra System Institute*
[1,2,3,4]*Czech Republic*
[5]*Germany*

1. Introduction

It is clear from the ever-growing evidence that human interference with vegetation cover and water flows have considerably impacted water circulation in the landscape and resulted in major changes in temperature distribution. Human changes in land use – extensive river channelization, forest clearance and land drainage – have greatly altered patterns of evapotranspiration over the landscape. To comprehend how the changes in evapotranspiration impact landscape sustainability it is necessary to take a holistic view of landscape functioning and gain understanding of the underlying natural processes.

The Earth's surface has been shaped by water - in interaction with geological processes - for billions of years. Water and the water cycle - along with living organisms - have been instrumental in the development of the Earth's atmosphere; free oxygen in the atmosphere is the result of the activity of autotrophic, photosynthetic organisms (stromatolites) that evolved in seawater some 3.5 billions years ago. This was the beginning of aerobic metabolism and enabled the evolution of higher organisms, including higher plants.

The emergence of terrestrial plants some 400 million years ago has played a major role in the amelioration of the climate. The process of evapotranspiration – evaporation from surfaces and transpiration by plants - is instrumental in temperature and water distribution in time and space. Whilst evaporation is a passive process driven solely by solar energy input, transpiration involves an active movement of water through the body of plants - transferring water from the soil to the atmosphere. The process of transpiration is also driven by solar energy but plants have the ability to control the rate of transpiration through their stomata and have developed many adaptations to conserve water when water is scarce.

Water vapour is the main greenhouse gas playing a protective role against heat loss from the Earth's surface; on average the earth is about 33°C warmer than it otherwise would be without water vapour and the other greenhouse gases in the atmosphere (water vapour's contribution being about 60 % on average, Schlesinger 1997). Water, thanks to its high heat-carrying capacity, is able to redistribute much of the solar heat energy received by the Earth through the water cycle: by evapotranspiration and condensation. Water evapotranspiration and condensation therefore plays an instrumental role in climate control with regard to temperature distribution in time and space, i.e. reducing the peaks and modulating the amplitudes of high and low temperatures on the land surface - making conditions on Earth suitable for life.

The natural vegetation cover that has developed over the Earth throughout millennia is best suited to utilize and dissipate the incoming solar energy, and to use the available water and matter in the most energy-efficient way. There is ample evidence for this. Since the time that human civilization begun greatly interfering with the landscape's natural vegetation cover - clearing forests, ploughing savannas and draining wetlands for agricultural use and urban settlements - many environmental problems have started to appear. More recently environmentally sustainable management systems have been sought - with various degrees of effort and understanding of the underlying problems.

In this chapter we will provide evidence of the role of water and vegetation in shaping the climate. Using data and observations from a virgin forest in Austria we will present and discuss the play rules of nature and offer a definition of landscape sustainability. We will present a living example of reduced precipitation over an area of 4000 square kilometres following the partial clearance of the Mau Forest in western Kenya and describe the situation in the de-watered landscape of the open-cast mining area of North-West Bohemia, Czech Republic. The connection between the disturbed water cycle and matter losses in the predominantly-agricultural Stör River catchment in Germany will be demonstrated and the role of evapotranspiration in maintaining landscape sustainability discussed.

2. The play rules of nature in search of sustainability

2.1 The energy-dissipative properties of water

Life on Earth depends on energy, water and a few basic elements (mainly C, H, O, N, P, S and about 20 others) that constitute living tissue. The biogeochemical cycles - the continuous cycles of matter and water - are essential for life to be sustained. The cycles are primarily powered by the energy received from the Sun. Driven by the sun's radiation water is cycled continuously: playing an instrumental role in energy dissipation and the cycling of matter. The dissipation of solar energy at the Earth's surface – i.e. the distribution of energy in time and space - creates suitable thermal conditions for natural processes and life on Earth.

To understand how the natural processes involved in energy dissipation are inter-related Ripl (1992, 1995) proposed a conceptual model based on the energy dissipative properties of water. In his Energy-Transport-Reaction Model (ETR Model), Ripl considered three essential processes (Fig. 1) that control the dissipation of energy:

- the process of water evaporation and condensation;
- the process of dissolution and precipitation of salts; and
- the process of disintegration and recombination of the water molecule within the biological cell

With water's high capacity for carrying energy in the form of latent heat, most energy is dissipated by the physical processor property of evaporation and condensation, making water a very efficient cooler or heater. When water changes from a liquid to its gaseous phase - as in evapotranspiration - energy is stored in the water vapour in the form of latent

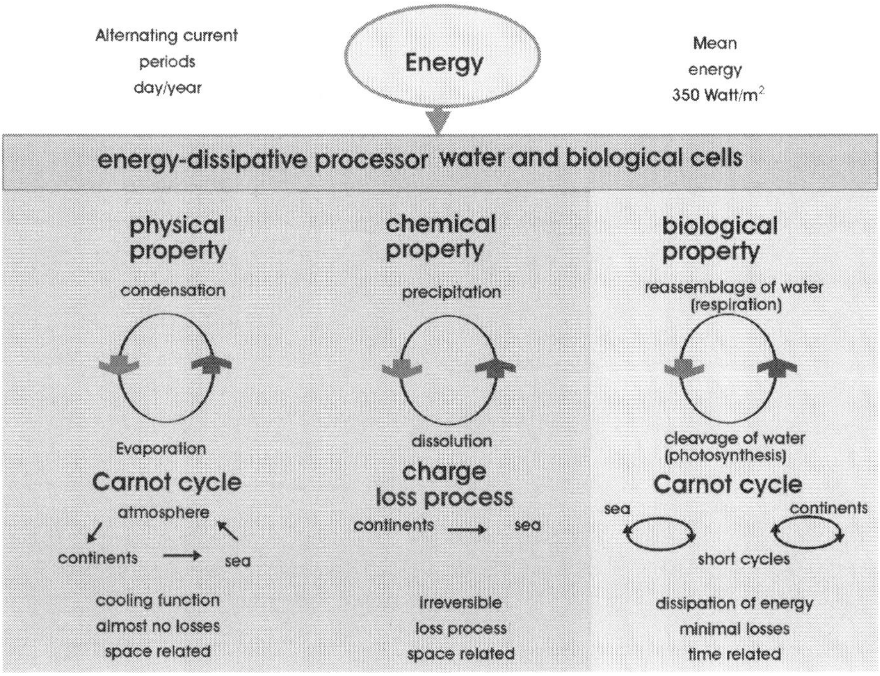

Fig. 1. Three processor properties of water

heat and the local area is cooled down. At night or early morning when water condenses on cooler surfaces, energy in the form of latent heat is released and the local area is warmed up. Without water, the energy of the incoming radiation is transformed into sensible heat and the local area becomes overheated during the day and likewise far cooler at night (as is well known from desert areas, with differences between day and night temperatures typically exceeding 50°C). Water-saturated landscapes provide much more stable environments than do dry terrestrial systems. In landscapes with water - abundant aquatic ecosystems, wetlands and soils with high water retention capacity - about 80 % of incoming solar energy is stored as latent heat of water vapour via evapotranspiration, whilst in de-watered landscapes (with a low-water retention capacity) the vast majority of solar energy is transformed into sensible heat (Pokorný et al. 2010b). In exceptional cases when, for example, hot air of low relative humidity moves across a wetland surrounded by dry areas, even more than equivalent of 100% of solar radiation can be stored safely in latent heat (Monteith 1975, Ryszkowski & Kedziora 1987, Kučerová et al. 2001). Below in Sections 3 and 4 we will show the high temperature differences measured between de-watered areas and sites with a good supply of water and high evapotranspiration.

Water has another important natural property - the ability to separate the charges in a given amount of molecules into protons and electrons. This chemical processor property of water is responsible for the dissolution of salts - using up the water's heat energy in the formation of ionic solutions – and then if concentrated by subsequent evaporation of the water crystals can be precipitated from the solute, releasing the same amount of energy as was required by the dissolution process. However, through dissolution and precipitation a much smaller fraction of energy is dissipated compared to evaporation and condensation.

In pure distilled water at 20°C, 10^{-7}moles of water are dissociated into protons (H^+) and electron-charged hydroxyl ions (OH^-). These electric charges represent chemical potentials, i.e. energy with the potential to be converted into chemical reactions. The number of charged parts (ions) per volume of water constitutes the concept of reactivity (pH, law of mass action). Importantly, reactivity is to a large part dependent on the temperature-, concentration- and pH gradients existing at various interfaces. Such interfaces between solid, liquid and gaseous phases are of special interest in all energy processes and provide sites for steady rates of change. Being essential tools for life processes, nature produces membranes and surfaces where life's important reactions can most readily take place. Even without there being differences in temperature at a liquid- (water-) solid interface, chemical reactions can still readily take place due to the singularity of charge distributions and the modulations of thermal motion (the thermal 'jiggling' of molecules / ions).

Kinetic energy ($mv^2/2$) consists of the frequency and amplitude of accelerated masses. At the interfaces between two phases (e.g. liquid-solid) a modulation of the mass movement of ions (molecules) can occur, especially in amplitude; reactivity is thus enhanced and reaction probabilities increased (in conditions of decreased pH and elevated proton density). As an example of this, take the distribution of highly-diluted, colloidal organic matter in a glass beaker of water. The organic colloids are coagulated at the glass wall, attracted and thus concentrated by the lowered pH conditions at the liquid-solid interface; this enables potential bacterial activity such as, for example, quicker growth of bacteria and decomposition of organic matter. Such phenomena are ubiquitous in nature: always occurring, for example, between the root membranes of plants and the interstitial water of the soil. Evapotranspiration by the leaves of plants lowers the water content in the capillary network of the soil interstitium, giving access to the oxygen of the air and thus exerting a positive feedback on root activity. If the 'water pump' of a productive growing plant should for some reason stop, then electron density (i.e. low redox conditions) will rise and decomposition processes will be severely retarded. Thus the activity of evapotranspiration - the switching on or off of the plant's water pump - controls soil bacterial activity and mineralization processes. In this way highly-efficient processes - control mechanisms closely connecting functioning plant systems and soil - are able to maintain loss-free conditions in the soil. Minerals and nutrients become 'available' only when the plant is actively growing and thus are readily 'used up'. The losses induced by the percolation of 'free' nutrients and minerals released by mineralization through to rivers via sub-surface groundwater flow are thus minimized. Such a mechanism is steadily optimizing the sustainable development of vegetation cover over the landscape by minimizing the irreversible losses from land sites to the sea (Ripl 2010).

Water is also the most important agent in the biological processes of production (photosynthesis) and decomposition (respiration) of organic matter. During photosynthesis water is split into reactive 2H and O. Oxygen is released to the atmosphere and hydrogen is used for the reduction of carbon dioxide to carbohydrates - organic compounds including sugars, starch and cellulose. The solar energy bound in organic matter is released again during mineralization (decomposition) when oxygen is used up to split sugars back into

CO_2 and H_2O. As the production and breakdown of organic matter generally occur within the same site, the biological process can be considered cyclic just like the physical dissipative process. However, considerably less incoming solar energy (about 1 - 2%) is bound by photosynthesis compared to that of water evaporation; the net efficiency of solar energy conversion into plant biomass is usually between 0.5 and a few percent of the incident radiation (for more details see, for example, Blankenship 2002).

The theories and ideas associated with dissipative structures, open dynamic systems operating far from equilibrium, and self-organization (Prigogine & Glansdorff 1971; Prigogine 1980; Prigogine & Stengers 1984) have given us a clearer understanding of how living organisms utilize a throughput of external energy to create new order and structures of increased complexity (Capra 1996). These theories cast light on how ecosystems have organized themselves during evolution: maximizing their sustainability through cycling water and matter and dissipating energy. The dissipation of energy takes place at various scales - from the micro-scale within cells to ecosystems and landscapes (Schneider & Sagan 2005). At the landscape level, evapotranspiration plays an essential role in energy dissipation and as such is highly dependent on the vegetation cover and water availability.

2.2 Plants and water availability

Water is supplied to the land and its vegetation through precipitation. The various sources of water contributing to precipitation differ in different regions of the Earth. In maritime regions, water derived from evaporation from the sea prevails whilst further inland precipitation may be derived equally from long-distance atmospheric transport of water from the sea and from evapotranspiration from within the basin itself (Schlesinger 1997). Availability of water is one of the most important factors determining the growth of plants: hence the distribution of plants on Earth coincides with the availability of water. Deserts are typically short of water and thus the vegetation is rather scarce or non-existent. Nevertheless, plants have developed a number of different strategies during evolution to cope with both conditions of water abundance on the one hand and water scarcity on the other. For the purpose of this chapter we will focus on mechanisms that plants use to control the local water cycle and why it is important.

There are several mechanisms that plants use to control the loss of water from their tissues. One of these is the operation of stomata, their intricate structure, position on plants, their size and numbers. Stomata are found in the leaf and stem epidermis of plants; they facilitate gas exchange and the passage of water from the leaf or stem tissues to the surrounding air by controlling the rate of transpiration. Stomata consist of a pair of guard cells, the opening between them providing the connection between the external air and the system of intercellular spaces. Plants adapted to dry conditions mostly have small stomata immersed within the epidermis. Numbers of stomata differ from about 50 to 1000 stomata per mm^2. Stomata respond to the amount of water in the leaf tissue and to air humidity: closing when the water content in leaf tissue is low and when ambient air humidity declines. In such cases only a small amount of water is transpired through the cuticle (a wax layer on the epidermis). In plants with a thin cuticle - most wetland plants (hygrophytes) belong to this category - the cuticle transpiration may amount to a considerable percentage of total transpiration. However, cuticle transpiration usually amounts to only a few percent of the water released by stomatal transpiration. The effectiveness of the cuticle in reducing loss of water is well seen in fruits, such as apples and pears, or potato tubers: if unpeeled they can stay many weeks without any great water loss (Harder et al. 1965).

Transpiration by plants can be seen as a water loss in such cases as water scarcity; managers of water reservoirs that supply drinking water would usually see it as a loss. For a plant, however, transpiration is a necessity by which a plant maintains its inner environment within the limit of optimal temperatures. And at the level of landscape, evapotranspiration is the most efficient air conditioning system developed by nature.

In addition to optimising temperature, through evapotranspiration plants control the optimum water balance in their root zone. The activity of plant roots in respect to water uptake regulates the redox conditions in the root zone, thus regulating the rate of organic matter decomposition that makes nutrients available for plants growth. It is therefore most likely that, through evapotranspiration, the vegetation cover controls the irreversible losses of matter: an efficient system where only so much organic matter is decomposed such that those mineral nutrients freed from organic bonds are rapidly taken up by plants for their nutrition.

In dry environments, plants have developed ways to attract water condensation. As water condensation takes place on surfaces, plants growing under the conditions of water-scarcity typically have a high surface-volume ratio. Spines and hairs on plants have developed to increase the plants' surface-volume ratio - thus providing more surfaces for water condensation (Fig. 2). Given the complex role of vegetation in maintaining a water balance, smooth temperature gradients and a control of matter cycles in the landscape, any potential economic profits expected from the destruction of natural vegetation cover need to be carefully weighed against the loss of the functioning role of vegetation.

Fig. 2. Spines and hairs on cacti enhance water condensation in arid environments (Photo: M. Marečková)

2.3 Water dynamics and matter losses

It is generally accepted that water is the most important transport and reaction medium - many chemical reactions can only take place in the presence of water and matter is transported mainly with water flow. Matter that is transported via rivers to the sea - both in a dissolved or particulate form - has to be seen as an irreversible matter loss for continents and their vegetation as it takes millions of years before the sea floor is lifted up to form a new continent. Equally, matter that is leached through the soil to the permanent groundwater is further unavailable for nutrition of the vegetation cover on land. Ripl (1992) used data from palaeolimnological studies of lake sediments in southern Sweden (Digerfeldt 1972) to demonstrate the role of vegetation cover in matter and water flows. Vegetation cover reconstruction and sediment dating has made it possible to document four distinctive stages in landscape and vegetation development in postglacial North European catchments and the relevant matter losses at each stage. During the first stage, the bare soils or soils with scarce pioneer vegetation that occurred after the retrieval of glaciers were prone to elevated soil erosion and high transport of dissolved matter. This was measured as a relatively high rate of matter deposition in lake sediments; analysis showed that sediment deposition rates were highly correlated with the deposition rates of base minerals, nutrients and organic material. When climax vegetation became established within catchments, rates of sediment deposition diminished some ten fold. With a fully developed vegetation cover in catchments, low deposition rates of approximately 0.1 to 0.2 mm per year remained rather constant right through until the second half of the 19th century. Since then increasing rates of sewage discharge to lakes, clearance of forest and intensification of agriculture have led to deposition rates increasing nearly a hundred fold to present levels of 8 to 10 mm per year.

The reduction in matter losses from catchments covered by climax vegetation is ascribed to the increased system efficiency of water and matter recycling. In catchments with a well-developed vegetation cover, water and matter are bound to short-circuited cycles and losses are minimal. In contrast, the increased clearance of forest, exposure of bare land, and drainage of agricultural land have accelerated matter losses from catchments. The lowering of the water table by humans has increased the rate of mineralization of organic matter and also enhanced water percolation through soils that carries away the dissolved mineral ions and nutrients. The increased inputs of nutrients to water bodies were documented by the much higher deposition rates of sediments - the beginning of eutrophication (Digerfeldt 1972, Björk 1988, Björk et al. 1972, 2010).

Ripl et al. (1995) confirmed by a laboratory lysimeter experiment that the water dynamics in a soil substrate has a major impact on the rate of organic matter decomposition; under the conditions of intermittent wet and dry phases more organic matter was mineralized and higher amounts of mineral ions leached through the soil than from the control soil substrate with a continuous water flow. The significance of interchanging dry/wet phases and its decisive role in matter losses can be documented also by many examples of drained lowland fens in northern Europe, where increased matter losses have been observed following fen drainage. The mineralization of organic matter accumulated throughout centuries has been of such dimensions that soil subsidence, for example in the fenland of Cambridgeshire, England, has amounted to more than 4.5 metres following the drainage that took place there in the 1650s (Purseglove 1989). By contrast, permanently moist soils slowly accumulate organic matter and matter losses are minimal.

2.4 Specific features of energy fluxes in wetland ecosystems – primary production and decomposition of organic matter

Wetlands which are eutrophic, i.e., well supplied with plant mineral nutrients, are highly productive because they do not suffer from water shortages. Individual types of wetlands differ significantly - not only in their production of plant biomass but also in their capability of long-term accumulation of dead organic matter (as detritus or peat). This capability depends on the ratio between average rates of primary production and decomposition. For example, bogs are distinguished by their low annual primary production of biomass (usually only 100 to 250 g m^{-2} of dry mass). Nonetheless, the strongly suppressed decomposition of organic matter that is produced in bogs results in a net annual accumulation of dead plant biomass that is eventually transformed into peat. As the peat layer grows upwards, the bog vegetation loses contact with the groundwater rich in minerals and its biomass production slows down. In contrast, though eutrophic fishponds have a typical primary production one order of magnitude higher than in bogs, they often hardly accumulate any dead biomass as the annual decomposition approaches or equals annual net primary production. In fishponds, however, like in other wetlands, the production to decomposition ratio depends on the supply of nutrients (especially P and N), i.e., on the trophic status of the water (Pokorný et al. 2010b). Thus any lake or fishpond, if oversupplied with nutrients, can accumulate a nutrient-rich organic sediment if the decomposition rate cannot keep pace with the extremely high primary production. Eventually, the fishpond becomes a source of nutrients; when oxygen gets depleted and anaerobic conditions at the sediment-water interface occur, phosphorus is released from the sediment enhancing the primary production even further.

2.5 Landscape sustainability

2.5.1 The dissipative-ecological-unit

The Earth's atmosphere has been described by Lovelock (1990) as an open system, far from equilibrium, characterized by a constant flow of energy and matter. Equally, living organisms are open systems with respect to continual flows of energy and matter. However, at a higher organisational level - such as an ecosystem - matter is continually recycled, i.e., what is a waste for one organism becomes a resource for another. Ripl & Hildmann (2000) termed the smallest functional unit that is capable of forming internalized cycles of matter and water while dissipating energy - the *dissipative-ecological unit* (DEU). The steadily increasing resource stability of DEUs is achieved by their reduction of water percolation through soils to the groundwater and instead their increase in local, short-circuited water cycling within ecosystems by enhancing their evapotranspiration.

The concept of the dissipative-ecological-unit is used to demonstrate how nature, when not disturbed by sudden changes in climatic conditions, tends to close cycles of matter, i.e. run an efficient local resource economy and maintain relatively even temperatures and moisture conditions.

2.5.2 Evapotranspiration and landscape sustainability

Results from a detailed study conducted in a predominantly agricultural catchment of the River Stör in NW Germany demonstrated how the destruction of natural vegetation cover over large areas has led to the opening up of cycles due to the disturbance of natural water flow dynamics (Ripl et al. 1995, Ripl & Eiseltová 2010). Water and matter no longer cycle within localized, short-circuited cycles; instead, reduced evapotranspiration has resulted in increased water percolation through the soil accompanied by increased losses of matter. The average losses of dissolved mineral ions measured within the Stör River catchment were alarmingly high, about 1,050 kg of mineral salts per ha and year (excluding NaCl). A detailed description of the measurements performed and methods used can be found in Ripl and Hildmann (2000). Such land management systems are unsustainable in the long-term as soil fertility will inevitably be gradually reduced.

A rather different situation can be observed in an undisturbed ecosystem, such as the rather unique virgin forest of Rothwald in Austria. Here the feedback control mechanism of this complex mature forest ecosystem is functioning according to the rules of nature. It is the interlinked vegetation cover that is in control of the processes. In this dolomitic bedrock area groundwater is very scarce - being present only in minor crevices. Oscillations of the water table within the thick debris layer are mainly controlled by the plants through their evapotranspiration. Despite the relatively high precipitation – over 1,000 mm a year – the run off from the virgin forest remains very low and is restricted mainly to the period of snow melt above frozen ground (February till May). The site does not suffer from shortage of water as can be deduced from the highly damped temperature distribution; the temperature amplitudes between day and night almost never exceed 8-9°C during summer (Ripl et al. 2004). The organic matter decomposition is rather slow due to the water-saturated conditions and the debris layer is rather high. The debris layer was 2-4 times higher in the Rothwald virgin forest that in the large areas of neighbouring managed forest (Splechtna, pers. comm., 2000). Water analyses of melted snow samples showed extremely low conductivity values (Table 1). This indicates that there is a much quicker turnover of water evaporated from the virgin forest in relation to precipitation brought from long distances away, as such precipitation water would have about 10 times higher conductivity. It is estimated that very short water cycles with a frequency of one day or less must be prevalent.

	Conductivity at 20° C mS m^{-1}	Alkalinity mmol l^{-1}	pH
Max	1.45	0.09	7.22
Min	0.26	0.00	4.73
Median	0.60	0.01	6.27
MW	0.72	0.03	6.49
no. of sites	17	16	16

Table 1. Conductivity, alkalinity and pH measured in melted snow from Rothwald virgin forest.

Based on the findings described above we can define landscape sustainability as the efficiency of the landscape to recycle water and matter, and to dissipate the incoming solar energy. We have provided evidence that matter losses increase with increased water percolation through soil – as a result of reduced evapotranspiration due to natural vegetation clearance. In the following sections we provide data from a thermal camera and satellite images. These data give supporting evidence that evapotranspiration plays a major role in the dissipation of the incoming solar energy and dampening temperature amplitudes.

3. Evapotranspiration as seen by thermal camera

Pictures of the landscape using a thermal camera show distinct differences in the temperatures of forest, grassland, bare soil and buildings. Even over relatively small areas of a few square metres, temperature differences can be over 20° C. Dry surfaces, such as concrete, when exposed to sunshine are the warmest, despite their higher albedo (higher reflection of solar radiation). This demonstrates that the surface temperatures in the landscape are controlled mainly by the process of water evapotranspiration while the albedo plays a less important role.

On a sunny day, dry surfaces such as the road show the highest temperature (up to 45° C), whilst meadows and forests have lower temperatures as they are cooled by evapotranspiration (Fig. 3). The cooling efficiency depends on water availability and vegetation type. The maize field (Fig. 4) shows a higher temperature over the bare soil (up to 47 °C) than on the top of the stand (32° C). Air heated by a warm soil ascends upwards and takes away water vapour. In hot air crops lose a high amount of water in the form of water vapour.

Fig. 3. Surface temperature of a drained meadow, road and forest as seen by thermovision camera on 17 July 2009 at 9.40 GMT+1 near the town of Třeboň, Czech Republic.

Fig. 4. Maize field and its surface temperature as seen by thermovision camera on 16 July 2010 at 14.19 GMT+1 in the vicinity of the town of Třeboň, Czech Republic.

Fig. 5. Surface temperature of forest canopy as seen by thermovision camera on 13 July 2010 at 14.15 GMT+1 in Novohradské hory, Czech Republic.

The vertical distribution of temperature in a forest canopy is opposite to that observed in a maize field. During a sunny day in a forest, a temperature inversion - a lower temperature at ground level in the shrub layer (23 - 26° C) than on tree crowns in the forest canopy (29.5° C) - has been observed (Fig. 5). A heavier cold air stays at the ground and hence the water vapour may condense on herb and shrub vegetation even during a sunny day. When temperatures go down at night the air becomes more saturated and condensation occurs above the tree canopy. Makarieva and Gorshkov (2010) have shown that intensive condensation is associated with the high evaporation from natural forest cover that is able to maintain regions of low atmospheric pressure on land - i.e. forests constitute acceptor regions for water condensation and precipitation.

4. Use of satellite images to assess cooling efficiency of vegetation cover

Satellite remote sensing data from the Landsat thermal infrared channel provide a suitable tool to evaluate the spatial and temporal distribution of land surface temperatures. This can be used to assess and compare the cooling efficiency of different vegetation cover types or land use. Two model sites (in central Europe and eastern Africa) were selected to demonstrate the role that a functioning vegetation cover plays in energy dissipation compared to the situation of bare or sparsely vegetated land characterized by highly reduced evapotranspiration and shortage of available water.

4.1 North-west Bohemia

4.1.1 Site description and methods

Landsat multispectral satellite data were used to analyze the effects of different land cover types on surface temperature. Two scenes - from 1 July 1995 and 10 August 2004 - were used to compare, firstly, the long-time change in vegetation cover and its effect on temperature distribution, and, secondly, the effect of seasonality of farming land. The selected model area of North-West Bohemia, Czech Republic and Saxony, Germany covers 8,722 km^2 (102 x 85 km). The site was selected to include different landscape types with a heterogeneous mix of land use - highly intensive agriculture, industrial areas and an open-cast brown coal mining area, small-scale farming lands, and broad-leaf and coniferous forests.

Supervised classification methods (Mather & Tso, 2009) were used to classify the land cover; five categories were defined - bare grounds, water, forest, non-forest vegetation and clouds. Surface radiation temperatures were calculated from the standard mono-window algorithm (Sobrino et al. 2004), using a conversion of thermal radiance values from the Landsat thermal infrared channel. As the satellite images selected differ in year and season, the temperature data were standardized by the normalization method of z-scores using the following equation:

$$Z = \frac{x_i - \overline{x}}{\sigma}$$

where x_i is the temperature value of a pixel, $\overline{x}$ is the mean average temperature, and σ is standard deviation.

Date	Min. temperature °C	Max. temperature °C	Mean average temperature °C	Standard deviation
1 July 1995	9.6	46	23.1	2.9
10 August 2004	12.3	46	24.1	4.8

Table 2. Temperature values calculated from Landsat thermal channel.

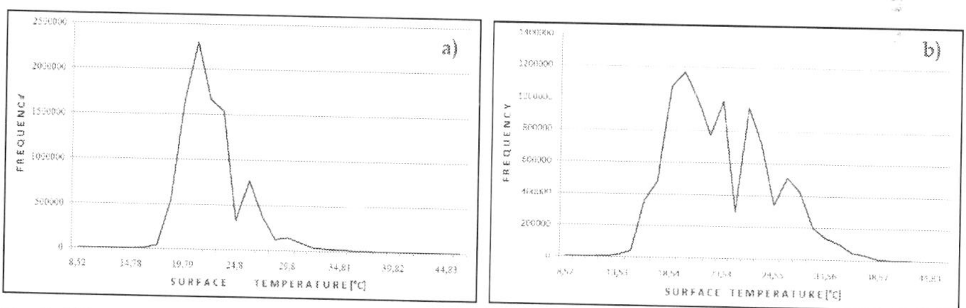

Fig. 6. Histograms showing frequencies of temperature distribution in satellite images of a) 1995 and b) 2004.

The result of normalization is a temperature image with a relative scale showing a range from minimum to maximum temperature. The real temperature values are given (Table 2) and their frequency of distribution displayed by histograms (Fig. 6).

4.1.2 Satellite data interpretation

The relationship between different land-cover types and their relative surface temperature distributions is shown in Figures 7 and 8. Surface temperature is an indicator of the system's

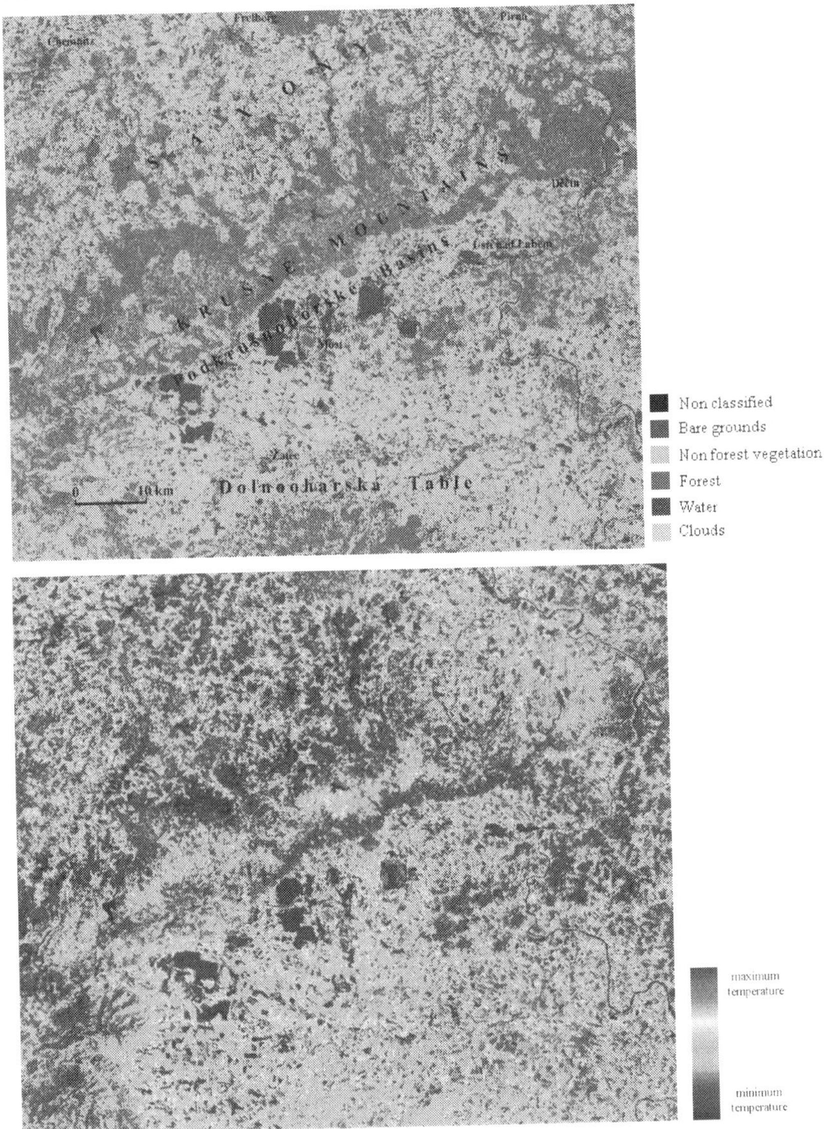

Fig. 7. Land cover (upper image) and temperature distribution (bottom image) over the model area of North-West Bohemia and Saxony obtained on 1 July 1995 at 9.40 GMT+1. The surface temperature data were obtained from Landsat thermal channel TM 6.

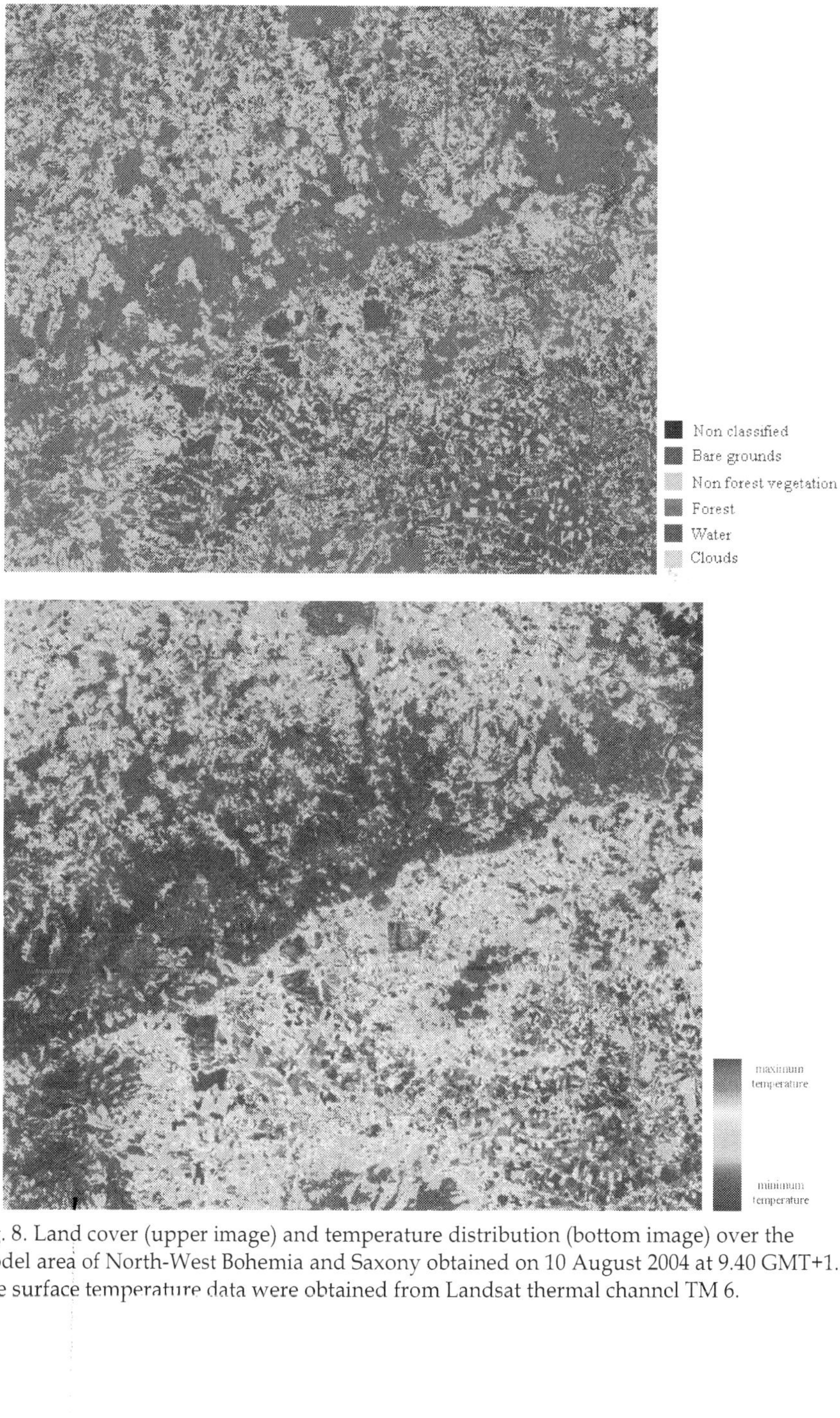

Fig. 8. Land cover (upper image) and temperature distribution (bottom image) over the model area of North-West Bohemia and Saxony obtained on 10 August 2004 at 9.40 GMT+1. The surface temperature data were obtained from Landsat thermal channel TM 6.

ability to convert (dissipate) the incoming solar energy; loss of vegetation is accompanied by changes in the distribution of solar radiation, resulting in a temperature rise. The lowest temperature range (20 - 22 °C) in both satellite scenes was obtained for a deciduous forest, followed by a coniferous forest (17 - 23°C) on the scene from the year 2004. In the satellite scene from 1995, the coniferous (spruce) forest across the top of the Krusne Hory mountains shows remarkably higher temperatures. This is explained by its lower cooling capacity, i.e. lower evapotranspiration rates, as, at that time, the forest was dying off due to the extreme depositions of sulphur dioxide emitted mainly in the 1980s. By 2004, the forest had mostly recovered and this can be seen through the recovery of the forest's cooling function and enhanced temperature damping. The highest temperatures (ranging between 31 - 45 °C) were found at the sites of open-cast brown coal mines and spoil heap tailings, but also on areas of arable fields after crop harvest (August 2004). Furthermore, this category of the highest temperatures, classified as bare grounds, also included urban areas, industrial and commercial zones, communication infrastructures (roads), as well as relatively natural surfaces such as rock outcrops, and peat bogs affected by drainage and/or peat mining. The temperature range of 23 - 32°C characterized the non-forest vegetation - a rather heterogeneous category of land cover that included meadows, areas with sparse vegetation, peatbogs and arable land in 1995 when this was still covered by green crops.

The temperature distribution over the prevailing different land uses depends on the water availability and rate of evapotranspiration. It illustrates the seasonal variability in the damping of temperatures over arable land: early in the season (satellite scene 1 July 1995; Fig. 7) when the crops are still green, the arable land belongs to the lower temperature class compared to the August 2004 scene (Fig. 8) when the crops had already been harvested and the sites fell into the highest temperature class, i.e. the sites were greatly overheated. Furthermore, we can observe the negative impact of the large arable fields of former state farms in the Czech Republic compared to smaller field sizes in Saxony (Germany) which show better temperature damping. The lack of functioning vegetation can be also seen in the shape of the temperature histograms (Fig. 6) - the histogram of 2004 is wider reflecting a higher spread of temperatures whilst the histogram of 1995 (with fields still mainly covered by crops) is narrower and shifted to lower values, reflecting the better cooling by vegetation. Sites with bare ground undoubtedly belong to the warmest places in the landscape; due to the lack of water evapotranspiration, more solar energy is transformed into sensible heat (raising the site's temperature) than into latent heat of water vapour. The higher albedo of bare ground (concrete, etc.) and the lower albedo of forests does not play such an important role when compared to the cooling effect of evapotranspiration (Pokorný et al. 2010a).

4.2 Mau Forest in Western Kenya

The Mau Forest complex, located about 150 km northwest of Nairobi, at an altitude between 1200 - 2600 m, is referred to as one of the largest remaining continuous blocks of indigenous forest in eastern Africa. With a high annual precipitation (reaching about 1000 mm on eastern slopes and more than 2000 mm on western ones) it is an area which includes the headwaters of many rivers feeding into the Rift Valley lakes (Lake Natron, Turkana, Victoria, Nakuru, Naivasha, Elmenteita and Baringo). In the last 25 years the site has been subject to extensive deforestation: forest cover of 5200 km^2 in 1986 was reduced to a mere 3400 km^2 in 2009 (for details, see Fig. 9). The availability of satellite images since the 1980s has enabled us to demonstrate the effect the forest clearance has had on temperature distribution over the whole area. Extreme rises in temperature (by more than 20° C; see Fig. 10) can be observed on sites of deforestation. Its consequences are also evident in the Rift Valley region, between lakes Nakuru and Naivasha. Areas that have been converted into fast-growing plantation forest show the opposite trend, i.e., temperature damping.

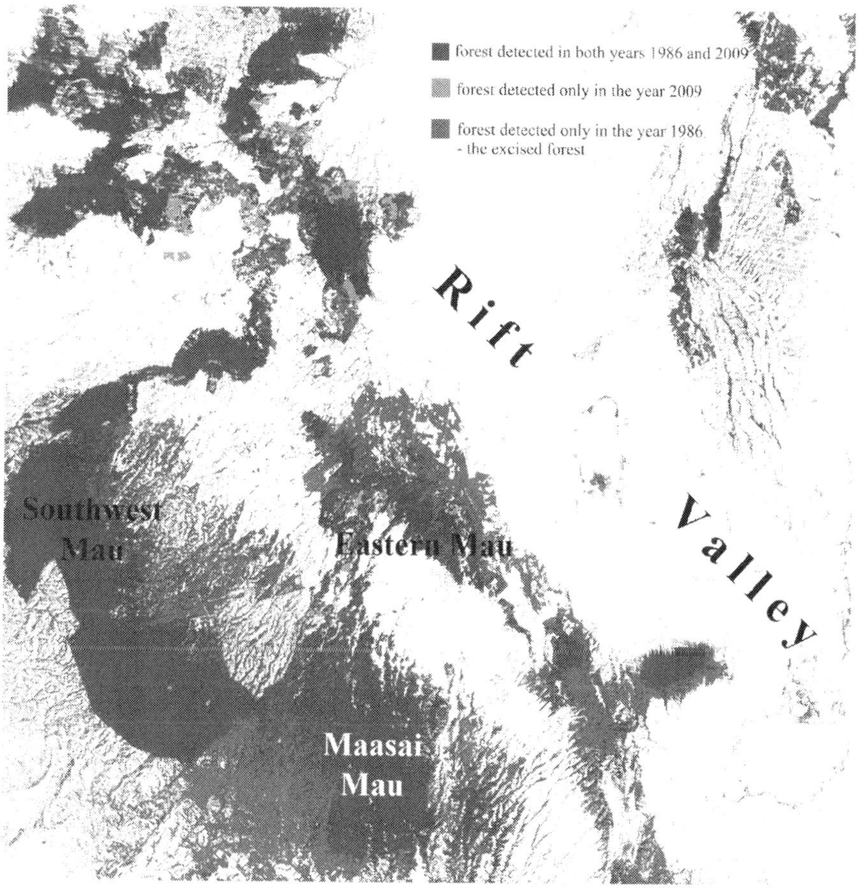

Fig. 9. Changes in the extent of the Mau Forest, East Africa, between the years 1986 and 2009.

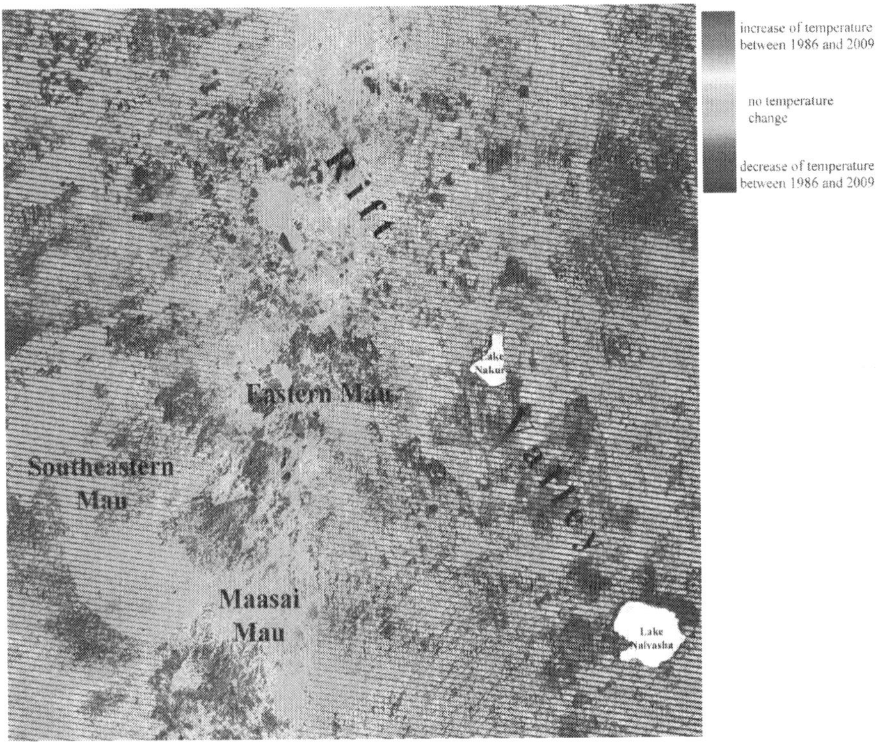

Fig. 10. Changes of temperature between the years 1986 and 2009 in the Mau Forest complex, East Africa, obtained as a difference of the standardized temperatures. The surface temperature data were obtained from Landsat thermal channel TM 6.

5. Discussion and conclusions

Deforestation and land drainage for agriculture or urbanisation has led to accelerated water discharge from catchments. From the self-regulating dissipative structures described earlier, loss of vegetation along with the water shortages has caused a shift to highly negative circumstances, with such consequences as temperature swings of increased amplitude and frequency leading to turbulent motion in warm dry air. The loss of functioning vegetation has been extensive for some time, as observed and reported in the Millennium Environmental Assessment (2005): every year some 60,000 km^2 of badly managed land is becoming a desert and about 200,000 km^2 of land loses agricultural productivity. The lack of water and ecosystem functionality now affects 30-40% of our global landmass.

Life on land is only possible when soil contains enough moisture for green plants to grow. The continuous runoff of river water discharge from land to ocean needs to be compensated by the opposite transport of water vapour from ocean to continent. Makarieva and Gorshkov (2010) have shown the role of the forest cover in a condensation-induced water cycle that maintains a flow of moist air from ocean to continent. They evaluated precipitation measured along transects from ocean to land on different continents; they revealed that along transects with continuous forest cover precipitation reaches as far as 1000 km inland (in hardly diminished amounts), whilst above deforested landscape precipitation rapidly diminishes when distance from the ocean exceeds 600 km. They identified two principles as to how the forest attracts and retains water. Firstly, the forest vertical architecture induces a temperature inversion: during the day, temperatures in the forest understorey are lower than that of the forest crown; in this way losses of water vapour to the atmosphere are reduced. Secondly, at night, water vapour condenses above the forest canopy causing a decrease in air pressure above the canopy; this 'sucks in' air horizontally and can bring moist air from the ocean enhancing further water condensation above the forest. Induced by living organisms, this atmospheric circulation maintaining the hydrological cycle on land has been termed the 'biotic pump' of atmospheric moisture (Makarieva & Gorshkov 2007).

In addition to the long water cycle (ocean to land), the role of the short water cycle has been emphasized by Ripl (1995, 2010) as playing an important role in climate amelioration and landscape sustainability. Figures 7, 8 and 10 clearly demonstrate the role of a healthy forest in modulating surface temperatures contributing to climate amelioration. In contrast, the bare or scarcely-vegetated land with low water supplies is often overheated as a result of the reduced evapotranspiration. Furthermore, matter cycles are dependent on water circulation; water being the most important transport and reaction medium. Not only does vegetation largely control water runoff and precipitation over land - with the help of evapotranspiration - but it also controls matter flows. By controlling soil moisture, vegetation governs the process of organic matter decomposition and hence the circulation of matter. Over millions of years of evolution biological communities have optimized their self-organization to run a highly efficient resource economy and hence maintain their sustainability.

We have provided evidence that the sustainability of river catchments has been seriously impaired by large-scale deforestation and drainage. The accelerated discharge of water via rivers to the sea - caused by extensive deforestation, land drainage and hence the landscape's reduced capacity for water retention - brings about overly high matter losses. Data provided for Germany have shown that areal matter losses have reached between 1 and 1.5 tons of dissolved matter per ha per year on average (Ripl & Eiseltová 2009). Areas of high matter losses correlate with areas of reduced evapotranspiration, and hence landscape overheating, as shown in the detailed study of the Stör River catchment in Germany (Ripl & Hildmann 2000) and three small sub-montane catchments in the Czech Republic (Procházka et al. 2001). The deforestation of large areas in tropical regions has resulted in a temperature increase of about 20°C (Hesslerová & Pokorný, 2010).

In arid zones, where irrigation is often a necessity for crop production, the situation is not better. Farmers may try to minimize the use of water for irrigation as water is scarce and costly; they will kill weeds by herbicides to reduce unwanted water losses through evapotranspiration. The ground thus remains void of an understorey or ground layer that would protect the soil from overheating, and rising hot air takes water vapour away. Even large irrigated areas, such as the cotton fields in Central Asia or irrigated farmland in Australia (e.g. the Murray-Darling Basin) do not achieve closed water cycles. Instead the excessive use of water for irrigation from rivers have had detrimental effects, such as the drying out of the Aral Sea due to water withdrawal from Amu-Darya and Syr-Darya (Central Asia) or the degradation of wetlands in the mouth of the Murray River in Australia. An additional problem is an increasing soil salinity in irrigated areas. There is an urgent need that agricultural research focuses on how to close water cycles in the landscape and the development of farming systems with a more vertically-layered vegetation structure keeping water and lower temperatures during a sunny day.

Observations of nature and studies of natural processes have offered us some understanding as to how nature tends to close the cycles of water and matter so that losses - water discharge and transport of matter via rivers to the sea - are kept to a minimum. We have provided evidence that evapotranspiration plays far the most important role in damping temperature amplitudes and helping to prevent large-scale overheating of land and atmosphere. Hence it is the vegetation cover that ameliorates the climate and can mitigate climate change. Studying the natural processes in a virgin forest in Austria has revealed how natural vegetation cover closes the cycles of water and matter and efficiently dissipates excesses in solar energy. An important question remains to be answered. How can we achieve such an efficient resource economy in a human-managed landscape - efficient water and matter recycling and energy dissipation as achieved by any undisturbed fully-functioning natural ecosystem? Below we offer some thoughts that we think are worth considering if society seriously wants to address landscape sustainability:

- An assembly of organisms that is ecologically-optimized will show the best local resource utilization in a given space; this ensemble will thus be the one that is able to grow and to expand over the area of that site. That is, at least, until some shift in the surrounding conditions immediately outside the given area should occur - and then another organism ensemble becomes the most efficient with respect to the available resources. According to the direct experience of farmers, the two mostly limiting factors for growth and expansion in our landscape are usually water and nitrogen.
- Farmers should be seen by society as the 'managers of our landscape': only their experience 'in tune' with their local environment - in direct feedback mode with the properties and harmonic patterns of their own locality - can rescue society's life-giving 'hardware', the land, and provide a sustainable management. The short water cycle -

with its inherent 'loss-free' matter flow - controlled by the land manager would appear to be the only way to a sustainable society. However, if intelligent land management is to be successful it has to be paid for: through appropriate rewards according to a land manager's achievements towards sustainability - such as low matter losses and efficient solar energy dissipation.

- All other conceptions of nature protection and conservation - that attempt some 'esoteric' protection of the landscape, with farmers trying to do 'nature conservation' by preserving structures in time and space - are in the long run deemed to fail. 'Fixed' structures cannot be sustainable within ecosystems that are living on dynamic changes towards keeping matter (biomass and soil) in place. Land management, as practiced today, that follows 'one rule fits all' centralized planning at the EU-level, must be seen as mismanagement; such planning results in an ever-growing disturbance of vegetation, climate, cooling and soil fertility, leading to steadily-growing desertification and loss of water, climate instabilities and increased food insecurities.
- There is not the slightest evidence for the belief that chasing the most necessary-for-life gas CO_2 through the trading of 'indulgence' certificates - and burying it deep down into what is mostly water-saturated zones - will change back a distorted climate. Neither has it been proved that, in an open atmosphere, CO_2 is acting as the driving greenhouse gas in the atmosphere as much as the far more dynamic water vapour under the aerodynamic conditions driving an ever-increasing number of wind-mills. To establish increasing areas of water evapotranspiration as the most desirable cooling mechanism, and dew formation as the most important process controlling air pressure in interaction with the vegetation cover of landscapes, would seem to be a far better strategy.

The water cycle is akin to the 'bloodstream' of the biosphere. Returning water to the landscape and restoring more natural vegetation cover is the only way to restore landscape sustainability. More attention in present-day science needs to be devoted to the study of the role of vegetation in the water cycle and climate amelioration. Restoration of a more natural vegetation cover over the landscape seems to be the only way forward.

Based on our current scientific knowledge, we can propose two criteria for assessing sustainable land management. These criteria are: the efficiency of an ecosystem to recycle water and matter; and its efficiency to dissipate solar energy. It is land managers that can substantially contribute to the restoration of the water cycle, climate amelioration and reduction of irreversible matter losses with river water flows to the sea. It is in the interest of society as a whole that land managers (farmers, foresters) be rewarded for their actions towards sustainable management of their land. Suitable tools to assess the achievements of individual land managers with respect to sustainable management of their land are: (1) continuous monitoring of conductivity – a measure of dissolved load - and flow rates in streams in order to estimate matter losses; and (2) the regular evaluation of satellite thermal channel images to assess temperature damping, i.e. the effectiveness of land use to dissipate solar energy. Restoration of natural 'cooling structures' – vegetation with its evapotranspiration and condensation-induced water circulation – is essential to renew landscape sustainability.

6. Acknowledgements

We would like to thank Mr. Steve Ridgill for improving the English text.

7. References

Björk, S. (1988). Redevelopment of lake ecosystems - a case study approach. *Ambio*, 17, 90-98.

Björk, S. et al. 1972. Ecosystem studies in connection with the restoration of lakes. *Verh. Internat. Verein. Limnol.* 18: 379-387.

Björk, S., Pokorný, J. & Hauser, V. (2010). Restoration of Lakes Through Sediment Removal, with Case Studies from Lakes Trummen, Sweden and Vajgar, Czech Republic. In: Eiseltová, M. (ed). *Restoration of Lakes, Streams, Floodplains, and Bogs in Europe: Principles and Case Studies*. Springer, Dordrecht, pp. 101-122.

Blankenship R.E. (2002). *Molecular Mechanisms of Phostosynthesis*. Blackwell Science, 336pp.

Capra F. (1996). *The Web of Life: A New Synthesis of Mind and Matter*. Harper Collins Publishers, New York, 320 pp.

Digerfeldt, G. (1972). The post-glacial development of lake Trumen. Regional vegetation history, water level changes and paleolimnology. *Folia Limnologica Scandinavica, 16*: 104.

Harder, R., Schumacher, W., Firbas, F. & von Denffer, D. (1965). *Strasburger's Textbook of Botany*. English translation from ed. 28 (1962) by Bell, P. and Coombe, D. Longmans. Green and Co. London.

Hesslerová, P. & Pokorný, J. (2010). Forest clearing, water loss and land surface heating as development costs. *Int. J. Water*, Vol. 5, No. 4, pp. 401 - 418.

Kučerová, A., Pokorný, J., Radoux, M., Němcová, M., Cadelli, D. & Dušek, J. (2001). Evapotranspiration of small-scale constructed wetlands planted with ligneous species. In: Vymazal, J. (Ed.): *Transformations of Nutrients in Natural and Constructed Wetlands*, Backhuys, Leiden, pp. 413-427.

Lovelock J. (1990): *The Ages of Gaia – A biography of Our Living Earth*. Oxford University Press, Oxford, 252 pp.

Makarieva, A. M. & Gorshkov, V. G. (2007). Biotic pump of atmospheric moisture as driver of the hydrological cycle on land. *Hydrology and Earth System Sciences*, Vol. 11, No. 2, pp. 1013-1033.

Makarieva, A. M. & Gorshkov, V. G. (2010). The Biotic Pump: Condensation, atmospheric dynamics and climate. *Int. J. Water*, Vol. 5, No. 4, pp. 365-385.

Mather, P.M. & Tso, B. (2009). *Classification Methods for Remotely Sensed Data*. CRC Press, Boca Raton, 332 pp.

Millennium Environmental Assessment (2005) *Ecosystems and Human Well-being: Desertification Synthesis*, World Resources Institute, Washington DC.

Monteith, J. L. (1975). *Vegetation and Atmosphere*, Academic Press, London.

Pokorný, J., Brom, J., Čermák, J. Hesslerová, P., Huryna, H., Nadyezhdina, N. & Rejšková, A. (2010a). Solar energy dissipation and temperature control by water and plants. *Int. J. Water*, Vol. 5, No. 4, pp. 311 - 336.

Pokorný, J., Květ, J., Rejšková, A. & Brom, J. (2010b). Wetlands as energy-dissipating systems. *J. Ind. Microbiol. Biotechol.*, Vol. 37, No. 12, pp. 1299 - 1305.

Prigogine, I. (1980). *From Being To Becoming: Time and Complexity in the Physical Sciences.* Freeman, San Francisco, 272 pp.

Prigogine, I. & Glansdorff, P. (1971). *Thermodynamic Theory of Structure, Stability and Fluctuations*, Wiley, New York.

Prigogine, I., Stengers, I. (1984). Order Out of Chaos. Bantam Books, New York, 349.

Procházka, J., Hakrová, P., Pokorný, J., Pecharová, E., Hezina, T., Wotavová, K., Šíma, M. & Pechar, L. (2001). Effect of different management practices on vegetation development, losses of soluble matter and solar energy dissipation in three small sub-mountain catchments. In: Vymazal, J. (ed.). *Transformations of nutrients in natural and constructed wetlands.* Backhuys, Leiden, pp. 143–175.

Purseglove, J. (1989). *Taming the Flood.* Oxford University Press, Oxford, 307pp.

Ripl, W. (1992). Management of Water Cycle: An Approach to Urban Ecology. *Water Pollution Journal Canada*, Vol. 27, No. 2, pp. 221-237.

Ripl, W. (1995). Management of water cycle and energy flow for ecosystem control: the energy-transport-reaction (ETR) model. *Ecological Modelling*, Vol. 78, No. 1-2, pp. 61-76.

Ripl, W. (2010). Losing fertile matter to the sea: How landscape entropy affects climate. *Int. J. Water*, Vol. 5, No. 4, pp. 353-364.

Ripl, W. & Eiseltová, M. (2009). Sustainable land management by restoration of shor water cycles and prevention of irreversible matter losses from topsoils. *Plant Soil Environ.*, Vol. 55, No. 9, pp. 404-410.

Ripl, W. & Eiseltová, M. (2010). Criteria for Sustainable Restoration of the Landscape. In: Eiseltová, M. (ed). *Restoration of Lakes, Streams, Floodplains, and Bogs in Europe: Principles and Case Studies.* Springer, Dordrecht, pp. 1-24.

Ripl, W. & Hildmann, C. (2000). Dissolved load transported by rivers as an indicator of landscape sustainability. *Ecological Engineering*, 14: 373–387.

Ripl, W., Hildmann, C., Janssen, T., Gerlach, I., Heller, S. & Ridgill, S. (1995). Sustainable redevelopment of a river and its catchment: the Stör River project. In: Eiseltová, M., Biggs, J. (eds.). *Restoration of Stream Ecosystems – An Integrated Catchment Approach.* IWRB Publishing, Slimbridge, Publ. No. 37: 76 112.

Ripl, W., Splechtna, K., Brande, A., Wolter, K.D., Janssen T., Ripl, W. jun. & Ohmeyer, C. (2004). *Funktionale Landschaftsanalyse im Albert Rothschild Wildnisgebiet Rothwald.* Im Auftrag von LIL (Verein zur Förderung der Landentwicklung und intakter Lebensräume) NÖ Landesregierung, Österreich, Final Report, 154 pp.

Ryszkowski, L. & Kedziora, A. (1987). Impact of agricultural landscape structure on energy flow and water cycling. *Landscape Ecology*, Vol. 1, No. 2, pp. 85-94.

Schlesinger, W. H. (1997). *Biochemistry: An Analysis of Global Change.* 2nd ed., Academic Press, San Diego, 588 pp.

Schneider, E. D. & Sagan, D. (2005). *Into the Cool: Energy Flow, Thermodynamics, and Life.* University of Chicago Press, Chicago.

Sobrino, J.A., Jiménez-Muñoz, J.C. & Paolini, L. (2004). Land surface temperature retrieval from LANDSAT TM 5, *Remote Sensing of Environment*, Vol. 90, pp. 434–440.

15

Critical Review of Methods for the Estimation of Actual Evapotranspiration in Hydrological Models

Nebo Jovanovic and Sumaya Israel
Council for Scientific and Industrial Research, Stellenbosch
South Africa

1. Introduction

The quantification of a catchment water balance is a fundamental requirement in the assessment and management of water resources, in particular under the impacts of human-induced land use and climate changes. The description and quantification of the water cycle is often very complex, particularly because of the spatial and temporal dimensions, variabilities and uncertainties inherent to the system. The advent of powerful computers, numerical modelling, and GIS is making it possible to describe the complexities of hydrological systems with statistically acceptable accuracy (Duan et al., 2004). Both local (e.g. on-farm) and catchment scale models, physically-based numerical models and simple conceptual balance models are now available to support water resource assessment, management, allocation as well as adaptation to climate change. In particular, the coupling of dedicated atmospheric, hydrological, unsaturated zone and groundwater models is becoming a powerful means of evaluating and managing water resources.

Evapotranspiration (ET) is a key process of the hydrological balance and arguably the most difficult component to determine, especially in arid and semi-arid areas where a large proportion of low and sporadic precipitation is returned to the atmosphere via ET. In these areas, vegetation is often subject to water stress and plant species adapt in different ways to prolonged drought conditions. This makes the process of ET very dynamic over time and variable in space. The focus of this chapter is on the methodologies used in hydrological models for the estimation of actual ET, which may be limited (adjusted) by water or other stresses. The chapter includes: i) a theoretical overview of ET processes, including the principle of atmospheric evaporative demand-soil water supply; ii) a schematic review of methods and techniques to measure and estimate ET; and iii) a review of methods for the estimation of ET in hydrological models.

2. Theoretical overview of evapotranspiration processes

ET is the combination of two separate processes, where liquid water is converted to water vapour (vaporization) from the soil, wet vegetation, open water or other surfaces, as well as from plants by transpiration through stomata (Allen et al., 1998). Evaporation and transpiration occur simultaneously and they are difficult to separate out. ET rate depends on weather conditions, water availability, vegetation characteristics, management and environmental constraints. The main weather variables affecting ET are temperature, solar radiation, wind speed and vapour pressure. The nature of the soil, its hydraulic properties and water retention capacity determine plant available water. Under natural conditions, water stored in the soil is replenished through precipitation, surface and groundwater. The type and developmental stage of vegetation, its adaptation to drought, structure and roughness, albedo, ground cover, root density and depth also affect ET rates. ET rates can be managed through different tillage practices, the establishment of windbreaks, different planting densities and thinning of vegetation, by reducing soil evaporation using, for example, localized irrigation targeting the root zone or mulching, and by reducing transpiration with herbicides or anti-transpirants (substances that induce closing of stomata, envelop vegetation with a surface film or change its albedo). Besides water stress, vegetation may be subject to other types of environmental stresses that are likely to result in a reduction of ET rates and plant growth, like for example pests, diseases, nutrient shortages, exposure to toxic substances and salinization (Allen et al., 1998).

Reference ET is the evaporation from a reference surface of the Earth and it depends on weather conditions. The reference surface can be an open water surface (open pan) or it can be related to weather variables (temperature, radiation, sunshine hours, wind speed, air humidity etc.). Many semi-empirical equations exist that relate reference ET to weather variables. Some of the most commonly adopted are Blaney-Criddle (Blaney and Criddle, 1950), Jensen-Haise (Jensen and Haise, 1963), Hargreaves (1983) and Thornthwaite (1948). Lu et al. (2005) compared the performance of three temperature-based methods, namely Thornthwaite (1948), Hamon (1963) and Hargreaves-Samani (1985), and three radiation-based methods, namely Turc (1961), Makkink (1957) and Priestley-Taylor (1972) for application in large scale hydrological studies in the south-eastern United States. Similarly, Oudin et al. (2005) tested the performance of 27 reference ET models in rainfall-runoff modelling of catchments located in France, Australia and the United States. Both Lu et al. (2005) and Oudin et al. (2005) proposed simple temperature-based methods for calculation of reference ET at catchment scale, in particular when availability of weather data sets is limited.

Theoretical equations that describe the mechanisms of the evaporation process are also available. For example, reference evaporation from an open water surface was first described by Penman (1948) and consisted of a radiation and a vapour pressure deficit term, representing the available energy for the endothermal evaporation process. Priestley and Taylor (1972) proposed the Priestley-Taylor equation, where the radiation term dominates over the advection term by a factor of 1.26, suitable for large forest catchments and humid environments. Based on decades of data and knowledge gained, the FAO (United Nations Food and Agricultural Organization) proposed a grass reference evapotranspiration (ETo) calculated with the Penman-Monteith equation (Monteith, 1965). The FAO Penman-Monteith ETo is defined as the evapotranspiration rate from a reference surface not short of water. The reference surface is a hypothetical grass reference crop with an assumed height of 0.12 m, a fixed surface resistance of 70 s m^{-1} and an albedo of 0.23 (Allen et al., 1998). The Penman-Monteith ETo is a function of the four main factors affecting evaporation, namely temperature, solar radiation, wind speed, and vapour pressure:

$$ETo = \frac{\frac{\Delta}{\lambda}(R_n - G) + \frac{\rho_a C_p}{\lambda}\frac{e_s - e_a}{r_a}}{\Delta + \gamma\left(1 + \frac{r_s}{r_a}\right)} \quad (1)$$

where λ is the latent heat of vaporization of water (MJ kg^{-1}); Δ is the gradient of the saturation vapour pressure-temperature function (kPa °C^{-1}); R_n is the net radiation (MJ m^{-2} d^{-1}); G is the soil heat flux (MJ m^{-2} d^{-1}); ρ_a is the air density (kg m^{-3}); C_p is the specific heat of the air at constant pressure = 1.013 kJ kg^{-1} K^{-1}; e_s is the saturated vapour pressure of the air (kPa), a function of air temperature measured at height z; e_a is the mean actual vapour pressure of the air measured at height z (kPa); r_a is the aerodynamic resistance to water vapour diffusion into the atmospheric boundary layer (s m^{-1}); γ is the psychrometric constant (kPa °C^{-1}); and r_s is the vegetation canopy resistance to water vapor transfer (s m^{-1}). Equation (1) uses standard climatic data that can be easily measured or derived from commonly collected weather data. Allen et al. (1998) also recommended procedures for the calculation of missing variables in equation (1).

In equation (1), the type of vegetation is accounted for through canopy resistance to gas exchange fluxes (r_s), vegetation height determining surface roughness (implicitly in r_a) and albedo (implicitly in R_n). Theoretically, the Penman-Monteith equation allows for direct calculation of actual ET through the introduction of canopy and air resistances to water vapour diffusion. However, this one-step approach is difficult to apply because canopy and air resistances are not known for many plant species and they are complex to measure. A two-step approach is then commonly used to determine actual ET, where the potential evapotranspiration (PET) is first calculated using a minimum value of canopy resistance for a specific crop/vegetation and the actual air resistance from weather data and vegetation height. In a second step, actual ET is calculated taking into account reduction in root water uptake due to water (and/or other) stress and reduction in soil evaporation due to drying of the top soil.

ET of crops or other vegetation differs distinctly from ETo because the ground cover, canopy properties, physiological adaptation and aerodynamic resistance of vegetation may be different from grass. These differences can be integrated into a factor Kc, commonly known as the crop coefficient because it is used to calculate crop water requirements (Allen et al., 1998). The FAO-56 model (Allen et al., 1998) provides a means of calculating reference and crop ET from meteorological data and crop coefficients. The effect of climate on crop water requirements is given by the reference evapotranspiration (ETo), and the effect of the crop by the crop coefficient Kc. Crop evapotranspiration under standard conditions (ETc) is the evapotranspiration from disease-free, well-fertilized crops, grown in large fields, under optimum soil water conditions, and achieving full production under the given climatic conditions. ETc can be calculated as:

$$ETc = Kc\,ETo \quad (2)$$

The Kc factor approach is applicable to uniform conditions, e.g. uniform crop fields with adequate fetch distance to minimize micrometeorological effects of field edges. Caution should therefore be exercised in the application of Kc under conditions where spatial variability of soil properties and crop management occur, in natural vegetation etc. The Kc factor can be split into two separate coefficients Kcb + Ke, where Kcb is the basal crop coefficient referred to crop transpiration and Ke is referred to direct evaporation from the soil (Allen et al., 1998).

The term ETc in equation (2) corresponds to evapotranspiration of vegetation at potential rates (PET) under given climatic conditions. In nature, PET seldom occurs, especially in semi-arid areas. When water is a limiting factor, physiological adaptation of plants occurs, stomata close and ET rates are below potential rates. This mechanism of stomatal control is described schematically in Figure 1.

In the soil-plant-atmosphere continuum (SPAC), water fluxes are driven by atmospheric evaporative demand and limited by soil water supply. Under wet soil conditions, the ratio of actual transpiration (T) and potential transpiration (PT), or relative transpiration (T/PT) is close to 1, showing that the root system is able to supply the canopy with water fast enough to keep up with the atmospheric evaporative demand and thereby preventing wilting. Under these conditions, transpiration is atmospheric demand-limited. As the soil dries beyond field capacity (FC) and beyond a threshold value of water content, T/PT drops below 1. Under these conditions, transpiration is soil water supply-limited as the root system can no longer supply water fast enough to keep up with demand and the soil water can be seen to be less available. Beyond soil water content at permanent wilting point (PWP), transpiration does not occur and T/PT = 0. The same mechanism can be represented for ratios of actual to potential evapotranspiration (ET/PET) as well as actual to maximum yield or productivity (Y/Ym). Plant available water depends on rooting depth, soil texture and structure. A similar mechanism occurs for direct evaporation from the soil surface. Canopy cover is generally used to split evaporation and transpiration, and approximates the available solar energy intercepted by the canopy compared to that reaching the soil surface (Ritchie, 1972).

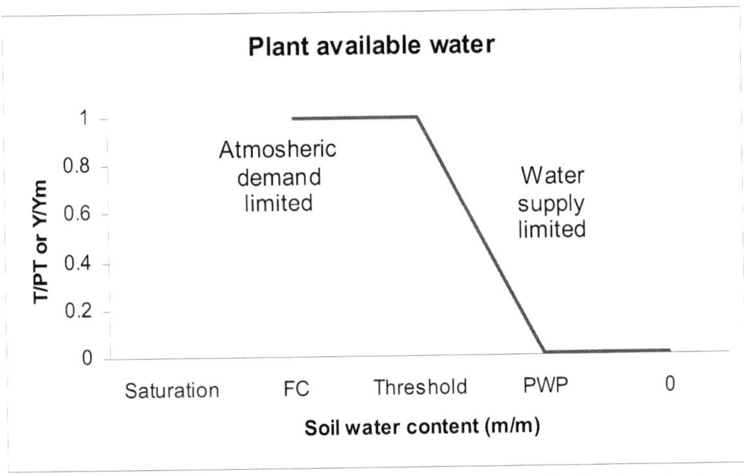

Fig. 1. Schematic representation of the plant available water graph. T - Actual transpiration; PT - Potential transpiration; Y - Actual yield or productivity; Ym - Maximum yield or productivity; FC - Soil water content at field capacity; PWP - Soil water content at permanent wilting point.

The original publication of Denmead and Shaw (1962) included the first scientific evidence of the concept of atmospheric evaporative demand-soil water supply (Figure 2) and this was followed in the last few decades by a large number of research studies on crop productivity-water functions (Doorenbos and Kassam, 1977; Hsiao et al., 2009; Raes et al., 2009; Steduto et al., 2009). This concept is applicable both to wet climates where the limiting factor for ET is generally atmospheric evaporative demand, and to dry climates where the dominant limiting factor is soil water supply.

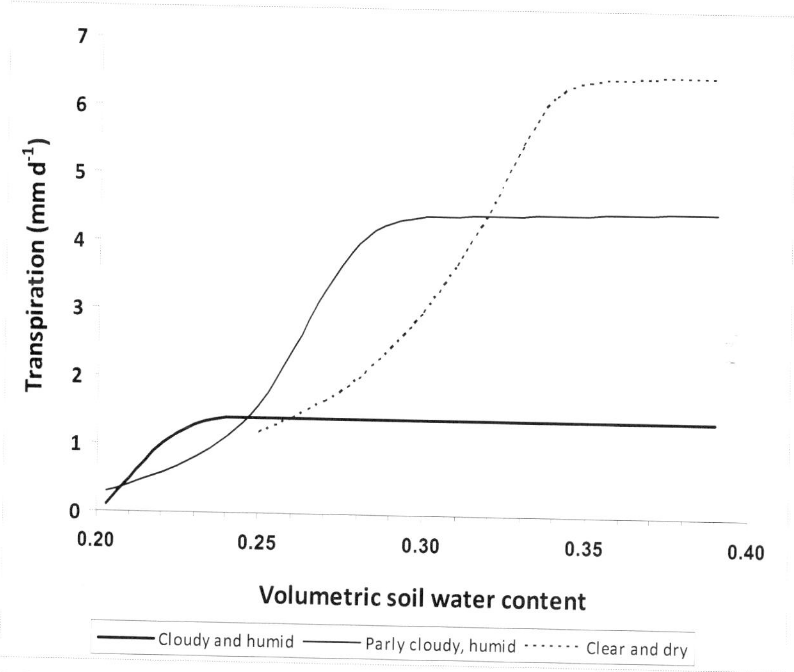

Fig. 2. Graph extracted from the original publication of Denmead and Shaw (1962), supplying scientific evidence of the dependence of transpiration on soil water supply and atmospheric demand.

3. Brief review of methods and techniques to measure and estimate actual evapotranspiration

A large number of methods and techniques for measurement and estimation of ET are available. These can be categorized into the following:

- Lysimeters (Allen et al., 1991): This is the only direct method to measure actual ET.
- Atmospheric measurements
    - Energy balance and micrometeorological methods: These methods are based on the computation of water fluxes based on measurements of atmospheric variables and they are therefore often referred to as direct measurements. Methods and techniques (e.g. Bowen ratio (Bowen, 1926), eddy correlation, scintillometry etc.) were widely discussed by Jarmain et al. (2008).

- Weather data: These methods are based on the calculation of ET from weather data (e.g. Penman-Monteith equation for reference grass ETo).
- Plant measurements
  - Remote sensing from aircraft and satellite: Reflected electromagnetic energy is measured using sensors to generate multi- or hyper-spectral digital images. These data can then be translated into spatial variables such as surface temperature, surface reflectance, and vegetation indices (e.g. the Normalized Difference Vegetation Index NDVI) that describe the vegetation activity and its energy status. These methods were not feasible in the past at large scale and high frequency; however, with the latest technological advances, these techniques show promise (e.g. SEBAL) (Bastiaanssen et al., 1998a and b).
- Soil measurements
  - Soil water balance:

$$ET = P - R - D + \Delta S \tag{3}$$

where P is precipitation; R is runoff or run-on (a component of lateral subsurface inflow/outflow can also be included); D is drainage (or capillary rise), it approximates vertical recharge; ΔS is the change in soil water content, usually measured continuously or manually with a variety of techniques like gravimetric method, soil water sensors, neutron probe, time domain reflectometry etc. (Hillel, 1982). All units are usually expressed in mm per time.

4. Estimation of actual evapotranspiration in hydrological models

Although methodologies for the estimation of ETo and PET are widely adopted, actual (below-potential) ET is difficult to quantify and it usually requires the reduction of PET through a factor that describes the level of stress experienced by plants (two-step approach). The level of stress can be mathematically expressed linearly (slope of line in Figure 1) or through more complex functions. Currently, many models developed for different purposes and operating at different scales apply different functions to reduce PET based on the concept of atmospheric evaporative demand-soil water supply limited ET.

4.1 Field scale models

One-dimensional, field (point) scale hydrological models generally use more detailed functions to predict ET compared to large scale catchment models. The Soil Water Balance (SWB) is an example of a one-dimensional crop model for uniform canopies (Annandale et al., 1999). It is a daily time step model that includes a multi-layer soil water reservoir, where infiltrating water cascades from the top soil layer towards the bottom of the soil profile. Actual transpiration is limited by the evaporative demand (T_{max}) and root water uptake determined by soil wetness (Figure 3). Soil water potential translates into leaf water potential taking into account resistances to water flow in the SPAC (parallel line intersecting the curve in Figure 3) (Annandale et al., 2000).

WATCROS is another example of one-dimensional, cascading water balance and dry matter production simulation model based on climate, soil and plant variables and parameters (Aslyng and Hansen, 1982). It calculates reference ET from grass using a modified formula of Makkink (1957), and it assumes that this grass reference represents any dense, green, growing agricultural crop under Nordic conditions. Such potential evapotranspiration is partitioned into potential evaporation from the soil and crop transpiration using Beer's law.

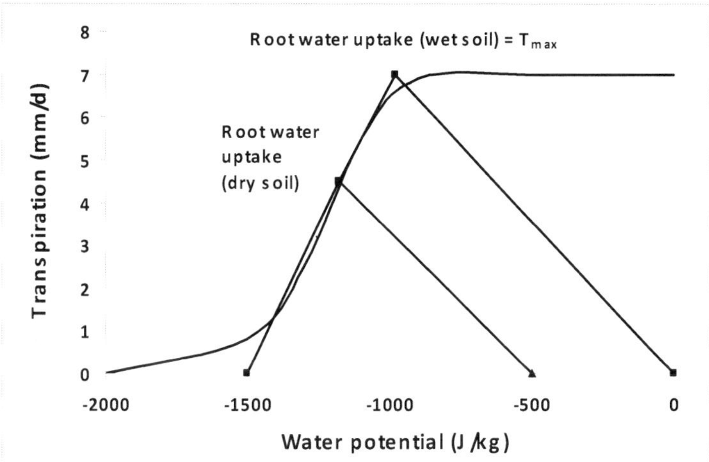

Fig. 3. Schematic representation of the root water uptake function adopted in SWB (adapted from Annandale et al., 2000). T_{max} – Maximum transpiration loss rate (mm d^{-1}).

In order to calculate actual transpiration, water is extracted at a potential rate when the actual soil water content is bigger than half the capacity of the root zone reservoir. Beyond this threshold, actual transpiration is decreased linearly as a function of the remaining water in the reservoir. If no water is left in the root zone reservoir, the transpiration rate equals 0. The size of the root zone reservoir depends on the soil and effective root depth (Hansen, 1984).

GLEAMS (Groundwater Loading Effects of Agricultural Management System) (Knisel, 1993) is also a one-dimensional, piston-flow water balance model used to simulate processes affecting water quality events in agricultural fields. It is the modified version of the well-validated CREAMS model (Knisel, 1980). PET is calculated with the Priestley and Taylor (1972) or with the Penman-Monteith equation (Allen et al., 1998). The model calculates actual soil evaporation and crop transpiration as a function of soil water content and leaf area index.

Cascading soil water balance models based on soil water reservoirs are often employed because of their conceptual simplicity and they are not data intensive. However, soil water movement in porous media can be best described physically with Richards' mass balance continuity equation for unsaturated water flow (Richards, 1931). Richards' equation equilibrates water between specified points (nodes) based on gradients in water energy and hydraulic conductivity:

$$\frac{\partial \theta}{\partial z} = \frac{d}{dz}\left[K(h,z)\frac{dh}{dz} - K(h,z)\right] - S(z,t) \quad (4)$$

where θ is the volumetric soil water content (m^3 m^{-3}); t is time (h); z is soil depth (m, assumed positive downward); h is the soil water pressure head (m); K is the unsaturated hydraulic conductivity (m h^{-1}), a function of h and z; S (z, t) is the sink term (h^{-1}). The conversion of soil water pressure heads into soil water contents and vice versa can be done using different forms of the soil water retention curve (van Genuchten, 1980). The unsaturated hydraulic conductivity-soil water pressure head functions were also described by van Genuchten (1980).

The sink term S(z, t) in equation (4) may include various sinks (or gains with a negative sign) like for example root water uptake. Root water uptake can be calculated with the approach of Nimah and Hanks (1973):

$$-S(z,t)=\frac{\left[H_r+(RRES\cdot z)-h(z,t)-s(z,t)\right]R(z)\cdot K(h)}{\Delta x\cdot \Delta z} \tag{5}$$

where H_r is the effective root water pressure head (m); RRES is a root resistance term; s is the osmotic pressure head (m); Δz is the soil depth increment (m); Δx is the horizontal distance increment; R is the proportion of the total root activity in the depth increment Δz. S cannot exceed potential transpiration.

Richards' equation (4) is non-linear and it can be solved iteratively through a finite-difference solution. It is adopted in several hydrological models to simulate water redistribution in the root zone and for accurate estimates of root water uptake and ET. For example, the RZWQM (Root Zone Water Quality Model) is a physically-based contaminant transport model that includes sub-models to simulate infiltration, runoff, water distribution and chemical movement in the soil (Ahuja et al., 2000). RZWQM simulates PET with a modified Penman-Monteith model and actual ET is constrained by water availability as estimated from Richards' equation.

Soil-Water-Atmosphere-Plant (SWAP) is a 2-D, transient model for water flow and solute transport in the unsaturated and saturated zones (Kroes and van Dam, 2003). It is applied to agrohydrological problems at field scale and it makes use of Richards' equation for soil water redistribution. The relative plant water uptake (T/PT) calculated with this model as a function of soil water potential is shown in Figure 4 (Feddes et al., 1978). The soil water potential values h_1, h_2, and h_4 are inputs. Threshold soil water potentials for reduction in T/PT vary in the range between h_{3h} and h_{3l} and they are applied depending on high (T_{high}) or low (T_{low}) transpiration demand. The h_4 input is wilting point. Reduction in T/PT occurs also in the wet soil range (close to saturation between h_2 and h_1) to simulate the effects of water-logging. The plant water uptake solution in SWAP (Feddes et al., 1978) is also adopted in the HYDRUS unsaturated flow and solute transport model (Simunek et al., 2007) as well as in the SIMGRO (SIMulation of GROundwater and surface water levels) catchment model (van Walsum et al., 2004).

MACRO (Jarvis, 1994) is a deterministic, one-dimensional, transient model for water and solute transport in field soils. It also uses the water uptake function proposed by Feddes et al. (1978). It accounts for conditions that are too wet (close to saturation h_1 in Figure 4) and too dry (close to wilting point h_4 in Figure 4). A dimensionless water stress index ω is used to calculate the ratio of actual to potential root water uptake. This stress index combines two functions describing the distribution of roots and water content in the multi-layered soil profile:

$$\omega=\sum_{i=1}^{i=k} r_i\,\omega_i \tag{6}$$

where k is the number of soil layers in the profile containing roots, and r_i and ω_i are the proportion of the total root length and a water stress reduction factor in layer i. Root length is distributed logarithmically with depth, whilst the stress factor ω_i depends on the soil water content in the particular layer. The root system is usually represented as an inverted cone and its distribution with depth is often non-linear (Yang et al., 2009). The shape of root distribution can therefore be represented with two inputs, namely root depth and an extractable water parameter (Gardner, 1991).

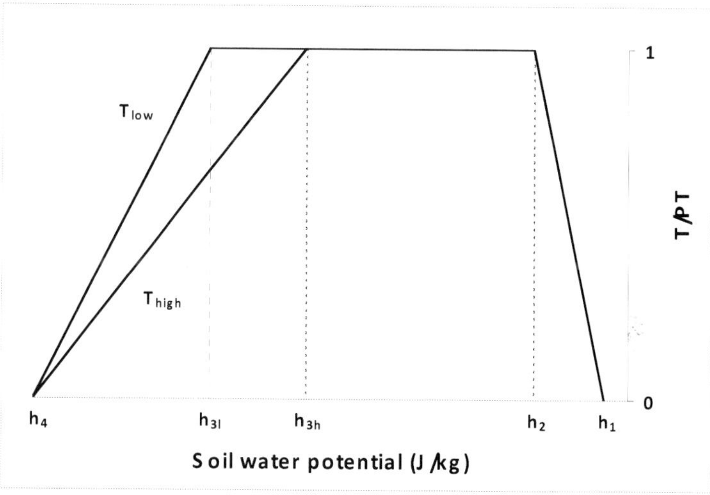

Fig. 4. Schematic representation of the plant available water graph adopted in SWAP (adapted from Feddes et al., 1978). T/PT – Relative plant water uptake; T_{low} – Low transpiration; T_{high} –High transpiration; h_n – Inputs of soil water potential.

The importance of knowing the root depth of vegetation in order to define the size of the soil reservoir and plant available water was underlined by Ritchie (1998) and illustrated in Figure 5. Ritchie (1998) proposed a linear relation between root water uptake and soil water content. Maximum, minimum and usual range of root water uptake are indicated in Figure 5. These depend on root length density Lv and the ability of plants to explore a certain volume of soil.

Another example of a model with a fairly detailed description of root distribution is WAVES (Dawes and Short, 1993; Zhang et al., 1996). WAVES is a water balance model that simulates surface runoff, soil infiltration, ET, soil water redistribution, drainage and water table interactions. Daily transpiration is calculated with the Penman-Monteith equation and reduced using weighting factors determined by the modelled root density and a normalized weighted sum of the matric and osmotic soil water potentials of each layer. The model has been parameterised and used to simulate the water use of various vegetation types in South Africa (Dye et al., 2008).

Feddes et al. (2001) discussed that deep-rooted vegetation and increased water availability may have an effect even on global climate. Deep rooting systems result in large volumes of soil being explored by the roots, large amounts of soil profile available water and large transpiration rates. This is even more prominent in the presence of shallow groundwater. Jovanovic et al. (2004) proved that the contribution of shallow water tables to root water uptake through capillary rise can be a substantial component of the water balance.

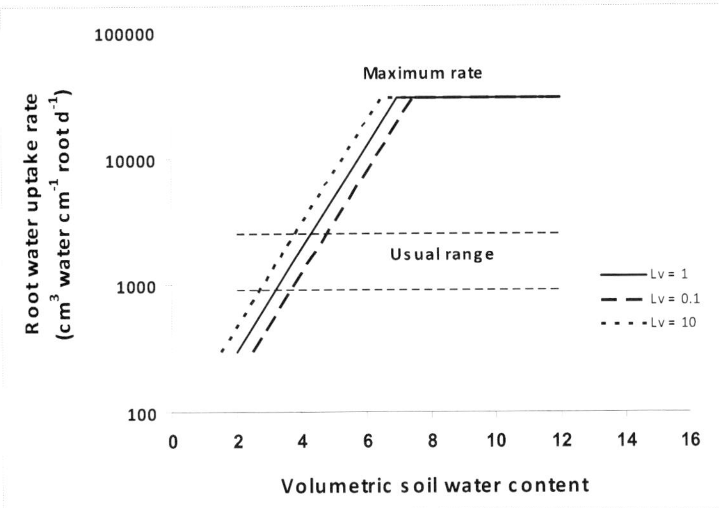

Fig. 5. Relationship between root water uptake rate, volumetric soil water content and root length density (Lv in cm cm-3) (adapted from Ritchie, 1998).

The DRAINMOD computer model was primarily developed to simulate the effects of drainage and associated water management practices on water table depths, the soil water regime and crop yields (Skaggs, 1978). ET is calculated according to the relationship of Norero (1969):

$$ET = \frac{PET}{1 + \left(\frac{h}{h^*}\right)^k} \tag{7}$$

where k is a constant that can be defined using methods given in Taylor and Ashcroft (1972) and Norero (1969), h is the soil water potential in the root zone which could be obtained from the soil water characteristics using the average root zone water content, and h^* is the value of h when ET = 0.5 PET. Equation (7) is graphically illustrated in Figure 6.

Given the purpose of the DRAINMOD model, direct evaporation from the soil can be estimated using the simplified Gardner (1958) equation relating maximum evaporation rate in terms of water table depth and unsaturated soil hydraulic conductivity:

$$\frac{d}{dz}\left[K(h,z)\frac{dh}{dz} - K(h,z)\right] = 0 \tag{8}$$

The symbols and units are the same as those defined for equation (4). Maximum soil evaporation rate for a given water table depth can be approximated by solving equation (8), using a large negative h value (for example h = -1000 cm) at the surface (z = 0) and h = 0 at the water table depth. An example of solution of equation (8) is shown in Figure 7 for a loamy sand.

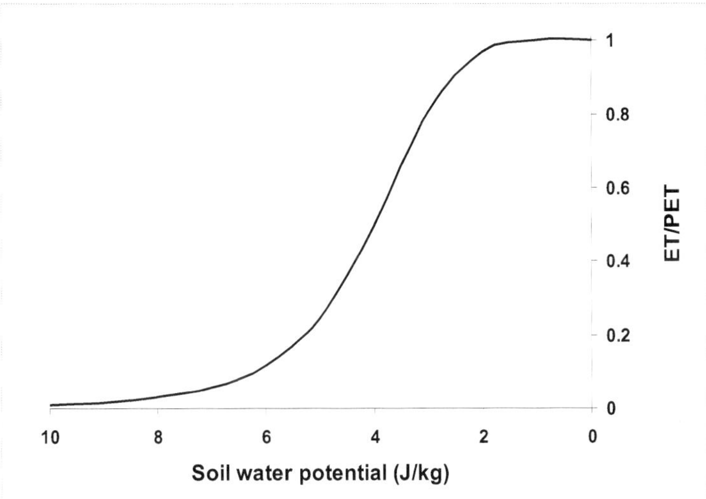

Fig. 6. Schematic of relative evapotranspiration (ET/PET), as affected by soil water potential in the root zone (adapted from Skaggs, 1978)

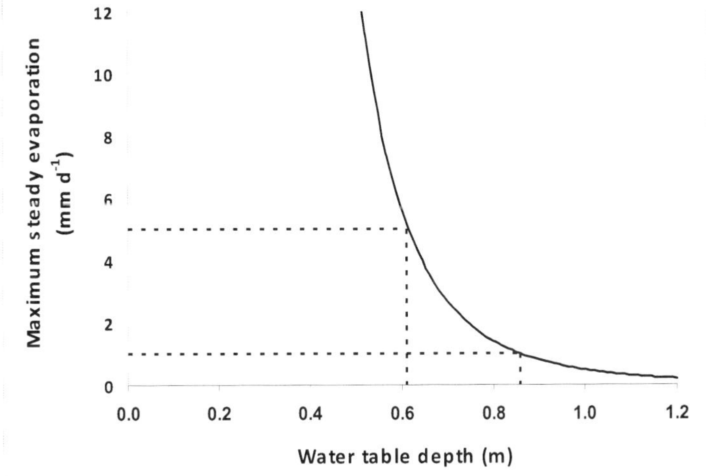

Fig. 7. Relationship between maximum upward movement of water versus water table depth for a loamy sand (adapted from Skaggs, 1978).

4.2 Catchment scale models

Many catchment scale models account for soil moisture in the estimate of ET (Viviroli et al., 2009) using more or less sophisticated approaches. For example, Zhang et al. (2001) developed a semi-empirical water balance model for forested and non-forested catchments in the Murray-Darling basin of Australia. This was based on the assumption that actual ET is equal to precipitation under very dry conditions, and that it equals PET under very wet conditions. On the other hand, Gurtz et al. (1999) applied the PREVAH (PRecipitation-Runoff-EVApotranspiration HRU Model) hydrological model in an alpine basin. They calculated ET using the Penman-Monteith equation by changing the canopy stomatal resistance (equation (1)) below a given threshold of soil moisture.

Barr et al. (1997) reviewed a number of studies where the dependence of ET on soil moisture was evidenced. In their study, they evaluated three methods for estimating ET in the SLURP mesoscale hydrological model (Kite, 1995), namely: i) the complementary relationship areal ET model (Morton, 1983), ii) the Granger (1991) modification of Penman's method and iii) the Spittlehouse (1989) energy-limited versus soil moisture-limited method. The method of Morton (1983) makes use of ET estimated with the Penman (1948) equation and reduced by an amount proportional to vapour pressure deficit, without taking into account the effects of soil moisture on ET. The method of Granger (1991) is a modification to the Penman (1948) equation that includes a relative evaporation variable in the vapour pressure deficit term. The Spittlehouse (1989) method takes into account soil moisture and it calculates actual ET withdrawal from the soil store as the lesser of the soil store and energy-limited rates. The energy-limited rate is calculated with the Priestley-Taylor equation (Priestley and Taylor, 1972) and the soil store-limited rate is calculated as a function of the fraction of extractable soil moisture (Spittlehouse, 1989). The formulation of all three methods was based on forests and grasslands in large catchments. Amongst the three methods tested over a 5-year period in the Kootenay Basin of eastern British Columbia, the Spittlehouse (1989) method including the soil moisture feedback to ET estimates gave the best agreement between simulated and recorded streamflow.

Zhou et al. (2006) used the Shuttleworth and Wallace (1985) model and NDVI to estimate ET from sparse canopies to feed in the BTOPMC distributed hydrological model (Takeuchi et al., 1999). The methodology adopted the Penman-Monteith ETo with an increase of stomatal resistance based on the generic equation:

$$r_s = \frac{r_{s\,\min}}{LAI\, F_i(X_i)} \tag{9}$$

where $r_{s\,\min}$ represents the minimal stomatal resistance of individual leaves under optimal conditions (s m^{-1}), LAI is the effective leaf area index and $F_i(X_i)$ is the stress function for a factor X_i (nutrients, pests, water etc.). The water stress function was expressed as a function of volumetric soil water content θ, in the range between field capacity θ_{fc} and residual soil water content θ_r:

$$f(\theta) = \frac{\theta - \theta_r}{\theta_{fc} - \theta_r} \tag{10}$$

The Agricultural Catchments Research Unit (ACRU) (Schulze, 1994) is a catchment scale agrohydrological modeling system. It calculates relative evapotranspiration (ET/PET) as a function of plant available water (Figure 8). The reduction of ET/PET on the left side of the graph in Figure 8 describes the effect of water-logging. The threshold f_s is user-specified, or it is calculated as a function of a critical leaf water potential ψ^{cr} and ET/PET:

$$f_s = F(\textit{user specified PAW}) \tag{11}$$

$$f_s = 0.94 + \frac{0.0026\,\psi^{cr}}{ET\,/\,PET} \tag{12}$$

Precipitation-Runoff Modular System (PRMS) is a hydrological modular modeling system for large scale basins (Leavesley et al., 1983, 1996). It calculates actual ET for four types of vegetation/land use and three types of soil texture (Figure 9).

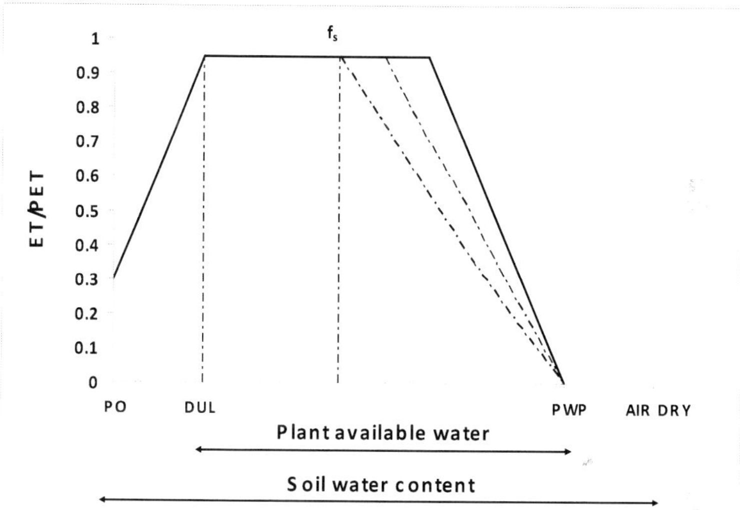

Fig. 8. Schematic representation of the plant available water graph adopted in ACRU (adapted from Schulze, 1994). ET – Actual evapotranspiration; PET – Potential evapotranspiration; PO – Soil water content at saturation; DUL – Drainage upper limit; PWP – Permanent wilting point; f_s – Threshold of reduction of relative evapotranspiration (ET/PET).

MIKE SHE is a physically-based, distributed, integrated hydrological and water quality modeling system (Abbott et al., 1986). ET is calculated based on PET, leaf area index and root depth, soil water content and physical characteristics as well as a set of empirical parameters (Kristensen and Jansen, 1975). Specifically, the ratio of ET to PET is calculated with two functions, the one describing the leaf area index and the other describing the soil water status.

More empirical approaches aimed at describing the hydrological cycle also take into consideration ET. A semi-empirical model called EARTH (Extended model for Aquifer Recharge and moisture Transport through unsaturated Hardrock; Department of Water

Affairs and Forestry, 2006) was developed in South Africa to estimate large scale groundwater recharge by accounting for the variables of the hydrological cycle. EARTH uses modules for vegetation, soil, linear reservoir and saturated flow. The soil module calculates ET as a linear function of soil moisture (Figure 10).

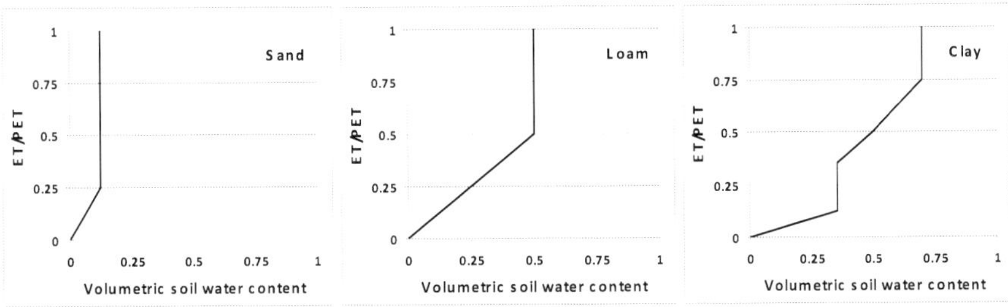

Fig. 9. Schematic representation of the plant available water graph adopted in PRMS for three types of soil texture (adapted from Leavesley et al., 1983). ET - Actual evapotranspiration; PET - Potential evapotranspiration.

The chloride mass balance (CMB) is another method commonly used to estimate groundwater recharge in semi-arid areas (Xu and Beekman, 2003). The estimates of groundwater recharge with CMB refer to long term annual averages, usually over hundreds of years. Implicitly, this technique accounts for the concentrating effects of water by ET in semi-arid regions. Groundwater recharge can be calculated with the following formula:

$$P Cl_p = R_T \, Cl_{gw} \tag{13}$$

where P is precipitation (mm a^{-1}); Cl_p is the chloride concentration in precipitation (mg L^{-1}); R_T is total groundwater recharge (mm a^{-1}), approximated with the term D in equation (3);

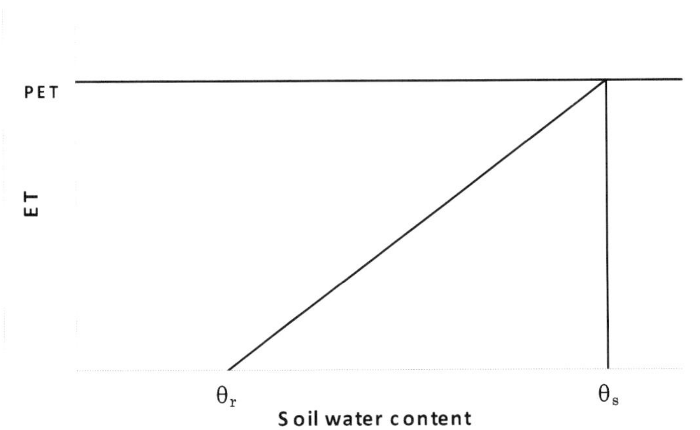

Fig. 10. Schematic representation of the plant available water graph adopted in EARTH (adapted from DWAF, 2006). PET - Potential evapotranspiration; ET - Actual evapotranspiration; θ_r -Soil moisture retained by the soil matrix; θ_s - Maximum soil moisture.

Cl_{gw} is the chloride concentration in groundwater (mg L^{-1}). The source of Cl has to be precipitation solely as other sources may intefere with the interpretation of Cl measurements. Other conservative tracers can also be used. As groundwater recharge can be approximated with term D and ΔS is negligible in the long term, equation (3) can be applied to calculate ET if mean annual runoff data are available.

4.3 Remote sensing applications in the estimate of actual evapotranspiration

The methods discussed above generate point estimates of ET. These values are usually applicable to uniform crop fields, hillslope transects or hydrologically homogeneous areas, and they often need to be upscaled (Oudin et al., 2005). Upscaling can be done through repetitive measurements in all representative areas of interest or through regionalization (Krause, 2002). Due to spatial variations in climate, vegetation, land use and physiographic characteristics, point methods for estimating ET are often too intensive to be applied at large catchment scales. A promising application that may overcome these shortcomings involves areal estimates of ET with remote sensing techniques.

The theory described in the canopy temperature-ET models of Hatfield et al. (1984) was the foundation for surface energy balance approaches based on remote sensing. In these approaches, each pixel of aircraft or satellite images is processed to determine the components of the energy balance equation:

$$\lambda E = R_n - H - G \tag{14}$$

where R_n is net radiation, λE is the latent heat of vaporization, H is the sensible heat flux, G is the soil heat flux and all terms are usually expressed in W m^{-2}. Algorithms such as the Surface Energy Balance Algorithm over Land (SEBAL) use remote sensing imagery, empirical relationships and physical modules to calculate the terms of the energy balance equation and estimate ET (converted from λE in equation (14)) (Bastiaanssen et al., 1998a, 1998b; Tasumi et al., 2005). In particular, SEBAL requires visible, near-infrared and thermal infrared input data obtained from satellite images. Instantaneous net radiation can be calculated from incoming solar radiation measured at ground stations and outgoing thermal radiation estimated from surface albedo, surface emissivity and temperature. Soil heat flux can be computed from surface temperature, albedo and NDVI. The sensible heat flux is calculated with an algorithm of standard heat and momentum transport equations including pixel-based Monin–Obukhov stability corrections. Both wet and dry surface pixels are required because these represent extreme limits in the studied domain at the specific time when the satellite images are taken. The sensible heat flux is constrained by a dry limit (surface with latent heat flux $\lambda E = 0$; sensible heat flux $H = R_n - G$) and wet limit (surface with sensible heat flux $H = 0$; vertical difference in air temperature $dT_a = 0$). A value of dT_a is assigned to all other pixels assuming it varies linearly between the dry and wet ranges. H is then calculated as a function of dT_a and λE computed as the residual of the energy balance. Instantaneous λE values are extrapolated over time assuming that the instantaneous evaporative fraction in equation (14) is stable for the given time period.

Other remote sensing based methods to estimate ET are also available. The Surface Energy Balance System (SEBS) is an energy balance algorithm for the estimation of ET (Su, 2002) that works on similar principles as SEBAL. The MODIS evapotranspiration (ET – MOD 16) algorithm is based on the Penman-Monteith equation (Allen et al., 1998). Land cover, fraction of absorbed photosynthetically active radiation, leaf area index and global surface meteorology information derived from MODIS are used to estimate daily ET and PET, which is then composited over an 8-day interval. ET is expressed in mm d^{-1} and calculated globally every day at 1 km resolution. METRIC (Mapping EvapoTranspiration at high Resolution with Internalized Calibration) is a computer model that uses LandSat data to compute and map ET. These ET maps (i.e. images) provide the means to quantify ET on a field by field basis in terms of both rates and spatial distribution (Allen et al., 2007). Sinclair and Pegram (2010) implemented a real time platform for supplying satellite-based information on ETo and soil moisture in South Africa. Wang et al. (2003) found a significant correlation between deseasonalized time series of NDVI and soil moisture, from where root zone depth can be indirectly estimated. This procedure, however, requires calibration for specific vegetation and climatic conditions.

Although some studies have been carried out in order to test and compare remote sensing methodologies to conventional methods for estimation of ET (Gibson et al., 2011; Kite & Droogers, 2000), more research is required in order to assess the feasibility of application of remote sensing techniques to improve water use efficiency, irrigation management on farms and catchment management, particularly in arid and semi-arid areas. Given the temporal dynamics of ET and its dependance on soil water supply conditions, the interpolation of instantaneous satellite information to estimate ET over a given time period may require verification (Olioso et al., 2005). Processed information from satellite images needs to be supplied at a required frequency for applications in water management on farms and in large catchments. In addition, cloud-free satellite images are required and these are not always available.

5. Conclusion

This chapter discussed the theoretical principles of some hydrological models as examples. It was not meant to provide a review of all models available. The models described here were extensively evaluated in specific studies. Wagener (2003) proposed models should be evaluated for performance (e.g. by minimizing the objective function which can be the difference between simulated and observed data), uncertainty (e.g. by analyzing reasonable ranges of model inputs, parameters and structure) and realism (e.g. by analyzing how consistent the model output is with our understanding of reality). No unique approach for model evaluation exists and, therefore, there is no easy answer to the question on which model is the most accurate. Models should be used for the purpose that they were developed and evaluated with different techniques and for different conditions.

The quantification of actual ET is of utmost importance for various applications in hydrology and water management, such as resource allocation, water footprinting, quantification of water use efficiency etc. This review has highlighted that a large number of both field (point) scale, one-dimensional models and catchment scale spatial GIS-based models adopt conceptually similar approaches to the estimation of actual ET. These approaches are based on the concept of atmospheric evaporative demand-soil water supply limited ET. Such a concept is applicable both to wet climates (limiting factor is atmospheric evaporative demand) and to dry climates (limiting factor is soil water supply). Some models make use of a one-step approach to increase canopy stomatal resistance directly in the Penman-Monteith equation, which represents a mechanistic and physically sound solution to the estimation of actual ET (e.g. BTOPMC). This methodology is, however, hampered by the difficulty in estimating the canopy resistance term. Other models adopt a more conventional two-step approach to calculate PET and reduce it using a water stress index generally based on soil water content (e.g. WATCROS). Some models make use of the data intensive and physically sound principles embedded in Richards' equation to redistribute water in the root zone (e.g. SWAP). Other models make use of a simplistic soil reservoir-based cascading water balance as finite differences are difficult to apply to complex and large scale systems (e.g. ACRU). In addition, abrupt and large changes in soil water content in space and time may lead to numerical instabilities in the finite difference solution of Richards' equation, or in longer simulation times compared to cascading soil water balance models because equilibrium conditions, usually solved through an iterative process, may not be reached easily.

When applying specific models, it is essential to be aware of the specific assumptions around which they were built, their advantages and limitations. Field scale models are generally more data intensive than catchment scale models. For example, dedicated crop and soil water balance models usually include moving thresholds in the atmospheric demand-soil water supply function (e.g. SWB). Models that estimate leaf area provide the opportunity to partition the energy available for soil evaporation and plant transpiration, and those that calculate root growth and depth facilitate the estimation of plant available water in the soil. If properly calibrated, such models are more accurate in predicting field (point) scale ET, but they are also more data intensive compared to large scale models. Large scale catchment models require ET-related inputs in the spatial domain and make use of less detailed ET calculation sub-routines as trade-off (e.g. PRMS).

Given the principles governing soil water redistribution, the soil water dynamics and ET, it is recommended that a daily time step be used in the calculation of water balance variables. Root depth is a very important variable that determines the volume of soil explored by plant roots. This is not often easily measured resulting in uncertainties in the estimation of ET and the water balance. Promising technologies for large scale spatial estimation of ET, soil moisture, and indirectly root depth include remote sensing. These techniques, however, need to be tested and validated for applicability to a wide range of water management conditions in arid and semi-arid areas. The purpose and applicability of remote sensing methods depend on the spatial resolution of the images and their temporal resolution (frequency).

6. Acknowledgment

The authors acknowledge the Water Research Commission (Pretoria, South Africa) for funding this study emanating from project No. K5/1909 on "Reducing Uncertainties of Evapotranspiration and Preferential Flow in the Estimation of Groundwater Recharge".

7. References

Abbott, M. B.; Bathurst, J. C.; Cunge, J. A.; O'onnell, P. E. & Rasmussen, J. (1986). An introduction to the European Hydrological System - Systeme Hydrologique Europeen (SHE). 1: History and Philosophy of a Physically-Based Distributed Modelling System. *Journal of Hydrology*, Vol. 87, pp. 45-59.

Ahuja, L.R.; Rojas, K.W.; Hanson, J.D.; Shaffer, M.J. & Ma, L. (2000). *Root Zone Water Quality Model. Modeling Management Effects on Water Quality and Crop Production,* Water Resources Publications, Colorado, USA, 356 pp.

Allen, R.G.; Howell, T.A.; Pruitt, W.O.; Walter, I.A. & Jensen, M.E. (Eds.) (1991). *Lysimeters for Evapotranspiration and Environmental Measurements*, American Society of Civil Engineers, New York, USA.

Allen, R.G.; Pereira, L.S.; Raes, D. & Smith, M. (1998). *Crop Evapotranspiration: Guidelines for Computing Crop Water Requirements*, United Nations Food and Agriculture Organization, Irrigation and Drainage Paper 56. Rome, Italy, 300 pp.

Allen, R.G.; Tasumi, M.; Morse, A.; Trezza, R.; Wright, J.L.; Bastiaanssen, W.; Kramber, W.; Lorite, I. & Robinson, C.W. (2007). Satellite-Based Energy Balance for Mapping Evapotranspiration with Internalized Calibration (METRIC) - Applications. *J. Irrig. and Drain. Eng.*, Vol. 133(4), pp. 395-406.

Annandale, J.G. ; Benade, N. ; Jovanovic, N.Z. ; Steyn, J.M.; Du Sautoy, N. & Marais, D. (1999). *Facilitating Irrigation Scheduling by Means of the Soil Water Balance Model*, Water Research Commission Report No. 753/1/99, Pretoria, South Africa.

Annandale, J.G.; Campbell, G.S.; Olivier, F.C. & Jovanovic, N.Z. (2000). Predicting crop water uptake under full and deficit irrigation: An example using pea (*Pisum sativum* cv. Puget). *Irrigation Science,* Vol. 19, pp. 65-72.

Aslyng, H.C. & Hansen, S. (1982). *Water Balance and Crop Production Simulation. Model WATCROS for Local and Regional Application.* Hydrotechnical Laboratory, The Royal Vet. and Agric. Univ., Copenhagen, 200 pp.

Barr, A. G.; Kite, G.W.; Granger, R. & Smith, C. (1997). Evaluating Three Evapotranspiration Methods in the SLURP Macroscale Hydrological Model. *Hydrological processes*, Vol. 11, pp. 1685-1705.

Bastiaanssen, W.G.M.; Menenti, M.; Feddes, R.A. & Holtslag, A.A.M. (1998a). A Remote Sensing Surface Energy Balance Algorithm for Land (SEBAL) 1. Formulation. *Journal of Hydrology*, Vol. 212-213, pp. 198-212.

Bastiaanssen, W.G.M.; Pelgrum, H.; Wang, J.; Ma, Y.; Moreno, J.F.; Roerink, G.J. & van der Wal, T. (1998b). A Remote Sensing Surface Energy Balance Algorithm for Land (SEBAL) 2. Validation. *Journal of Hydrology*, Vol. 212-213, pp. 213-229.

Blaney, H.F. & Criddle, W.D. (1950). *Determining Water Requirements in Irrigated Areas from Climatological and Irrigation Data*. USDA Soil Conserv. Serv. SCS-TP96, 44 pp.

Bowen, I. S. (1926). The Ratio of Heat Losses by Conduction and by Evaporation from any Water Surface. *Phys. Rev.*, Vol. 27, pp. 779--787.

Dawes, W.R. & Short, D.L. (1993). *The Efficient Numerical Solution of Differential Equations for Coupled Water and Solute Dynamics: the WAVES model.* CSIRO Division of Water Resources Technical Memorandum 93/18, Canberra, ACT, Australia.

Denmead, O.T. & Shaw, R.H. (1962). Availability of soil water to plants as affected by soil moisture content and meteorological conditions. *Agronomy Journal*, Vol. 54, 385-390.

Department of Water Affairs and Forestry. (2006). *Groundwater Resource Assessment II: Recharge Literature Review Report 3aA,* Project No. 2003-150, Department of Water Affairs and Forestry, Pretoria, South Africa.

Doorenbos, J. & Kassam, A.H. (1979) *Yield Response to Water*. Irrigation and Drainage Paper n. 33. FAO, Rome, Italy, 193 pp.

Duan, Q.; Gupta, H.V.; Sorooshian, S.; Rousseau, A. & Turcotte, R. (Eds.) (2004). *Calibration of Watershed Models*, AGU Monograph Series, Water Science and Application 6, American Geophysical union, Washington, D.C., 345 pp.

Dye, P.J.; Jarmain, C.; Le Maitre, D.; Everson, C.S.; Gush, M. & Clulow, A. (2008). *Modelling Vegetation Water Use for General Application in Different Categories of Vegetation.* Water Research Commission Report No.1319/1/08, ISBN 978-1-77005-559-9, Pretoria, South Africa.

Feddes, R.A.; Hoff, H.; Bruen, M.; Dawson, T.; de Rosnay, P.; Dirmeyer, P.; Jackson, R.B.; Kabat, P.; Kleidon, A.; Lilli, A. & Pitman, A.J. (2001). Modeling Root Water Uptake in Hydrological and Climate Models. *Bulletin of the American Meteorological Society,* Vol. 82(12), 2797-2809.

Feddes, R.A.; Kowalik, R.J. & Zaradny, H. (1978). *Simulation of Field Water Use and Crop Yield. Simulation Monographs,* Pudoc, Wageningen, The Netherlands, 189 pp.

Gardner, W.R. (1958). Some Steady-State Solutions of the Unsaturated Moisture Flow Equation with Application to Evaporation from a Water Table. *Soil Science,* Vol. 85, pp. 228–232.

Gardner, W.R. (1991). Modeling Water Uptake by Roots. *Irrigation Science,* Vol. 12, 109-114.

Gibson, L.A., Munch, Z. & Engelbrecht, J. (2011). Particular Uncertainties Encountered in Using a Pre-Packaged SEBS Model to Derive Evapotranspiration in a Heterogeneous Study Area in South Africa. *Hydrology and Earth System Science,* Vol. 15, pp. 295-310.

Granger, R.J. (1991). *Evaporation from Natural Non-Saturated Surfaces,* PhD Thesis, Department of Agricultural Engineering, University of Saskatchewan, Saskatoon, 140 pp.

Gurtz, J.; Baltensweiler, A. & Lang, H. (1999). Spatially Distributed Hydrotope-Based Modelling of Evapotranspiration and Runoff in Mountainous Basins. *Hydrological Processes,* Vol. 13, pp. 2751–2768.

Hamon, W.R. (1963). Computation of Direct Runoff Amounts from Storm Rainfall. *Int. Assoc. Sci. Hydrol. Pub.,* Vol. 63, pp. 52-62.

Hansen, S. (1984). Estimation of Potential and Actual Evapotranspiration. *Nordic Hydrology,* Vol. 15, pp. 205-212.

Hargreaves, G.H. (1983). Discussion of 'Application of Penman Wind Function' by Cuenca, R.H. and Nicholson, M.J. *J. Irrig. and Drain. Eng. ASCE,* Vol. 109(2), pp. 277-278.

Hargreaves, G.H. & Samani, Z.A. (1985). Reference Crop Evapotranspiration from Temperature. *Applied Eng. in Agric.,* Vol. 1(2), pp. 96-99.

Hatfield, J.L.; Reginato, R.J. & Idso, S.B. (1984). Evaluation of Canopy Temperature-Evapotranspiration Models over Various Crops. *Agricultural and Forest Meteorology,* Vol. 32, pp. 41-53.

Hillel, D. (1982). *Introduction to Soil Physics,* Academic Press Inc., New York, USA.

Hsiao, T.C.; Heng, L.; Steduto, P.; Rojas-Lara, B.; Raes, D. & Fereres, E. (2009). AquaCrop The FAO Crop Model to Simulate Yield Response to Water: III. Parameterization and Testing for Maize. *Agronomy Journal,* Vol. 101(3), pp. 448-459.

Jarmain, C.; Everson, C.S.; Savage, M.J.; Mengistu, M.G.; Clulow, A.D.; Walker, S. & Gush, M.B. (2008). *Refining Tools for Evaporation Monitoring in Support of Water Resources Management,* Water Research Commission Report No. K5/1567/1/08, Pretoria, South Africa.

Jarvis, N.J. (1994). *The MACRO Model (Version 3.1). Technical Description and Sample Simulations,* Reports and Dissertations 19, Dept. Soil Sci., Swedish Univ. Agric. Sci., Uppsala, 51 pp.

Jensen, M.E. & Haise, H.R. (1963). Estimating Evapotranspiration from Solar Radiation. *J.Irrig. and Drain. Div. ASCE*, Vol. 89, pp. 15-41.

Jovanovic, N.Z.; Ehlers, L.; Bennie, A.T.P.; Du Preez, C.C. & Annandale, J.G. (2004). Modelling the Contribution of Root Accessible Water Tables towards Crop Water Requirements. *South African Journal of Plant and Soil*, Vol. 21(3), pp. 171-182.

Kite, G.W. (1995). *Manual for the SLURP Hydrological Model*, NHRI, Saskatoon, 111 pp.

Kite, G.W. & Droogers, P. (2000). Comparing Evapotranspiration Estimates from Satellites, Hydrological Models and Field Data. *Journal of Hydrology*, Vol. 229, pp. 3–18.

Knisel, W.G. (Ed.) (1980). *CREAMS: A Field-Scale Model for Chemical, Runoff, and Erosion from Agricultural Management Systems*, Conservation Research Report 26, U.S. Department of Agriculture, Washington, D.C.

Knisel, W.G. (1993). *GLEAMS: Groundwater Loading Effects of Agricultural Management Systems, V.2.10*, University of Georgia, Coastal Plain Experiment Station, Biological and Agricultural Engineering Department (Publication No. 5).

Krause, P. (2002). Quantifying the Impact of Land Use Changes on the Water Balance of Large Catchments using the J2000 Model. *Physics and Chemistry of the Earth*, Vol. 27, pp. 663–673.

Kristensen, K. J. & Jensen, S.E. (1975). A Model for Estimating Actual Evapotranspiration from Potential Evapotranspiration. *Nordic Hydrology*, Vol. 6, pp. 70-88.

Kroes, J.G. & van Dam, J.C. (2003). *Reference Manual SWAP Version 3.0.3.*, Wageningen, Alterra, Green World Research. Alterra Rep. No. 773, pp. 211.

Leavesley, G.H.; Lichty, R.W.; Troutman, B.M. & Saindon, L.G. (1983). *Precipitation-Runoff Modelling System – User's Manual*, US Geological Survey Water Resources Investigation Report 83-4238, Denver, Colorado, USA.

Leavesley, G.J.; Restrepo, P.J.; Markstrom, S.L.; Dixon, M. & Stannard, L.G. (1996). *The Modular Modeling System (MMS): User's Manual*, Open-File Report 96-151, US Geological Survey, Denver, Colorado, USA.

Lu, J.; Sun, G.; McNulty, S.G. & Amatya, D.M. (2005). A Comparison of Six Potential Evapotranspiration Methods for Regional Use in the South-Eastern United States. *Journal of the American Water Resources Association*, Vol. 41(3), pp. 621-633.

Makkink, G.F. (1957). Testing the Penman Formula by Means of Lysimeters. *J. Inst. of Water Eng.*, Vol. 11, pp. 277-288.

Monteith, J.L. (1965). *Evaporation and the Environment, The State and Movement of Water in Living Organisms, XIXth symposium*, Cambridge University Press, Swansea.

Morton, F.I. (1983). Operational Estimate of Aerial Evapoutranspiration and their Significance to the Science and Practice of Hydrology. *Journal of Hydrology*, Vol. 66, pp. 77-100.

Nimah, M. & Hanks, R.J. (1973). Model for Estimating Soil-Water-Plant-Atmospheric Interrelation: I. Description and Sensitivity. *Soil Science Society of America Proceedings*, Vol. 37, pp. 522-527.

Norero, A.L. (1969). *A Formula to Express Evapotranspiration as a Function of Soil Moisture and Evaporative Demand of the Atmosphere*, PhD Thesis., Utah State University, Logan/Utah, USA.

Olioso, A.; Inoue, Y.; Ortega-Farias, S.; Demarty, J.; Wigneron, J.-P.; Braud, I.; Jacob, F.; Lecharpentier, P.; Ottle, C.; Calvet, J.-C. & Brisson, N. (2005). Future Directions for Advanced Evapotranspiration Modeling: Assimilation of Remote Sensing Data into

Crop Simulation Models and SVAT Models. *Irrigation and Drainage Systems*, Vol. 19, 377–412.

Oudin, L.; Hervieu, F.; Michel, C.; Perrin, C.; Andreassian, V.; Anctil, F. & Loumagne, C. (2005). Which Potential Evapotranspiration Input for a Lumped Rainfall-Runoff Model? Part 2 - Towards a Simple and Efficient Potential Evapotranspiration Model for Rainfall–Runoff Modelling. *Journal of Hydrology*, Vol. 303, pp. 290–306.

Penman, H.L. (1948). Natural Evaporation from Open Water, Bare Soil and Grass. *Proc. Roy. Soc. London A(194), S.* pp. 120-145.

Priestley, C.H.B. & Taylor, R.J. (1972). On the Assessment of Surface Heat Flux and Evaporation Using Large Scale Parameters. *Mon. Weath. Rev.*, Vol. 100, pp. 81-92.

Raes, D.; Steduto, P.; Hsiao, T.C. & Fereres, E. (2009) AquaCrop-The FAO Crop Model to Simulate Yield Response to Water: II. Main Algorithms and Software Description. *Agronomy Journal*, Vol. 101(3), pp. 438-447.

Richards, L.A. (1931). Capillary Conduction of Liquids through Porous Mediums. *Physics*, Vol. 1(5), pp. 318-333.

Ritchie, J.T. (1972). Model for Predicting Evaporation from a Row Crop with Incomplete Cover. *Water Resources Research*, Vol. 8, pp. 1204-1213.

Ritchie, J.T. (1998). Soil Water Balance and Plant Water Stress, In: *Understanding Options for Agricultural Production*, G.Y. Tsuji, G. Hoogenboom & P.K. Thornton (Eds.), pp. 41-53, Kluwer Academic Publishers, UK.

Schulze, R.E. (1994). *Hydrology and Agrohydrology: A Text to Accompany the ACRU-300 Agrohydrological Modelling System*, Agricultural Catchments Research Unit, Department of Agricultural Engineering, University of Natal.

Shuttleworth, W.J. & Wallace, J.S. (1985). Evaporation from Sparse Crops - An energy Combination Theory. *Quart. J. Roy. Meteorol. Soc.*, Vol. 111, pp. 839-855.

Simunek, J.; Sejna, M. & van Genuchten, M.Th. (2007). The HYDRUS Software Package for Simulating Two- and Three-Dimensional Movement of Water, Heat, and Multiple Solutes in Variably-Saturated Media, User Manual, Version 1.0, PC Progress, Prague, Czech Republic.

Sinclair, S. & Pegram, G.G.S. (2010). A Comparison of ASCAT and Modelled Soil Moisture over South Africa, Using TOPKAPI in Land Surface Mode. *Hydrology and Earth System Science*, Vol. 14, pp. 613–626.

Skaggs, R.W. (1978). *A Water Management Model for Shallow Water Tables*, Report No. 134, Water Resources Research Institute, The University of North Carolina, Raleigh, NC, USA.

Spittlehouse, D.L. (1989). Estimating Evapotranspiration from Land Surfaces in British Columbia, In: *Estimation of Areal Evapotranspiration*, pp. 245-253, IAHS, Publ., 177.

Steduto, P.; Hsiao, T.C.; Raes, D. & Fereres, E. (2009) AquaCrop-The FAO Crop Model to Simulate Yield Response to Water: I. Concepts and Underlying Principles. *Agronomy Journal*, Vol. 101(3), pp. 426-437.

Su, Z. (2002). The Surface Energy Balance System (SEBS) for Estimation of Turbulent Heat Fluxes. *Hydrology and Earth System Sciences*, Vol. 6, pp. 85-99.

Takeuchi, A.; Ao, T. & Ishidaira, H. (1999). Introduction of Block-Wise Use of TOPMODEL and Muskingum-Cunge Method for the Hydroenvironmental Simulation of a Large Ungauged Basin. *Hydrol. Sci. J.*, Vol. 44(4), pp. 633–646.

Tasumi, M.; Trezza, R.; Allen, R.G. & Wright, J.L. (2005). Operational Aspects of Satellite Based Energy Balance Models for Irrigated Crops in the Semi-Arid US. *Irrigation and Drainage Systems*, Vol. 19, pp. 355-376.

Taylor, S.A. & Ashcroft, G.L. (1972). *Physical Edaphology. The Physics of Irrigated and Non-Irrigated Soils*. Freeman, San Francisco, pp. 533.

Thornthwaite, C.W. (1948). An approach toward a rational classification of climate. *Geograph. Rev.*, Vol. 38, pp. 55.

Turc, L. (1961). Evaluation de Besoins en Eau d'Irrigation, ET Potentielle. *Ann. Agron.*, Vol. 12, pp. 13-49.

Van Genuchten, M.Th. (1980). A Closed-Form Equation for Predicting the Hydraulic Conductivity of Unsaturated Soils. *Soil Science Society of America Journal*, Vol. 44(5), pp. 892–898.

Van Walsum, P.E.V.; Veldhuizen, A.A.; van Bakel, P.J.T.; van der Bolt, F.J.E.; Dik, P.E.; Groenendijk, P.; Querner, E.P. & Smit, M.R.F. (2004). *SIMGRO 5.0.1, Theory and Model Implementation*, Report No. 913.1, Alterra, Wageningen, The Netherlands.

Viviroli, D.; Zappa, M.; Gurtz, J. & Weingartner, R. (2009). An Introduction to the Hydrological Modelling System PREVAH and its Pre- and Post-Processing Tools. *Environmental Modelling & Software*, Vol. 24, pp. 1209-1222.

Wang, X.; Xie, H.; Guan, H. & Zhou, X. (2003). Different Responses of MODIS-Derived NDVI to Root-Zone Soil Moisture in Semi-Arid and Humid Regions. *Journal of Hydrology*, Vol. 340, pp. 12-24.

Wapener, T. (2003) Evaluation of Catchment Models. *Hydrological Processes*, Vol. 17, pp. 3375-3378.

Xu, Y. & Beekman, H.E. (Eds.) (2003). *Groundwater Recharge Estimation in Southern Africa*, ISBN 92-9220-000-3, UNESCO IHP Series no. 64, UNESCO, Paris.

Yang, D.; Zhang, T.; Zhang, K.; Greenwood, D.; Hammond, J.P. & White P.J. (2009). An Easily Implemented Agro-hydrological Procedure with Dynamic Root Simulation for Water Transfer in the Crop–Soil System: Validation and Application. *Journal of Hydrology*, Vol. 370, 177–190

Zhang, L.; Dawes, W.R. & Hatton, T.J. (1996). Modelling Hydrologic Processes Using a Biophysically Based Model – Application of WAVES to FIFE and HAPEX-MOBILHY. *Journal of Hydrology*, Vol. 185, pp. 147-169.

Zhang, L.; Dawes, W.R. & Walker, G.R. (2001). The Response of Mean Annual Evapotranspiration to Vegetation Changes at Catchment Scale. *Water Resources Research*, Vol. 37, pp. 701-708.

Zhou, M.C.; Ishidaira, H.; Hapuarachchi, H.P.; Magome, J.; Kiem, A.S. & Takeuchi, K. (2006). Estimating Potential Evapoutranspiration Using Shuttleworth-Wallace Model and NOAA-AVHRR NDVI Data to Feed a Distributed Hydrological Model over the Mokeng River Basin. *Journal of Hydrology*, Vol. 327, pp. 151-173.

16

Development of Hybrid Method for the Modeling of Evaporation and Evapotranspiration

Sungwon Kim
*Dongyang University,
Republic of Korea*

1. Introduction

Evaporation is the process whereby liquid water is converted to water vapor and removed from the evaporating surface. In hydrological practice, the estimation of evaporation can be achieved by direct or indirect methods. Direct method is based on the field measurements. Evaporation pan have also been used and compared to estimate evaporation by researchers (Choudhury, 1999; McKenzie and Craig, 2001; Vallet-Coulomb et al., 2001). The Class A evaporation pan is one of the most widely used instruments for the measurements of evaporation from a free water surface. The pan evaporation (PE) is widely used to estimate the evaporation of lakes and reservoirs (Finch, 2001). Many researchers have tried to estimate the evaporation through the indirect methods using the climatic variables, but some of these techniques require the data which cannot be easily obtained (Rosenberry et al., 2007).

Evapotranspiration (ET) is the sum of volume of water used by vegetation, evaporated from the soil, and intercepted precipitation (Singh, 1988). ET plays an important role in our environment at global, regional, and local scales. ET is observed using a lysimeter directly or can be calculated using the water balance method or the climatic variables indirectly. Because the measurements of ET using a lysimeter directly, however, requires much unnecessary time and needs correct and careful experience, it is not always possible in field measurements. Thus, an empirical approach based on the climatic variables is generally used to calculate ET (Penman, 1948; Allen et al., 1989). In the early 1970s, the Food and Agricultural Organization of the United Nations (FAO), Rome, developed practical procedures to calculate the crop water requirements (Doorenbos & Pruitt, 1977), which have become the widely accepted standard for irrigation studies. A common practice for estimating ET from a well-watered agricultural crop is to calculate the reference crop ET such as the grass reference ET (ET_o) or the alfalfa reference ET (ET_r) from a standard surface and to apply an appropriate empirical crop coefficient, which accounts for the difference between the standard surface and the crop ET.

Recently, the outstanding results using the neural networks model in the fields of PE and ET modeling have been obtained (Bruton et al., 2000; Sudheer et al., 2003; Trajkovic et al., 2003; Trajkovic, 2005; Keskin and Terzi, 2006; Kisi, 2006; Kisi, 2007; Zanetti et al., 2007; Jain et al., 2008; Kumar et al., 2008; Landeras et al., 2008; Kisi, 2009; Kumar et al., 2009; Tabari et al.,

2009; Chang et al., 2010; Guven and Kisi, 2011). Sudheer et al. (2002) investigated the prediction of Class A PE using the neural networks model. They used the neural networks model for the evaporation process using proper combinations of the observed climate variables such as temperature, relative humidity, sunshine duration, and wind speed for the neural networks model. Shiri and Kisi (2011) investigated the ability of genetic programming (GP) to improve the accuracy of daily evaporation estimation. They used proper combinations of air temperature, sunshine hours, wind speed, relative humidity, and solar radiation for GP. Kumar et al. (2002) developed the neural networks models to calculate the daily ET. They used proper combinations of the observed climatic variables such as solar radiation, temperature, relative humidity, and wind speed for the neural networks models. Kisi & Ozturk (2007) used the neuro-fuzzy models to calculate FAO-56 PM ET_0 using the observed climatic variables. They used proper combinations of the observed climatic variables such as air temperature, solar radiation, wind speed, and relative humidity for the neuro-fuzzy models. Kim & Kim (2008) developed the neural networks model embedding the genetic algorithm for the modeling of the daily PE and ET_r simultaneously, and constructed the optimal neural networks model using the uncertainty analysis of the input layer nodes/variables. Furthermore, they suggested the 2-dimensional and 3-dimensional maps for PE and ET_r to provide the reference data for irrigation and drainage system, Republic of Korea. And, the recent researches combining the stochastic models and the neural networks models in the fields of hydrology and water resources have been accomplished. Mishra et al. (2007) developed a hybrid model, which combined a linear stochastic model and a nonlinear neural networks model, for drought forecasting. Kim (2011) investigated the modeling of the monthly PE and ET_r simultaneously using the specific method, which combined the stochastic model with the neural networks models.

The purpose of this study is to develop the hybrid method for the modeling of the monthly PE and FAO-56 PM ET_0 simultaneously. The hybrid method represents the combination of Univariate Seasonal periodic autoregressive moving average (PARMA) model and support vector machine neural networks model (SVM-NNM). For this research, first, the stochastic model, Univariate Seasonal PARMA(1,1) model, is used for the generation of the reliable data, which are considered as the training dataset. Therefore, the observed data are considered as the testing dataset. Second, the neural networks model, SVM-NNM, is used for the modeling of the monthly PE and FAO-56 PM ET_0 simultaneously. Homogeneity evaluation using the One-way ANOVA and Mann-Whitney *U* test, furthermore, is carried out for the observed and calculated PE and FAO-56 PM ET_0 data. And, the correlation relationship between the observed PE and FAO-56 PM ET_0 data can be derived using the bivariate linear regression analysis model (BLRAM), respectively.

2. Calculation of FAO-56 PM ET_0

Penman (1948) combination method links evaporation dynamics with the flux of net radiation and aerodynamic transport characteristics of the natural surface. Based on the observations that latent heat transfer in plant stem is influenced not only by these abiotic factors, Monteith (1965) introduced a surface conductance term that accounted for the response of leaf stomata to its hydrologic environment. This modified form of the Penman-Monteith (PM) ET_0 model. Jensen et al. (1990) measured ET_0 using the lysimeters at 11 stations located in the different climatic zones of various regions around the world. They compared the results of the lysimeters with those of 20 different empirical equations and methodologies for ET_o measurements. It was found that PM ET_o model showed the optimal results over all the climatic zones. If the observed/measured data for ET_o does not exist, therefore, PM ET_o model can be considered as a standard methodology to calculate ET_o. In Gwangju and Haenam stations which were selected for this study, there are no observed data for ET_o using a lysimeter. The data calculated using PM ET_o model can be assumed as the observed ET_o, whose reliability was verified by many previous studies. All calculation procedures as used in PM ET_o model are based on the FAO guidelines as laid down in the publication No. 56 of the Irrigation and Drainage Series of FAO "Crop Evapotranspiration–Guidelines for Computing Crop Water Requirements" (1998). Therefore, FAO-56 PM ET_o equation means PM ET_o equation suggested by the Irrigation and Drainage Paper No. 56, FAO. FAO-56 PM ET_o equation is given by Allen et al. (1998) and can be shown as the following equation (1).

$$\text{FAO-56 PM ET}_o = \frac{0.408\Delta(R_n - G) + \gamma(900/(T+273))u_2(e_s - e_a)}{\Delta + \gamma(1+0.34u_2)} \tag{1}$$

where FAO-56 PM ET_o = the grass reference evapotranspiration (mm/day); R_n = the net radiation at the crop surface (MJ/m^2 day); G = the soil heat flux density (MJ/m^2 day); T = the mean air temperature at 2m height (◦C); u_2 = the wind speed at 2m height (m/sec); e_s = the saturation vapor pressure (kPa); e_a = the actual vapor pressure (kPa); e_s - e_a = the saturation vapor pressure deficit (kPa); Δ = the slope vapor pressure curve (kPa/◦C); and γ = the psychometric constant (kPa/◦C). FAO CROPWAT 8.0 computer program has been used to calculate FAO-56 PM ET_o and extraterrestrial radiation (R_a). FAO CROPWAT 8.0 computer program allows the user to enter the climatic data available including maximum temperature (T_{max}), minimum temperature (T_{min}), mean relative humidity (RH_{mean}), mean wind speed (WS_{mean}), and sunshine duration (SD) for calculating FAO-56 PM ET_o. On the base of climatic data available, FAO CROPWAT 8.0 computer program calculates the solar radiation reaching soil surface. Fig. 1(a)-(b) show the calculation of FAO-56 PM ET_o using FAO CROPWAT 8.0 computer program in Gwangju and Haenam stations, respectively.

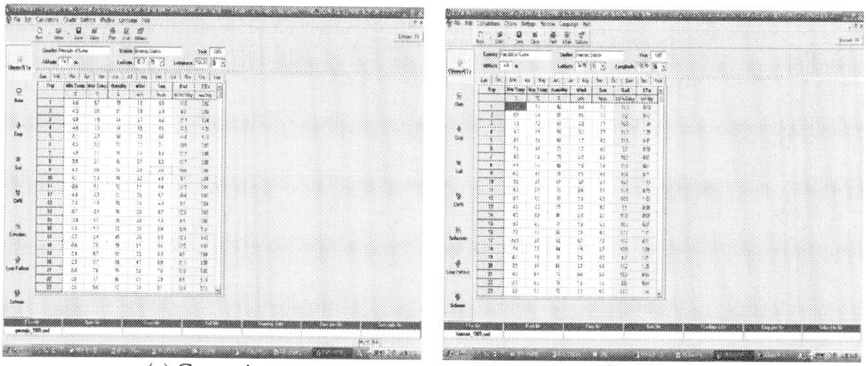

(a) Gwangju (b) Haenam

Fig. 1. Calculation of FAO-56 PM ET_o using FAO CROPWAT 8.0 Computer Program

3. Stochastic model

3.1 Univariate seasonal periodic autoregressive moving average model

Stationary ARMA models have been widely applied in stochastic hydrology for modeling of annual time series where the mean, variance, and the correlation structure do not depend on time. For seasonal hydrologic time series, such as monthly series, seasonal statistics including the mean and standard deviation may be reproduced by a stationary ARMA model by means of standardizing the underlying seasonal series. Hydrologic time series such as monthly streamflows, PE, and FAO-56 PM ET_o are usually characterized by different dependence structure depending on the season (Salas, 1993). One may extend Univariate Seasonal periodic autoregressive (PAR) model to include periodic moving average (MA) parameters. Such a model is Univariate Seasonal periodic autoregressive moving average (PARMA) model and is expressed as Univariate Seasonal PARMA(p,q) model. The stochastic models are generally simple to use. When the errors, however, happen in model identification and parameter estimation, the generated data using the stochastic models cannot reconstruct the statistical properties of the observed data exactly. Furthermore, the high-order PARMA(p,q) models have generally many parameters, and the calculation process is much complex (Salas et al., 1980). In this study, the author determined in advance 4 kinds of Univariate Seasonal PARMA(p,q) models including PARMA(1,1), PARMA(1,2), PARMA (2,1), and PARMA(2,2), which are the low-order models and contain the seasonal properties. In general, the low-order Univariate Seasonal PARMA(p,q) models are useful for the periodic hydrologic time series modeling (Salas et al., 1982). Furthermore, the author generated 100 years data in advance using each Univariate Seasonal PARMA(p,q) model for the climatic variables of the neural networks models, respectively. As a result, the author selected Univariate Seasonal PARMA(1,1) model, which shows the best statistical properties and is simple in parameter estimation. Univariate Seasonal PARMA(1,1) model has been applied to monthly streamflow time series from the previous studies (Tao and Delleur, 1976; Hirsch, 1979), and is shown as the following equation (2).

$$y_{v,\tau} = \mu_\tau + \phi_{1,\tau}(y_{v,\tau-1} - \mu_{\tau-1}) + \varepsilon_{v,\tau} - \theta_{1,\tau}\varepsilon_{v,\tau-1} \tag{2}$$

where $y_{v,\tau} / y_{v,\tau-1}$ = the monthly PE and FAO-56 PM ET_o for year= v and month= $\tau / \tau-1$; $\mu_\tau / \mu_{\tau-1}$ = the means for month= $\tau / \tau-1$; $\phi_{1,\tau}$ = the seasonal autoregressive parameter for month= τ; $\theta_{1,\tau}$ = the seasonal moving average parameter for month= τ; $\varepsilon_{v,\tau} / \varepsilon_{v,\tau-1}$ = uncorrelated noise terms; v = year; τ = month $(1,2,\ldots,\omega)$; and ω =12. Furthermore, Univariate Seasonal PARMA(2,1), PARMA(2,2) models, and more complex multiplicative PARMA(p,q) models may be needed for hydrologic modeling and simulation when the preservation of both the seasonal and the annual statistics is desired (Salas and Abdelmohsen, 1993).

3.2 Construction of Univariate Seasonal PARMA(p,q) model

The author used Univariate Seasonal PARMA(1,1) model to generate the sufficient training dataset, and obtained two generated samples. They included the input nodes/variables including mean temperature (T_{mean}), maximum temperature (T_{max}), minimum temperature (T_{min}), mean dew point temperature (DP_{mean}), minimum relative humidity (RH_{min}), mean relative humidity (RH_{mean}), mean wind speed (WS_{mean}), maximum wind speed (WS_{max}), and sunshine duration (SD) in mean values and the output nodes/variables including PE and

FAO-56 PM ET_0 in total values, respectively. Furthermore, they were generated for 100 years (Short-term), 500 years (Mid-term), and 1000 years (Long-term), respectively. The author selected the second generated sample, and the first 50 years of the 100, 500, and 1000 years was abandoned to eliminate the biases, respectively. The parameters of Univariate Seasonal PARMA(1,1) model were determined using the method of approximate least square, respectively.

4. Support Vector Machine Neural Networks Model (SVM-NNM)

SVM-NNM has found wide application in several areas including pattern recognition, regression, multimedia, bio-informatics and artificial intelligence. Very recently, SVM-NNM is gaining recognition in hydrology (Dibike et al., 2001; Khadam & Kaluarachchi, 2004). SVM-NNM implements the structural risk minimization principle which attempts to minimize an upper bound on the generalization error by striking a right balance between the training performance error and the capacity of machine. The solution of traditional neural networks models such as MLP-NNM may tend to fall into a local optimal solution, whereas global optimum solution is guaranteed for SVM-NNM (Haykin, 2009). SVM-NNM is a new kind of classifier that is motivated by two concepts. First, transforming data into a high-dimensional space can transform complex problems into simpler problems that can use linear discriminant functions. Second, SVM-NNM is motivated by the concept of training and using only those inputs that are near the decision surface since they provide the most information about the classification. The first step in SVM-NNM is transforming the data into a high-dimensional space. This is done using radial basis function (RBF) that places a Gaussian at each sample data. Thus, the feature space becomes as large as the number of sample data. RBF uses backpropagation to train a linear combination of the gaussians to produce the final result. SVM-NNM, however, uses the idea of large margin classifiers for the training performance. This decouples the capacity of the classifier from the input space and at the same time provides good generalization. This is an ideal combination for classification (Vapnik, 1992, 2000; Principe et al., 2000; Tripathi et al., 2006).

In this study, the basic ideas of SVM-NNM are reviewed. Consider the finite training pattern (x_i, y_i). where $x_i \in \Re^n$ = a sample value of the input vector x considering of N training patterns; and $y_i \in \Re^n$ = the corresponding value of the desired model output. A nonlinear transformation function $\phi(\cdot)$ is defined to map the input space to a higher dimension feature space, $\Re^{n_h}$. According to Cover's theorem (Cover, 1965), a linear function, $f(\cdot)$, could be formulated in the high dimensional feature space to look for a nonlinear relationship between inputs and outputs in the original input space. It can be written as the following equation (3).

$$\bar{y} = f(x) = w^T \phi(x) + b \tag{3}$$

where $\bar{y}$ = the actual model output. The coefficient w and b are adjustable model parameters. In SVM-NNM, we aim at minimizing the empirical risk. It can be written as the following equation (4).

$$R_{emp} = \frac{1}{N} \sum_{i=1}^{N} \left| y_i - \bar{y}_i \right|_{\varepsilon} \tag{4}$$

where R_{emp} = the empirical risk; and $\left|y_i - \bar{y}_i\right|_\varepsilon$ = the Vapnik's ε-insensitive loss function. Following regularization theory (Haykin, 2009), the parameters w and b are calculated by minimizing the cost function. It can be written as the following equation (5).

$$\psi_\varepsilon(w,\xi,\xi^*) = \frac{1}{2}w^T w + C\sum_{i=1}^{N}(\xi_i + \xi_i^*) \tag{5}$$

subject to the constraints: 1) $y_i - \bar{y}_i \le \varepsilon + \xi_i$ i = 1, 2, ... , N, 2) $-y_i + \bar{y}_i \le \varepsilon + \xi_i^*$ i = 1, 2, ... , N, and 3) $\xi_i, \xi_i^* \ge 0$ i = 1, 2, ... , N. where $\psi_\varepsilon(w,\xi,\xi^*)$ = the cost function; ξ_i, ξ_i^* = positive slack variables; and C = the cost constant. The first term of the cost function, which represents weight decay, is used to regularize weight sizes and to penalize large weights. This helps in improving generalization performance (Hush and Horne, 1993). The second term of the cost function, which represents penalty function, penalizes deviations of $\bar{y}$ from y larger than $\pm\varepsilon$ using Vapnik's ε-insensitive loss function. The cost constant C determines the amount up to which deviations from ε are tolerated. Deviations above ε are denoted by ξ_i, whereas deviations below - ε are denoted by ξ_i^*. The constrained quadratic optimization problem can be solved using the method of Lagrangian multipliers (Haykin, 2009). From this solution, the coefficient w can be written as the following equation (6).

$$w = \sum_{i=1}^{N}(\alpha_i - \alpha_i^*)\phi(x_i) \tag{6}$$

where α_i, α_i^* = the Lagrange multipliers, which are positive real constants. The data points corresponding to non-zero values for $(\alpha_i - \alpha_i^*)$ are called support vectors. In SVM-NNM to calculate PE and FAO-56 PM ET_o, there are several possibilities for the choice of kernel function, including linear, polynomial, sigmoid, splines and RBF. In this study, RBF is used to map the input data into higher dimensional feature space. RBF can be written as the following equation (7).

$$k(x,x_j) = \Phi_1 = \exp(-B_1 R_j^2) = \exp\left(-\frac{\left\|x_i - x_j\right\|^2}{2\sigma^2}\right) \tag{7}$$

where i, j = the input layer and the hidden layer; $K(x,x_j) = \Phi_1$ = the inner product kernel function; $B_1 = \frac{1}{2\sigma^2}$, and has a constant value; and σ = the width/spread of RBF, which can be adjusted to control the expressivity of RBF. The function for the single node of the output layer which receives the calculated results of RBF can be written as the following equation (8).

$$G_k = [\sum_{j=1}^{N}(\alpha_j - \alpha_j^*)\cdot K(x,x_j)] + B \tag{8}$$

where k = the output layer; G_k = the calculated value of the single output node; and B = the bias in the output layer. Equation (8), finally, takes the form of equation (9) and (10), which represents SVM-NNM for the modeling of PE and FAO-56 PM ET_o.

$$PE = W_1 \cdot \Phi_2(G_k) = W_1 \cdot \Phi_2[[\sum_{j=1}^{N}(\alpha_j - \alpha_j^*) \cdot K(x,x_j)] + B] \tag{9}$$

$$\text{FAO-56 PM ET}_o = W_2 \cdot \Phi_2(G_k) = W_2 \cdot \Phi_2[[\sum_{j=1}^{N}(\alpha_j - \alpha_j^*) \cdot K(x,x_j)] + B] \tag{10}$$

where $\Phi_2(\cdot)$ = the linear sigmoid transfer function; W_1 = the specific weights connected to the output variable of PE; and W_2 = the specific weights connected to the output variable of FAO-56 PM ET_o. A number of SVM-NNM computer programs are now available for the modeling of PE and FAO-56 PM ET_o. NeuroSolutions 5.0 computer program was used to develop SVM-NNM structure. Fig. 2 shows the developed structure of SVM-NNM. From the Fig. 2, the input nodes/variables of climatic data are mean temperature (T_{mean}), maximum temperature (T_{max}), minimum temperature (T_{min}), mean dew point temperature (DP_{mean}), minimum relative humidity (RH_{min}), mean relative humidity (RH_{mean}), mean wind speed (WS_{mean}), maximum wind speed (WS_{max}), and sunshine duration (SD) in mean values (01/1985-12/1990). The output nodes/variables of climatic data are PE and FAO-56 PM ET_o in total values (01/1985-12/1990).

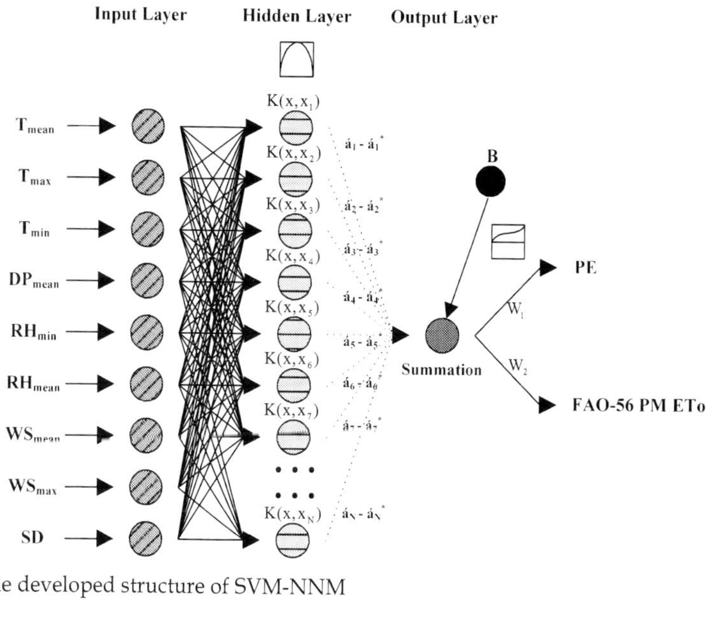

Fig. 2. The developed structure of SVM-NNM

5. Study scope and data

In this study, Gwangju and Haenam stations from the Yeongsan River catchment are selected among the 71 weather stations including Jeju-do under the control of the Korea meteorological administration (KMA). They have possessed long-term climatic data dating back over at least 30 years. The Yeongsan River catchment covers an area of 3455 km^2, and lies between latitudes 34.4°N and 35.2°N, and between longitudes 126.2°E and 127.0°E. Fig. 3 shows the Yeongsan River catchment including Gwangju and Haenam stations. The climatic data, which was necessary for the modeling of PE and FAO-56 PM ET$_o$, were collected from the Internet homepage of water management information system (www.wamis.go.kr) and the Korea meteorological administration (www.kma.go.kr).

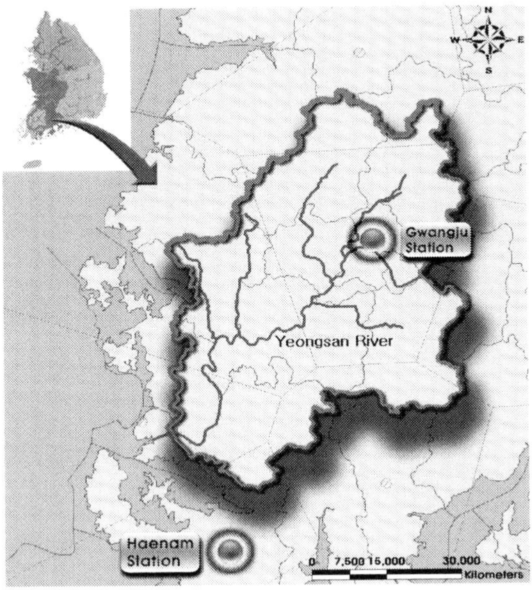

Fig. 3. The Yeongsan River Catchment including Gwangju and Haenam stations

6. SVM-NNM performance

6.1 Performance statistics

The performance of SVM-NNM to account for calculating the monthly PE and FAO-56 PM ET$_o$ was evaluated using a wide variety of standard statistics index. A total of 3 different standard statistics indexes were employed; the coefficient of correlation (CC), root mean square error (RMSE), and Nash-Sutcliffe coefficient (R^2) (Nash & Sutcliffe, 1970; ASCE, 1993). Table 1 shows summary of the statistics index in this study. where $\overline{y}_i(x)$ = the calculated PE and FAO-56 PM ET$_o$ (mm/month); $y_i(x)$ = the observed PE and FAO-56 PM ET$_o$ (mm/month); $\overline{u}_y$ = mean of the calculated PE and FAO-56 PM ET$_o$ (mm/month); u_y = mean of the observed PE and FAO-56 PM ET$_o$ (mm/month); and n = total number of the monthly PE and FAO-56 PM ET$_o$ considered. A model which is effective in the modeling of PE and FAO-56 PM ET$_o$ accurately, and efficient in capturing the complex relationship among the various inputs and output variables involved in a particular problem, is considered the best. CC, RMSE, and R^2 statistics quantify the efficiency of SVM-NNM in capturing the extremely complex, dynamic, and nonlinear relationships (Kim, 2011).

Statistics Indexes	Equation
CC	$\dfrac{\frac{1}{n}\sum_{i=1}^{n}[y_i(x)-u_y][\bar{y}_i(x)-\bar{u}_y]}{\sqrt{\frac{1}{n}\sum_{i=1}^{n}[y_i(x)-u_y]^2}\sqrt{\frac{1}{n}\sum_{i=1}^{n}[\bar{y}_i(x)-\bar{u}_y]^2}}$
RMSE	$\sqrt{\frac{1}{n}\sum_{i=1}^{n}[\bar{y}_i(x)-y_i(x)]^2}$
R^2	$1-\dfrac{\sum_{i=1}^{n}[y_i(x)-\bar{y}_i(x)]^2}{\sum_{i=1}^{n}[y_i(x)-u_y]^2}$

Table 1. Summary of statistics indexes

6.2 Data normalization

The climatic data used in this study including mean temperature (T_{mean}), maximum temperature (T_{max}), minimum temperature (T_{min}), mean dew point temperature (DP_{mean}), minimum relative humidity (RH_{min}), mean relative humidity (RH_{mean}), mean wind speed (WS_{mean}), maximum wind speed (WS_{max}), and sunshine duration (SD) were normalized for preventing and overcoming problem associated with the extreme values. An important reason for the normalization of input nodes is that each of input nodes represents an observed value in a different unit. Such input nodes are normalized, and the input nodes in dimensionless unit are relocated. The similarity effect of input nodes is thus eliminated (Kim et al., 2009). According to Zanetti et al. (2007), by grouping the daily values into averages, ET_o may be calculated due to their highest stabilization. For data normalization, the data of input and output nodes were scaled in the range of [0 1] using the following equation (11).

$$Y_{norm} = \frac{Y_i - Y_{min}}{Y_{max} - Y_{min}} \tag{11}$$

where Y_{norm} = the normalized dimensionless data of the specific input node/variable; Y_i = the observed data of the specific input node/variable; Y_{min} = the minimum data of the specific input node/variable; and Y_{max} = the maximum data of the specific input node/variable.

6.3 Training performance

The method for calculating parameters is generally called the training performance in the neural networks model category. The training performance of neural networks model is iterated until the training error is reached to the training tolerance. Iteration means one completely pass through a set of inputs and target patterns or data. In general, it is assumed that the neural networks model does not have any prior knowledge about the example problem before it is trained (Kim, 2004). A difficult task with the neural networks model is to choose the number of hidden nodes. The network geometry is problem dependent. This study adopted one hidden layer for the construction of SVM-NNM since it is well known that one hidden layer is enough to represent PE and FAO-56 PM ET_o nonlinear complex relationship (Kumar et al., 2002; Zanetti et al., 2007). The testing performance in the modeling of PE and FAO-56 PM ET_o, therefore, is carried out using the optimal parameters, which are calculated during the training performance.

The hybrid method, which was developed in this study, consisted of the following training patterns. First, the stochastic model was selected. As explained previously, Univariate Seasonal PARMA(1,1) model, which consisted of 1 pattern only, was used to generate the training dataset. Second, the data, which were generated by Univariate Seasonal PARMA(1,1) model, consisted of 3 patterns including 100 years (Short-term), 500 years (Mid-term), and 1000 years (Long-term), respectively. Finally, the neural networks model, which consisted of 1 pattern only including SVM-NNM, was used for the training and testing performances, respectively. Therefore, the hybrid method consisted of 3 training patterns including 100/PARMA(1,1)/SVM-NNM, 500/PARMA(1,1)/SVM-NNM, and 1000 /PARMA(1,1)/SVM-NNM, respectively. For Gwangju and Haenam stations, the training dataset including the climatic, PE, and FAO-56 PM ET_o data were generated by Univariate Seasonal PARMA(1,1) model using observed data (01/1985-12/1990) for 100 years (Short-term), 500 years (Mid-term), and 1000 years (Long-term), respectively. After the first 50 years of the generated data for 100, 500, and 1000 years was abandoned to eliminate the biases, the training performance should be carried out using SVM-NNM. Therefore, the total amount of data used for the training performance consisted of 600, 5400, and 11400, respectively. For the training performance of SVM-NNM, NeuroSolutions 5.0 computer program was used to carry out the training performance. Fig. 4(a)-(b) show SVM-NNM training performance using NeuroSolutions 5.0 computer program.

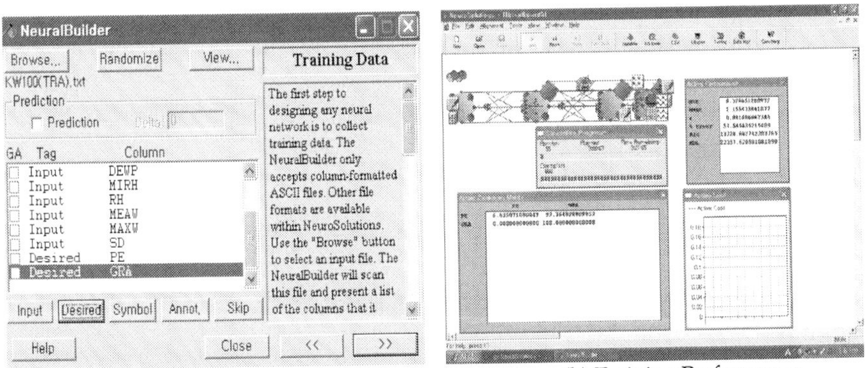

(a) Training Data (b) Training Performance

Fig. 4. SVM-NNM training performance using NeuroSolution 5.0 Computer Program

Table 2 shows the statistics results of the training performances for 3 training patterns of PE modeling. In PE of Gwangju station, from the Table 2, 100/PARMA(1,1)/SVM-NNM training pattern produced the statistics results with CC value of 0.929, RMSE value of 16.360 mm and R^2 value of 0.858. 500/PARMA(1,1)/SVM-NNM training pattern produced the statistics results with CC value of 0.919, RMSE value of 17.942 mm and R^2 value of 0.833. 1000/PARMA(1,1)/SVM-NNM training pattern produced the statistics results with CC value of 0.905, RMSE value of 20.489 mm and R^2 value of 0.781, respectively. In PE of Haenam station, from the Table 2, 100/PARMA(1,1)/SVM-NNM training pattern produced the statistics results with CC value of 0.965, RMSE value of 9.643 mm and R^2 value of 0.930. 500/PARMA(1,1)/SVM-NNM training pattern produced the statistics results with CC value of 0.958, RMSE value of 10.673 mm and R^2 value of 0.914. 1000/PARMA(1,1)/SVM-NNM training pattern produced the statistics results with CC value of 0.962, RMSE value of 10.105 mm and R^2 value of 0.922, respectively. From the above results, the statistics results of the training performance for 100/PARMA(1,1)/SVM-NNM training pattern were better than those of the training performances for 500/PARMA(1,1)/SVM-NNM and 1000/PARMA(1,1)/SVM-NNM training patterns for PE of Gwangju and Haenam stations, respectively.

Station	Statistics Indexes	100/PARMA(1,1)/ SVM-NNM	500/PARMA(1,1)/ SVM-NNM	1000/PARMA(1,1) SVM-NNM
Gwangju	CC	0.929	0.919	0.905
	RMSE (mm)	16.360	17.942	20.489
	R^2	0.858	0.833	0.781
Haenam	CC	0.965	0.958	0.962
	RMSE (mm)	9.643	10.673	10.105
	R^2	0.930	0.914	0.922

Table 2. Statistics results of the training performances (PE)

Table 3 shows the statistics results of the training performances for 3 training patterns of FAO-56 PM ET_o modeling. In FAO-56 PM ET_o of Gwangju station, from the Table 3, 100/PARMA(1,1)/SVM-NNM training pattern produced the statistics results with CC value of 0.975, RMSE value of 8.517 mm and R^2 value of 0.948. 500/PARMA(1,1)/SVM-NNM training pattern produced the statistics results with CC value of 0.966, RMSE value of 10.237 mm and R^2 value of 0.924. 1000/PARMA(1,1)/SVM-NNM training pattern produced the statistics results with CC value of 0.963, RMSE value of 11.061 mm and R^2 value of 0.911, respectively. In FAO-56 PM ET_o of Haenam station, from the Table 3, 100/PARMA(1,1) /SVM-NNM training pattern produced the statistics results with CC value of 0.965, RMSE value of 9.935 mm and R^2 value of 0.926. 500/PARMA(1,1)/SVM-NNM training pattern produced the statistics results with CC value of 0.956, RMSE value of 10.822 mm and R^2 value of 0.912. 1000/PARMA(1,1)/SVM-NNM training pattern produced the statistics results with CC value of 0.962, RMSE value of 10.014 mm and R^2 value of 0.925, respectively. From the above results, the statistics results of the training performance for 100/PARMA(1,1)/SVM-NNM training pattern were better than those of the training performances for 500/PARMA(1,1)/SVM-NNM and 1000/PARMA(1,1)/SVM-NNM training patterns for FAO-56 PM ET_o of Gwangju and Haenam stations, respectively. Therefore, the data length has less effect on the training performance of PE and FAO-56 PM ET_o in this study. Tokar and Johnson (1999) suggested that the data length has less effect on the neural networks model performance than the data quality. Sivakumar et al. (2002) suggested that it is imperative to select a good training dataset from the available data series. They indicated that the best way to achieve a good training performance seems to be to include most of the extreme events such as very high and very low values in the training dataset. Furthermore, Kim (2011) did not carry out the statistics analysis of the training performance since the training dataset of 6 training patterns consisted of the generated (not observed) data only.

Station	Statistics Indexes	100/PARMA(1,1)/ SVM-NNM	500/PARMA(1,1)/ SVM-NNM	1000/PARMA(1,1) SVM-NNM
Gwangju	CC	0.975	0.966	0.963
	RMSE (mm)	8.517	10.237	11.061
	R^2	0.948	0.924	0.911
Haenam	CC	0.965	0.956	0.962
	RMSE (mm)	9.935	10.822	10.014
	R^2	0.926	0.912	0.925

Table 3. Statistics results of the training performances (FAO-56 PM ET_o)

6.4 Testing performance

The neural networks model is tested by determining whether the model meets the objectives of modeling within some preestablished criteria or not. Of course, the optimal parameters, which are calculated during the training performance, are applied for the testing performance of the neural networks model (Kim, 2004). For the testing performance, the monthly climatic data (01/1985-12/1990) in Gwangju and Haenam stations were used. The total amount of data used for the testing performance consisted of 72 data for the monthly time series. The testing performance applied the cross-validation method in order to overcome the over-fitting problem of SVM-NNM. The cross-validation method is not to train all the training data until SVM-NNM reaches the minimum RMSE, but is to cross-validate with the testing data at the end of each training performance. If the over-fitting problem occurs, the convergence process over the mean square error of the testing data will not decrease but will increase as the training data are still trained (Bishop, 1994; Haykin, 2009).

Table 4 shows the statistics results of the testing performances for 3 training patterns of PE modeling. In PE of Gwangju station, from the Table 4, 100/PARMA(1,1)/SVM-NNM training pattern produced the statistics results with CC value of 0.955, RMSE value of 12.239 mm and R^2 value of 0.908. 500/PARMA(1,1)/SVM-NNM training pattern produced the statistics results with CC value of 0.956, RMSE value of 13.501 mm and R^2 value of 0.888. 1000/PARMA(1,1)/SVM-NNM training pattern produced the statistics results with CC value of 0.953, RMSE value of 15.103 mm and R^2 value of 0.860, respectively. In PE of Haenam station, from the Table 4, 100/PARMA(1,1)/SVM-NNM training pattern produced the statistics results with CC value of 0.966, RMSE value of 9.581 mm and R^2 value of 0.932. 500/PARMA(1,1)/SVM-NNM training pattern produced the statistics results with CC value of 0.968, RMSE value of 9.370 mm and R^2 value of 0.935. 1000/PARMA(1,1)/SVM-NNM training pattern produced the statistics results with CC value of 0.953, RMSE value of 11.313 mm and R^2 value of 0.905, respectively. From the above results, the statistics results of the testing performance for 100/PARMA(1,1)/SVM-NNM training pattern were better than those of the testing performances for 500/PARMA(1,1)/SVM-NNM and 1000/ PARMA(1,1)/SVM-NNM training patterns for PE of Gwangju station. And, the statistics results of the testing performance for 500/PARMA(1,1)/SVM-NNM training pattern were better than those of the testing performances for 100/PARMA(1,1)/SVM-NNM and 1000/ PARMA(1,1)/SVM-NNM training patterns for PE of Haenam station, respectively.

Station	Statistics Indexes	100/PARMA(1,1)/ SVM-NNM	500/PARMA(1,1)/ SVM-NNM	1000/PARMA(1,1) SVM-NNM
Gwangju	CC	0.955	0.956	0.953
	RMSE (mm)	12.239	13.501	15.103
	R^2	0.908	0.888	0.860
Haenam	CC	0.966	0.968	0.953
	RMSE (mm)	9.581	9.370	11.313
	R^2	0.932	0.935	0.905

Table 4. Statistics results of the testing performances (PE)

Table 5 shows the statistics results of the testing performances for 3 training patterns of FAO-56 PM ET_o modeling. In FAO-56 PM ET_o of Gwangju station, from the Table 5, 100/PARMA(1,1)/SVM-NNM training pattern produced the statistics results with CC value of 0.981, RMSE value of 7.300 mm and R^2 value of 0.962. 500/PARMA(1,1)/SVM-NNM training pattern produced the statistics results with CC value of 0.974, RMSE value of 8.803 mm and R^2 value of 0.944. 1000/PARMA(1,1)/SVM-NNM training pattern produced the statistics results with CC value of 0.976, RMSE value of 9.448 mm and R^2 value of 0.936, respectively. In FAO-56 PM ET_o of Haenam station, from the Table 5, 100/PARMA(1,1)/ SVM-NNM training pattern produced the statistics results with CC value of 0.971, RMSE value of 9.007 mm and R^2 value of 0.939. 500/PARMA(1,1)/SVM-NNM training pattern produced the statistics results with CC value of 0.970, RMSE value of 8.882 mm and R^2 value of 0.941. 1000/PARMA(1,1)/SVM-NNM training pattern produced the statistics results with CC value of 0.972, RMSE value of 8.581 mm and R^2 value of 0.945, respectively. From the above results, the statistics results of the testing performance for 100/PARMA(1,1)/SVM-NNM training pattern were better than those of the testing performances for 500/PARMA(1,1)/SVM-NNM and 1000/PARMA(1,1)/SVM-NNM training patterns for FAO-56 PM ET_o of Gwangju station. And, the statistics results of the testing performance for 1000/PARMA(1,1)/SVM-NNM training pattern were better than those of the testing performances for 100/PARMA(1,1)/SVM-NNM and 500/PARMA(1,1)/SVM-NNM training patterns for FAO-56 PM ET_o of Haenam station, respectively. Kim (2011) suggested, however, that the statistics results of testing performance for 1000/ PARMA(1,1)/GRNNM-GA training pattern were better than those of testing performances for 100/PARMA(1,1)/GRNNM-GA and 500/PARMA(1,1)/GRNNM-GA training patterns for the modeling of PE and ET_r, South Korea. The continuous research will be needed to establish the neural networks models available for the optimal training patterns and modeling of PE and FAO-56 PM ET_o.

From the statistics results of the testing performances for PE and FAO-56 PM ET_o, the statistics results of FAO-56 PM ET_o were better than those of PE. PE is the observed data as a reason and represents the natural phenomenon including strong nonlinear patterns and various uncertainties, whereas ET_0 is calculated by FAO-56 PM equation with the constant operation processes. In FAO-56 PM ET_0, furthermore, the strong nonlinear patterns of the natural phenomena are transformed into linear patterns including the constant uncertainty. The author can consider that the modeling of FAO-56 PM ET_0 has significantly less uncertainty compared to that of PE. Kim (2011) suggested the similar results for the modeling of PE and ET_r using the neural networks models. He suggested that the statistics results of ET_r were better than those of PE for the modeling of PE and ET_r using GRNNM-BP and GRNNM-GA, South Korea. Fig. 5(a)-(f) show comparison plots of observed and calculated PE/FAO-56 PM ET_0 for the testing performances of 3 training patterns for Gwangju station, respectively. Fig. 6(a)-(f) show comparison plots of observed and calculated PE/FAO-56 PM ET_0 for the testing performances of 3 training patterns for Haenam station, respectively.

Station	Statistics Indexes	100/PARMA(1,1)/ SVM-NNM	500/PARMA(1,1)/ SVM-NNM	1000/PARMA(1,1) SVM-NNM
Gwangju	CC	0.981	0.974	0.976
	RMSE (mm)	7.300	8.803	9.448
	R^2	0.962	0.944	0.936
Haenam	CC	0.971	0.970	0.972
	RMSE (mm)	9.007	8.882	8.581
	R^2	0.939	0.941	0.945

Table 5. Statistics results of the testing performances (FAO-56 PM ET_0)

6.5 Homogeneity evaluation

FAO-56 PM ET_0 equation, which has been unanimously accepted by the FAO consultation members for ET_0 calculation (Allen et al., 1998), was used to calculate ET_0 since there are no observed data for ET_0 using a lysimeter, South Korea. Even if FAO-56 PM ET_0 is not observed data, the reliability for the calculated FAO-56 PM ET_0 is adequate and proper. Homogeneity evaluation was performed to compare the observed PE/FAO-56 PM ET_0 with the calculated PE/FAO-56 PM ET_0 for the results of the test performances, respectively. In this study, homogeneity evaluation consisted of the One-way ANOVA and the Mann-Whitney *U* test, respectively.

6.5.1 The One-way ANOVA

The One-way ANOVA is a class of statistical analysis that is widely used because it encourages systematic decision making for the underlying problems that involve considerable uncertainty. It enables inferences to be made in such a way that sample data can be combined with statistical theory. It supposedly removes the effects of the biases of the individual, which leads to more rational and accurate decision making. The One-way ANOVA is the formal procedure for using statistical concepts and measures in performing decision making. The following six steps can be used to make statistical analysis of the One-way ANOVA on the means and variances: 1) Formulation of hypotheses 2) Define the test statistic and its distribution 3) Specify the level of significance 4) Collect data and compute test statistic 5) Determine the critical value of the test statistic 6) Make a decision (McCuen, 1993; Salas et al., 2001; Ayyub and McCuen, 2003).

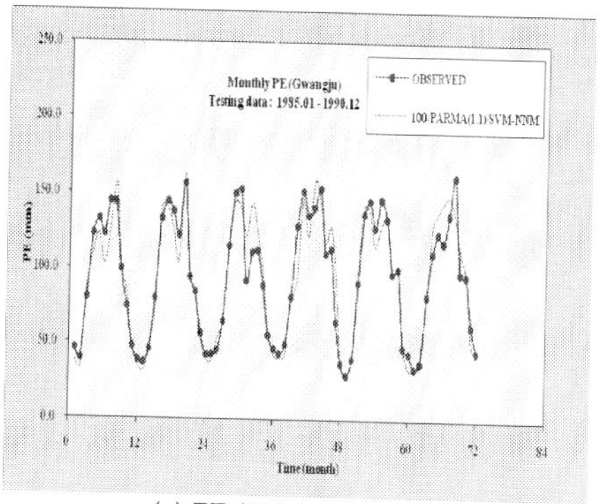

(a) PE (100 year)

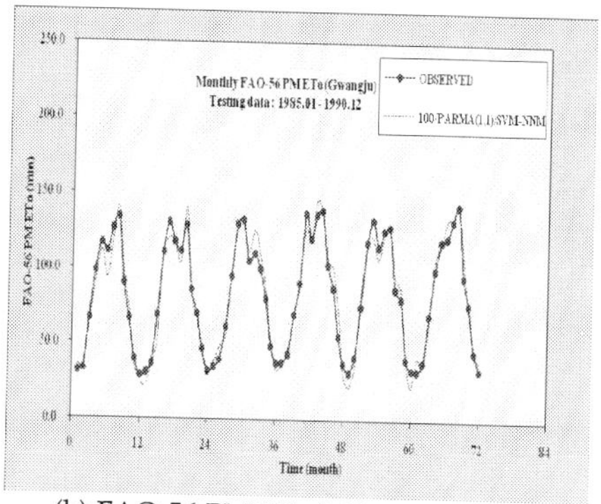

(b) FAO-56 PM ET_o (100 year)

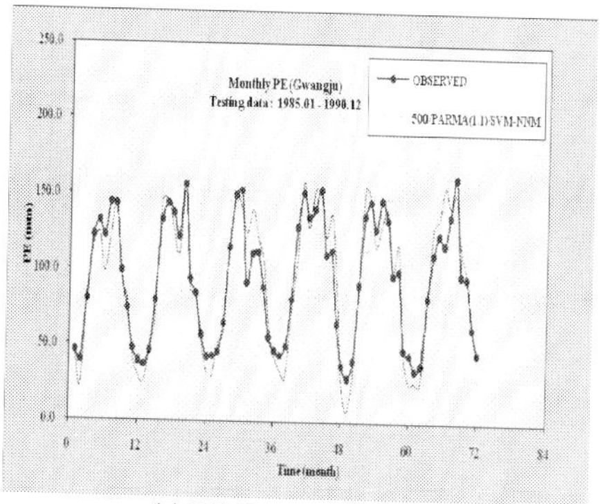

(c) PE (500 year)

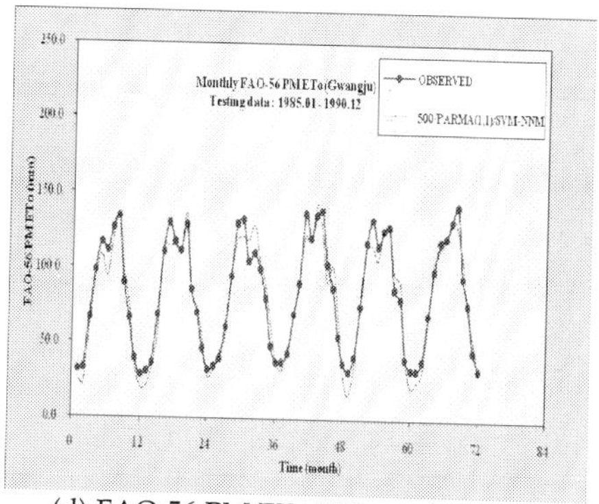

(d) FAO-56 PM ET_o (500 year)

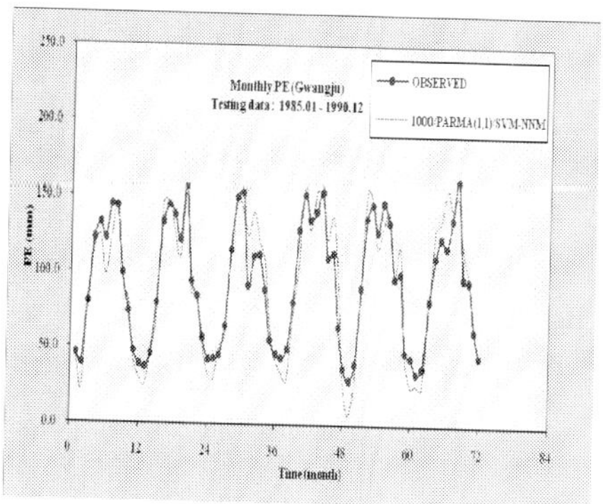

(e) PE (1000 year)

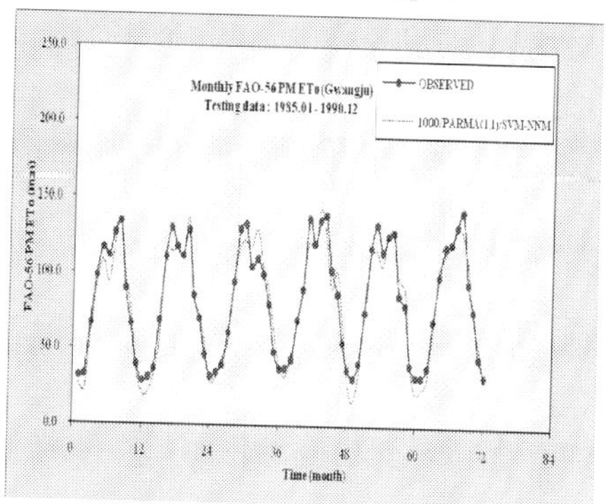

(f) FAO-56 PM ET_o (1000 year)

Fig. 5. Comparison plots of observed and calculated PE and FAO-56 PM ET_o (Gwangju)

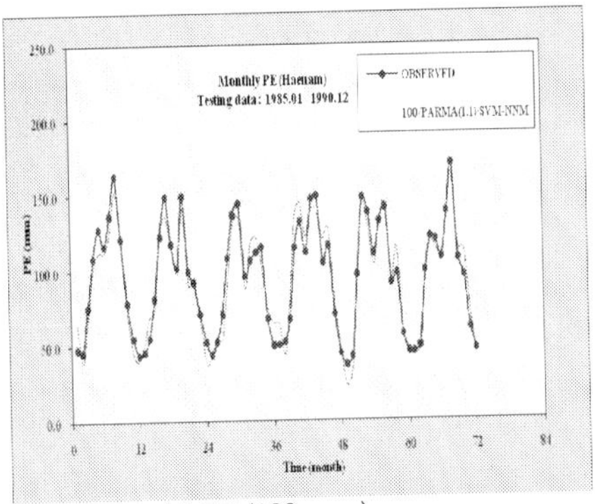

(a) PE (100 year)

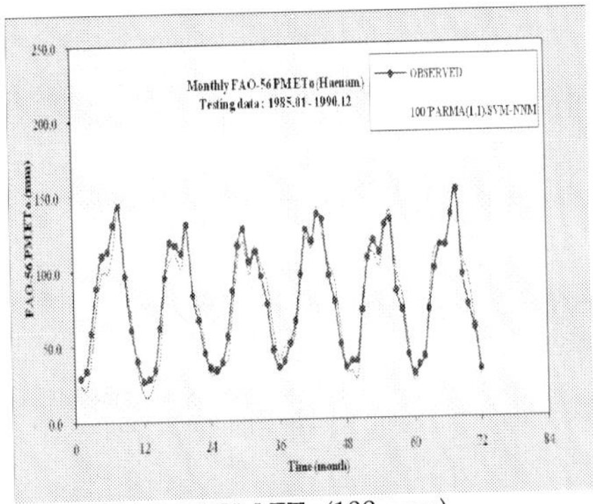

(b) FAO-56 PM ET_o (100 year)

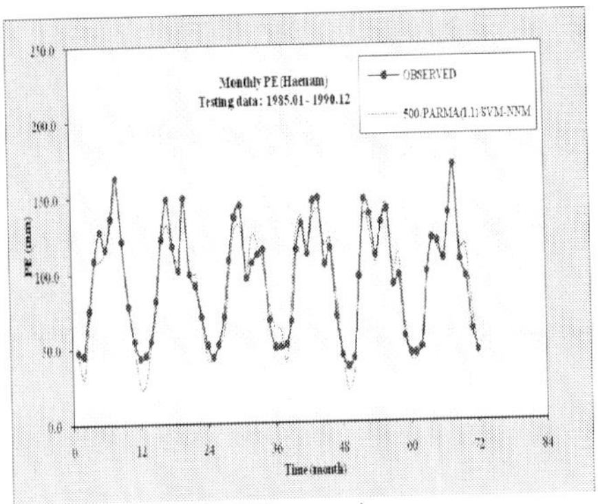

(c) PE (500 year)

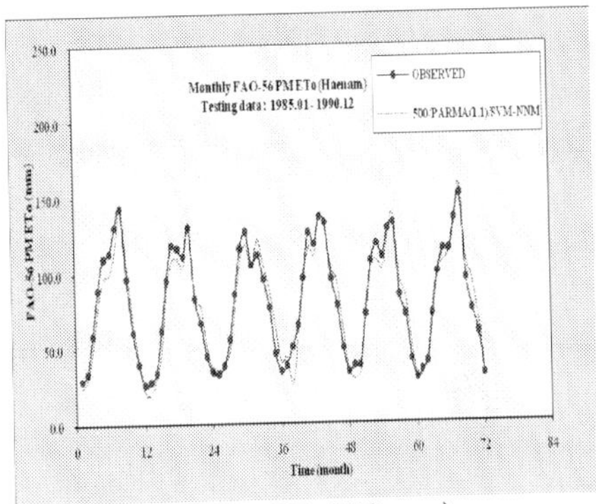

(d) FAO-56 PM ET_o (500 year)

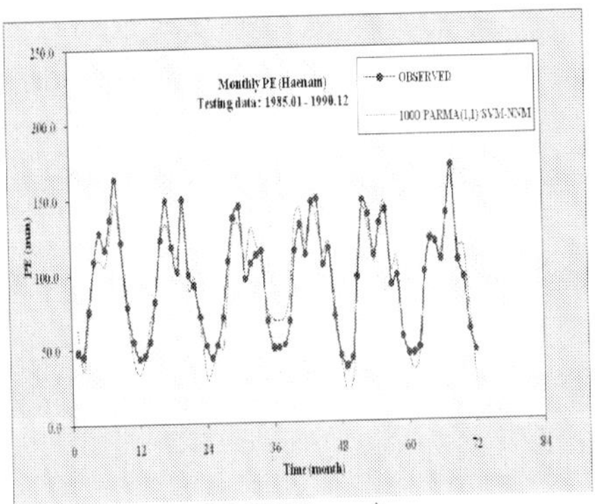

(e) PE (1000 year)

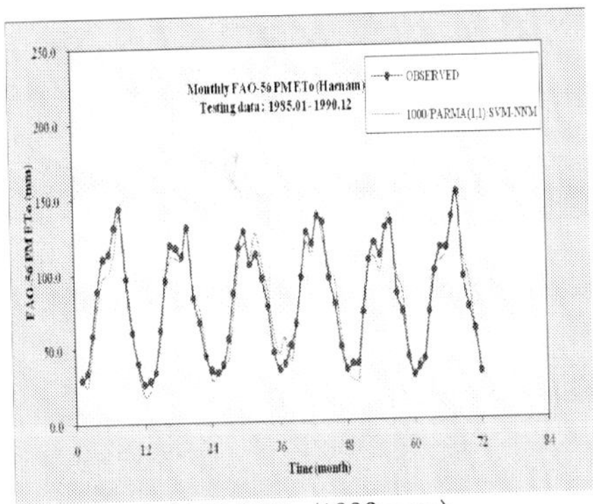

(f) FAO-56 PM ET_o (1000 year)

Fig. 6. Comparison plots of observed and calculated PE and FAO-56 PM ET_o (Haenam)

The One-way ANOVA on the means was performed and computed t statistic using two-sample t test between the observed PE/FAO-56 PM ET_o and the calculated PE/FAO-56 PM ET_o, respectively. The critical value of t statistic was computed for the level of significance 5 percent (5%) and 1 percent (1%). If the computed value of t statistic is greater than the critical value of t statistic, the null hypothesis, which is the means are equal, should be rejected and the alternative hypothesis should be accepted. The One-way ANOVA on the variances was performed and computed F statistic using two-sample F test between the observed PE/FAO-56 PM ET_o and the calculated PE/FAO-56 PM ET_o, respectively. The critical value of F statistic was computed for the level of significance 5 percent (5%) and 1 percent (1%). If the computed value of F statistic is greater than the critical value of F statistic, the null hypothesis, which is the population variances are equal, should be rejected and the alternative hypothesis should be accepted.

Table 6 shows the results of the One-way ANOVA on the means of PE. The critical value of t statistic was computed as $t_{0.05}$ =1.981 and $t_{0.01}$=2.620 for the level of significance 5 percent (5%) and 1 percent (1%), respectively. The computed values of t statistic with 0.045 for 100/PARMA/SVM-NNM training pattern, 0.111 for 500/PARMA/SVM-NNM training pattern, 0.390 for 1000/PARMA/SVM-NNM training pattern were not significant for PE of Gwangju station. So, the null hypothesis, which is the means are equal, was accepted for PE of Gwangju station. Furthermore, the computed values of t statistic with 0.145 for 100/PARMA/SVM-NNM training pattern, 0.169 for 500/PARMA/SVM-NNM training pattern, 0.103 for 1000/PARMA/SVM-NNM training pattern were not significant for PE of Haenam station. So, the null hypothesis, which is the means are equal, was accepted for PE of Haenam station.

Station	Training Pattern	Level of Significance	Two-sample t test		
			Critical t Statistic	Computed t Statistic	Null Hypothesis
Gwangju	100/PARMA(1,1)/SVM-NNM	0.05/0.01	1.981/2.620	0.045	Accept/Accept
	500/PARMA(1,1)/SVM-NNM	0.05/0.01	1.981/2.620	0.111	Accept/Accept
	1000/PARMA(1,1)/SVM-NNM	0.05/0.01	1.981/2.620	0.390	Accept/Accept
Haenam	100/PARMA(1,1)/SVM-NNM	0.05/0.01	1.981/2.620	0.145	Accept/Accept
	500/PARMA(1,1)/SVM-NNM	0.05/0.01	1.981/2.620	0.169	Accept/Accept
	1000/PARMA(1,1)/SVM-NNM	0.05/0.01	1.981/2.620	0.103	Accept/Accept

Table 6. Results of the One-way ANOVA on the means of PE

Table 7 shows the results of the One-way ANOVA on the means of FAO-56 PM ET_o. The critical value of t statistic was computed as $t_{0.05}$ =1.981 and $t_{0.01}$=2.620 for the level of significance 5 percent (5%) and 1 percent (1%), respectively. The computed values of t statistic with 0.040 for 100/PARMA/SVM-NNM training pattern, 0.283 for 500/PARMA /SVM-NNM training pattern, 0.483 for 1000/PARMA/SVM-NNM training pattern were not significant for FAO-56 PM ET_o of Gwangju station. So, the null hypothesis, which is the means are equal, was accepted for FAO-56 PM ET_o of Gwangju station. Furthermore, the computed values of t statistic with 0.231 for 100/PARMA/SVM-NNM training pattern, 0.071 for 500/PARMA/SVM-NNM training pattern, 0.018 for 1000/PARMA/SVM-NNM training pattern were not significant for FAO-56 PM ET_o of Haenam station. So, the null hypothesis, which is the means are equal, was accepted for FAO-56 PM ET_o of Haenam station.

Station	Training Pattern	Level of Significance	Two-sample *t* test		
			Critical *t* Statistic	Computed *t* Statistic	Null Hypothesis
Gwangju	100/PARMA(1,1)/SVM-NNM	0.05/0.01	1.981/2.620	0.040	Accept/Accept
	500/PARMA(1,1)/SVM-NNM	0.05/0.01	1.981/2.620	0.283	Accept/Accept
	1000/PARMA(1,1)/SVM-NNM	0.05/0.01	1.981/2.620	0.483	Accept/Accept
Haenam	100/PARMA(1,1)/SVM-NNM	0.05/0.01	1.981/2.620	0.231	Accept/Accept
	500/PARMA(1,1)/SVM-NNM	0.05/0.01	1.981/2.620	0.071	Accept/Accept
	1000/PARMA(1,1)/SVM-NNM	0.05/0.01	1.981/2.620	0.018	Accept/Accept

Table 7. Results of the One-way ANOVA on the means of FAO-56 PM ET_o

Table 8 shows the results of the One-way ANOVA on the variances of PE. The critical value of *F* statistic was computed as $F_{0.05}$ =1.981 and $F_{0.01}$=2.620 for the level of significance 5 percent (5%) and 1 percent (1%), respectively. The computed values of *F* statistic with 1.040 for 100/PARMA/SVM-NNM training pattern, 1.249 for 500/PARMA/SVM-NNM training pattern, 1.343 for 1000/PARMA/SVM-NNM training pattern were not significant for PE of Gwangju station. So, the null hypothesis, which is the variances are equal, was accepted for PE of Gwangju station. Furthermore, the computed values of *F* statistic with 1.033 for 100/PARMA/SVM-NNM training pattern, 1.030 for 500/PARMA/SVM-NNM training pattern, 1.036 for 1000/PARMA/SVM-NNM training pattern were not significant for PE of Haenam station. So, the null hypothesis, which is the variances are equal, was accepted for PE of Haenam station.

Station	Training Pattern	Level of Significance	Two-sample *F* test		
			Critical *F* Statistic	Computed *F* Statistic	Null Hypothesis
Gwangju	100/PARMA(1,1)/SVM-NNM	0.05/0.01	1.981/2.620	1.040	Accept/Accept
	500/PARMA(1,1)/SVM-NNM	0.05/0.01	1.981/2.620	1.249	Accept/Accept
	1000/PARMA(1,1)/SVM-NNM	0.05/0.01	1.981/2.620	1.343	Accept/Accept
Haenam	100/PARMA(1,1)/SVM-NNM	0.05/0.01	1.981/2.620	1.033	Accept/Accept
	500/PARMA(1,1)/SVM-NNM	0.05/0.01	1.981/2.620	1.030	Accept/Accept
	1000/PARMA(1,1)/SVM-NNM	0.05/0.01	1.981/2.620	1.036	Accept/Accept

Table 8. Results of the One-way ANOVA on the variances of PE

Table 9 shows the results of the One-way ANOVA on the variances of FAO-56 PM ET_o. The critical value of *F* statistic was computed as $F_{0.05}$ =1.981 and $F_{0.01}$=2.620 for the level of significance 5 percent (5%) and 1 percent (1%), respectively. The computed values of *F* statistic with 1.055 for 100/PARMA/SVM-NNM training pattern, 1.045 for 500/PARMA/ SVM-NNM training pattern, 1.154 for 1000/PARMA/SVM-NNM training pattern were not significant for FAO-56 PM ET_o of Gwangju station. So, the null hypothesis, which is the variances are equal, was accepted for FAO-56 PM ET_o of Gwangju station. Furthermore, the computed values of *F* statistic with 1.033 for 100/PARMA/SVM-NNM training pattern, 1.021 for 500/PARMA/SVM-NNM training pattern, 1.031 for 1000/PARMA/SVM-NNM training pattern were not significant for FAO-56 PM ET_o of Haenam station. So, the null hypothesis, which is the variances are equal, was accepted for FAO-56 PM ET_o of Haenam station.

Station	Training Pattern	Level of Significance	Two-sample F test		
			Critical F Statistic	Computed F Statistic	Null Hypothesis
Gwangju	100/PARMA(1,1)/SVM-NNM	0.05/0.01	1.981/2.620	1.055	Accept/Accept
	500/PARMA(1,1)/SVM-NNM	0.05/0.01	1.981/2.620	1.045	Accept/Accept
	1000/PARMA(1,1)/SVM-NNM	0.05/0.01	1.981/2.620	1.154	Accept/Accept
Haenam	100/PARMA(1,1)/SVM-NNM	0.05/0.01	1.981/2.620	1.033	Accept/Accept
	500/PARMA(1,1)/SVM-NNM	0.05/0.01	1.981/2.620	1.021	Accept/Accept
	1000/PARMA(1,1)/SVM-NNM	0.05/0.01	1.981/2.620	1.031	Accept/Accept

Table 9. Results of the One-way ANOVA on the variances of FAO-56 PM ET_o

6.5.2 The Mann-Whitney *U* test

The Mann-Whitney U test is a nonparametric alternative to the two-sample t test for two independent samples and can be used to test whether two independent samples have been taken from the same population. It is the most powerful alternative to the two-sample t test. Therefore, when the assumptions of the two-sample t test are violated or are difficult to evaluate such as with small samples, the Mann-Whitney U test should be applied. The Mann-Whitney U test is to be used in the case of two independent samples, and the Kruskal-Wallis test is an extension of the Mann-Whitney U test for the case of more than two independent samples (McCuen, 1993; Salas et al., 2001; Ayyub and McCuen, 2003).

The Mann-Whitney U test was performed and computed z statistic between the observed PE/FAO-56 PM ET_o and the calculated PE/FAO-56 PM ET_o, respectively. The critical value of z statistic was computed for the level of significance 5 percent (5%) and 1 percent (1%). If the computed value of z statistic is greater than the critical value of z statistic, the null hypothesis, which is the two independent samples are from the same population, should be rejected and the alternative hypothesis should be accepted in this study.

Table 10 shows the results of the Mann-Whitney U test of PE. The critical value of z statistic was computed as $z_{0.05}$ =1.645 and $z_{0.01}$=2.327 for the level of significance 5 percent (5%) and 1 percent (1%), respectively. The computed values of z statistic with -0.196 for 100/PARMA/ SVM-NNM training pattern, -0.136 for 500/PARMA/SVM-NNM training pattern, -0.288 for 1000/PARMA/SVM-NNM training pattern were not significant for PE of Gwangju station. So, the null hypothesis, which is the two independent samples are from the same population, was accepted for PE of Gwangju station. Furthermore, the computed values of z statistic with -0.172 for 100/PARMA/SVM-NNM training pattern, -0.124 for 500/PARMA/SVM-NNM training pattern, -0.076 for 1000/PARMA/SVM-NNM training pattern were not significant for PE of Haenam station. So, the null hypothesis, which is the two independent samples are from the same population, was accepted for PE of Haenam station.

Table 11 shows the results of the Mann-Whitney U test of FAO-56 PM ET_o. The critical value of z statistic was computed as $z_{0.05}$ =1.645 and $z_{0.01}$=2.327 for the level of significance 5 percent (5%) and 1 percent (1%), respectively. The computed values of z statistic with -0.056 for 100/PARMA/SVM-NNM training pattern, -0.515 for 500/PARMA/SVM-NNM training pattern, -0.711 for 1000/PARMA/SVM-NNM training pattern were not significant for FAO-56 PM ET_o of Gwangju station. So, the null hypothesis, which is the two independent samples are from the same population, was accepted for FAO-56 PM ET_o of Gwangju station. Furthermore, the computed values of z statistic with -0.380 for 100/PARMA/SVM-NNM training pattern, -0.212 for 500/PARMA/SVM-NNM training pattern, -0.176 for 1000/PARMA/SVM-NNM training pattern were not significant for FAO-56 PM ET_o of Haenam station. So, the null hypothesis, which is the two independent samples are from the same population, was accepted for FAO-56 PM ET_o of Haenam station.

Station	Training Pattern	Level of Significance	Mann-Whitney U test		
			Critical z Statistic	Computed z Statistic	Null Hypothesis
Gwangju	100/PARMA(1,1)/SVM-NNM	0.05/0.01	1.645/2.327	-0.196	Accept/Accept
	500/PARMA(1,1)/SVM-NNM	0.05/0.01	1.645/2.327	-0.136	Accept/Accept
	1000/PARMA(1,1)/SVM-NNM	0.05/0.01	1.645/2.327	-0.288	Accept/Accept
Haenam	100/PARMA(1,1)/SVM-NNM	0.05/0.01	1.645/2.327	-0.172	Accept/Accept
	500/PARMA(1,1)/SVM-NNM	0.05/0.01	1.645/2.327	-0.124	Accept/Accept
	1000/PARMA(1,1)/SVM-NNM	0.05/0.01	1.645/2.327	-0.076	Accept/Accept

Table 10. Results of the Mann-Whitney U test of PE

Station	Training Pattern	Level of Significance	Mann-Whitney U test		
			Critical z Statistic	Computed z Statistic	Null Hypothesis
Gwangju	100/PARMA(1,1)/SVM-NNM	0.05/0.01	1.645/2.327	-0.056	Accept/Accept
	500/PARMA(1,1)/SVM-NNM	0.05/0.01	1.645/2.327	-0.515	Accept/Accept
	1000/PARMA(1,1)/SVM-NNM	0.05/0.01	1.645/2.327	-0.711	Accept/Accept
Haenam	100/PARMA(1,1)/SVM-NNM	0.05/0.01	1.645/2.327	-0.380	Accept/Accept
	500/PARMA(1,1)/SVM-NNM	0.05/0.01	1.645/2.327	-0.212	Accept/Accept
	1000/PARMA(1,1)/SVM-NNM	0.05/0.01	1.645/2.327	-0.176	Accept/Accept

Table 11. Results of the Mann-Whitney U test of FAO-56 PM ET_o

7. Construction of the Bivariate Linear Regression Analysis Model

The bivariate linear regression analysis model (BLRAM) was adopted to calculate FAO-56 PM ET_o simply using the observed PE and compare the observed PE and the calculated FAO-56 PM ET_o. The BLRAM is a conventional and universal model, which can calculate FAO-56 PM ET_o simply using the observed PE for Gwangju and Haenam stations, respectively. The BLRAM consists of two variables; PE as the independent variable X_t, and FAO-56 PM ET_o as the dependent variable Y_t. The mathematical expression can be written as the following equation (12) (McCuen, 1993; Salas et al., 2001; Ayyub and McCuen, 2003).

$$Y_t = b_0 + b_1 \cdot X_t \tag{12}$$

where b_1 = the slope coefficient, which is also known as the regression coefficient because it is calculated by the result of a regression analysis. Using the BLRAM, the correlation relationship was investigated between the observed PE and FAO-56 PM ET_o for 3 training patterns. A very good relationship was found with the BLRAM, which could calculate FAO-56 PM ET_o in this study. The results of the BLRAM were the same for 3 training patterns. Therefore, it can be considered that the observed PE and FAO-56 PM ET_o for 3 training patterns are homogeneous groups. It can be inferred from the homogeneity evaluation of the previous chapter.

Table 12 shows the BLRAM, goodness-of-fit test, and regression coefficient between the observed PE and FAO-56 PM ET_o for 3 training patterns of Gwangju and Haenam stations, respectively. From the Table 12, for Gwangju station, the regression coefficient of the BLRAM indicates that FAO-56 PM ET_o increases 0.9060 mm as each 1 mm increase in PE. R^2 = 0.966 indicates that the total variance of FAO-56 PM ET_o corresponds to 96.6%. The ratio of S(e)/S(y) = 0.187 suggests a very good level of accuracy. In addition, the standard error of estimate (SEE) decreases from 37.471 mm (S(y)) to 7.007 mm (S(e)). where S(y) = the standard deviation of FAO-56 PM ET_o; and S(e) = the standard error of FAO-56 PM ET_o using the equation (12). The overall deviations are nearly zero, and this tendency always occurs for the BLRAM. The standard error ratios of the regression coefficient (b_1) and the intercept (b_0) are 0.023 and 0.911, which indicates that the regression coefficient is relatively more accurate than the intercept. For Haenam station, the regression coefficient of the BLRAM indicates that FAO-56 PM ET_o increases 0.9574 mm as each 1 mm increase in PE. R^2 = 0.919 indicates that the total variance of FAO-56 PM ET_o corresponds to 91.9%. The ratio of S(e)/S(y) = 0.287 suggests a very good level of accuracy. In addition, the standard error of estimate (SEE) decreases from 36.794 mm (S(y)) to 10.560 mm (S(e)). where S(y) = the standard deviation of FAO-56 PM ET_o; and S(e) = the standard error of FAO-56 PM ET_o using the equation (12). The overall deviations are nearly zero, and this tendency always occurs for the BLRAM. The standard error ratios of the regression coefficient (b_1) and the intercept (b_0) are 0.035 and 0.345, which indicates that the regression coefficient is relatively more accurate than the intercept. If a large and negative intercept exists, it can create some problems for forecasting or modeling (McCuen, 1993). Fig. 7(a)-(b) show comparison plots of the observed PE and FAO-56 PM ET_o for 1000/PARMA(1,1)/SVM-NNM training pattern of Gwangju and Haenam station, respectively.

	BLRAM (3 Training Patterns)	Goodness-of-fit test			Regression coefficient analysis	
		S(e) (mm)	S(e)/S(y)	R^2	$Se(b_1)/b_1$	$Se(b_0)/b_0$
Gwangju	$ET_o = 0.9060\ PE - 2.2901$	7.007	0.187	0.966	0.023	0.911
Haenam	$ET_o = 0.9574\ PE - 9.9968$	10.560	0.287	0.919	0.035	0.345

Table 12. Regression analysis between the observed PE and FAO-56 PM ET_o

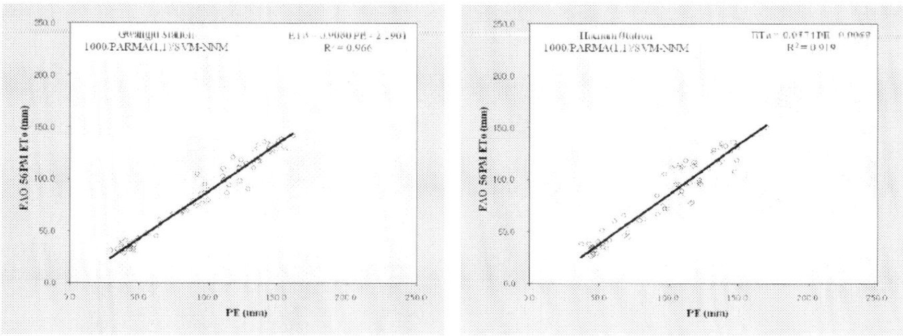

(a) Gwangju (b) Haenam

Fig. 7. Comparison plots of the observed PE and FAO-56 PM ET_o

8. Conclusions

The hybrid method was developed for the modeling of the monthly PE and FAO-56 PM ET_0 simultaneously. The author determined in advance 4 kinds of Univariate Seasonal PARMA(p,q) models including PARMA(1,1), PARMA(1,2), PARMA (2,1), and PARMA(2,2), which are the low-order models and contain the seasonal properties. As a result, the author selected Univariate Seasonal PARMA(1,1) model, which show the best statistical properties and is simple in parameter estimation. The data which were generated by Univariate Seasonal PARMA(1,1) model consisted of 100 years (Short-term), 500 years (Mid-term), and 1000 years (Long-term), respectively. The following conclusions can be drawn from this study.

[1] The statistics results of the testing performance for 100/PARMA(1,1)/SVM-NNM training pattern were better than those of the testing performances for 500/PARMA(1,1) /SVM-NNM and 1000/PARMA(1,1)/SVM-NNM training patterns for PE of Gwangju station. And, the statistics results of the testing performance for 500/PARMA(1,1)/SVM-NNM training pattern were better than those of the testing performances for 100 /PARMA(1,1)/SVM-NNM and 1000/ PARMA(1,1)/SVM-NNM training patterns for PE of Haenam station, respectively

[2] The statistics results of the testing performance for 100/PARMA(1,1)/SVM-NNM training pattern were better than those of the testing performances for 500/PARMA(1,1) /SVM-NNM and 1000/PARMA(1,1)/SVM-NNM training patterns for FAO-56 PM ET_0 of Gwangju station. And, the statistics results of the testing performance for 1000 /PARMA(1,1)/SVM-NNM training pattern were better than those of the testing performances for 100/PARMA(1,1)/SVM-NNM and 500/PARMA(1,1)/SVM-NNM training patterns for FAO-56 PM ET_0 of Haenam station, respectively

[3] Homogeneity evaluation consisted of the One-way ANOVA and the Mann-Whitney *U* test. The null hypothesis, which is the means are equal, was accepted using the One-way ANOVA on the means for PE and FAO-56 PM ET_0 of Gwangju and Haenam stations, respectively. And, the null hypothesis, which is the variances are equal, was accepted using the One-way ANOVA on the variances for PE and FAO-56 PM ET_0 of Gwangju and Haenam stations, respectively. The null hypothesis, which is the two independent samples are from the same population, was accepted using the Mann-Whitney *U* test for PE and FAO-56 PM ET_0 of Gwangju and Haenam stations, respectively.

[4] The BLRAM was adopted to calculate FAO-56 PM ET_0 simply using the observed PE and compare the observed PE and the calculated FAO-56 PM ET_0 of Gwangju and Haenam stations, respectively. A very good relationship was found with the BLRAM, which could calculate FAO-56 PM ET_0.

As PE and FAO-56 PM ET_0 are relatively important for the design of irrigation facilities and agricultural reservoirs, the spread of an automatic measuring system for PE and FAO-56 PM ET_0 is important and urgent to ensure the reliable and accurate data from the measurements of PE and FAO-56 PM ET_0. Furthermore, the continuous research will be needed to establish the neural networks models available for the optimal training patterns and modeling of PE and FAO-56 PM ET_0.

9. References

Allen, R.G.; Jensen, M.E.; Wright, J.L. & Burman, R.D. (1989). Operational estimates of reference evapotranspiration. *Agronomy Journal*, Vol. 81, No. 4, pp. 650-662, ISSN: 0002-1962.

Allen, R.G.; Pereira, L.S.; Raes, D. & Smith, M. (1998). *Crop evapotranspiration. Guidelines for computing crop water requirement,* Irrigation and Drainage Paper No. 56, FAO, Rome, Italy.

ASCE Task Committee on Definition of Criteria for Evaluation of Watershed Models. (1993). Criteria for evaluation of watershed models. *Journal of Irrigation and Drainage Engineering, ASCE,* Vol. 119, No. 3, pp. 429-442, ISSN: 0733-9437.

Ayyub, B.M. & McCuen, R.H. (2003). *Probability, Statistics, and Reliability for Engineers and Scientists 2nd Edition,* Taylor & Francis, ISBN: 1584882867, Boca Raton, FL, USA.

Bishop, C.M. (1994). Neural networks and their applications. *Review of Scientific Instruments,* Vol. 65, pp. 1803-1832, ISSN: 0034-6748.

Bruton, J.M.; McClendon, R.W. & Hoogenboom, G. (2000). Estimating daily pan evaporation with artificial neural networks. *Transactions of the American Society Agricultural Engineers, ASAE,* Vol. 43, No. 2, pp. 491-496, ISSN:0001-2351.

Chang, F.J.; Chang, L.C.; Kao, H.S. & Wu, G.R. (2010). Assessing the effort of meteorological variables for evaporation estimation by self-organizing map neural network. *Journal of Hydrology,* Vol. 384, pp. 118-129, ISSN: 0022-1694.

Choudhury, B.J. (1999). Evaluation of empirical equation for annual evaporation using field observations and results from a biophysical model. *Journal of Hydrology,* Vol. 216, pp. 99-110, ISSN: 0022-1694.

Cover, T.M. (1965). Geometrical and statistical properties of systems of linear inequalities with applications in pattern recognition. *IEEE Transactions on Electronic Computers EC-14,* pp. 326-334.

Dibike, Y.B.; Velickov, S.; Solomatine, D. & Abbott, M.B. (2001). Model induction with support vector machines: introductions and applications. *Journal of Computing in Civil Engineering, ASCE,* Vol. 15, No. 3, pp. 208-216, ISSN: 0887-3801.

Doorenbos, J. & Pruitt, W.O. (1977). *Guidelines for predicting crop water requirement,* Irrigation and Drainage Paper No. 24 2nd Edition, FAO, Rome, Italy.

Finch, J.W. (2001). A comparison between measured and modeled open water evaporation from a reservoir in south-east England. *Hydrological Processes,* Vol. 15, pp. 2771-2778, ISSN: 0885-6087.

Guven, A. & Kisi, O. (2011). Daily pan evaporation modeling using linear genetic programming technique. *Irrigation Science,* Vol. 29, No. 2, pp. 135-145, ISSN: 0342-7188.

Haykin, S. (2009). *Neural networks and learning machines 3rd Edition,* Prentice Hall, ISBN: 0131471392, NJ, USA.

Hirsch R.M. (1979). Synthetic hydrology and water supply reliability. *Water Resources Research,* Vol. 15, pp.1603-1615, ISSN: 0043-1397.

Hush, D.R. & Horne, B.G. (1993). Progress in supervised neural network : What's new since Lippmann ?. *IEEE Signal Processing Magazine,* Vol. 10, pp. 8-39, ISSN: 1053-5888.

Jain, S.K.; Nayak, P.C. & Sudheer, K.P. (2008). Models for estimating evapotranspiration using artificial neural networks, and their physical interpretation. *Hydrological Processes,* Vol. 22, pp. 2225-2234, ISSN: 0885-6087.

Jensen, M.E.; Burman, R.D. & Allen, R.G. (1990). *Evapotranspiration and irrigation water requirements,* ASCE Manual and Report on Engineering Practice No. 70, ASCE, NY, USA.

Keskin, M.E. & Terzi, O. (2006). Artificial neural networks models of daily pan evaporation. *Journal of Hydrologic Engineering, ASCE,* Vol. 11, No. 1, pp. 65-70, ISSN: 1084-0699.

Khadam, I.M. & Kaluarachchi, J.J. (2004). Use of soft information to describe the relative uncertainty of calibration data in hydrologic models. *Water Resources Research,* Vol. 40, No. 11, W11505, ISSN: 0043-1397.

Kim, S. (2004). Neural Networks Model and Embedded Stochastic Processes for Hydrological Analysis in South Korea. *KSCE Journal of Civil Engineering,* Vol. 8, No. 1, pp. 141-148, ISSN: 1226-7988.

Kim, S. (2011). Nonlinear hydrologic modeling using the stochastic and neural networks approach. *Disaster Advances,* Vol. 4, No. 1, pp. 53-63, ISSN: 0974-262X.

Kim, S.; Kim, J.H. & Park, K.B. (2009). Neural networks models for the flood forecasting and disaster prevention system in the small catchment. *Disaster Advances,* Vol. 2, No. 3, pp. 51-63, ISSN: 0974-262X.

Kim, S. & Kim, H.S. (2008). Neural networks and genetic algorithm approach for nonlinear evaporation and evapotranspiration modeling. *Journal of Hydrology,* Vol. 351, pp. 299-317, ISSN: 0022-1694.

Kisi, O. (2006). Generalized regression neural networks for evapotranspiration modeling. *Hydrological Sciences Journal,* Vol. 51, No. 6, pp. 1092-1105, ISSN: 0262-6667.

Kisi, O. (2007). Evapotranspiration modeling from climatic data using a neural computing technique. *Hydrological Processes,* Vol. 21, pp. 1925-1934, ISSN: 0885-6087.

Kisi, O. (2009). Modeling monthly evaporation using two different neural computing technique. *Irrigation Science,* Vol. 27, No. 5, pp. 417-430, ISSN: 0342-7188.

Kisi, O. & Ozturk, O. (2007). Adaptive neurofuzzy computing technique for evapotranspiration estimation. *Journal of Irrigation and Drainage Engineering, ASCE,* Vol. 133, No. 4, pp. 368-379, ISSN: 0733-9437.

Kumar, M.; Bandyopadhyay, A.; Raghuwanshi, N.S. & Singh, R. (2008). Comparative study of conventional and artificial neural network-based ET_0 estimation models. *Irrigation Science,* Vol. 26, No. 6, pp. 531-545, ISSN: 0342-7188.

Kumar, M.; Raghuwanshi, N.S. & Singh, R. (2009). Development and validation of GANN model for evapotranspiration estimation. *Journal of Hydrologic Engineering, ASCE,* Vol. 14, No. 2, pp. 131-140, ISSN: 1084-0699.

Kumar, M.; Raghuwanshi, N.S.; Singh, R.; Wallender, W.W. & Pruitt, W.O. (2002). Estimating evapotranspiration using artificial neural network. *Journal of Irrigation and Drainage Engineering, ASCE,* Vol. 128, No. 4, pp. 224-233, ISSN: 0733-9437.

Landeras, G.; Ortiz-Barredo, A. & Lopez, J.J. (2008). Comparison of artificial neural network models and empirical and semi-empirical equations for daily reference evapotranspiration estimation in the Basque country (Northern Spain). *Agricultural Water Management,* Vol. 95, No. 5, pp. 553-565, ISSN: 0378-3774.

McCuen, R.H. (1993). *Microcomputer applications in statistical hydrology,* Prentice Hall, ISBN: 0135852900, Englewood Cliffs, NJ, USA.

McKenzie, R.S. & Craig, A.R. (2001). Evaluation of river losses from the Orange River using hydraulic modeling. *Journal of Hydrology,* Vol. 241, pp. 62-69, ISSN: 0022-1694.

Mishra, A.K.; Desai, V.R. & Singh, V.P. (2007). Drought forecasting using a hybrid stochastic and neural network model. *Journal of Hydrologic Engineering, ASCE,* Vol. 12, No. 6, pp. 626-638, ISSN: 1084-0699.

Monteith, J.L. (1965). The state and movement of water in living organism, *Proceedings of Evaporation and Environment,* pp. 205-234, Swansea, Cambridge University Press, NY, USA.

Nash, J.E. & Sutcliffe, J.V. (1970). River flow forecasting through conceptual models, Part 1 - A discussion of principles. *Journal of Hydrology*, Vol. 10, No. 3, pp. 282-290, ISSN: 0022-1694.

Penman, H.L. (1948). Natural evaporation from open water, bare soil and grass, *Proceedings of the Royal Society of London*, Vol. 193, pp. 120-146, London, England.

Principe, J.C.; Euliano, N.R. & Lefebvre, W.C. (2000). *Neural and adaptive systems: fundamentals through simulation*, Wiley, John & Sons, ISBN: 0471351679, NY, USA.

Rosenberry, D.O.; Winter, T.C.; Buso, D.C. & Likens, G.E. (2007). Comparison of 15 evaporation methods applied to a small mountain lake in the northeastern USA. *Journal of Hydrology*, Vol. 340, pp. 149-166, ISSN: 0022-1694.

Salas, J.D. (1993). Analysis and modeling of hydrologic time series. In: *Handbook of Hydrology, Chapter 19*, Maidment, D.R., (Ed.), pp. 19.1-19.72, McGraw-Hill, ISBN: 0070397325, NY, USA.

Salas, J.D. & Abdelmohsen, M. (1993). Initialization for generating single site and multisite low order PARMA processes. *Water Resources Research*, Vol.29, pp.1771-1776, ISSN: 0043-1397.

Salas J.D.; Boes D.C.; & Smith, R.A. (1982). Estimation of ARMA models with seasonal parameters. *Water Resources Research*, Vol. 18, pp.1006-1010, ISSN: 0043-1397.

Salas J.D.; Delleur J.R.; Yevjevich, V & Lane, W.L. (1980). *Applied modeling of hydrologic time series*, Water Resources Publication, ISBN: 0918334373, CO, USA.

Salas, J.D.; Smith, R.A., Tabios III, G.O. & Heo, J.H. (2001). *Statistical computing techniques in water resources and environmental engineering*, Unpublished book in CE622, Colorado State University, Fort Collins, CO, USA.

Shiri, J. & Kisi, O. (2011). Application of artificial intelligence to estimate daily pan evaporation using available and estimated climatic data in the Khozestan Province (Southwestern Iran). *Journal of Irrigation and Drainage Engineering, ASCE*, Vol. 137, No. 7, pp. 412-425, ISSN: 0733-9437.

Singh, V.P. (1988). *Hydrologic system rainfall-runoff modeling. Vol. 1*, Prentice Hall, ISBN: 0134480511, NJ, USA.

Sivakumar B.; Jayawardena, A.W. & Fernando, T.M.K.G. (2002). River flow forecasting : use of phase-space reconstruction and artificial neural networks approaches. *Journal of Hydrology*, Vol. 265, pp. 225-245, ISSN: 0022-1694

Sudheer, K.P.; Gosain, A.K.; Rangan, D.M. & Saheb, S.M. (2002). Modeling evaporation using an artificial neural network algorithm. *Hydrological Processes*, Vol. 16, pp. 3189-3202, ISSN: 0885-6087.

Sudheer, K.P.; Gosain, A.K. & Ramasastri, K.S. (2003). Estimating actual evapotranspiration from limited climatic data using neural computing technique. *Journal of Irrigation and Drainage Engineering, ASCE*, Vol. 129, No. 3, pp. 214-218, ISSN: 0733-9437.

Tabari, H.; Marofi, S. & Sabziparvar, A.A. (2009). Estimation of daily pan evaporation using artificial neural network and multivariate non-linear regression. *Irrigation Science*, Vol. 28, No. 5, pp. 399-406, ISSN: 0342-7188.

Tao, P.C. & Delleur, J.W. (1976). Seasonal and nonseasonal ARMA models in hydrology. *Journal of Hydraulic Division, ASCE*, Vol. 102, pp.1591-1599, ISSN: 0733-9429.

Tokar, A.S. & Johnson, P.A. (1999). Rainfall-runoff modeling using artificial neural networks. *Journal of Hydrologic Engineering, ASCE*, Vol. 4, No. 3, pp. 232-239, ISSN: 1084-0699.

Trajkovic, S. (2005). Temperature-based approaches for estimating reference evapotranspiration. *Journal of Irrigation and Drainage Engineering, ASCE,* Vol. 131, No. 4, pp. 316-323, ISSN: 0733-9437.

Trajkovic, S; Todorovic, B. & Stankovic, M. (2003). Forecasting reference evapotranspiration by artificial neural networks. *Journal of Irrigation and Drainage Engineering, ASCE,* Vol. 129, No. 6, pp. 454-457, ISSN: 0733-9437.

Tripathi, S.; Srinivas, V.V. & Nanjundish, R.S. (2006). Downscaling of precipitation for climate change scenarios: A support vector machine approach. *Journal of Hydrology,* Vol. 330, pp. 621-640, ISSN: 0022-1694.

Vallet-Coulomb, C.; Legesse, D.; Gasse, F.; Travi, Y. & Chernet, T. (2001). Lake evaporation estimates in tropical Africa (Lake Ziway, Ethiopia). *Journal of Hydrology,* Vol. 245, pp. 1-18, ISSN: 0022-1694.

Vapnik, V.N. (1992). Principles of risk minimization for learning theory. In: *Advances in Neural Information Processing Systems Vol. 4,* Moody, Hanson & Lippmann, (Ed.), pp. 831-838, Elsevier, ISBN: 1558602224, NY, USA.

Vapnik, V.N. (2010). *The nature of statistical learning theory 2nd Edition,* Springer-Verlag, ISBN: 0387987800, NY, USA.

Zanetti, S.S.; Sousa, E.F.; Oliveira, V.P.S.; Almeida, F.T. & Bernardo, S. (2007). Estimating evapotranspiration using artificial neural network and minimum climatological data. *Journal of Irrigation and Drainage Engineering, ASCE,* Vol. 133, No. 2, pp. 83-89, ISSN: 0733-9437.

17

Modelling Evapotranspiration and the Surface Energy Budget in Alpine Catchments

Giacomo Bertoldi[1], Riccardo Rigon[2] and Ulrike Tappeiner[1,3]
[1] *Institute for Alpine Environment, EURAC Research, Bolzano.*
[2] *University of Trento.*
[3] *Institute of Ecology, University of Innsbruck.*
[1,2] *Italy*
[3] *Austria.*

1. Introduction

Accurate modelling of evapotranspiration (ET) is required to predict the effects of climate and land use changes on water resources, agriculture and ecosystems. Significant progress has been made in estimating ET at the global and regional scale. However, further efforts are needed to improve spatial accuracy and modeling capabilities in alpine regions (Brooks & Vivoni, 2008b). This chapter will point out the components of the energy budget needed to model ET, to discuss the fundamental equations and to provide an extended review of the parametrizations available in the hydrological and land surface models literature. The second part of the chapter will explore the complexity of the energy budget with special reference to mountain environments.

2. The energy budget components

Evapotranspiration is controlled by the surface water and energy budget. In this section the single components of the energy budget will be discussed: radiation, soil heat flux, sensible and latent heat fluxes.
The surface energy budget inside a control volume can be written as:

$$\Delta t\,(R_n - H - LE - G) = \Delta E \tag{1}$$

where the energy fluxes concerning the soil-atmosphere interface in the time interval Δt are:

- R_n net radiative flux;
- H sensible heat flux;
- LE latent heat flux;
- G heat flux in the soil;
- ΔE internal energy variation in the control volume;

The control volume is assumed with a thickness of some meters, so as to include the soil layer close to the surface and the first meters of the atmosphere, including the possible vegetation cover.

Besides the surface energy budget also the mass budget must be considered in order to quantify ET, namely the water conservation inside the control volume.

$$\frac{\Delta S}{\Delta t} = P - ET - R - R_G - R_S \tag{2}$$

where ΔS is the storing of the various supplies (underground and surface storage, soil moisture, vegetation interception, storing in channels); P is precipitation; $ET = \lambda LE$ is evapotranspiration; λ is the latent heat of vaporization; R is the surface runoff; R_G is the runoff towards the deep water table; R_S is the sub-surface runoff.

In the next sections it will be explained how the different components of the energy budget are usually described in hydrological and Land Surface Models (LSMs): radiation, sensible and latent heat flux, soil heat flux.

2.1 Radiation

The radiation is usually divided in short-wave components - indicated here as SW (including visible light, part of the ultraviolet radiation and the close infrared) and long-wave components - indicated here as LW (infrared radiation), with wavelength λ ranging respectively from 0.1 to 3 μm (98% of extraterrestrial radiation) and from 3 to 100 μm (2% of extraterrestrial radiation). In photosynthetic processes the short-wave radiation is further divided in photosynthetically active radiation ($0.4 < \lambda < 0.7 \mu m$) and near infrared (Bonan, 1996).

Moreover, there is a further distinction between diffuse radiation D (coming by diffusion in the atmosphere from any direction) and direct radiation P (coming only from the sun), and also between radiation from the sky downwards (*"down"* $\downarrow$) and radiation from the soil upwards (*"up"* $\uparrow$).

The net radiation at the soil level can be factorized as follows:

$$R_n = sw \cdot (R\downarrow_{SW\,P} + R\downarrow_{LW\,P}) + V \cdot (R\downarrow_{SW\,D} + R\downarrow_{LW\,D}) - R\uparrow_{SW} - R\uparrow_{LW\,R} - R\uparrow_{LW} + R\downarrow_{SW\,O} + R\downarrow_{LW\,O} \tag{3}$$

with:

$R\downarrow_{SW\,P}$	short-wave direct radiation;
$R\downarrow_{LW\,P}$	long-wave direct radiation;
$R\downarrow_{SW\,D}$	short-wave diffuse radiation;
$R\downarrow_{LW\,D}$	long-wave diffuse radiation;
$R\uparrow_{SW}$	short-wave reflected radiation;
$R\uparrow_{LW\,R}$	long-wave reflected radiation;
$R\uparrow_{LW}$	radiation emitted by surface;
$R\downarrow_{SW\,O} + R\downarrow_{LW\,O}$	reflected radiation emitted by the surrounding surfaces;

The effects of topography on the diffused radiation can be expressed through the sky view factor V, indicating the sky fraction visible in one point.

In the presence of reliefs only part of the horizon is visible and this has consequences on the radiative exchanges.

V is defined as:

$$V = \omega / 2\pi \tag{4}$$

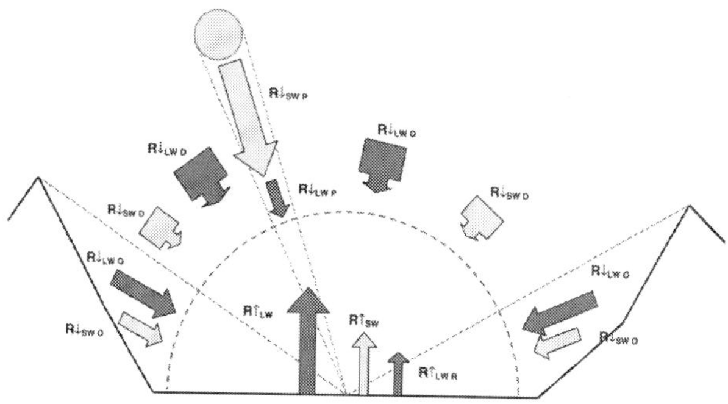

Fig. 1. Scheme of the solar radiation components.

where ω is the solid angle seen from the point considered.
The presence of shadows due to surrounding mountains can be expressed through a factor sw, a function of topography and sun position, defined as:

$$sw = \begin{cases} 1 \text{ if the point is in the sun} \\ 0 \text{ if the point is in the shadow} \end{cases} \tag{5}$$

All direct radiation terms have to be multiplied by this factor.
In the next paragraphs we analyze in detail the parametrization of the single terms composing the radiation flux.

2.1.1 Direct radiation $R\downarrow_{SW\ P}$ and $R\downarrow_{LW\ P}$

The direct long-wave radiation $R\downarrow_{LW\ P}$ is emitted directly by the sun and therefore it is negligible at the soil level (differently from the long-wave diffuse radiation).
Usually the short-wave radiation $R\downarrow_{SW\ P}$ is assumed as an input variable, measured or calculated by an atmospheric model. The direct radiation can be written as the product of the extraterrestrial radiation R_{Extr} by an attenuation factor varying in time and space.

$$R\downarrow_{SW\ P} = F_{att} R_{Extr} \tag{6}$$

The extraterrestrial radiation can be easily calculated on the basis of geometric formulas (Iqbal, 1983). The atmospheric attenuation is due to Rayleigh diffusion, to the absorption on behalf of ozone and water vapor, to the extinction (both diffusion and absorption) due to atmospheric dust and shielding caused by the possible cloud cover. Moreover the absorption entity depends on the ray path length through the atmosphere, a function of the incidence angle and of the measurement point elevation. The effect of the latter can be very important in a mountain environment, where it is necessary to consider the shading effects.
Part of the dispersed radiation is then returned as short-wave diffuse radiation ($R\downarrow_{SW\ D}$) and part of the energy absorbed by atmosphere is then re-emitted as long-wave diffuse radiation ($R\downarrow_{LW\ D}$).
From a practical point of view, according to the application type and depending on the measured data possessed, the attenuation coefficient can be calculated with different degrees of complexity. The radiation transfer through the atmosphere is a well studied phenomenon

and there exist many models providing the soil incident radiation spectrum in a detailed way, considering the various attenuation effects separately (Kondratyev, 1969).

2.1.2 Diffuse downward short-wave radiation $R\downarrow_{SW\ D}$

This term is a function of the atmospheric radiation due to Rayleigh dispersion and to the aereosols dispersion, as well as to the presence of cloud cover. The $R\downarrow_{SW\ D}$ actually is not isotropic and it depends on the sun position above the horizon. For its parametrization, see, for example, Paltrige & Platt (1976).

2.1.3 Diffuse downward long-wave radiation $R\downarrow_{LW\ D}$

Often this term in not provided by standard meteorological measurements, and many LSMs provide expressions to calculate it. This term indicates the long-wave radiation emitted by atmosphere towards the earth. It can be calculated starting from the knowledge of the distribution of temperature, humidity and carbon dioxide of the air column above. If this information is not available, various formulas, based only on ground measurements, can be found in literature with expressions as follows:

$$R\downarrow_{LW\ D}= \epsilon_a \sigma T_a^4 \tag{7}$$

with:

T_a air temperature [K];
ϵ_a atmosphere emissivity $f(e_a, T_a,$ cloud cover$)$;
e_a vapor pressure [mb];

Usually for ϵ_a empirical formulas have been used, but it is also possible to provide a derivation based on physical topics like in Prata (1996). The cloud cover effect on this term is significant and not easy to consider in a simple way. Cloud cover data can be provided during the day by ground or satellite observations but, especially on night, is difficult to collect.

2.1.4 Reflected short-wave radiation $R\uparrow_{SW}$

This term indicates the short-wave energy reflection.

$$R\uparrow_{SW}= a(R\downarrow_{SW\ P} + R\downarrow_{SW\ D}) \tag{8}$$

where a is the albedo.
The albedo depends strongly on the wave length, but generally a mean value is used for the whole visible spectrum. Besides its dependance on the surface type, it is important to consider its dependence on soil water content, vegetation state and surface roughness. The albedo depends moreover on the sun rays inclination, in particular for smooth surfaces: for small angles it increases. There is very rich literature about albedo description, it being a key parameter in the radiative exchange models, see for example Kondratyev (1969). Albedo is often divided in visible, near infrared, direct and diffuse albedo, as in Bonan (1996).

2.1.5 Long-wave radiation emitted by the surface $R\uparrow_{LW}$

This term indicates the long-wave radiation emitted by the earth surface, considered as a grey body with emissivity ε_s (values from 0.95 to 0.98). The surface temperature T_s $[K]$ is unknown and must be calculated by a LSM. $\sigma = 5.6704 \cdot 10^{-8}\ W/(m^2K^4)$ is Stefan-Boltzman constant.

$$R\uparrow_{LW} = \varepsilon_s \sigma T_s^4 \tag{9}$$

2.1.6 Reflected long-wave radiation $R\uparrow_{LW\ R}$

This term is small and can be subtracted by the incoming long-wave radiation, assuming surface emissivity ε_s equal to surface absorptivity:

$$R\downarrow_{LW\ D} = \varepsilon_s \cdot \epsilon_a \sigma T_a^4 \tag{10}$$

2.1.7 Radiation emitted and reflected by surrounding surfaces $R\downarrow_{SW\ O} + R\downarrow_{LW\ O}$

It indicates the radiation reflected ($R\uparrow_{SW} + R\uparrow_{LW\ R}$) and emitted ($R\uparrow_{LW}$) by the surfaces adjacent to the point considered. This term is important at small scale, in the presence of artificial obstructions or in the case of a very uneven orography. To calculate it with precision it is necessary to consider reciprocal orientation, illumination, emissivity and the albedo of every element, through a recurring procedure (Helbig et al., 2009). A simple solution is proposed for example in Bertoldi et al. (2005).

If the intervisible surfaces are hypothesized to be in radiative equilibrium, i.e. they absorb as much as they emit, these terms can be quantified in a simplified way:

$$\begin{aligned} R\downarrow_{SW\ O} &= (1-V)R\uparrow_{SW} \\ R\downarrow_{LW\ O} &= (1-V)(R\uparrow_{LW} + R\uparrow_{LW\ R}) \end{aligned} \tag{11}$$

2.1.8 Net radiation

Inserting expressions (7) and (9) in the (3), the net radiation is:

$$R_n = [sw \cdot R\downarrow_{SW\ P} + V \cdot R\downarrow_{SW\ D}](1 - V \cdot a) + V \cdot \varepsilon_s \cdot \varepsilon_a \cdot \sigma \cdot T_a^4 - V \cdot \varepsilon_s \cdot \sigma \cdot T_s^4 \tag{12}$$

with $\varepsilon_a = f(e_a, T_a, \text{cloud cover})$ as for example in Brutsaert (1975).

Equation (12) is not invariant with respect to the spatial scale of integration: indeed it contains non-linear terms in T_a, T_s, e_a, consequently the same results are not obtained if the local values of these quantities are substituted by the mean values of a certain surface. Therefore, the shift from a treatment valid at local level to a distributed model valid over a certain spatial scale must be done with a certain caution.

2.1.9 Radiation adsorption and backscattering by vegetation

Expression (12) needs to be modified to take into account the radiation adsorption and backscattering by vegetation, as shown in Figure 2. This effect is very important to obtain a correct soil surface skin temperature (Deardorff, 1978). From Best (1998) it is possible to derive the following relationship:

$$\begin{aligned} R_n = {} & [sw \cdot R\downarrow_{SW\ P} + V \cdot R\downarrow_{SW\ D}](1 - V \cdot a) * (f_{trasm} + a_v) \\ & + (1-\varepsilon_v) \cdot V \cdot \varepsilon_s \cdot \varepsilon_a \cdot \sigma \cdot T_a^4 + \varepsilon_v \cdot \varepsilon_s \cdot \sigma \cdot T_v^4 \end{aligned} \tag{13}$$

where T_v is vegetation temperature, ε_v vegetation emissivity (supposed equal to absorption), a_v vegetation albedo (downward albedo supposed equal to upward albedo) and f_{trasm}

vegetation transmissivity, depending on plant type, leaf area index and photosynthetic activity.
Models oriented versus ecological applications have a very detailed parametrization of this term (Dickinson et al., 1986). Bonan (1996) uses a two-layers canopy model. Law et al. (1999) explicit the relationship between leaf area distribution and radiative transfer. A first energy budget is made at the canopy cover layer, and the energy fluxes are solved to find the canopy temperature, then a second energy budget is made at the soil surface. Usually a fraction of the grid cell is supposed covered by canopy and another fraction by bare ground.

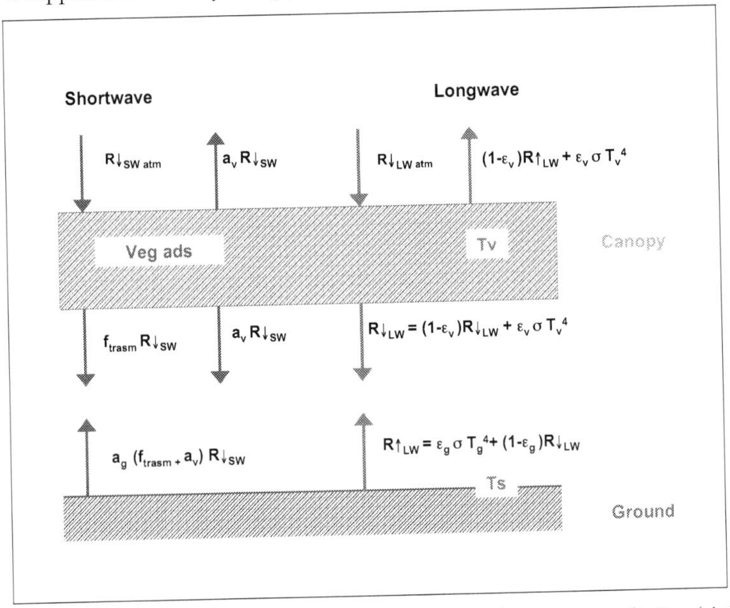

Fig. 2. Schematic diagram of short-wave radiation (left) and long-wave radiation (right) absorbed, transmitted and reflected by vegetation and ground , as in equation 13 (from Bonan (1996), modified).

2.2 Soil heat flux

The soil heat flux G at a certain depth z depends on the temperature gradient as follows:

$$G = -\lambda_S \frac{\partial T_S}{\partial z} \tag{14}$$

where λ_S is the soil thermal conductivity ($\lambda_S = \rho_S c_S \kappa_S$ with ρ_S density, c_S specific heat and κ_S soil thermal diffusivity) depending strongly on the soil saturation degree. The heat transfer inside the soil can be described in first approximation with Fourier conduction law:

$$\frac{\partial T_S}{\partial t} = \kappa_S \frac{\partial^2 T_S}{\partial z^2} \tag{15}$$

Equation (14) neglects the heat associated to the vapor transportation due to a vertical gradient of the soil humidity content as well as the horizontal heat conduction in the soil. The vapor transportation can be important in the case of dry climates (Saravanapavan & Salvucci, 2000). The soil heat flux can be calculated with different degrees of complexity. The most simple assumption (common in weather forecast models) is to calculate G as a fraction of net radiation (Stull (1988) suggests $G = 0.1R_n$). Another simple approach is to use the analytical solution for a sinusoidal temperature wave. A compromise between precision and computational work is the force restore method (Deardorff, 1978; Montaldo & Albertson, 2001), still used in many hydrological models (Mengelkamp et al., 1999). The main advantage is that only two soil layers have to be defined: a surface thin layer, and a layer getting down to a depth where the daily flux is almost zero. The method uses some results of the analytical solution for a sinusoidal forcing and therefore, in the case of days with irregular temperature trend, it provides less precise results.

The most general solution is the finite difference integration of the soil heat equation in a multilayered soil model (Daamen & Simmonds, 1997). However, this method is computationally demanding and it requires short time steps to assure numerical stability, given the non-linearity and stationarity of the surface energy budget, which is the upper boundary condition of the equation.

2.2.1 Snowmelt and freezing soil

In mountain environments snow-melt and freezing soil should be solved at the same time as soil heat flux. A simple snow melt model is presented in Zanotti et al. (2004), which has a lumped approach, using as state variable the internal energy of the snow-pack and of the first layer of soil. Other models consider a multi-layer parametrization of the snowpack (e.g. Bartelt & Lehning, 2002; Endrizzi et al., 2006). Snow interception by canopy is described for example in Bonan (1996). A state of the art freezing soil modeling approach can be found in Dall'Amico (2010) and Dall'Amico et al. (2011).

2.3 Turbulent fluxes

A modeling of the ground heat and vapor fluxes cannot leave out of consideration the schematization of the atmospheric boundary layer (ABL), meant as the lower part of atmosphere where the earth surface properties influence directly the characteristics of the motion, which is turbulent. For a review see Brutsaert (1982); Garratt (1992); Stull (1988).

A flux of a passive tracer x in a turbulent field (as for example heat and vapor close to the ground), averaged on a suitable time interval, is composed of three terms: the first indicates the transportation due to the mean motion v, the second the turbulent transportation $\overline{x'\,\mathbf{v}'}$, the third the molecular diffusion k.

$$\overline{\mathbf{F}} = \overline{x}\,\overline{\mathbf{v}} + \overline{x'\,\mathbf{v}'} - k\nabla x \tag{16}$$

The fluxes parametrization used in LSMs usually only considers as significant the turbulent term only. The molecular flux is not negligible only in the few centimeters close the surface, and the horizontal homogeneity hypothesis makes negligible the convective term.

2.3.1 The conservation equations

The first approximation done by all hydrological and LSMs in dealing with turbulent fluxes is considering the Atmospheric Boundary Layer (ABL) as subject to a stationary, uniform motion, parallel to a plane surface.

This assumption can become limitative if the grid size becomes comparable to the vertical heterogeneity scale (for example for a grid of 10 m and a canopy height of 10 m). In this situation horizontal turbulent fluxes become relevant. A possible approach is the Large Eddy Simulation (Albertson et al., 2001).

If previous assumptions are made, then the conservation equations assume the form:

- Specific humidity conservation, failing moisture sources and phase transitions:

$$k_v \frac{\partial^2 \overline{q}}{\partial z^2} - \frac{\partial}{\partial z}(\overline{w'q'}) = 0 \tag{17}$$

where:

k_v is the vapor molecular diffusion coefficient $[m^2/s]$

$q = \frac{m_v}{m_v + m_d}$ is the specific humidity [vapor mass out of humid air mass].

- Energy conservation:

$$k_h \frac{\partial^2 \overline{\theta}}{\partial z^2} - \frac{\partial}{\partial z}(\overline{w'\theta'}) - \frac{1}{\rho c_p}\frac{\partial H_R}{\partial z} = 0 \tag{18}$$

where:

k_h is the thermal diffusivity $[m^2/s]$

H_R is the radiative flux $[W/m^2]$

θ is the potential temperature [K]

ρ is the air density $[kg/m^3]$

w is the vertical velocity $[m/s]$.

- The horizontal mean motion equations are obtained from the momentum conservation by simplifying Reynolds equations (Stull, 1988; Brutsaert, 1982 cap.3):

$$-\frac{1}{\rho}\frac{\partial \overline{p}}{\partial x} + 2\omega \, sin\, \phi \, \overline{v} + \nu \frac{\partial^2 \overline{u}}{\partial z^2} - \frac{\partial}{\partial z}(\overline{w'u'}) = 0 \tag{19}$$

$$-\frac{1}{\rho}\frac{\partial \overline{p}}{\partial y} - 2\omega \, sin\, \phi \, \overline{u} + \nu \frac{\partial^2 \overline{v}}{\partial z^2} - \frac{\partial}{\partial z}(\overline{w'v'}) = 0 \tag{20}$$

where:

ν is the kinematic viscosity $[m^2/s]$

ω is the earth angular rotation velocity [rad/s]

ϕ is the latitude [rad] .

The vertical motion equation can be reduced to the hydrostatic equation:

$$\frac{\partial p}{\partial z} = -\rho g. \tag{21}$$

In a turbulent motion the molecular transportation terms of the momentum, heat and vapor quantity, respectively ν, k_h and k_v, are several orders of magnitude smaller than Reynolds fluxes and can be neglected.

2.3.2 Wind, heat and vapor profile at the surface

Inside the ABL we can consider, with a good approximation, that the decrease in the fluxes intensity is linear with elevation. This means that in the first meters of the air column the fluxes and the friction velocity u^* can be considered constant. Considering the momentum flux constant with elevation implies that also the wind direction does not change with elevation (in the layer closest to the soil, where the geostrofic forcing is negligible). In this way the alignment with the mean motion allows the use of only one component for the velocity vector, and the problem of mean quantities on uniform terrain becomes essentially one-dimensional, as these become functions of the only elevation z.

In the first centimeters of air the energy transportation is dominated by the molecular diffusion. Close to the soil there can be very strong temperature gradients, for example during a hot summer day. Soil can warm up much more quickly than air. The air temperature diminishes very rapidly through a very thin layer called *micro layer*, where the molecular processes are dominant. The strong ground gradients support the molecular conduction, while the gradients in the remaining part of the surface layer drive the turbulent diffusion. In the remaining part of the surface layer the potential temperature diminishes slowly with elevation.

The effective turbulent flux in the interface sublayer is the sum of molecular and turbulent fluxes. At the surface, where there is no perceptible turbulent flux, the effective flux is equal to the molecular one, and above the first cm the molecular contribution is neglegible. According to Stull (1988), the turbulent flux measured at a standard height of 2 m provides a good approximation of the effective ground turbulent flux.

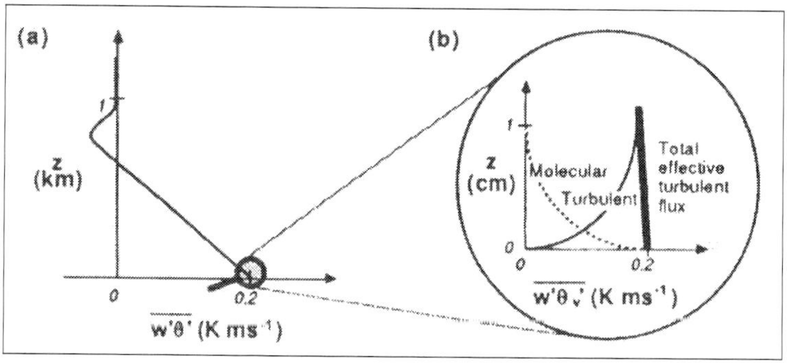

Fig. 3. (a) The effective turbulent flux in diurnal convective conditions can be different from zero on the surface. (b) The effective flux is the sum of the turbulent flux and the molecular flux (from Stull, 1988).

Applying the concept of effective turbulent flux, the molecular diffusion term can be neglected, while the hypothesis of uniform and stationary limit layer leads to neglect the convective terms due to the mean vertical motion and the horizontal flux. The vertical flux at the surface can then be reduced to the turbulent term only:

$$\overline{\mathbf{F_z}} = \overline{x'\, w'} \tag{22}$$

In the case of the water vapor, equation (17) shows that, if there is no condensation, the flux is:

$$ET = \lambda\rho\overline{w'q'} \tag{23}$$

where ET is the evaporation quantity at the surface, ρ the air density and λ is the latent heat of vaporization.
Similarly, as to sensible heat, equation (18) shows that the heat flux at the surface H is:

$$H = \rho c_p \overline{w'\theta'} \tag{24}$$

where c_p is the air specific heat at constant pressure.
The entity of the fluctuating terms $\overline{w'u'}$, $\overline{w'\theta'}$ and $\overline{w'q'}$ remains unknown if further hypotheses (called closing hypotheses) about the nature of the turbulent motion are not introduced. The closing model adopted by the LSMs is Bousinnesq model: it assumes that the fluctuating terms can be expressed as a function of the vertical gradients of the quantities considered (diffusive closure).

$$\tau_x = -\rho\overline{u'w'} = \rho K_M \partial\overline{u}/\partial z \tag{25}$$
$$H = -\rho c_p \overline{w'\theta'} = -\rho c_p K_H \partial\overline{\theta}/\partial z \tag{26}$$
$$ET = -\lambda\rho\overline{w'q'} = -\rho K_W \partial\overline{q}/\partial z \tag{27}$$

where K_M is the turbulent viscosity, K_H and K_W $[m^2/s]$ are turbulent diffusivity. Moreover a logarithmic velocity profile in atmospheric neutrality conditions is assumed:

$$\frac{k\,u}{u_{*0}} = \ln(\frac{z}{z_0}) \tag{28}$$

where k is the Von Karman constant, z_0 is the aerodynamic roughness, evaluated in first approximation as a function of the height of the obstacles as $z_0/h_c \simeq 0.1$ (for more precise estimates see Stull (1988) p.379; Brutsaert (1982) ch.5; Garratt (1992) p.87). In the case of compact obstacles (e.g. thick forests), the profile can be thought of as starting at a height d_0, and the height z can be substituted with a fictitious height $z - d_0$.

Surface type	z_0 [cm]
Large water surfaces	0.01-0.06
Grass, height 1 cm	0.1
Grass, height 10 cm	2.3
Grass, height 50 cm	5
Vegetation, height 1-2 m	20
Trees, height 10-15 m	40-70
Big towns	165

Table 1. Values of aerodynamic roughness length z_0 for various natural surfaces (from Brutsaert, 1982).

Also the other quantities θ and q have an analogous distribution. Using as scale quantities $\theta_{*0} = -\overline{w'\theta'}_0/u_{*0}$ e $q_{*0} = -\overline{w'q'}_0/u_{*0}$ and substituting them in the (25), the following integration is obtained:

$$\frac{k(\theta - \theta_0)}{\theta_{*0}} = \ln(\frac{z}{z_T}) \tag{29}$$

$$\frac{k(q - q_0)}{q_{*0}} = \ln(\frac{z}{z_q}). \tag{30}$$

The boundary condition chosen is $\theta = \theta_0$ in $z = z_T$ and $q = q_0$ in $z = z_q$. The temperature θ_0 then is not the ground temperature, but that at the elevation z_T. The roughness height z_T is the height where temperature assumes the value necessary to extrapolate a logarithmic profile. Analogously, z_q is the elevation where the vapor concentration assumes the value necessary to extrapolate a logarithmic profile.
Indeed, close to the soil (interface sublayer) the logarithmic profile is not valid and then, to estimate z_T and z_q, it would be necessary to study in a detailed way the dynamics of the heat and mass transfer from the soil to the first meters of air.
If we consider a real surface instead of a single point, the detail requested to reconstruct accurately the air motion in the upper soil meters is impossible to obtain. Then there is a practical problem of difficult solution: on the one hand, the energy transfer mechanisms from the soil to the atmosphere operate on spatial scales of few meters and even of few cm, on the other hand models generally work with a spatial resolution ranging from tens of m (as in the case of our approach) to tens of km (in the case of mesoscale models). Models often apply to local scale the same parametrizations used for mesoscale. Therefore a careful validation test, even for established theories, is always important.
Observations and theory (Brutsaert, 1982, p.121) show that z_T and z_q generally have the same order of magnitude, while the ratio $\frac{z_T}{z_0}$ is roughly included between $\frac{1}{5} - \frac{1}{10}$.

2.3.3 The atmospheric stability

In conditions different from neutrality, when thermal stratification allows the development of buoyancies, Monin & Obukhov (1954) similarity theory is used in LSMs. The similarity theory wants to include the effects of thermal stratification in the description of turbulent transportation. The stability degree is expressed as a function of Monin-Obukhov length, defined as:

$$L_{MO} = -\frac{u_{0*}^3 \theta_0}{kg\overline{w'\theta'}} \tag{31}$$

where θ_0 is the potential temperature at the surface.
Expressions of the stability functions can be found in many texts of Physics of the Atmosphere, for example Katul & Parlange (1992); Parlange et al. (1995). The most known formulation is to be found in Businger et al. (1971). Yet stability is often expressed as a function of bulk Richardson number Ri_B between two reference heights, expresses as:

$$Ri_B = \frac{g\ z\ \Delta\theta}{\overline{\theta} u^2} \tag{32}$$

where $\Delta\theta$ is the potential temperature difference between two reference heights, and $\overline{\theta}$ is the mean potential temperature.

If $Ri_B > 0$ atmosphere is steady, if $Ri_B < 0$ atmosphere is unsteady. Differently from L_{MO}, Ri_B is also a function of the dimensionless variables z/z_0 e z/z_T. The use of Ri_B has the advantage that it does not require an iterative scheme.
Expressions of the stability functions as a function of Ri_B are provided by Louis (1979) and more recently by Kot & Song (1998). Many LSMs use empirical functions to modify the wind profile inside the canopy cover.
From the soil up to an elevation $h_d = f(z_0)$, limit of the interface sublayer, the logarithmic universal profile and Reynolds analogy are no more valid. For smooth surfaces the interface sublayer coincides with the viscous sublayer and the molecular transport becomes important. For rough surfaces the profile depends on the distribution of the elements present, in a way which is not easy to parametrize. Particular experimental relations can be used up to elevation h_d, to connect them up with the logarithmic profile (Garratt, 1992, p. 90 and Brutsaert, 1982, p. 88). These are expressions of non-easy practical application and they are still little tested.

2.3.4 Latent and sensible heat fluxes

As consequence of the theory explained in the previous paragraph, the turbulent latent and sensible fluxes H and LE can be expressed as:

$$H = \rho c_p \overline{w'\theta'} = \rho c_p C_H u(\theta_0 - \theta) \tag{33}$$

$$ET = \lambda \rho \overline{w'q'} = \lambda \rho C_E u(q_0 - q), \tag{34}$$

where $\theta_0 - \theta$ and $q_0 - q$ are the difference between surface and measurement height of potential temperature and specific humidity respectively. C_H and C_E are usually assumed to be equal and depending on the bulk Richardson number (or on Monin-Obukhov lenght):

$$C_H = C_{Hn} F_H(Ri_B), \tag{35}$$

where C_{Hn} is the heat bulk coefficient for neutral conditions:

$$C_{Hn} = C_{En} = \frac{k^2}{[\ln(z/z_0)][\ln(z_a/z_T)]} \tag{36}$$

derived on Eq. 29 and depending on the wind speed u, the measurement height z, the temperature (or moisture) measurement height z_a, the momentum roughness length z_0 and the heat roughness length z_T.
A common approach is the 'electrical resistance analogy' (Bonan, 1996), where the atmospheric resistance is expressed as:

$$r_{aH} = r_{aE} = (C_H\, u)^{-1} \tag{37}$$

3. Evapotranspiration processes

In order to convert latent heat flux in evapotranspiration the energy conservation must be solved at the same time as water mass budget. In fact, there must be a sufficient water quantity available for evaporation. Moreover, vegetation plays a key role.

3.1 Unsaturated soil evaporation

If the availability of water supply permits to reach the surface saturation level, then evaporation is potential $ET = EP$ and then we have air saturation at the surface $q(T_s) = q^*(T_s)$ (the superscript * stands here for saturation). If the soil is unsaturated, $q(T_s) \neq q^*(T_s)$ and different approaches are possible to quantify the water content at the surface, in dependance of the water budget scheme adopted.

1. A first possibility is to introduce then the concept of surface resistance r_g to consider the moisture reduction with respect to the saturation value. As it follows from equation (34):

$$ET = \lambda \rho C_E u(q_0 - q) = \lambda \rho \frac{1}{r_a}(q_0 - q) = \lambda \rho \frac{1}{r_a + r_g}(q_0^* - q) \tag{38}$$

2. As an alternative, we can define a soil-surface relative moisture

$$r_h = q_0 / q_0^* \tag{39}$$

and then the expression for evaporation becomes:

$$ET = \lambda \rho \frac{1}{r_a}(r_h\, q_0^* - q) \tag{40}$$

An expression of r_h as a function of the potential ψ_s [m] (work required to extract water from the soil against the capillarity forces) and of the ratio of the soil water content η to the saturation water content η_s is given in Philip & Vries (1957):

$$r_h = exp(-(g/R_v T_s)\psi_s(\eta/\eta_s)^{-b}) \tag{41}$$

where $R_v = 461.53\ [J/(kg\ K)]$ is the gas constant for water vapor, T_s is the soil temperature, b an empirical constant. Tables of these parameters for different soil types can be found in Clapp & Hornberger (1978).

Another more simple expression frequently applied in models to link the value r_h with the soil water content η is provided by Noilhan & Planton (1989):

$$r_h = \begin{cases} 0.5(1 - \cos(\frac{\eta}{\eta_k}\pi)) & se\ \eta < \eta_k \\ 1 & if\ \eta \geq \eta_k \end{cases} \tag{42}$$

where η is the moisture content of a soil layer with thickness d_1, and η_k is a critical value depending on the saturation water content: $\eta_k \simeq 0.75\eta_s$.

3. A third possibility, very used in large-scale models, is that of expressing the potential/real evaporation ration through a simple coefficient:

$$ET = x\, EP = x\, \lambda \rho \frac{1}{r_a}(q_0^* - q) \tag{43}$$

The value of x can be connected to the soil water content η through the expression (Deardorff, 1978) (see Figure 4):

$$x = min(1, \frac{\eta}{\eta_k}) \tag{44}$$

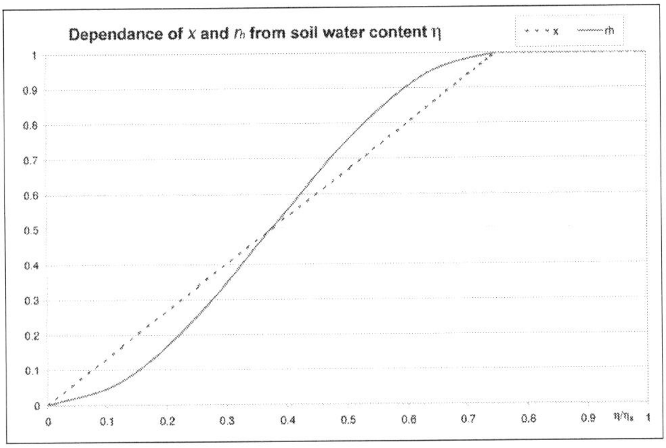

Fig. 4. Dependence of x and r_h on the soil water content η (Eq. 44-42)

3.2 Transpiration

Usually transpiration takes into account the canopy resistance r_c to add to the atmospheric resistance r_a:

$$ET = x\,EP = x\,\lambda\rho\frac{1}{r_a + r_c}(q_0^* - q) \tag{45}$$

The canopy resistance depends on plant type, leaf area index, solar radiation, vapor pressure deficit, temperature and water content in the root layer. There is a wide literature regarding such dependence, see for example Feddes et al. (1978); Wigmosta et al. (1994).
Canopy interception and evaporation from wet leaves are important processes modeled that should be modelled, according to Deardorff (1978). It is possible to distinguish two fundamental approaches: single-layer canopy models and multi-layer canopy models.

Single-layer canopy models (or "big leaf" models)

The vegetation resistance is entirely determined by stomal resistance and only one temperature value, representative of both vegetation and soil, is considered. Moreover a vegetation interception function can be defined so as to define when the foliage is wet or when the evaporation is controlled by stomal resistance.

Multi-layer canopy models

These are more complex models in which a soil temperature T_g, different from the foliage temperature T_f, is considered. Therefore, two pairs of equations of latent and sensible heat flux transfer, from the soil level to the foliage level, and from the latter to the free atmosphere, must be considered (Best, 1998). Moreover the equation for the net radiation calculation must consider the energy absorption and the radiation reflection by the vegetation layer.

Deardorff (1978) is the first author who presents a two-layer model with a linear interpolation between zones covered with vegetation and bare soil, to be inserted into atmosphere general circulation models. Over the last years many detailed models have been developed, above all with the purpose of evaluating the CO_2 fluxes between vegetation and atmosphere. Particularly complex is the case of scattered vegetation, where evaporation is due to a combination of soil/vegetation effects, which cannot be schematized as a single layer (Scanlon & Albertson, 2003).

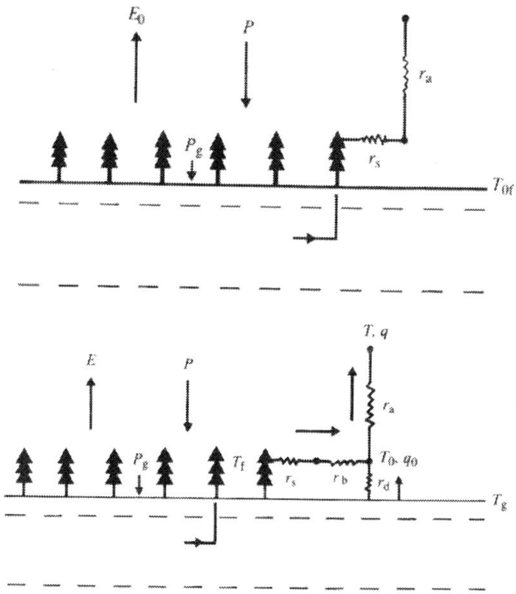

Fig. 5. **Above**: scheme representing a single-layer vegetation model. Linked both with atmosphere (with resistance r_a) and with the deep soil (through evapotranspiration with resistance r_s), vegetation and soil surface layer are assumed to have the same temperature T_{0f}. **Below**: scheme representing a multilayer vegetation model. Linked both with atmosphere (with resistances r_b and r_a), and with the deep soil (through evapotranspiration with resistance r_s), as well as with the soil under the vegetation (r_d), vegetation and soil surface layer are assumed to have different temperature T_f and T_g. P_g is the rainfall reaching the soil surface (from Garratt, 1992).

Given the many uncertainties regarding the forcing data and the components involved (soil, atmosphere), and the numerous simplifying hypotheses, the detail requested in a vegetation cover scheme is not yet clear.

A single-layer description of vegetation cover (big-leaf) and a two-level description of soil represent probably the minimum level of detail requested. In general, if the horizontal scale is far larger than the vegetation scale, a single-layer model is sufficient (Garratt, 1992, p. 242), as in the case of the general circulation atmospheric models or of mesoscale hydrologic models for large basins. These models determine evaporation as if the vegetation cover were but a partially humid plane at the atmosphere basis. In an approach of this kind surface resistance, friction length, albedo and vegetation interception must be specified. The surface resistance must include the dependence on solar radiation or on soil moisture, as transpiration decreases when humidity becomes smaller than the withering point (Jarvis & Morrison, 1981). For the soil, different coefficients depending on moisture are requested, together with a functional relation of evaporation to the soil moisture.

4. Water in soils

Real evaporation is coupled to the infiltration process occurring in the soil, and its physically-based estimate cannot leave the estimation of soil water content consideration.
The most simple schemes to account water in soils used in LSMs single-layer and two-layer methods. The most general approach, which allows water transport for unsaturated stratified soil, is based on the integration of Richards (1931) equation, under different degrees of approximations.

4.1 Single layer or bucket method

In this method the whole soil layer is considered as a bucket and real evaporation E_0 is a fraction x of potential evaporation E_p, with x proportional to the saturation of the whole soil.

$$E_0 = xE_p \tag{46}$$

with x expressed by Eq. (44). The main problem of this method is that evaporation does not respond to short precipitation, leading to surface saturation but not to a saturation of the whole soil layer (Manabe, 1969).

4.2 Two-layer or force restore method

This method is analogous to the one developed to calculate the soil heat flux, but it requires calibration parameters which are unlikely to be known. With this method it is possible to consider the water quantity used by plants for transpiration, considering a water extraction by roots in the deepest soil layer (Deardorff, 1978).

4.3 Multilayer methods and Richards equation

Richards (1931) equation and Darcy-Buckingham law govern the unsaturated water transport in isobar and isothermal conditions:

$$\vec{q} = -K\nabla\,(z + \psi) \tag{47}$$

$$\frac{\partial \psi}{\partial \eta}\,\nabla \cdot (K\nabla\psi) - \frac{\partial K}{\partial z} = \frac{\partial \psi}{\partial t} \tag{48}$$

where $\vec{q} = (q_x, q_y, q_z)$ is the specific discharge, K is the hydraulic conductivity tensor, z is the upward vertical coordinate and ψ is the suction potential or matrix potential.
The determination of the suction potential allows also a more correct schematization of the plant transpiration and it lets us describe properly flow phenomena from the water table to the surface, necessary to the maintenance of evaporation from the soils.
Richards equation is, rightfully, an energy balance equation, even if this is not evident in the modes from which it has been derived. Then the solutions of the equation (48) must be searched by assigning the water retention curve which relates ψ with the soil water content η and an explicit relation of the hydraulic conductivity as a function of ψ (or η). Both relationships depend on the type of terrain and are variable in every point. K augments with η, until it reaches the maximum value K_s which is reached at saturation.

Although the integration of the Richards equation is the only physically based approach, it requires remarkable computational effort because of the non linearity of the water retention curve. It is difficult to find a representative water retention curve because of the high degree of spatial variability in soil properties (Cordano & Rigon, 2008).

4.4 Spatial variability in soil moisture and evapotranspiration

Topography controls the catchment-scale soil moisture distribution (Beven & Freer, 2001) and therefore water availability for ET. Two methods most frequently used to incorporate sub-grid variability in soil moisture and runoff production SVATs models are the variable infiltration capacity approach (Wood, 1991) and the topographic index approach (Beven & Kirkby, 1979). They represent computationally efficient ways to represent hydrologic processes within the context of regional and global modeling. A review and a comparison of the two methods can be found in Warrach et al. (2002).

More detailed approaches need to track surface or subsurface flow within a catchment explicitly. Such approaches, which require to couple the ET model with a distributed hydrological model, are particularly useful in mountain regions, as presented in the next section.

5. Evapotranspiration in Alpine Regions

In alpine areas, evapotranspiration (ET) spatial distribution is controlled by the complex interplay of topography, incoming radiation and atmospheric processes, as well as soil moisture distribution, different land covers and vegetation types.

1. Elevation, slope and aspect exert a direct control on the incoming solar radiation (Dubayah et al., 1990). Moreover, elevation and the atmospheric boundary layer of the valley affect the air temperature, moisture and wind distribution (e.g., Bertoldi et al., 2008; Chow et al., 2006; Garen & Marks, 2005).
2. Vegetation is organized along altitudinal gradients, and canopy structural properties influence turbulent heat transfer processes, radiation divergence (Wohlfahrt et al., 2003), surface temperature (Bertoldi et al., 2010), therefore transpiration, and, consequently, ET.
3. Soil moisture influences sensible and latent heat partitioning, therefore ET. Topography controls the catchment-scale soil moisture distribution (Beven & Kirkby, 1979) in combination with soil properties (Romano & Palladino, 2002), soil thickness (Heimsath et al., 1997) and vegetation (Brooks & Vivoni, 2008a).

Spatially distributed hydrological and land surface models (e.g., Ivanov et al., 2004; Kunstmann & Stadler, 2005; Wigmosta et al., 1994) are able to describe land surface interactions in complex terrain, both in the temporal and spatial domains. In the next section we show an example of the simulation of the ET spatial distribution in an Alpine catchment simulated with the hydrological model GEOtop (Endrizzi & Marsh, 2010; Rigon et al., 2006).

6. Evapotranspiration in the GEOtop model

The GEOtop model describes the energy and mass exchanges between soil, vegetation and atmosphere. It takes account of land cover, soil moisture and the implications of topography on solar radiation. The model is open-source, and the code can be freely obtained from

the web site: http://www.geotop.org/. There, we provide a brief description of the 0.875 version of the model (Bertoldi et al., 2005), used in this example. For details of the most recent numerical implementation, see Endrizzi & Marsh (2010).
The model has been proved to simulate realistic values for the spatial and temporal dynamics of soil moisture, evapotranspiration, snow cover (Zanotti et al., 2004) and runoff production, depending on soil properties, land cover, land use intensity and catchment morphology (Bertoldi et al., 2010; 2006).
The model is able to simulate the following processes: (i) coupled soil vertical water and energy budgets, through the resolution of the heat and Richard's equations, with temperature and water pressure as prognostic variables (ii) surface energy balance in complex topography, including shadows, shortwave and longwave radiation, turbulent fluxes of sensible and latent heat, as well as considering the effects of vegetation as a boundary condition of the heat equation (iii) ponding, infiltration, exfiltration, root water extraction as a boundary condition of Richard's equation (iv) subsurface lateral flow, solved explicitly and considered as a source/sink term of the vertical Richard's equation (v) surface runoff by kinematic wave, and (vi) multi-layer glacier and snow cover, with a solution of snow water and energy balance fully integrated with soil.
The incoming direct shortwave radiation is computed for each grid cell according to the local solar incidence angle, including shadowing (Iqbal, 1983). It is also split into a direct and diffuse component according to atmospheric and cloud transmissivity (Erbs et al., 1982). The diffuse incoming shortwave and longwave radiation is adjusted according to the theory described in Par. 2.1. The soil column is discretized in several layers of different thicknesses. The heat and Richards' equations are written respectively as:

$$C_t(P)\frac{\partial T}{\partial t} - \frac{\partial}{\partial z}\left[K_t(P)\frac{\partial T}{\partial z}\right] = 0 \tag{49}$$

$$C_h(P)\frac{\partial P}{\partial t} - \frac{\partial}{\partial z}\left[K_h(P)\left(\frac{\partial P}{\partial z}+1\right)\right] - q_s = 0 \tag{50}$$

Where T is soil temperature, P the water pressure, C_t the thermal capacity, K_t the thermal conductivity, C_h the specific volumetric storativity, K_h the hydraulic conductivity, and q_s the source term associated with lateral flow. The variables C_t, K_t, C_h, and K_h depend on water content, and, in turn, on water pressure, and are therefore a source of non-linearity. At the bottom of the soil column a boundary condition of zero fluxes has been imposed.
The boundary conditions at the surface are consistent with the infiltration and surface energy balance, and are given in terms of surface fluxes of water (Q_h) and heat (Q_t) at the surface, namely:

$$Q_h = min\left[p_{net}, K_{h1}\frac{(h-P_1}{dz/2} + K_{h1}\right] - E(T_1, P_1) \tag{51}$$

$$Q_t = SW_{in} - SW_{out} + LW_{in} - LW_{out}(T_1) - H(T_1) - LE(T_1) \tag{52}$$

Where p_{net} is the net precipitation, K_{h1} and P_1 are the hydraulic conductivity and water pressure of the first layer, h is the pressure of ponding water, dz the thickness of the first layer, T_1 the temperature of the first layer. E is evapotranspiration (as water flux), SW_{in} and SW_{out} are the incoming and outgoing shortwave radiation, LW_{in} and LW_{out} the incoming and outgoing longwave radiation, H the sensible heat flux and LE the latent heat flux. H and LE are calculated taking into consideration the effects of atmospheric stability (Monin & Obukhov, 1954).

E is partitioned by evaporation or sublimation from the soil or snow surface E_G, transpiration from the vegetation E_{TC}, evaporation of the precipitation intercepted by the vegetation E_{VC}. Every cell has a fraction covered by vegetation and a fraction covered by bare soil. In the 0.875 version of the model, a one-level model of vegetation is employed, as in Garratt (1992) and in Mengelkamp et al. (1999): only one temperature is assumed to be representative of both soil and vegetation. In the most recent version, a two-layer canopy model has been introduced.

Bare soil evaporation E_g is related to the water content of the first layer through the soil resistance analogy (Bonan, 1996):

$$E_G = (1 - cop)\, E_P \,\frac{r_a}{r_a + r_s} \tag{53}$$

where cop is fraction of soil covered by the vegetation EP is the potential ET calculated with equation 34 and r_a the aerodynamic resistance:

$$r_a = 1/\,(\rho\, C_E\, \hat{u}) \tag{54}$$

The soil resistance r_s is function of the water content of the first layer.

$$r_s = r_a\,\frac{1.0 - (\eta_1 - \eta_r)/(\eta_s - \eta_r)}{(\eta_1 - \eta_r)/(\eta_s - \eta_r)} \tag{55}$$

where η_1 is the water content of the first soil layer close the surface, η_r is the residual water content (defined following Van Genuchten, 1980) and η_s is the saturated water content, both in the first soil layer.

The evaporation from wet vegetation is calculated following Deardorff (1978):

$$E_{VC} = cop\; E_P\; \delta_W \tag{56}$$

where δ_W is the wet vegetation fraction.

The transpiration from dry vegetation is calculated as:

$$E_{TC} = cop\; E_P(1 - \delta_W)\sum_i^n \frac{f_{root}^i\, r_a}{r_a + r_c^i} \tag{57}$$

The root fraction f_{root}^i of each soil layer i is calculated decreasing linearly from the surface to a maximum root depth, depending from the cover type. The canopy resistance r_c depends on solar radiation, vapor pressure deficit, temperature as in Best (1998) and on water content in the root zone as in Wigmosta et al. (1994).

6.1 The energy balance at small basin scale: application to the Serraia Lake.

An application of the model to a small basin is shown here, in order to bring out the problems arising when passing from local one-dimensional scale to basin-scale. The Serraia Lake basin is a mountain basin of 9 km^2, with an elevation ranging from 900 to 1900 m, located in Trentino, Italy. Within the basin there is a lake of about 0.5 km^2. During the year 2000 a study to calculate the yearly water balance was performed (Bertola & al., 2002).

The model was forced with meteorological measurement of a station located in the lower part of the basin at about 1000 m, and the stream-flow was calibrated for the sub-catchment of Foss Grand, of about 4 km^2. Then the model was applied to the whole basin. Further details on the calibration can be found in Salvaterra (2001). Meteorological data are assumed to be constant across the basin, except for temperature, which varies linearly with elevation (0.6 ^oC / 100 m) and solar radiation, which slightly increases with elevation and is affected by shadow and aspect.

With the GEO_{TOP} model it is possible to simulate the water and energy balance, aggregated for the whole basin (see figure 6 and 7) and its distribution across the basin. Figure 7 shows the map of the seasonal latent heat flux (ET) in the basin. During winter and fall ET is low (less than 40 W/m^2), with the lowest values in drier convex areas. During summer and spring ET increases (up to 120 W/m^2), with highest values in the bottom of the main valley (where indeed there are a lake and a wetland) and lowest values in north-facing, high-elevation areas.

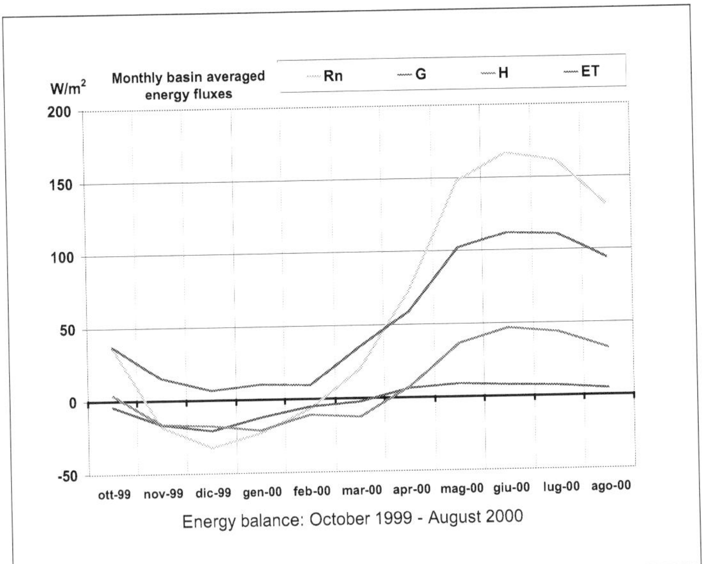

Fig. 6. Monthly energy balance for the Serraia basin (TN, Italy).

The main factors controlling the ET pattern in a mountain environment (see Figure 8) are also: elevation, which controls temperature, aspect, which influences radiation, soil thickness, which determines storage capacity, topographic convergence, which controls the moisture availability. In particular, aspect has a primary effect on net radiation and a secondary effect more on sensible rather than on latent heat flux, as in Figure 9, where south aspect locations have larger R_n and H, but similar behavior for the other energy budget components). Water content changes essentially the rate between latent and heat flux, as in Figure 10 where wet locations have larger ET and lower H.

Therefore, the surface fluxes distribution seems to agree with experience and current hydrology theory, but the high degree of variability poses some relevant issues because the hypothesis of homogenous turbulence at the basis of the fluxes calculation is no more valid

Garen, D. C. & Marks, D. (2005). Spatially distributed energy balance snowmelt modeling in a mountainous river basin: estimation of meteorological inputs and verification of model results, *J. Hydrol.* 315: 126–153.

Garratt, J. R. (1992). *The Atmospheric Boundary Layer*, Cambridge University Press.

Heimsath, M. A., Dietrich, W. E., Nishiizumi, K. & Finkel, R. (1997). The soil production function and landscape equilibrium, *Nature* 388: 358–361.

Helbig, N., Lowe, H.&Lehning, M. (2009). Radiosity approach for the short wave surface radiation balance in complex terrain., *J.Atmos.Sci.* 66(doi:10.1175/2009JAS2940.1): 2900–2912.

Iqbal, M. (1983). *An Introduction to Solar Radiation*, Academic Press.

Ivanov, V. Y., Vivoni, E. R., Bras, R. L. & Entekhabi, D. (2004). Catchment hydrologic response with a fully distributed triangulated irregular network model, *Water Resour. Res.* 40: doi:10.1029/2004WR003218.

Jarvis, P. & Morrison, J. (1981). The control of transpiration and photosynthesis by the stomata., *in* P. Jarvis & T. Mansfield (eds), *Stomatal Physiology*, Cambridge Univ. Press, UK, pp. 247-279.

Katul, G. G. & Parlange, M. B. (1992). A penman-brutsaert model for wet surface evaporation, *Water Resour. Res.* 28(1): 121-126.

Kondratyev, K. Y. (1969). *Radiation in the atmosphere*, Academic Press, New York.

Kot, S. C. & Song, Y. (1998). An improvement of the Louis scheme for the surface layer in an atmospheric modelling system, *Bound. Layer Meteorol.* 88(2): 239–254.

Kunstmann, H. & Stadler, C. (2005). High resolution distributed atmospheric-hydrological modelling for alpine catchments, *J. Hydrol.* 314: 105–124.

Law, B. E., Cescatti, A. & Baldocchi, D. D. (1999). Leaf area distribution and radiative transfer in open-canopy forests: Implications to mass and energy exchange., *Tree Physiol.* 21: 287–298.

Louis, J. F. (1979). A parametric model of vertical eddy fluxes in the atmosphere, *Bound. Layer Meteorol.* 17: 187–202.

Manabe, S. (1969). Climate and ocean circulation. i. the atmospheric circulation and the hydrology of the earth's surface., *Monthly Weather Review* 97: 739–774.

Mengelkamp, H.,Warrach, K. & Raschke, E. (1999). SEWAB - a parametrization of the surface energy and water balance for atmospheric and hydrologic models, *Adv.Water Resour.* 23: 165–175.

Monin, A. S. & Obukhov, A. M. (1954). Basic laws of turbulent mixing in the ground layer of the atmosphere, *Trans. Geophys. Inst. Akad.* 151: 163–187

Montaldo, N. & Albertson, J. (2001). On the use of the force-restore svat model formulation for stratified soils, *J. Hydromet.* 2(6): 571–578.

Noilhan, J. & Planton, S. (1989). A simple parametriztion of land surface processes for meteorological models, *Mon. Wea. Rev* 117: 536–585.

Paltrige, G. W. & Platt, C. M. R. (1976). *Radiative Processes in Meteorology and Climatology*, Elsevier.

Parlange, M. B., Eichinger,W. E. & Albertson, J. D. (1995). Regional scale evaporation and the atmospheric boundary layer, *Reviews of Geophysic* 33(1): 99–124.

Philip, J. R. & Vries, D. A. D. (1957). Moisture movement in porous materials under temperature gradients, *Trans. Am. Geophys. Union* 38(2): 222–232.

Prata, A. J. (1996). A new long-wave formula for estimating downward clear-sky radiation at the surface., *Quarterly Journal of the Royal Meteorological Society* 122(doi: 10.1002/qj.49712253306): 1127–1151.

Richards, L. A. (1931). Capillary conduction of liquids in porous mediums, *Physics* 1: 318–333.

Rigon, R., Bertoldi, G. & Over, T. M. (2006). GEOtop: a distributed hydrological model with coupled water and energy budgets, *Journal of Hydrometeorology* 7: 371–388.

Romano, N. & Palladino, M. (2002). Prediction of soil water retention using soil physical data and terrain attributes, *J. Hydrol.* 265: 56–75.

Salvaterra, M. (2001). *Applicazione di un modello di bilancio idrologico al bacino del Lago di Serraia (TN)*, Tesi di diploma, Corso di diploma in Ingegneria per l'Ambiente e le Risorse.

Saravanapavan, T. & Salvucci, G. D. (2000). Analysis of rate-limiting processes in soil evaporation with implications for soil resistance models, *Adv. Water Resour.* 23: 493–502.

Scanlon, T. M. & Albertson, J. D. (2003). Water availability and the spatial complexity of co2, water, and energy fluxes over a heterogeneous sparse canopy, *J. Hydromet.* 4(5): 798–809.

Stull, R. B. (1988). *An Introduction to Boundary Layer Meteorology*, Kluwer Academic Publisher.

Van Genuchten, M. T. (1980). A closed-form equation for predicting the hydraulic conductivity of unsaturated soils., *Soil Sci. Soc. Am. J.* 44: 892–898.

Warrach, K., Stieglitz, M., Mengelkamp, H. & Raschke, E. (2002). Advantages of a topographically controlled runoff simulation in a SVAT model, *J. Hydromet.* 3: 131–148.

Wigmosta, M. S., Vail, L. & Lettenmaier, D. (1994). A Distributed Hydrology-Vegetation Model for complex terrain, *Water Resour. Res.* 30(6): 1665–1679.

Wohlfahrt, G., Bahn, M., Newesely, C. H., Sapinsky, S., Tappeiner, U. & Cernusca, A. (2003). Canopy structure versus physiology effects on net photosynthesis of mountain grasslands differing in land use., *Ecological modelling.* 170: 407–426.

Wood, E. F. (1991). Land-surface-atmosphere interactions for climate modelling. observations, models and analysis, *Surv. in Geophys.* 12: 1–3.

Zanotti, F., Endrizzi, S., Bertoldi, G. & Rigon, R. (2004). The geotop snow module, *Hydrol. Proc.* 18: 3667–3679. DOI:10.1002/hyp.5794.

18

Stomatal Conductance Modeling to Estimate the Evapotranspiration of Natural and Agricultural Ecosystems

Giacomo Gerosa[1], Simone Mereu[2], Angelo Finco[3] and Riccardo Marzuoli[1,4]
[1]Dipartimento Matematica e Fisica, Università Cattolica del Sacro Cuore, via Musei 41, Brescia
[2]Dipartimento di Economia e Sistemi Arborei, Università di Sassari,
[3]Ecometrics s.r.l., Environmental Monitoring and Assessment, via Musei 41, Brescia
[4]Fondazione Lombardia per l'Ambiente, Piazza Diaz 7, Milano
Italy

1. Introduction

This chapter presents some of the available modelling techniques to predict stomatal conductance at leaf and canopy level, the key driver of the transpiration component in the evapotranspiration process of vegetated surfaces. The process-based models reported, are able to predict fast variations of stomatal conductance and the related transpiration and evapotranspiration rates, e.g. at hourly scale. This high-time resolution is essential for applications which couple the transpiration process with carbon assimilation or air pollutants uptake by plants.

2. Stomata as key drivers of plant's transpiration

Evapotranspiration from vegetated areas, as suggested by the name, has two different components: evaporation and transpiration. Evaporation refers to the exchange of water from the liquid to the gaseous phase over living and non-living surfaces of an ecosystem, while transpiration indicates the process of water vaporisation from leaf tissues, i.e. the mesophyll cells of leaves. Both processes are driven by the available energy and the drying potential of the surrounding air, but transpiration depends also on the capacity of plants to replenish the leaf tissues with water coming from the roots through their hydraulic conduction system, the xylem. This capacity depends directly on soil water availability (i.e. soil water potential), which contributes to the onset of the water potential gradient within the soil-plant-atmosphere continuum.

Moreover, since the cuticle -a waxy coating covering the leaf surface- is nearly impermeable to water, the main part of leaf transpiration (about 95%) results from the diffusion of water vapour through the stomata. Stomata are little pores in the leaf lamina which provide low-resistance pathways to the diffusional movement of gases (CO_2, H_2O, air pollutants) from

outside to inside the leaf and vice versa. Following complex signal pathways, environmental, osmotic and hormonal, stomata regulate their opening area and thus the water vapour loss from leaves. When the evaporative demand is bigger than the water replenishing capability from the xylem, stomata closes partially or even totally. High evaporative demands can be due to elevated air temperature, high leaf-to-air vapour pressure deficit (VPD), and intense winds. Stomatal closure can also be caused by high concentration of carbon dioxide in the mesophyll space.
Stomata, thus, directly control plant transpiration preventing plants from excessive drying, and acting as key drivers of water vapour movements from vegetated surfaces to the atmosphere.
This chapter illustrates the modelling techniques to predict the stomatal behaviour of vegetation at high-resolution time scale, and the related water fluxes.

3. Modelling stomatal behaviour: The Jarvis-Stewart model and the Ball-Berry model

Stomata play an essential role in the regulation of both water losses by transpiration and CO_2 uptake for photosynthesis and plant growth. Stomatal aperture is controlled by the turgor pressure difference between the guard cells surrounding the pore and the bulk leaf epidermis. In order to optimize CO_2 uptake and water losses in rapidly changing environmental conditions, plants have evolved the ability to control stomatal aperture in the order of seconds. Stomatal aperture responds to multiple environmental factors such as, solar radiation, temperature, drought, VPD, wind speed, and sub-stomatal CO_2 concentrations.
The availability of modern physiological instrumentation (diffusion porometers, gas-exchange analyzers) has allowed to measure leaf stomatal conductance (g_s) in field conditions and to study how environmental variables influence this parameter.
However, measurements of g_s by porometers and gas-exchange analyzers can be made only when foliage is dry, and long-term enclosure in measuring chamber may lead to changes in the physiological state of the leaves. Consequently measurements in the field are usually made intensively over selected periods of a few hours in selected days.
Furthermore, stomatal conductance values depend also upon the physiological condition of the plant, which relates to the weather of the previous days as well as to the previous season for perennial species.
Therefore it is important to have continuous g_s measurements over the whole vegetative season in order to improve the interpretation of other physiological data such as photosynthesis rate and carbon assimilation.
An alternative to very frequent measurements of g_s in the field is to predict them from models that describe its dependency on environmental factors. These models can be parameterized using the available field measurements conducted on occasional periods.
Furthermore, modeling appears the most effective tool for integration, simulation and prediction purposes concerning the effects of climatic global change on vegetation.
Stomatal conductance is among the processes that have been most extensively modeled during the last decades. In their excellent review, Damour et al. (2010) describe 35 stomatal conductance models classified as:

1. models based on climatic control only
2. models mainly based on the gs-photosynthesis relationship
3. models mainly based on an Abscisic Acid (ABA) control
4. models mainly based on the turgor regulation of guard cell.

The next paragraphs provides information on two early developed g_s models which are currently among the most widely used: the multiplicative model of Jarvis (1976) based on climatic control and later modified by Stewart (1988), and the Ball Berry model (1988), based on g_s-photosynthesis relationship.

3.1 The Jarvis-Stewart model

The stomatal conductance model developed by Jarvis (1976) can be defined as an empirical multiplicative model based on the observed responses of g_s to environmental factors. The assumption of this model is that the influence of each environmental factor on g_s is independent of the others and can be determined by boundary line analysis (Webb 1972).
The Jarvis model, in its first form, integrates the responses of g_s to light intensity, leaf temperature, vapour pressure deficit, ambient CO_2 concentration and leaf water potential, according to the following equation:

$$g_s = f(Q) \cdot f(T_l) \cdot f(VPD_l) \cdot f(C_a) \cdot f(\Psi) \tag{1}$$

where Q is the quantum flux density (μE $m^{-2}s^{-1}$), T_l is the leaf temperature (°C), VPD_l is the leaf-to-air vapour pressure deficit calculated at leaf temperature (kPa), C_a is the ambient CO_2 concentration (ppm) and Ψ is leaf water potential (MPa).
Stewart (1988) further implemented this model adopting the assumption that the functions of environmental variables have values between zero and unit and exert their influence reducing the maximum stomatal conductance of the plant (g_{smax}), a species-specific value depending on leaf stomatal density, that can be defined as the largest value of conductance observed in fully developed leaves - but not senescent - of well-watered plants under optimal climatic conditions (Körner et al., 1979). This value can be derived from field measurements conducted under the above mentioned optimal conditions.
Furthermore, in Stewart formulation, quantum flux density is replaced by global solar radiation, leaf temperature by air temperature, leaf-to-air vapour pressure deficit by air vapour pressure deficit and leaf water potential by soil moisture deficit measured in the first meter of soil (i.e. soil water content, SWC). Stewart also omitted $f(C_a)$ because the effect of CO_2 ambient concentrations was considered negligible. This simplification allows for an easier data collection to run the model, but it must be kept in mind that C_a change considerably among seasons and thus the simplification may lead to a considerable error, especially when the model is used for annual g_s behavior of evergreen species.
The model is defined by the following equation:

$$g_s = g_{smax} \cdot f(Q) \cdot f(T_a) \cdot f(VPD_a) \cdot f(SWC) \tag{2}$$

It is important to notice that f(Q) can also be replaced by the more specific f(PAR), based on the photosynthetically active radiation.
Each function has a characteristic shape described by the following equations:

$$f(Q) - 1 - exp^{-aQ} \tag{3}$$

$$f(T_a) = \frac{(T - T_{\min})}{(T_{opt} - T_{\min})} \left[\frac{(T_{\max} - T)}{(T_{\max} - T_{opt})} \right]^b \tag{4}$$

where $b = \frac{(T_{\max} - T_{opt})}{(T_{opt} - T_{\min})}$, $f(T_a) = 0.1$ when $T \leq T_{\min}$ or $T \geq T_{\max}$ and $f(T_a) = 1$ when $T = T_{opt}$

$$f(VPD) = \frac{(1 - 0.1) \cdot (c - VPD)}{(c - d)} \tag{5}$$

where $f(VPD) = 1$ when VPD $\leq d$ and $f(VPD) = 0.1$ when VPD $\geq c$

$$f(SWC) = 1 - exp\left[k(SWC - SWC_{\max})\right] \tag{6}$$

where $f(SWC) = 1$ when $SWC = SWC_{max}$

Since g_s depends on four major variables, field measurements do not usually show a clear relationship with any of the considered variables. Often, g_s is reduced below the value expected for a value of a single independent variable, as the result of the influences of the other variables. As a consequence, the coefficients of each function must be derived with boundary-line analysis, plotting all field measurements of relative g_s ($g_{srel} = g_s/g_{smax}$) against each environmental variable considered separately.

Provided that enough measurements have been adequately performed to cover variable space, the upper limit of the scatter diagram indicates the response of g_s to the particular independent variable, when the other variables are not limiting.

An example of boundary-line analysis is reported in figure taken from Gerosa et al. (2009):

The main criticism formulated against this kind of approach is that the interactive effects between environmental factors are not properly taken into account, since interactions are only partially explained by the multiplicative nature of the model which simply multiplies concomitant effects, avoiding any synergistic interaction.

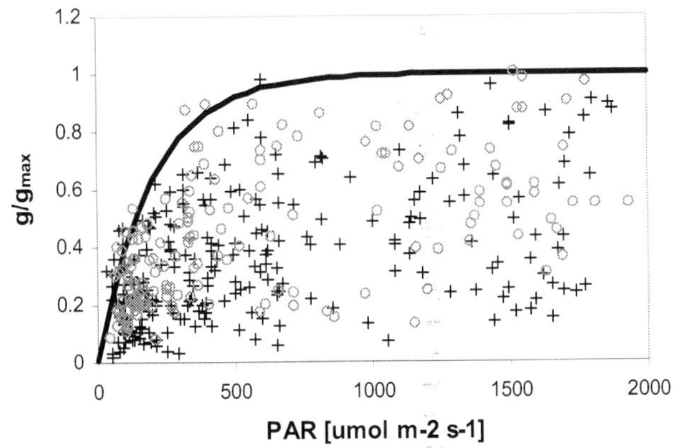

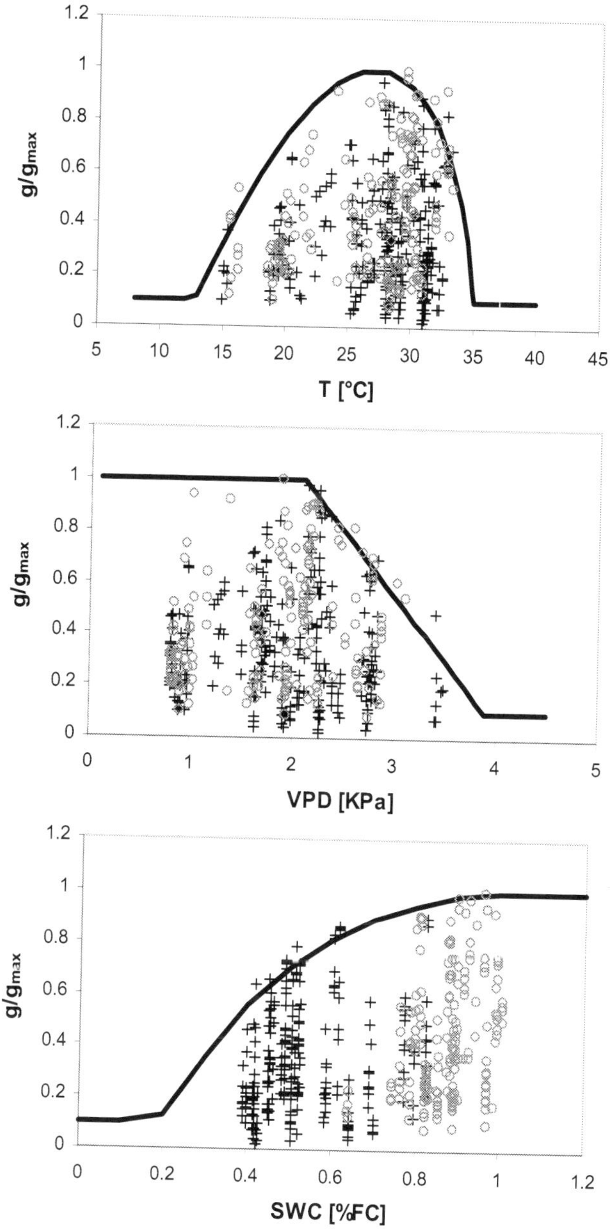

Fig. 1. Boundary-line analysis for the definition of g_s limiting function parameters (modified from Gerosa et al. 2008)

3.2 The Ball-Berry model

The Ball-Berry empirical model describes the behaviour of g_s as a function of environmental conditions and net photosynthetic rate. In its simplest form (Ball et al., 1987) the model states:

$$g_s = g_0 + a_1 An \frac{RH}{C_s} \tag{7}$$

Where g_s is the stomatal conductance to water vapour, g_0 is the stomatal conductance at the light compensation point, a_1 is a fitting parameter representing the slope of the equation, An is photosynthesis, RH is relative humidity and C_s is the molar fraction of CO_2 at the leaf surface. The model takes advantage of the feedback loop that exists between A and g_s (Farquhar at al., 1978) implying that they are interdependent. Additionally, An and g_s can respond independently to environmental variables and so they cannot be considered driving variables but rather state variables. The empirical relationship emerges from optimized vegetation behaviour that maximizes productivity (Patwardhan et al., 2006): the relationship, actually, corresponds roughly to the value of maximum surface conductance that maximizes productivity.

In order to derive g_s, the model needs to be coupled with a photosynthesis model (most often the Farquhar biochemical model) from which An is calculated. In order to derive g_s two equations must be solved simultaneously:

$$g_s = g_0 + a_1 An \frac{RH}{C_s} \tag{8}$$

$$A = (C_s - C_i) g_s \tag{9}$$

The problem is often solved by reiteration of the two equations where g_s at time t_{n+1} is computed with An at time t_n and An at time t_{n+2} is computed using g_s at time $t+1$. The reiteration approach however can give birth to oscillations in time of g_s and An due to chaotic solution in particular conditions. However, Baldocchi et al. (1994) found an analytical solution for the set of equations that bypasses this problem.

The original Ball-Berry model (Ball et al., 1987) was further implemented by Leuning (1995), considering that stomata respond to vapour pressure deficit (VPD) rather than humidity. In its modified version the equation takes the form:

$$g_s = g_0 + \frac{a_1 An}{(C_s - \Gamma)\left(1 + \frac{VPD_0}{VPD_s}\right)} \tag{10}$$

Where, Γ is the CO_2 compensation point, Cs and $VPDs$ are the CO_2 concentration and vapour pressure deficit at the leaf surface, and VPD_0 is an empirical coefficient.

This model encapsulates two empirical trends reported in the literature. First, through the correlation between g_s and An the equation predicts that the ratio $(C_i-\Gamma)/(C_s-\Gamma)$ is largely independent of leaf irradiance and C_s, except near the light and CO_2 compensation points. It also predicts that g_s declines linearly as VPD_s increases, in fact through the relation

$$E = 1.6 \cdot g_s \cdot VPD_s \tag{11}$$

for the transpiration rate (E), the hyperbolic function of VPD_s is equivalent to a linear decline of g_s with increasing E.

The main limitation of the Ball-Berry-Leuning (BBL) model is its failure in describing stomatal closure in drought conditions. The model has been further implemented by Dewar (2002) to take SWC in consideration by coupling the BBL model with Tardieu model for stomatal response to drought. The coupled model takes the form:

$$g_s = \frac{a_1(An + Rd)}{Ci\left(1 + \frac{VPD_0}{VPDs}\right)} exp\{-[ABA]\beta exp(\delta\Psi)\} \tag{12}$$

Where Rd is dark respiration, $[ABA]$ is the concentration of abscisic acid in the leaf xylem, Ψ is the leaf water potential, β is the basal sensitivity of ion diffusion to $[ABA]$ at zero leaf water potential, and δ describes the increase in the sensitivity of ion diffusion to $[ABA]$ as Ψ declines.

The model has the advantage of describing stomatal responses to both atmospheric and soil variables and has proven to reproduce a number of common water use trends reported in the literature as, for example, isohydric and anisohydric behaviour.

4. Modelling water vapour exchange between leaves and atmosphere and scaling it up to plant and ecosystem level: The *big-leaf* approach and the resistive analogy

The exchange of water vapour through stomata is a molecular diffusion process since air in the sub-stomatal cavities is motionless as well as the air in the first layer outside the stomata directly in contact with the outer leaf surface, i.e. the leaf boundary-layer,. Outside the leaf boundary-layer, it is the turbulent movement of air that removes water vapour, and this process is two orders of magnitude more efficient than the molecular diffusion. The exchange of water between the plant and the atmosphere is further complicated by the physiological control that stomatal resistance exerts on the diffusion of water vapour to the atmosphere.

Transpiration is modelled through an electric analogy (Ohm's law) introduced by Chamberlain and Chadwick (1953). Transpiration behaves analogously to an electric current, which originates from an electric potential difference and flows through a conductor of a given resistance from the high to the low potential end (Figure 2).

The driving potential of the water flux E is assumed to be the difference between the water vapour pressure in ambient air $e(T_a)$ and the water vapour pressure inside the sub-stomatal cavity $e_s(T_l)$, the latter being considered at saturation. The resistances that water vapour encounters from within the leaf to the atmosphere is given by the resistance of the stomatal openings (r_s) and the resistance of the leaf boundary laminar sub-layer (r_b). This process can be represented by the following equation:

$$E = \frac{[e_s(T_l) - e(T_a)]}{r_b + r_s} \cdot \frac{\rho c_p}{\lambda \gamma} \tag{13}$$

where T_a is air temperature (°K), T_l is leaf temperature (°K), e is water vapour pressure in the ambient air (Pa), e_s is water vapour pressure of saturated air (Pa) and the term $\rho c_p/\lambda\gamma$ is a factor to express E in mass density units (kg m^{-2} s^{-1}), equivalent to mm of water per second, being c_p the heat capacity of air at constant pressure (1005 J K^{-1} kg^{-1}), ρ the air density (kg m^{-3}), λ the vaporisation heat of water (2.5x10^6 J kg^{-1}), and $\gamma = c_p/\lambda$ the psychrometric constant (67 Pa K^{-1}). Despite the apparent difference with the well-known Penman-Monteith equation (Monteith, 1981), Eq. 13 is an equivalent formulation of this latter, as demonstrated by Gerosa et al. (2007).

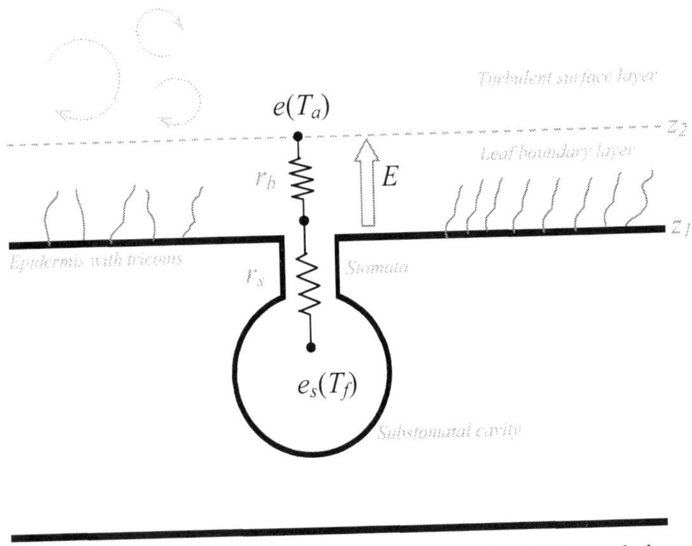

Fig. 2. Schematic picture of the transpirative process form a leaf. The symbols are explained in the text.

While the water vapour pressure deficit $[e_s(T_l)-e(T_a)]$ driving the water exchange is determined by temperature difference, the amount of water flux is regulated by the resistances along the path of the flux.
The stomatal resistance r_s, reciprocal of the stomatal conductance g_s, is obtained applying one of the stomatal prediction models presented in the previous paragraph, which are fed by meteorological and agrometeorological data.
The quasi-laminar sub-layer resistance r_b depends on the molecular properties of the diffusive substance and on the thickness of the layer. The resistance against the diffusion of a gas through air is defined as:

$$r = \int_{z_1}^{z_2} \frac{1}{D_{H_2O}} dz \tag{14}$$

for the leaf boundary-layer the equation gives:

$$r_b = (z_2 - z_1)/D_{H2O} \tag{15}$$

where D_{H2O} is the diffusion coefficient of water vapour in the air, z_1 and z_2 representing the lower and upper height of the leaf boundary-layer.

However, the thickness of the leaf boundary-layer depends on leaf geometry, wind intensity and atmospheric turbulence. In order to take these factors in consideration, a more practical formulation, proposed by Unsworth et al. (1984), can be used:

$$r_b = k\left(d / u\right)^{1/2} \tag{16}$$

where k is an empirical coefficient set to a value of 132 (Thom 1975), d is the downwind leaf dimension, and u is the horizontal wind speed near the leaves.

The transpiration of a whole plant, or of a vegetated surface with closed canopy, may be modelled using a similar approach referred to as the *big-leaf*. The *big-leaf* assumes the canopy vegetation as an ideal big-leaf lying at a virtual height $z=d+z_0$ above ground (Figure 3). The d parameter is the displacement height, i.e. the height of the zero-plane of the canopy, equal to 2/3 of the canopy height, z_0 is the roughness length, i.e. the additional height above d where the wind extinguishes inside the canopy (sink for momentum), around 1/10 of the canopy height, and $d+z_0'$ is the apparent height of water vapour source.

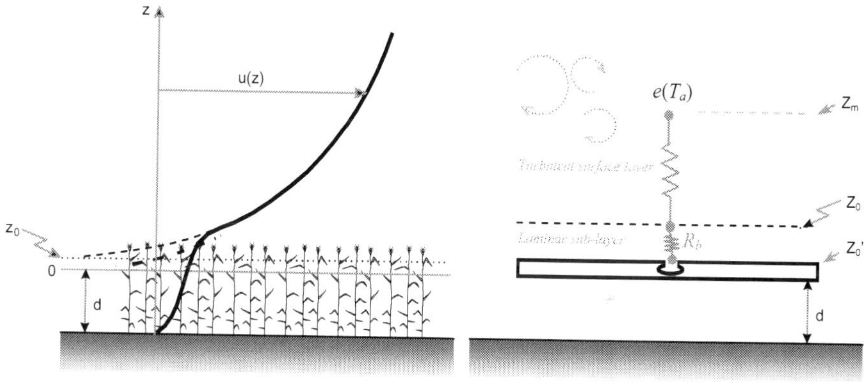

Fig. 3. The *big-leaf* approach to model water vapour exchange of a vegetated surface. Left side a real canopy; right side its big-leaf representation. The laminar sub-layer has been enlarged and the stomatal resistance is not shown. Please note the upper case notation of the resistances.

This transpiring big-leaf has a bulk stomatal resistance R_s equal to the sum of the stomatal resistances r_s of all the n leaves of the canopy. Recalling the rules of composition for parallel resistances:

$$1 / R_s = \sum_{1}^{n} 1 / r_s = n / r_s \tag{17}$$

Since the number of leaf is rarely known, a practical way of upscaling r_s is to consider the thickness of the "big-leaf" equal to the leaf area index of the canopy (LAI= $m^{-2}_{leaf}/m^{-2}_{ground}$)

i.e. the square meters of leaf area projected on each square meter of ground surface. This assumption is equivalent to stating that the light extinction coefficient of the big-leaf is equal to the light extinction of the canopy.

The transpiration rate of the "big-leaf", the whole canopy, is then obtained in a way very similar to those above developed for the leaves:

$$E = \frac{\left[e_s(T_l) - e(T_a)\right]}{R_a + R_b + R_s} \cdot \frac{\rho c_p}{\lambda \gamma} \tag{18}$$

It is worth noticing the upper case notation for the "bulk" resistances and the introduction of the aerodynamic resistance R_a.

The aerodynamic resistance depends on the turbulent features of the atmospheric surface layer, and it is introduced to account for the distance z_m at which the atmospheric water potential is measured above the canopy. It is formally the vertical integration of the reciprocals of the turbulent diffusion coefficients for all scalars, which in turn depends on the friction velocity u^* and the atmospheric stability. The integrated version of R_a is given by

$$R_a = \frac{1}{k \cdot u^*}\left[\ln\left(\frac{z_m - d}{z_0}\right) - \Psi_M\right] \tag{19}$$

where k is the von Kármán dimensionless constant (0.41), u^* is the friction velocity (m s^{-1}), a quantity indicating the turbulent characteristic of the atmosphere, and Ψ_M is the integrated form of the atmospheric stability function for momentum (non-dimensional).

The friction velocity, if not available, can be derived with the following equation:

$$u^* = \frac{k u_{zm}}{\ln\left(\frac{z_m - d}{z_0}\right) - \Psi_M} \tag{20}$$

where u_{zm} is the wind velocity measured at z_m, and Ψ_M is a function defined as:

$$\Psi_M = \begin{cases} 2\ln\left(\frac{1+y^2}{2}\right) \quad \textit{with} \quad y = (1 - 16(z-d)/L)^{1/4} & \textit{if } (z-d)/L < 0 \quad \textit{unstable condition} \\ 0 & \textit{if } (z-d)/L = 0 \quad \textit{neutral condition} \\ -5(z-d)/L & \textit{if } (z-d)/L > 0 \quad \textit{stable condition} \end{cases} \tag{21}$$

L is the length (m) of Monin-Obukhov (1954) indicating the atmospheric stability:

$$L = \frac{-\rho\, c_p\, T_0\, u_*^3}{k g H} \tag{22}$$

with T_0 the reference temperature (273.16 K), g the gravity acceleration (9.81 m s^{-2}) and H the sensible heat flux (W m^{-2}).

Since L is a function of u^* and H, and vice versa, concurrent determination of u^* and Ψ_M from routine weather data would normally require an iterative procedure (Holtslag and van Ulden, 1983).

If the atmospheric stability is not known as well as the sensible heat flux, and the water potential in the atmosphere is measured near the canopy, a neutral stability can be assumed by setting Ψ_M=0 in the u^* equation with fairly good approximation.
The laminar sub-layer resistance R_b can by computed with a general purpose formulation proposed by Hicks et al. (1987) which involves the Schmidt and Prandtl numbers, being Sc=0.62 for water vapour and Pr=0.72 respectively:

$$R_b = \frac{2}{ku^*}(Sc / \Pr)^{2/3} \tag{23}$$

where k is here the von Kármán constant.
Modelling canopy transpiration using only three resistances in series might seem an oversemplification; however the approach has proven valid in different cases in predicting fast variations of water exchange over a vegetated surface following the stomatal behaviour, as well as to predict the total amount of transpired water (Grunhage et al., 2000).
To obtain a higher modelling performance, the resistive network of the "big-leaf" model can be implemented for specific needs. For example, multiple vegetation layers can be included in order to account for the transpiration of the understory vegetation below a forest, or the canopy can be decomposed in several layers, each with its own properties (De Pury and Farquhar, 1997) In such cases the models take the name of multi-layer models. Other improvements are required when multiple sources of water vapour have to be considered, for example when the evaporation from a water catchment, or evaporation from bare soil in ecosystems with sparse vegetation. .
All these models are collectively known as *1-D* SVAT models (one-dimensional Soil Vegetation Atmosphere Transfer models).
In the following paragraph a multi-layer dual-source model to predict the evapotranspiration from a poplar plantation ecosystem with understory vegetation is presented.

5. Example and applications - a multi-layer model for the transpiration of a mature poplar plantation ecosystem - comparison with eddy covariance measurements

The poplar plantation used for this modelling exercise was located in the Po valley near the city of Pavia. The ecosystem was made by mature poplar trees of about 27 m height with the soil below the plant mainly covered by poplar saplings and perennial grasses. Since the canopy was completely closed, most of the evapotranspiration was due to plants transpiration i.e. evaporation from other surfaces can be considered negligible. According to Choudhury and Monteith (1988), less than 5% of the water vapour flux is due to evaporation from soil for a closed canopy. In this case study evaporation from soil was strongly limited by the absence of tillage and by the coverage of understory vegetation. Moreover the upper soil layer resulted very dry and acted as a screen against water vapour transport from wetter underlying soil layers.
The water exchange was modelled using only two water sources, both of them transpirative: the poplar crown and the understory vegetation. Thus this example model includes only two layers (Figure 4).
The model is composed of three different sub-models: one stomatal sub-model for the stomatal conductance of the transpiring plants, one soil sub-model for the soil water content, and one atmospheric sub-model to describe the water vapour exchange dynamic at canopy level following the adopted resistive network.

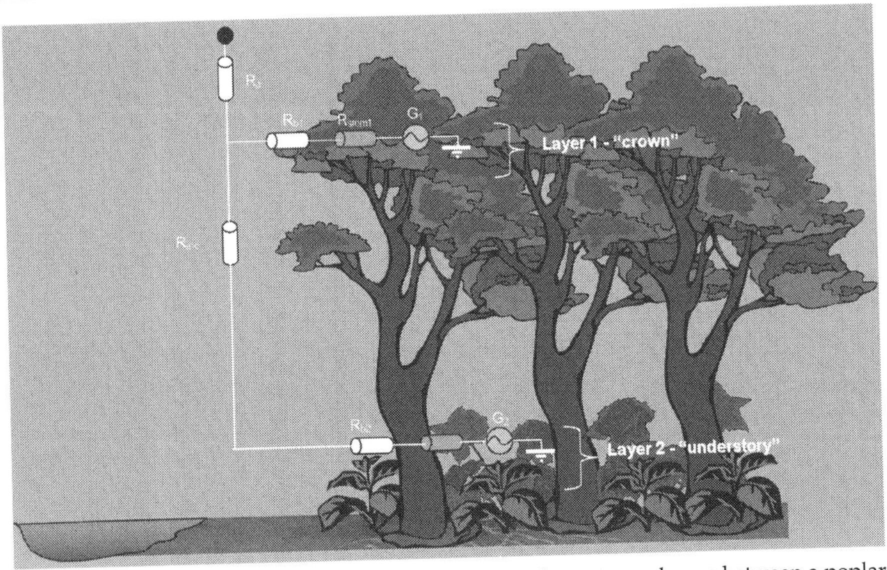

Fig. 4. A multi-layer multiple source model to estimate the water exchange between a poplar plantation ecosystem and the atmosphere.

5.1 The stomatal conductance sub-model

To describe the physiological behaviour of the bulk stomatal conductance (G_s) a Jarvis-Stewart multiplicative model was used, according to the following formulation:

$$G_s = g_{smax} \cdot [f(PHEN) \cdot f(T) \cdot f(PAR) \cdot f(VPD) \cdot f(SWC)] \quad (24)$$

where g_{smax} is the maximum stomatal conductance expressed by the poplar trees in non-limiting conditions. A maximum value of 1.87 cm s^{-1} (referred to the Projected Leaf Area) has been found in the literature for g_{smax} of poplar leaves located at 2 meter of height in Italian climatic condition (Marzuoli et al., 2009). This value has been reduced to 57% to account for the decreasing of g_{smax} with the canopy height, as proposed by Schafer et al. (2000). Thus a g_{smax} value of 0.8 cm s^{-1} was assumed for the canopy.

The phenology function *f(PHEN)* has been assumed equal to zero when the vegetation was without leaves and equal to one after the leaf burst when the leaves were fully expanded. This was fixed to the 110th day of the year (*DOY*).

Compared to Eq. 2, Eq. 24 includes a limiting function based on phenology *f(PHEN)* which grows linearly from 0 to 1 during the first 10 days after leaves emergence, and decreases linearly in the last 10 days, starting from *DOY* 285th, simulating leaf's senescence:

$$f(PHEN) = \begin{cases} 0 & \forall DOY \le SGS \quad or \quad \forall DOY \ge EGS \\ 1 & \forall (SGS + DayUp) < DOY < (EGS - DayDown) \\ ((\text{DOY} - \text{SGS}) / \text{DayUp}) & \forall\, SGS < \text{DOY} < (SGS + DayUp) \\ ((\text{EGS} - \text{DOY}) / \text{DayDown}) & \forall\, (EGS - DayDown) < \text{DOY} < \text{EGS} \end{cases} \quad (25)$$

SGS and *EGS* are the days for the start and the end of the growing season respectively. . *DayUp* and *DayDown* are the number of days necessary to complete the new leaves expansion and to complete the leaves senescence, respectively.
The G_s dependence on light was modelled according to Eq. 3 form:

$$f(PAR) = 1 - exp^{-aPAR} \tag{26}$$

where a represents a specie-specific coefficient (0.006 in this study) and *PAR* is the Photosynthetically Active Radiation expressed as μmol photons m^{-2} s^{-1}.
Eq. 4 and Eq. 5 were used for *Gs* dependence on temperature and VPD, respectively.
For soil water content *SWC* a different limiting function, from that reported by Sterwart (1988), was used. The boundary-line analysis revealed that SWC exerted its influence on g_{smax} according to the following equation:

$$f(SWC) = \max\left\{0.1; \min\left[1; g \cdot SWC^{(h/SWC)}\right]\right\} \tag{27}$$

where *SWC* is expressed as fraction of soil field capacity while g and h are two coefficients whose values are respectively 1.0654 and 0.2951.
The bulk stomatal conductance of the understory vegetation was modelled using the same parameterization but assuming a g_{smax} value equal to 1.87 cm s^{-1}. The inherent approximation is that the understory vegetation was entirely composed of young poplar plantlets.

	Parameter	*Value*	*Unit*
	g_{max} (H_2O)	0.8	cm/s
f_{PHEN}	*SGS*	110	DOY
	EGS	285	DOY
	DayUp	10	Days
	DayDown	10	Days
f_{PAR}	a	0.006	adim.
f_T	T_{opt}	27	°C
	T_{max}	36	°C
	T_{min}	12	°C
	b	0.5625	adim.
f_{VPD}	c	3.7	KPa
	d	2.1	KPa
f_{SWC}	g	1.0654	adim.
	h	0.2951	adim.

Table 1. Values of the f limiting functions coefficients and g_{smax} for the stomatal conductance model of *Populus nigra*.

5.2 The soil sub-model

The water availability in the soil was modelled using a simple "bucket" model. In this paradigm the soil is considered as a bucket and the water content is assessed dynamically, step by step, via the hydrological balance between the water inputs (rains) and outputs (plant consumption) occurred in the previous time step. The model was initialised assuming the soil water saturated at the beginning of the season and assuming a root depth for soil exploitation of 3 m:

$$AWHC = (\theta_{FC}-\theta_{WP}) \cdot 1000 \cdot RootDepth = 243 \text{ mm H}_2\text{O / m}^3 \text{ soil} \quad (28)$$

$$AW_t\text{=0} = AWHC \text{ (mm)} \quad (29)$$

where *AWHC* is the available water holding capability of the sandy soil between the wilting point (θ_{WP}= 0.114 m^3 m^{-3} for our sandy loam soil) and the field capacity (θ_{FC}=0.195 m^3 m^{-3}). The running equations were:

$$ET_{t-1} = F_{H20,\,t-1} \cdot 3600 / \lambda \text{ (mm)} \quad (30)$$

$$AW_t = AW_{t-1} + Rain_{t-1} - ET_{t-1} \text{ (mm)} \quad (31)$$

$$SWC_t = AW_t / AWHC \text{ (\% of FC)} \quad (32)$$

Eq. 32 represents the water loss of plant ecosystem through the transpiration of the two layers ($F_{H20,\,t-1}$) in the previous time step. Since water fluxes are expressed as rates (mm s^{-1}), for an hourly time step, as in our cases, their values must be multiplied by 3600 in order to get the water consumed in one hour.
AW_t is the available water in the soil after water inputs and consumptions. The effects of runoff and groundwater level rising have been neglected due to the flatness of the ecosystem and the groundwater level which were deeper than the root exploration depth.
SWC represents the soil water content expressed as percentage of field capacity, as requested by the *f(SWC)* function of the stomatal sub-models.

5.3 The atmospheric sub-model and the resistive network

The resistance R_a was calculated by using Eq. 19 and Eq. 21, with z_m=33 m the measurement height, h= 26.3 m the canopy height, u^* the friction velocity, u the horizontal wind speed, L the Monin-Obhukhov length, $d=2/3 \cdot h$ the zero-plane displacement height and $z_0=1/10 \cdot h$ the roughness length.
The laminar sub-layer resistances of the layers 1 and 2 (R_{b1} and R_{b2}) were both calculated using the Eq. 23 given u^*.
The stomatal resistances of the layers 1 and 2 (R_{stom1} and R_{stom2}) were calculated using the stomatal sub-model after having estimated the leaf temperatures from the air temperature T and the heat fluxes H:

$$T_l = T + H \cdot (R_a + R_{b,\,heat}) / (\rho \cdot c_p) \quad (33)$$

where $R_{b,heat}$ was calculated using the Eq. 23 with $Sc=0.67$ and $Pr=0.71$.
Then the vapour pressure deficit $VPD = e_s(T_l) - e(T)$ was derived from the T_l for the calculation of $e_s(T_l)$ and from the air temperature T and the relative humidity RH for the actual e: $e(T)=UR \cdot e_s(T)$.
The vapour pressure of the saturated air can be calculated from the well-known Teten-Murray empirical equation:

$$e_s(T) = 0.611 \cdot exp(17.269 \cdot (T - 273) / (T - 36)) \quad (34)$$

which gives e_s in kPa when T is expressed as °K.
The stomatal resistance of the crown R_{stom1} was obtained as the reciprocal of the stomatal conductance obtained by the Jarvis–Stewart sub-model fed with PAR, T_{leaf}, VPD and SWC_t, the latter being the soil water content calculated with the Eq. 32 .

The understory R_{stom2} was obtained in a similar way but considering a understory g_{max} (=1.87 cm s^{-1}) and the *PAR fraction* reaching the below canopy vegetation instead of the original *PAR*:

$$PAR_{fraction} = \exp(-k \cdot LAI_1) \quad (35)$$

where k is the light extinction factor within the canopy, set to 0.54, and LAI_1 is the leaf area index of the crown, assumed to be equal to 2 at maximum leaf expansion.
The in-canopy resistance R_{inc} was calculated following Erisman et al. (1994):

$$R_{inc} = (14 \cdot LAI_1 \cdot h) / u^* \quad (36)$$

where h is the canopy height and LAI_1 the leaf area index of the crown.
The stomata of the big leaves of the two layers of Figure 4 (G_1 and G_2) were assumed as water generators driven by the difference of water concentration between the leaves (χ_{sat}), assumed water saturated al leaf temperature T_l, and the air (χ_{air}):

$$G_1 = G_2 = \chi_{sat} - \chi_{air} \quad (g\ m^{-3}) \quad (37)$$

where

$$\chi_{sat} = 2.165 \cdot e_s(T_l) / T_l \quad (g\ m^{-3})$$

$$\chi_{air} = 2.165 \cdot e(UR, T) / T \quad (g\ m^{-3})$$

being 2.165 the ratio between the molar weight of water molecules M_w (18 g mol^{-1}) and the gas constant R (8.314 J mol^{-1} K^{-1}) if e and e_s are expressed in Pa (multiplied by 1000 if expressed in kPa).
Then the total water flux of the ecosystem F_{H2O} could be calculated by composing all the resistances and the generators within the modelled resistive network, following the electrical composition rules for resistances and generators in series and in parallel, and applying the scaling strategy according to the *LAI*:

$$R_1 = (R_{b1} + R_{stom1} / LAI_1) \quad (s/m) \quad (38)$$

$$R_2 = (R_{b2} + R_{stom2} / LAI_2) \quad (s/m) \quad (39)$$

$$R_3 = R_{inc} + R_2 \quad (s/m) \quad (40)$$

$$G_{eq} = G_2 - (G_2 - G_1) \cdot R_3 / (R_1 + R_3) \quad (g\ m^{-3}) \quad (41)$$

$$R_{eq} = R_1 \cdot R_3 / (R_1 + R_3) \quad (s/m) \quad (42)$$

$$F_{H2O} = G_{eq} / (R_{eq} + R_a) / 1000 \quad (kg\ m^{-2}\ s^{-1} = mm\ s^{-1}) \quad (43)$$

where LAI_2 is the leaf area index of the understory vegetation (=0.5)

5.4 Comparison with EC measurements

Concurrent measurements of λE were performed over the same ecosystem by means of eddy covariance technique with instrumentation set-up according to Gerosa et al. (2005).

The comparison between the direct λE measurements and the modelled ones allowed the evaluation of model performance.
The model performance was very good in predicting the hourly variation of λE both during the summer season (*Modeled* = 0.885 · *Measured* + 8.4389; R^2=0.85, p<0.001, n=1872) with a slight tendency to underestimate the peaks.
An example of the comparison exercise for a summer week is shown in Figure 5

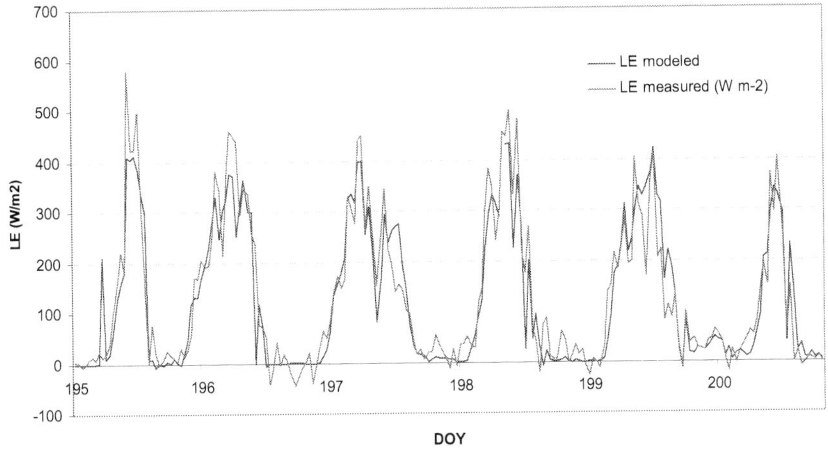

Fig. 5. Comparison between modelled and measured λE, expressed as W m^{-2} unit

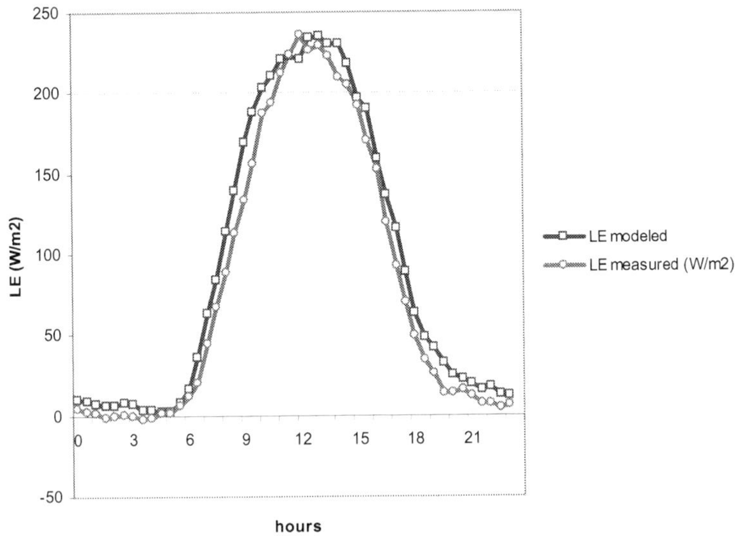

Fig. 6. Mean daily course of the modeled λE compared to the measured one. All the available hourly measurements were considered (n=3914)

For the whole year the performance was less good (*Modeled* = 0.876 · *Measured* + 21.443; R^2=0.68, p<0.001, n=3914) but still acceptable, especially in reproducing the average daily course of λE (Figure 6).

6. Conclusions

Scientific literature provides many ways (e.g. FAO) to estimate the evapotranspiration of a vegetated surface. Sometimes there is the need to predict this process at a very high-time resolution (e.g. hourly means). Hourly estimations of evapotranspiration, for example, are important in all the applications and the methodologies which couple transpiration process with carbon assimilation or air pollutants uptake by plants.

In these cases, the *big-leaf* approach, together with the resistive analogy which simulates the gas-exchange between vegetation and atmosphere, is a simple but valid example of a process-based model which includes the stomatal conductance behaviour, as well as a basic representation of the canopy features.

7. Acknowledgements

This publication was partially funded by the Catholic University's program for promotion and divulgation of scientific research.

8. References

Ball J.T., Woodrow I.E., Berry J.A., 1987. A model predicting stomatal conductance and its contribution to the control of photosynthesis under different environmental conditions. In: Biggins J, ed. *Progress in photosynthesis research*. Dordrecht: Martinus Nijhoff Publishers, 221-224

Chamberlain A.C., Chadwick R.C., 1953. Deposition of airborne radioiodine vapour. *Nucleonics* 11, 22–25

Choudhury B.J., Monteith J.L., 1988. A four-layer model for heat budget of homogeneous land surfaces. *Quarterly Journal of the Royal Meteorological Society* 114, 373–398

Damour G., Simonneau T., Cochard H., Urban L., 2010. An overview of models of stomatal conductance at leaf level. *Plant, Cell & Environment* 33, 1419-1438.

De Pury D.D.G., Farquhar G.D., 1997. Simple scaling of photosynthesis from leaves to canopies without the errors of big-leaf models. *Plant, Cell & Environment* 20, 537-557.

Dewar R.C., 2002. The Ball-Berry-Leuning and Tardieu-Davies stomatal models: synthesis and extension within a spatially aggregated picture of guard cell function. *Plant, Cell & Environment* 25: 1383-1398

Erisman J.W., Van Pul A., Wyers P., 1994. Parameterization of surface-resistance for the quantification of atmospheric deposition of acidifying pollutants and ozone. *Atmospheric Environment* 28, 2595–2607

Farquhar G.D., Dubbe D.R., Raschke K., 1978. Gain of the feedback loop involving carbon dioxide and stomata: theory and measurement. *Plant physiology* 62: 406-412

Gerosa G., Derghi F., Cieslik S., 2007. Comparison of Different Algorithms for Stomatal Ozone Flux Determination from Micrometeorological Measurements. *Water Air & Soil Pollution* 179, 309-321.

Gerosa G., Marzuoli R., Desotgiu R., Bussotti F., Ballarin-Denti A., 2008. Visible leaf injury in young trees of *Fagus sylvatica* L. and *Quercus robur* L. in relation to ozone uptake and ozone exposure. An Open-Top Chambers experiment in South Alpine environmental conditions. *Environmental Pollution* 152, 274–284.

Gerosa G., Vitale M., Finco A., Manes F., Ballarin Denti A. and Cieslik S., 2005. Ozone uptake by an evergreen Mediterranean forest (*Quercus ilex*) in Italy. Part I: Micrometeorological flux measurements and flux partitioning. *Atmospheric Environment* 39, 3255-3266.

Grünhage L., Haenel H.D., Jager H.J., 2000. The exchange of ozone between vegetation and atmosphere: micrometeorological measurement techniques and models. *Environmental Pollution* 109, 373–392.

Hicks B.B., Baldocchi, D.D., Meyers T.P., Hosker R.P., Matt D.R., 1987. A Preliminary multiple resistance routine for deriving dry deposition velocities from measured quantities. *Water, Air and Soil Pollution* 36, 311–330.

Holtslag A.A.M., van Ulden A.P., 1983. A simple scheme for daytime estimates of the surface fluxes from routine weather data. *Journal of Climate and Applied Meteorology* 22, 517–529.

Jarvis P.G., 1976. The interpretation of the variations in leaf water potential and stomatal conductance found in canopies in the field. *Philosophical Transactions of the Royal Society of London*, Series B 273, 593–610.

Körner C., Scheel J., Bauer H., 1979. Maximum leaf diffusive conductance in vascular plants. Photosynthetica 13, 45-82.

Leuning R., 1995. A critical appraisal of a combined stomatal photosynthesis model for C3 plants. *Plant, Cell & Environment*. 18, 339–355.

Marzuoli R., Gerosa G., Desotgiu R., Bussotti F., Ballarin-Denti A., 2008. Ozone fluxes and foliar injury development in the ozone-sensitive poplar clone Oxford *(Populus maximowiczii x Populus berolinensis)*: a dose–response analysis. Tree Physiology 29, 67-76.

Monin A.S., Obukhov A.M., 1954. Basic laws of turbulent mixing in the atmosphere near the ground. *Translation in Aerophysics of Air Pollution* (In: Fay, J.A., Hoult D.P. (Eds.), AIAA, New York, 1969, pp. 90–119). Akademija Nauk CCCP, Leningrad, TrudyGeofizich eskowo Instituta 151(24), 163–187.

Monteith J.L., 1981. Evaporation and surface temperature. *Quarterly Journal of the Royal Meteorological Society* 107, 1–27.

Patwardhan S., Pavlick, R. Kleidon A., 2006. Does the empirical Ball-Berry law of stomatal conductance emerge from maximization of productivity? *American Geophysical Union*, Fall Meeting 2006, abstract #H51C-0498.

Schafer K.V.R., Oren R., Tenhunen J.D., 2000. The effect of tree height on crown level stomatal conductance. *Plant, Cell & Environment* 23, 4 365-375.

Stewart J.B. (1988) Modelling surface conductance of pine forest. *Agricultural and Forest Meteorology* 43, 19–35.

Thom A.S., 1975. Momentum, mass and heat exchange of plant communities. In Vegetation and Atmosphere. Ed. J.L. Monteith. Academic Press, London.

Unsworth M.H., Heagle A.S., Heck W.W., 1984. Gas Exchange in open field chambers – I. Measurement and analysis of atmospheric resistance to gas exchange. Atmospheric *Environment* 18, 373–380.

Webb R.A., 1972. Use of the boundary line in the analysis of biological data. *Journal of Horticultural Science* 47, 309-319.

19

A Distributed Benchmarking Framework for Actual ET Models

Yann Chemin
International Water Management Institute
Sri Lanka

1. Introduction

With the various types of actual ET models being developed in the last 20 years, it becomes necessary to inter-compare methods. Most of already published ETa models comparisons address few number of models, and small to medium areas (Chemin et al., 2010; Gao & Long, 2008; García et al., 2007; Suleiman et al., 2008; Timmermans et al., 2007). With the large amount of remote sensing data covering the Earth, and the daily information available for the past ten years (i.e. Aqua/Terra-MODIS) for each pixel location, it becomes paramount to have a more complete comparison, in space and time.
To address this new experimental requirement, a distributed computing framework was designed, and created. The design architecture was built from original satellite datasets to various levels of processing until reaching the requirement of various ETa models input dataset. Each input product is computed once and reused in all ETa models requiring such input. This permits standardization of inputs as much as possible to zero-in variations of models to the models internals/specificities.

2. Theoretical points of observation

2.1 Net radiation and soil heat flux

In the two-source energy balance approach, like TSEB and SEBS differ from the single-source concept of SEBAL and METRIC in the sense that the radiation and energy balances have separate formulations for either bare soil or canopy. The energy balance at any instantaneous moment is expressed by equation Eq. 1.

$$Rn = G + H + LE \tag{1}$$

Where Rn is Net Radiation, G is soil heat flux, H is sensible heat flux and LE is latent heat of vaporization. This is what is appearing in single-source models like SEBAL and METRIC. Single source models concentrate on identifying Rn and G from astronomical and semi-empirical equations respectively, while H is being iteratively solved based on thermodynamically exceptional geographical locations, often referred in literature (Bastiaanssen, 1995) as *wet* and *dry* pixels, also the technique to identify them is referred in more recent literature as *end-members selection/identification* (Timmermans et al., 2007).

In two-source models, it is separated into bare soil and canopy energy balances as in Eq. 2 and 3, respectively:

$$Rns = G + Hs + LEs \tag{2}$$

Where Rns is the net radiation of bare soil surface, Hs is the sensible heat flux from bare soil, LEs is the latent heat of vaporization from soil surface.

$$Rnc = Hc + LEc \tag{3}$$

Where Rnc is the net radiation from canopy of crop, Hc is the sensible heat flux from canopy, LEc is the latent heat of vaporization of crop. Once the elements of those two equations are found, the fraction of vegetation cover (fc) is used to combine them into the area of a satellite remote sensing pixel, which is inherently a mixel of bare soil and canopy.

The Net Radiation is partitioned according to the formulation commonly used in two-sources model (Eq. 4 and 5), where the soil partition of Rn is an LAI-based extinction coefficient (Choudhury, 1989) with a coefficient C ranging from 0.3 to 0.7 (Friedl, 2002), depending on the arrangement of the canopy elements. Friedl (2002) mentions that a canopy with spherical (random) leaf angle distribution would lead to a C value of 0.5.

$$Rns = Rn\, e^{\frac{-C\, LAI}{Cos(sunza)}} \tag{4}$$

$$Rnc = Rn - Rns \tag{5}$$

Where LAI is the leaf area index, sunza is the sun zenith angle. Friedl (2002) mentions that he derived his soil heat flux formulation from his previous work (Friedl, 1996). It takes the already available soil fraction of net radiation and the cosine of the sun zenith angle (Eq. 6). A coefficient is then multiplied to those whereby soil type and moisture conditions are taken into consideration after (Choudhury et al., 1987).

$$G = Kg\, Rns\, Cos(sunza) \tag{6}$$

Where Kg is the soil type and moisture condition coefficient in the soil heat flux. The Fraction of Vegetation cover is necessary to split the two-sources of heat transfer studied in such models. They are the soil surface (bare soil) and the vegetation canopy surface. The fraction of vegetation cover from Jia et al. (2003) quoting Baret et al. (1995) is developed as in Eq. 7:

$$fc = 1 - [\frac{(NDVI - NDVI_{min})}{(NDVI_{min} - NDVI_{max})}]^K \tag{7}$$

with K being taken as 0.4631 in Jia et al. (2003) and NDVImin at LAI=0 and NDVImax at LAI = +INF. As can be seen, a very large weight of potential deviation from the expected result is resting in the proper assessment of fc (Eq. 7). There are also uncertainties in the LAI raster input (Yang, Huang, Tan, Stroeve, Shabanov, Knyazikhin, Nemani & Myneni, 2006; Yang, Tan, Huang, Rautiainen, Shabanov, Wang, Privette, Huemmrich, Fensholt, Sandholt, Weiss, Ahl, Gower, Nemani, Knyazikhin & Myneni, 2006).

The soil heat flux computed for Bastiaanssen (1995), is what could be called a *partial contribution* of soil heat flux to the energy balance of the pixel, as the semi-empirical relationship is proportional to various elements of thermodynamic forcing within each pixel (Eq. 8).

$$G = \frac{Rn}{Albedo}\, T_c\, (0.0032(\frac{Albedo}{r_0}) + 0.0062(\frac{Albedo}{r_0})^2)\,(1 - 0.978NDVI^4) \tag{8}$$

with T_c the temperature in Celsius and r_0 the Albedo to apparent Albedo correction ranging 0.9 to 1.1 depending on the time of the day.
SEBS uses a two-source Albedo anchors stretching equation multiplied by the soil fraction of the pixel to extract a percentage of the net radiation as soil heat flux (Eq. 9).

$$G = Rn\,(Albedo_{dark} + (1 - f_c)\,(Albedo_{bright} - Albedo_{dark})) \tag{9}$$

Generic values are $Albedo_{dark} = 0.05$ and $Albedo_{bright} = 0.35$, while adjustements are made when concentrating on a specific land use, eventually.

2.2 Monin-Obukhov Similarity Theory

The Monin-Obukhov Similarity Theory (Monin & Obukhov, 1954) is being used in single source and two-source energy balance models. It is interesting to note that Monin & Obukhov (1954), in the development of their Monin-Obukhov Similarity Theory (MOST) considered the friction velocity to be about 5% of the geostrophic wind velocity having an average speed of 10m/s results in the friction velocity being around 0.5 m/s, and with the Coriolis parameter $l = 10^{-4}s^{-1}$ and a tolerance of 20%, an estimate of the height of the surface layer is found at h=50m, that is also the DisALEXI blending height for air temperature (Norman et al., 2003). The dynamic velocity within this layer can be considered near to constant and the effect of Coriolis Force neglected (Monin & Obukhov, 1954). Under those conditions of neutral stratification the processes of turbulent mixing in the surface layer can be described by the logarithmic model of the boundary layer (Eq. 10).

$$\begin{aligned}
L &= \frac{-1004\,\rho u^3 T}{kgH} \\
most_x &= (1 - 16\,\frac{h}{L})^{\frac{1}{4}} \\
\psi_h &= 2\,log(\frac{1 + most_x^2}{2}) \\
\psi_m &= 2\,log(\frac{1 + most_x}{2}) + log(\frac{1 + most_x^2}{2}) - 2\,atan(most_x) + 0.5\pi
\end{aligned} \tag{10}$$

with ψ_m, ψ_h the diabatic correction of momentum and heat through their changes of states, $most_x$ a MOST internal parameter, L the Monin-Obukhov Length (MOL), k is the von Karman constant, g the gravity acceleration, u is the wind speed, ρ is the air density, T is the temperature and h is the height of interest (measurement height of the wind speed, roughness length, etc.).
Constraints to MOST as found in Bastiaanssen (1995) are of two types, first avoiding the latent heat flux input to be nil as its input location is in the denominator of the MOL equation (Equation Eq. 11), the second constraint is when the MOL is becoming positive, to force ψ_m and ψ_h to a ranged negative value (Bastiaanssen, 1995).

$$\begin{aligned}
&if(H = 0.0) : L = -1000.0 \\
&if(L > 0.0) : \psi_h = \psi_m = -5\frac{2}{L}
\end{aligned} \tag{11}$$

It turns out that Su (2002), extending the reach of his SEBS model to the GCM community has included a dual model for the convective processes within the Atmospheric Boundary Layer (ABL). Su (2002) followed the observations of Brutsaert (1999) that the ABL lower layer is either stable, either unstable and that the thickness of this lower layer is α = 10-15% of the ABL height, which is about β = 100-150 times the surface roughness. SEBS takes the highest from both as its estimation of h_{st}, the height of ABL sublayer separation. If the reference height is lower than that, then the lower sublayer model is run, otherwise the upper sublayer models is used.

The cutline between the two sublayers of the ABL permits SEBS to process the lower layer (Atmospheric Surface Layer, ASL) under the MOST paradigm (Eq. 12), whether it proves unstable (generally in the day) or stable (generally in the night). The momentum and heat functions for the upper sublayer of the ABL where flow is laminar (free convection) can then be merged with the ASL by what Su (2002) calls a Bulk ABL Similarity (BAS) stability correction set of functions called here ξ_m and ξ_h for momentum and heat respectively (Eq. 13).

$$
\begin{aligned}
&if(\frac{-z_0}{L} < 0.0) : \xi_m = -2.2\, alog(1 + \frac{z_0}{L}) \\
&if(z_0 < \frac{\alpha}{\beta}h_i) : \xi_m = -alog(\alpha) + \psi_m(\frac{-\alpha h_i}{L}) - \psi_m(-\frac{z_0}{L}) \\
&if(z_0 \geq \frac{\alpha}{\beta}h_i) : \xi_m = alog(\frac{h_i}{\beta z_0}) + \psi_m(\frac{-\beta z_0}{L}) - \psi_m(-\frac{z_0}{L})
\end{aligned}
\tag{12}
$$

$$
\begin{aligned}
&if(\frac{-z_0}{L} < 0.0) : \xi_h = -7.6\, alog(1 + \frac{z_0}{L}) \\
&if(z_0 < \frac{\alpha}{\beta}h_i) : \xi_h = -alog(\alpha) + \psi_h(\frac{-\alpha h_i}{L}) - \psi_h(-\frac{z_0}{L}) \\
&if(z_0 \geq \frac{\alpha}{\beta}h_i) : \xi_h = alog(\frac{h_i}{\beta z_0}) + \psi_h(\frac{-\beta z_0}{L}) - \psi_h(-\frac{z_0}{L})
\end{aligned}
\tag{13}
$$

with z_0 the surface roughness for momentum, h_i the height of ABL or Planetary Boundary Layer (PBL). The formulation of ψ_m and ψ_h functions are inherited from Beljaars & Holtslag (1991) and include either correction weights inside the standard equations in some cases (unstable conditions of ψ_m and ψ_h), either a polynomial with exponential in other cases (stable conditions of ψ_m and ψ_h). However, Beljaars & Holtslag (1991) stated categorically that the data described are characteristic for grassland and agricultural land *with sufficient water supply*.

2.3 Roughness height

Allen et al. (2005) mentions that METRIC and SEBAL do not require knowledge of crop type (no satellite based crop classification is needed). SEBAL relies on a type of semi-empirical equation relating NDVI to the roughness length (also called roughness height) for momentum and heat (Eq. 14).

$$z_{0m} = e^{a+b\,NDVI} \tag{14}$$

Among many others, Chandrapala & Wimalasuriya (2003) proposed $a = -5.5$ and $b = 5.8$ for Sri Lanka using AVHRR NDVI images (sensor response curves, atmospheric correction and pixel size are all influencing NDVI response). Bastiaanssen (1995) preferred using extrema conditions to define a and b.

SEBS takes a ground truth added to mapping point of view and uses a look up table to translate land use raster maps into roughness length.

Land Cover	**NDVI**	z_{0m}
Vegetation	$NDVI_{max}$	$\frac{h_v}{7}$
Desert	0.02	0.002

Table 1. Boundary conditions on z_{0m} from NDVI satellite data after Bastiaanssen (1995)

While SEBAL and METRIC use a fixed conversion rate between roughness length for momentum and roughness length for heat, Su (2002) is introducing in SEBS the use of the exponential of kB^{-1} (Eq. 15).

$$z_{0m} = 0.136\, h_{vegetation}$$
$$z_{0h} = \frac{z_{0m}}{e^{kB^{-1}}} \tag{15}$$

with $h_{vegetation}$ the height of the vegetation relating to the roughness length for momentum z_{0m} from Brutsaert (1982), z_{0h} the roughness length for heat, Su (2002) refers to the work of Massman (1999) on the combined *von Karman constant - sublayer Stanton number* (kB^{-1}), where B^{-1} is defined by Gieske (2007) as in Eq. 16:

$$B^{-1} = St_k^{-1} - C_d^{-\frac{1}{2}}$$
$$\frac{St_k^{-1}}{u_*} = \frac{\rho C_p \Delta T}{H} \tag{16}$$

C_d is the drag coefficient, St_k^{-1} is the roughness Stanton number, u_* is the friction velocity, ρ the air density, C_p the specific heat and ΔT the temperature difference.

2.4 Aerodynamic roughness for heat

The aerodynamic resistance (roughness) for heat r_{ah} is an input to the sensible heat flux and is often a source of concentration in ET models based on energy balance. For logical reasons, as the parameterization of r_{ah} needs prior knowledge of the state of the sensible heat flux to enable knowledge of the MOL to parameterize ψ_m and ψ_h the diabatic correction of momentum and heat through their changes of states. In turn, ψ_m and ψ_h, offset the logarithmic relation of the observation height to the respective roughness lengths (z_{0m} and z_{0h}) being the driving force to curve the relation to the wind shear profile.

SEBS resistance at the wet limiting case re_{wet} (Eq. 17) is using the MOL as L_{wet} configured to use all the energy available for evaporation, which in turn is used in conjunction with the reference height (z_{ref}) either in the computation of the ψ_h or χ_h whether ASL or BAS models are at work.

$$re_{wet} = \frac{alog(\frac{z_{ref}}{L_{wet}} - [\psi|\chi]_h)}{ku^*} \tag{17}$$

2.5 Ground to air temperature difference

Allen et al. (2005) mentions that METRIC is a variant of the important model SEBAL and that it has been extended to provide tighter integration with ground-based reference ET. SEBAL formulation for the ground to air temperature difference (dT) is estimated (Eq. 18) as an affine function with two extreme conditions found in the satellite image processed (Bastiaanssen,

1995).

$$
\begin{aligned}
&pixel_{cold} \mapsto dT = 0 \\
&pixel_{hot} \mapsto dT = \frac{(Rn - G) \times r_{ah}}{\rho C_p} \\
&dT = a + b \times T
\end{aligned}
\tag{18}
$$

with r_{ah} the aerodynamic resistance for heat, $pixel_{cold}$ and $pixel_{hot}$ the end-members defined in Bastiaanssen (1995); Bastiaanssen et al. (1998); Timmermans et al. (2007) and that are the *signet ring* of the model.
METRIC formulation (Eq. 19) includes the reference ET paradigm found in Allen et al. (1998), the $Kc \times ETo$ crop ET, also called ETc, being translated into METRIC as $k \times ET_r$ (Allen et al., 2005), practically using Alfalfa at full growth as anchor point in their Idaho study. Some extra metorological data being available when using METRIC, permits a daily surface soil water balance to be run to enforce conditions on the *dry/hot* pixel energy balance and effective dT. Selection of the extreme pixels is focused on cropped area as much as possible.

$$
\begin{aligned}
&pixel_{cold} \mapsto dT = \frac{(Rn - G - k \times ET_r) \times r_{ah}}{\rho C_p} \\
&pixel_{hot} \mapsto dT = \frac{(Rn - G) \times r_{ah}}{\rho C_p} \\
&dT = a + b \times T
\end{aligned}
\tag{19}
$$

SEBS (Su, 2002) is computing dT (Eq. 20) from a surface skin virtual temperature (T_0) and PBL virtual temperature (T_{pbl}).

$$
\begin{aligned}
T_v &= \frac{log(h_{pbl} - h_{disp})}{log(h_u - h_{disp})} \\
T_{pbl} &= \frac{T_s \times (1 - f_c) + T_v \times f_c}{\frac{1-DEM}{44331.0}^{1.5029}} \\
T_0 &= \frac{T_c}{\frac{1-DEM}{44331.0}^{1.5029}} \\
dT &= T_0 - T_{pbl}
\end{aligned}
\tag{20}
$$

Where T_s is the soil temperature, T_v is the (virtual) vegetation canopy temperature, f_c is the fraction of vegetation cover, DEM is the elevation and T_c is the satellite sensed temperature in Celsius, h_u is the wind speed measurement height. h_{disp} is the displacement height being 0.65 of the canopy height in SEBS, Monin & Obukhov (1954) mention that for observations made at height superior to 1 meter, the displacement height can be nullified. The blending height (h_{pbl}) is given by an external mean or if no data is available, default value of 200m is used (same as in SEBAL) or 1000m.

2.6 Actual ET

Single sources models (METRIC and SEBAL) have promoted a particular way of closing the energy-balance (Eq. 21), using a ratio of fluxes called the evaporative fraction (Λ; Eq. 21).

$$\Lambda = \frac{Rn - G - H}{Rn - G} \tag{21}$$

This instantaneous ratio (could be called an efficiency, since it is unitless) is multiplied with a potential value of ET in order to provide with an Actual ET (equation Eq. 22).

$$ET_a = \Lambda \, ET_{potential} \tag{22}$$

METRIC (Allen et al., 2007) evaporative fraction is used with the potential ET based on Penman-Monteith method as published in Allen et al. (1998) in order to produce an actual ET estimation. In contrast SEBAL (Bastiaanssen, 1995; Bastiaanssen et al., 1998) evaporative fraction is used with the potential ET computed from exo-atmospheric solar radiation, a single-way atmospheric transmissivity, and the reflected proportion from Albedo.
SEBS (Su, 2002) is computing an instantaneous latent heat flux (LE) from the evaporative fraction and the subtraction of net radiation and soil heat flux (Eq. 23). However, using what is called in SEBS the *relative evaporation*, the pixel value of H the sensible heat flux is allowed to vary only between H_{wet} and H_{dry}.

$$\begin{aligned}
H_{dry} &= Rn - G \\
H_{wet} &= \frac{(Rn - G) - \frac{\left(\frac{\rho C_p}{r_{e_wet}} e_{sat}\right)}{\gamma}}{\frac{1+slope}{\gamma}} \\
Enforce &: H_{wet} \leq H \leq H_{dry} \\
E_{relative} &= 1 - \frac{H - H_{wet}}{H_{dry} - H_{wet}} \\
\Lambda &= E_{relative} \frac{Rn - G - H_{wet}}{Rn - G}
\end{aligned} \tag{23}$$

H_{wet} is derived from the Priestley-Taylor equation (Priestley & Taylor, 1972), from which γ and $slope$ come from, r_{e_wet} is the resistance at the wet limiting case. $E_{relative}$ is the relative evaporation.
SSEB (Senay et al., 2007) is evaluating a regional approximation of the evaporative fraction Λ as a ranging of satellite-based temperature products (Eq. 24) and uses the reference ET found in Allen et al. (1998) as a mean to compute the actual ET.

$$\Lambda = \frac{T_{hot} - T}{T_{hot} - T_{cold}} \tag{24}$$

3. Methodology

3.1 Conceptual design and processing flow

The distributed framework is a Linux system based on GDAL library (GDAL, 2011) and C programming, enhanced with a distributed language called OpenMP (OpenMP, 2011), used essentially for data distribution as seen in (Chemin, 2010).
As the conceptual architecture of the framework (Fig. 1) signifies, there are several layers of processing involved. Initially, downloaded satellite imagery is located in a single directory (referred as *RS data* in Fig. 1).

During the pre-processing phase, a parsing agent will select images of the same kind and same date and stitch them together, then reproject them to the projection system selected. Upon completion, a second agent will perform a relatively conservative quality check and assign null value to failed pixels according to the satellite imagery information available.

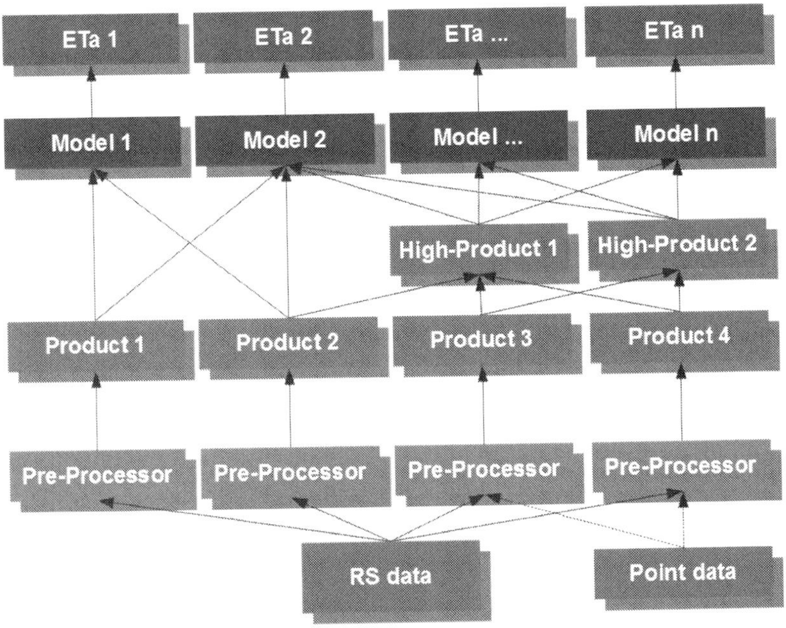

Fig. 1. Architecture Concept of the Framework

Once these two steps are performed, the raster maps are tagged as *products*. Those products will be shared in between any of the ET models that require such type of input.
Some ET models require some higher-level input raster maps, by this, we define *higher-level product* as a raster that requires at least one *product* as defined above as precursor to its creation.

3.2 Meteorological data

Meteorological data (referred as *Point data* in Fig. 1) is encoded with Fourier Transforms (FT) in function of cumulative day of year from the beginning of the satellite imagery data set. This insures both faithfulness of the data and high-portability as well as an elegant way to summarize a complex and variable non-spatial dataset.
Actual state of the research in meteorological data time-series encoding for this framework is to develop an array of geo-tagged FT. Also under consideration are Wavelet Transforms (WT), for reasons of time tagging. This array will be used to interpolate on-the-fly from an appropriate number of neighbours and with an interpolation algorithm suiting best the operator requirement. The reason for this, is that it transfers the load from storage requirements (which can be heavy for daily meteorological raster datasets) to thread computing that is benefiting from distributed speed-up as only a marginal number of equations will be added by such process compared to the ET models processing load.

4. Implementation

Practically (Fig. 2), MODIS datasets are grouped by products and by day and batch processed each in one core of the computer in parallel. This involves format changing, merging tiles, reprojecting, renaming outputs according to the nomenclature of the processing system. The tools involved in that step are either standard Linux Shell tools, either part of GDAL (2011) standard tools (i.e. *gdalwarp* and *gdal_translate*). Both of these tools are still essentially sequential programs at this time, thus, they are being sent to each core in a distributed manner through the Shell with a check loop to ensure that there is at all time the same number of programs running as there are cores/threads available in the CPU architecture. It becomes clear that for each new leap in number of cores in future commercial offerings, the framework will automatically increase its processing capacity to the new enlarged number of cores/threads available, thus also reducing by the same factor the time needed to process a given number of satellite images.

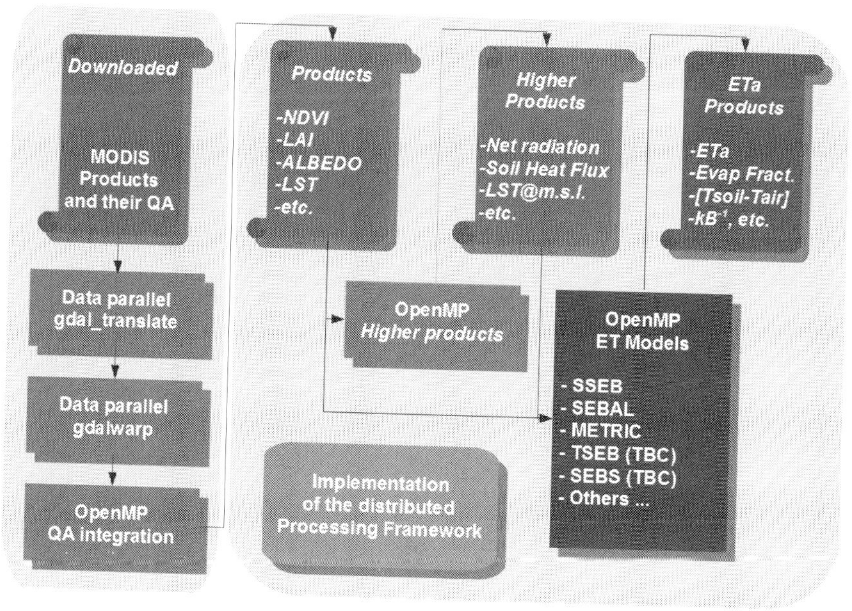

Fig. 2. Architecture Implementation of the Framework

Models that are already inside the framework account to SSEB from Senay et al. (2007), METRIC from Allen et al. (2007), SEBAL from Bastiaanssen et al. (1998) using the work from Alexandridis et al. (2009), in progress are SEBS from Su (2002) and TSEB from both Kustas & Norman (1999) and Norman et al. (1995).

One of the many candidate for inclusion was the Two-Source Algorithm (TSA) from Yunhao et al. (2005). After extensive calculus and referring to the mathematic academia, it became clear that both temperature equations below when combined to extract $T_{vegetation}$ and T_{soil} (Eq. 25) have a large amount of solutions even within the constraining dimension of surface skin temperature range. The available information to possibly close the equations unknown parameters are relating to satellite T after Price (1984) and the satellite-based emissivity (ϵ), partitioned into $\epsilon_{vegetation} = 0.93$ & $\epsilon_{soil} = 0.97$ quoting Sui et al. (1997). Unless missing information or other sources of solution dimension constraints are published, the proposition of the model is so far considered impossible and not included in the framework until such condition is being met through publication.

$$\begin{aligned} T &= f_c\, T_{vegetation} + (1 - f_c)\, T_{soil} \\ \epsilon\sigma T^4 &= f_c\, \epsilon_{vegetation}\sigma T^4_{vegetation} + (1 - f_c)\, \epsilon_{soil}\sigma T^4_{soil} \\ \epsilon &= f_c\, \epsilon_{vegetation} + (1 - f_c)\, \epsilon_{soil} \end{aligned} \tag{25}$$

With ϵ the satellite-based emissivity, $\epsilon_{vegetation}$ & ϵ_{soil} assumed fixed emissivity values for vegetation and bare soil (satellite response dependent), σ the Stephan-Boltzmann constant, T the satellite-based land surface temperature, and $T_{vegetation}$ & T_{soil} the pixel vegetation and soil fractions temperatures. The first equation answers to the geographical two-source proportion within the pixel, while the second equation answers to the two-source flux merging according to Yunhao et al. (2005).

Reference ET models included are Allen et al. (1998) from Cannata (2006), Priestley and Taylor (Priestley & Taylor, 1972) and Hargreaves (Hargreaves et al., 1985), Modified Hargreaves (Droogers & Allen, 2002), Hargreaves-Samani (Hargreaves & Samani, 1985). Only the reference ET from Allen et al. (1998) is being used as a precursor of SSEB (Senay et al., 2007) and METRIC (Allen et al., 2007) actual ET. It was found preponderant to have a minimum group of reference ET models available as baseline for all the work, especially when looking into geographical areas where meteorological data has always been dominant in agricultural literature.

Some models requiring operator intervention (SEBAL, METRIC) have add there internals modified with specially designed heuristics acting as operators. Initial developments were not looking into heuristics but stochastic algorithms. Some efforts using a genetic algorithm were eventually too expensive in processing time, while at the same time end-member selection information were becoming more common (Chandrapala & Wimalasuriya, 2003; Timmermans et al., 2007). Thus heuristics were designed and implemented on a regional basis, initially studied under the Greek conditions for the purpose of Alexandridis et al. (2009) and Chemin et al. (2010). Eventually, the heuristics are extended to fit data sources, continent/climate combinations and model types on an adhoc basis as new regions are included into the geographical scope of research.

5. Initial results

In the case of SEBAL heuristic, the convergence reached 82% of the images processed for the Australian Murray-Darling Basin (1 Million Km^2), enabling the automatic processing of 3635 MODIS multi-tiles images within a single day of computing. Fig. 3 is the output from SEBAL with such heuristic for some irrigated areas in Australia, the total area being processed amounts to more than 5 Billions pixels of ETa values, being multiplied by as many temporary rasters and original data as required for each of the ET models. The Australian irrigation system (less than 100,000 ha) has a sharp, contrasted and well-defined pattern of water depletion, characteristic of continental dry climate with high water supply control for defined periods of the year where crops are in the field.

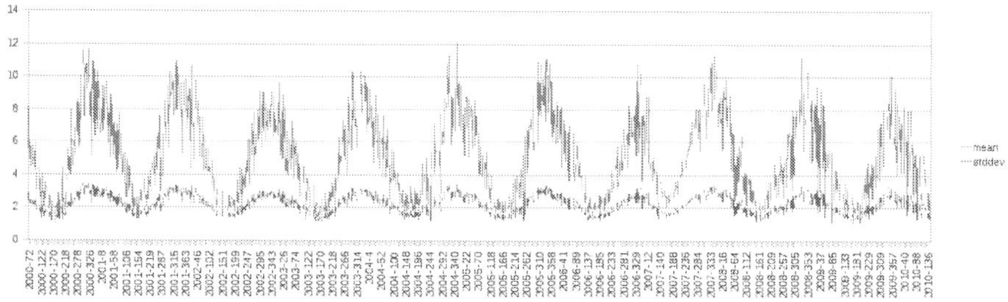

Fig. 3. Daily RS-based ETa (mm/day) in an Australian irrigation system (2000-2010)

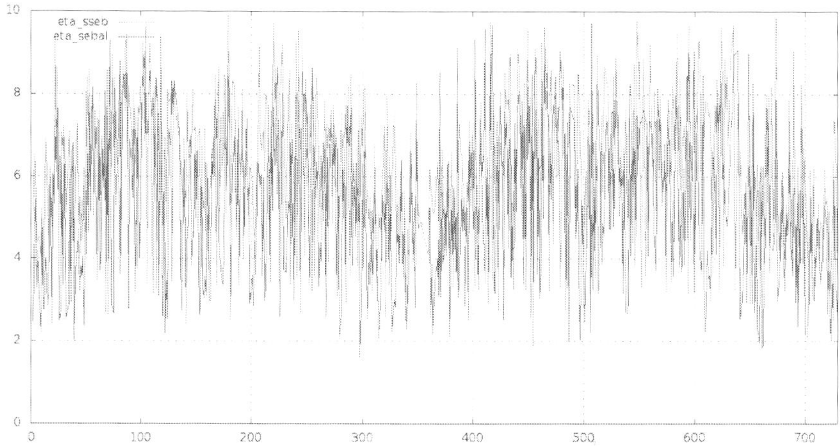

Fig. 4. Daily ETa (mm/day) averages for Sri Lanka (2003-2004)

Looking into the matter of comparing ETa results from different ETa models, Fig. 4 is the averaged ETa output from two models (SEBAL and SSEB) over the tropical island of Sri Lanka in 2003 and 2004. It turns out that the relatively small island of Sri Lanka has an average ETa that is changing much more on a day to day basis than our previous example in Australia. Scale, climate, topography yield exposure to ocean events frequently, having drastic impact on thermodynamics of the island surface as the Fig. 5 also confirms. Changes between models of actual ET from SEBAL and SSEB are relatively constant throughout the RS modeling period. Actual ET from SSEB is in the upper range of SEBAL's one. The work of de Silva (1999) in the dry zone of Sri Lanka and the work of Hemakumara et al. (2003) in the wet zone of Sri Lanka are falling within the expected results found here. Likewise the average evaporative fractions found for Sri Lanka in Fig. 5 are especially leveraging the larger dry zone area of the island with value in the range of 0.3 to 0.5.

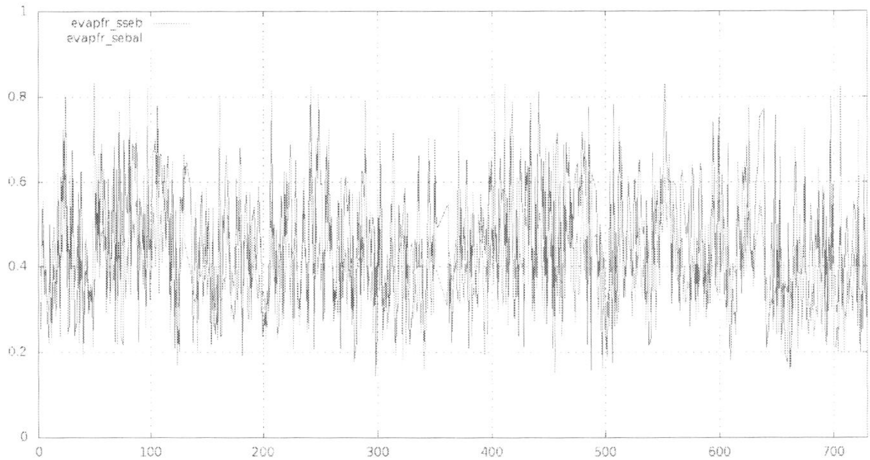

Fig. 5. Instantaneous Evaporative Fractions for Sri Lanka (2003-2004)

6. Conclusion

Challenges to experimentally compare ET models are immense, the theoretical points of comparison are sometimes clear, sometimes rather difficult to pinpoint. To try and address this situation, a framework for benchmarking ET actual models has been designed. Its implementation has embedded parallel data distribution at the base of each parts of the framework to remove the resistance of the data size to process large areas, high frequency and large time period with commonly available computers.

Future work includes the finalization of SEBS (Su, 2002) and TSEB (Kustas & Norman, 1999) integration in the framework, looking for other ETa model candidates to add to existing ones. Also there is a need for designing and creating statistical tools to cross-compare several depths and layers of ETa models processing datasets. Finally, the use of OpenMPI (OpenMPI, 2011) is envisaged for concurrently running several ET models diagnostics in different multi-core machines or OpenCL (Khronos.org, 2011) kernel-based data distributed language to process all analysis as one large computation on a Graphical Processing Unit (GPU).

7. References

Alexandridis, T. K., Cherif, I., Chemin, Y., Silleos, G. N., Stavrinos, E. & Zalidis, G. C. (2009). Integrated methodology for estimating water use in mediterranean agricultural areas, *Remote Sensing* 1(3): 445–465.
URL: *http://www.mdpi.com/2072-4292/1/3/445/*

Allen, R. G., Peirera, L. S., Raes, D. & Smith, M. (1998). *Crop evapotranspiration - Guidelines for computing crop water requirements - FAO Irrigation and Drainage Paper 56*, FAO - Food and Agriculture Organization of the United Nations.
URL: *http://www.fao.org/docrep/x0490e/x0490e00.htm*

Allen, R. G., Tasumi, M., Morse, A., Kramber, W. J. & Bastiaanssen, W. G. M. (2005). Computing and mapping evapotranspiration, *in* U. Aswathanarayana (ed.), *Advances in Water Science Methodologies*, Taylor & Francis, pp. 73–85.

Allen, R. G., Tasumi, M. & Trezza, R. (2007). Satellite-based energy balance for mapping evapotranspiration with internalized calibration (metric) model, *Journal of Irrigation and Drainage Engineering* 133(6): 380–394.
URL: *http://ascelibrary.org/iro/resource/1/jidedh/v133/i4/p380*

Baret, F., Clevers, J. G. P. W. & Steven, M. D. (1995). The robustness of canopy gap fraction estimates from red and near-infrared reflectances: A comparison of approaches, *Remote Sensing of Environment* 54(2): 141–151.
URL: *http://www.sciencedirect.com/science/article/B6V6V-3YYMS35-10/2/dcb889861d7a768d454934f074fe487a*

Bastiaanssen, W. G. M. (1995). *Regionalization of surface flux densities and moisture indicators in composite terrain. A remote sensing approach under clear skies in mediterranean climates*, PhD thesis, Agricultural University of Wageningen.
URL: *http://library.wur.nl/WebQuery/clc/918192*

Bastiaanssen, W. G. M., Pelgrum, H., Wang, J., Ma, Y., Moreno, J. F., Roerink, G. J. & van der Wal, T. (1998). A remote sensing surface energy balance algorithm for land (sebal).: Part 2: Validation, *Journal of Hydrology* 212-213: 213–229.
URL: *http://www.sciencedirect.com/science/article/B6V6C-4CYFRND-H/2/6bc33f78398c9c74ffabe611b3ed1b6b*

Beljaars, A. C. M. & Holtslag, A. A. M. (1991). Flux parameterization over land surfaces for atmospheric models, *Journal of Applied Meteorology* 30(3): 327–341.
URL: *http://journals.ametsoc.org/doi/abs/10.1175/1520-0450%281991%29030%3C0327%3AFPOLSF%3E2.0.CO%3B2*

Brutsaert, W. (1982). *Evaporation in the Atmosphere - Theory, History and Applications*, D. Reidel Publishing Company.
URL: *http://www.springer.com/978-90-277-1247-9*

Brutsaert, W. (1999). Aspects of atmospheric boundary layer similarity under free-convective conditions, *Reviews of geophysics* 37(4): 439–451.
URL: *http://www.agu.org/journals/ABS/1999/1999RG900013.shtml*

Cannata, M. (2006). *GIS embedded approach for Free & Open Source Hydrological Modelling*, PhD thesis, Department of Geodesy and Geomatics, Polytechnic of Milan, Italy.
URL: *http://istgis.ist.supsi.ch:8001/geomatica/index.php?id=1*

Chandrapala, L. & Wimalasuriya, M. (2003). Satellite measurements supplemented with meteorological data to operationally estimate evaporation in sri lanka, *Agricultural Water Management* 58(2): 89–107.
URL: *http://www.sciencedirect.com/science/article/B6T3X-47PPC5B-1/2/d0844a2f9391840f54c00ff0da926a75*

Chemin, Y. H. (2010). *Remote Sensing Raster Programming*, Lulu Eds.
URL: *http://www.lulu.com/product/paperback/remote-sensing-raster-programming/16217224*

Chemin, Y. H., Alexandridis, T. K. & Cherif, I. (2010). Grass image processing environment - application to evapotranspiration direct readout, *OSGEO Journal* 6: 27–31.
URL: *http://www.osgeo.org/ojs/index.php/journal/article/viewFile/131/132*

Choudhury, B. (1989). Estimating evaporation and carbon assimilation using infrared temperature data: vistas in modeling, *in* G. Asrar (ed.), *Theory and applications of remote sensing*, New York Wiley, pp. 628–690.

Choudhury, B., Idso, S. & Reginato, R. (1987). Analysis of an empirical model for soil heat flux under a growing wheat crop for estimating evaporation by an infrared-temperature based energy balance equation, *Agricultural and Forest Meteorology* 39(4): 283–297. URL: *http://www.sciencedirect.com/science/article/B6V8W-4894NGY-9B/2/5668c2204c44352f9fca7d9aef38216c*

de Silva, R. (1999). A comparison of different models of estimating actual evapotranspiration from potential evapotranspiration in the dry zone of sri lanka, *Sabaragamuwa University Journal* 2: 87–100. URL: *http://www.sab.ac.lk/journal/1999/1999A10.pdf*

Droogers, P. & Allen, R. G. (2002). Estimating reference evapotranspiration under inaccurate data conditions, *Irrigation and Drainage Systems* 16: 33–45. URL: *http://dx.doi.org/10.1023/A:1015508322413*

Friedl, M. A. (1996). Relationships among remotely sensed data, surface energy balance, and area-averaged fluxes over partially vegetated land surfaces, *Journal of Applied Meteorology* 35(11): 2091–2103. URL: *http://journals.ametsoc.org/doi/abs/10.1175/1520-0450%281996%29035%3C2091%3ARARSDS%3E2.0.CO%3B2*

Friedl, M. A. (2002). Forward and inverse modeling of land surface energy balance using surface temperature measurements, *Remote Sensing of Environment* 79(2-3): 344–354. URL: *http://www.sciencedirect.com/science/article/B6V6V-44R1BH4-K/2/57215e156bb3b7681a6460684503a761*

Gao, Y. & Long, D. (2008). Intercomparison of remote sensing-based models for estimation of evapotranspiration and accuracy assessment based on swat, *Hydrological Processes* 22: 4850–4869. URL: *http://onlinelibrary.wiley.com/doi/10.1002/hyp.7104/abstract*

García, M., Villagarcía, L., Contreras, S., Domingo, F. & Puigdefábregas, J. (2007). Comparison of three operative models for estimating the surface water deficit using aster reflective and thermal data, *Sensors* 7(6): 860–883. URL: *http://www.mdpi.com/1424-8220/7/6/860/*

GDAL (2011). Gdal - geospatial data abstraction library. URL: *http://www.gdal.org*

Gieske, A. (2007). Numerical modeling of heat and water vapor transport through the interfacial boundary layer into a turbulent atmosphere, *in* B. J. Geurts, H. Clercx & W. Uijttewaal (eds), *Particle-Laden Flow*, Vol. 11 of *ERCOFTAC Series*, Springer Netherlands, pp. 71–83. 10.1007/978-1-4020-6218-6_6. URL: *http://dx.doi.org/10.1007/978-1-4020-6218-6_6*

Hargreaves, G. H. & Samani, Z. A. (1985). Reference crop evapotranspiration from temperature, *Applied Engineering in Agriculture* 1: 96–99.

Hargreaves, G. L., Hargreaves, G. H. & Riley, J. P. (1985). Agricultural benefits for senegal river basin, *Journal of Irrigation and Drainage Engineering* 111: 113–124.

Hemakumara, H., Chandrapala, L. & Moene, A. (2003). Evapotranspiration fluxes over mixed vegetation areas measured from large aperture scintillometer, *Agricultural Water Management* 58: 109–122.

Jia, L., Su, Z., van den Hurk, B., Menenti, M., Moene, A., De Bruin, H. A. R., Yrisarry, J. J. B., Ibanez, M. & Cuesta, A. (2003). Estimation of sensible heat flux using the surface energy balance system (sebs) and atsr measurements, *Physics and Chemistry of the Earth* 28(1-3): 75–88. Applications of Quantitative Remote Sensing to Hydrology.

URL: *http://www.sciencedirect.com/science/article/B6X1W-483SMNY-5/2/735667f1cf3fef ea27e001736c7d728a*

Khronos.org (2011). Opencl: Open source computing language.
URL: *http://www.khronos.org/opencl/*

Kustas, W. P. & Norman, J. M. (1999). Evaluation of soil and vegetation heat flux predictions using a simple two-source model with radiometric temperatures for partial canopy cover, *Agricultural and Forest Meteorology* 94(1): 13–29.
URL: *http://www.sciencedirect.com/science/article/B6V8W-3W7XBGD-2/2/8c44f505c27c3 9f01073e096dcef8b8d*

Massman, W. J. (1999). A model study of kbh-1 for vegetated surfaces using 'localized near-field' lagrangian theory, *Journal of Hydrology* 223(1-2): 27–43.
URL: *http://www.sciencedirect.com/science/article/B6V6C-3XG1T11-3/2/b77eb802830b880 f3946171e0adfb12a*

Monin, A. & Obukhov, A. (1954). Basic laws of turbulent mixing in the surface layer of the atmosphere, *Tr. Akad. Nauk. SSSR Geophiz. Inst.* 24: 163–187.

Norman, J. M., Anderson, M., Kustas, W. P., French, A. N., Mecikalski, J., Torn, R., Diak, G. R., Schmugge, T. J. & Tanner, B. C. W. (2003). Remote sensing of energy fluxes at 10îm pixel resolution, *Water Resources Research* 39(8): 1221–1229.

Norman, J. M., Kustas, W. P. & Humes, K. S. (1995). Source approach for estimating soil and vegetation energy fluxes in observations of directional radiometric surface temperature, *Agricultural and Forest Meteorology* 77(3-4): 263–293.
URL: *http://www.sciencedirect.com/science/article/B6V8W-4031C00-9/2/a0a99cd55406a97 5e6f7b6ab446006f0*

OpenMP (2011). The openmp api specification for parallel programming.
URL: *http://www.openmp.org*

OpenMPI (2011). Open mpi: Open source high performance computing.
URL: *http://www.open-mpi.org*

Price, J. C. (1984). Land surface temperature measurements from the split window channels of the noaa 7 advanced very high resolution radiometer, *Journal of Geophysical Research* 89(D5): 7231–7237.
URL: *www.agu.org/journals/ABS/JD089iD05p07231.shtml*

Priestley, C. H. B. & Taylor, R. J. (1972). On the assessment of surface heat flux and evaporation using large-scale parameters, *Monthly Weather Review* 100(2): 81–92.
URL: *http://journals.ametsoc.org/doi/abs/10.1175/1520-0493%281972%29100%3C0081% 3AOTAOSH%3E2.3.CO%3B2*

Senay, G. B., Budde, M., Verdin, J. P. & Melesse, A. M. (2007). A coupled remote sensing and simplified surface energy balance approach to estimate actual evapotranspiration from irrigated fields, *Sensors* 7(6): 979–1000.
URL: *http://www.mdpi.com/1424-8220/7/6/979/*

Su, Z. (2002). The surface energy balance system (sebs) for estimation of turbulent heat fluxes, *Hydrology and Earth System Sciences* 6(1): 85–100.
URL: *http://www.hydrol-earth-syst-sci.net/6/85/2002/*

Sui, H. Z., Tian, G. L. & Li, F. Q. (1997). Two-layer model for monitoring drought using remote sensing, *Journal of Remote Sensing* 1: 220–224.

Suleiman, A., Al-Bakri, J., Duqqah, M. & Crago, R. (2008). Intercomparison of evapotranspiration estimates at the different ecological zones in jordan, *Journal of*

Hydrometeorology 9(5): 903–919.
URL: *http://journals.ametsoc.org/doi/abs/10.1175/2008JHM920.1*

Timmermans, W. J., Kustas, W. P., Anderson, M. C. & French, A. N. (2007). An intercomparison of the surface energy balance algorithm for land (sebal) and the two-source energy balance (tseb) modeling schemes, *Remote Sensing of Environment* 108(4): 369 – 384.
URL: *http://www.sciencedirect.com/science/article/B6V6V-4MV19YK-3/2/02bab62a1b4dc9f9298946f723c196b3*

Yang, W., Huang, D., Tan, B., Stroeve, J. C., Shabanov, N. V., Knyazikhin, Y., Nemani, R. R. & Myneni, R. B. (2006). *IEEE Transactions on Geoscience and Remote Sensing* 44: 1829–1842.

Yang, W., Tan, B., Huang, D., Rautiainen, M., Shabanov, N. V., Wang, Y., Privette, J. L., Huemmrich, K. F., Fensholt, R., Sandholt, I., Weiss, M., Ahl, D. E., Gower, S. T., Nemani, R. R., Knyazikhin, Y. & Myneni, R. B. (2006). *IEEE Transactions on Geoscience and Remote Sensing* 44: 1885–1898.

Yunhao, C., Xiaobing, L., Jing, L., Peijun, S. & Wen, D. (2005). Estimation of daily evapotranspiration using a two-layer remote sensing model, *International Journal of Remote Sensing* 26(8): 1755–1762.
URL: *http://www.informaworld.com/10.1080/01431160512331314074*

20

Possibilities of Deriving Crop Evapotranspiration from Satellite Data with the Integration with Other Sources of Information

Gheorghe Stancalie and Argentina Nertan
National Meteorological Administration 97, Soseaua Bucuresti-Ploiesti, Bucharest Romania

1. Introduction

After precipitation, evapotranspiration is one of the most significant components in terrestrial water budgets.

Evapotranspiration (ET) describes the transport of water into the atmosphere from surfaces (including soil - soil evaporation) and from vegetation (transpiration). Those are often the most important contributors to evapotranspiration. Other contributors to evapotranspiration are the e from wet canopy surface (wet-canopy evaporation) and evaporation from vegetation-covered water surface in wetlands the process of evapotranspiration is one of the main consumers of solar energy at the Earth's surface. The energy used for evapotranspiration is generally referred to as latent heat flux. The term latent heat flux includes other related processes unrelated to transpiration including condensation (e.g., fog, dew), and snow and ice sublimation.

There are several factors that affect the evapotranspiration processes: energy availability; the humidity gradient away from the surface(the rate and quantity of water vapor entering into the atmosphere are higher in drier air); the wind speed at the soil level (wind affects evapotranspiration by bringing heat energy into an area); Water availability (it is well known that the evapotranspiration cannot occur if water is not available); Vegetation biophysical parameters (many physical parameters of the vegetation, like cover plant height, leaf area index and leaf shape and the reflectivity of plant surfaces can affect evapotranspiration); Stomatal resistance (the transpiration rate is dependent on the diffusion resistance provided by the stomatal pores, and also on the humidity gradient between the leaf's internal air spaces and the outside air); soil characteristics which includes its heat capacity, and soil chemistry and albedo. For a given climatic region the evapotranspiration follows the seasonal declination of solar radiation and the resulting air temperatures: minimum evapotranspiration rates generally occur during the coldest months of the year and maximum rates generally coincide with the summer season (Burba, 2010). Even so evapotranspiration depends on solar energy; the availability of soil moisture and plant maturity, the seasonal maximum evapotranspiration actually may precede or follow the seasonal maximum solar radiation and air temperature by several weeks (Burba, 2010). If the moisture is available, evapotranspiration is dependent mainly on the availability of solar energy to vaporize water: evapotranspiration varies with latitude, season, time of day, and cloud cover. Most of the evapotranspiration of water at the Earth's surface level occurs in the subtropical regions (Fig.1). In these areas, high quantities of solar radiation provide the energy necessary to convert liquid water into a gas. Usually, evapotranspiration exceeds precipitation on middle and high latitude large areas during the summer season. As a result of climate change it is expected to induce a further intensification of the global water cycle, including ET (Huntington, 2006). Therefore accurate estimates of evapotranspiration are needed for weather forecasting and projecting the long-term effects of land use change and global climate change, irrigation scheduling and watershed management.

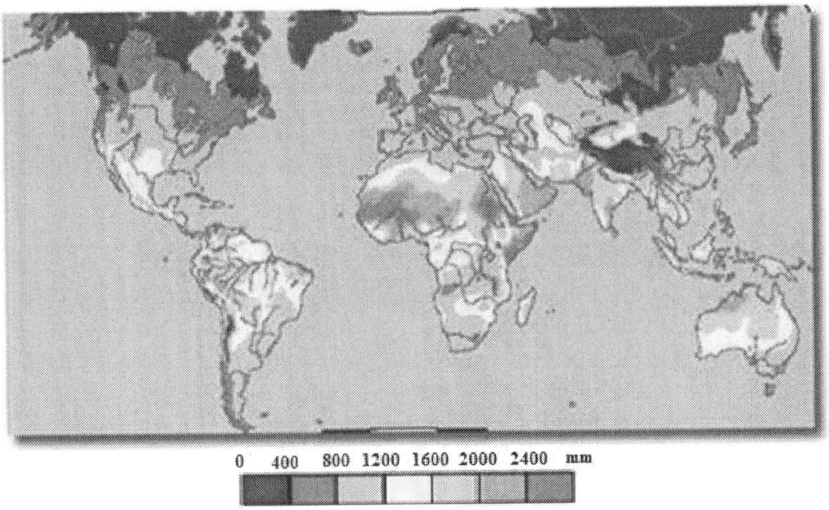

Fig. 1. Mean Annual Potential Evapotranspiration (UNEP World Atlas of Desertification)

In this regard, remote sensing data with the increasing imagery resolution is a useful tool to provide ET information over different temporal and spatial scales. During the last decades important progresses were made in the determination of ET using remote sensing techniques. Some studies have classified the methods of ET estimation in two categories: semi- empirical methods - use empirical relationship and a minimum set of meteorological data; analytical methods – consist in the establishment of the physical process at the scale of interest. A study done by Courault (2007) proposed a few methods which can be classified as follows: empirical direct methods, residual methods of the energy budget, deterministic methods, and vegetation index methods.

In agriculture, an accurate quantification of ET is important for effective and efficient irrigation management. When evaporative demand exceeds precipitation, plant growth and quality may be unfavorably affected by soil water deficit. A large part of the irrigation water applied to agricultural lands *(Fig. 2)* is consumed by evaporation and transpiration. In a given crop, evapotranspiration process is influenced by several factors: plant species, canopy characteristics, plant population, degree of surface cover, plant growth stage, irrigation regime (over irrigation can increase ET due to larger evaporation), soil water availability, planting date, tillage practice, etc. As it can be observed from Fig. 2 the movement of the water vapor from the soil and plant surface, a t a field level is influenced mainly by wind speed and direction although other climatic factors also can play a role. Evapotranspiration increases with increasing air temperature and solar radiation. Wind speed can cause ET increasing. For high wind speed values the plant leaf stomata (the small pores on the top and bottom leaf surfaces that regulate transpiration) close and evapotranspiration is reduced. There are situations when wind can cause mechanical damage to plants which can decrease ET due to reduced leaf area. Hail can reduce also leaf area and evapotranspiration. Higher relative humidity decreases ET as the demand for water vapor by the atmosphere surrounding the leaf surface decreases. If relative humidity (dry air) has lower values, the ET increases due to the low humidity which increases the vapor pressure deficit between the vegetation surface and air. On rainy days, incoming solar radiation decreases, relative humidity increases, and air temperature usually decreases, generation ET decreasing. But, depending on climatic conditions, actual crop water use usually increases in the days after a rain event due to increased availability of water in the soil surface and crop root zone.

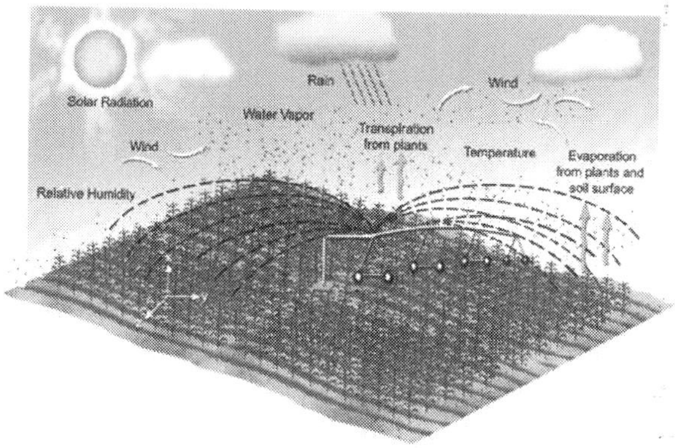

Fig. 2. Evaporation and transpiration and the factors that impact these processes in an irrigated crop.

2. Evapotranspiration and energy budget

The estimation of ET parameter, corresponding to the latent heat flux (λE) from remote sensing is based on the energy balance evaluation through several surface properties such as albedo, surface temperature (T_s), vegetation cover, and leaf area index (LAI). Surface energy balance (SEB) models are based on the surface energy budget equation. To estimate regional crop ET, three basic types of remote sensing approaches have been successfully applied (Su, 2002).

The first approach computes a surface energy balance (SEB) using the radiometric surface temperature for estimating the sensible heat flux (H), and obtaining ET as a residual of the energy balance. The single-layer SEB models implicitly treat the energy exchanges between soil, vegetation and the atmosphere and compute latent heat flux (λE) by evaluating net (all-wave) radiant energy (R_n), soil heat flux (G) and H. For instantaneous conditions, the energy balance equation is the following:

$$\lambda E = R_n - H - G \tag{1}$$

where: R_n = net radiant energy (all-wave); G = soil heat flux; H = sensible heat flux (Wm^{-2}); λE = latent energy exchanges (E = the rate of evaporation of water (kg m^{-2} s^{-1}) and λ = the latent heat of vaporization of water (J kg^{-1})). λE is obtained as the residual of the energy balance contain biases from both H and (R_n - G). There are several factors which affect the performance of single-source approaches, like the uncertainties about atmospheric and emissivity effects. LST impacts on all terms of the energy balance in particular on long wave radiation. The radiative surface temperatures provided by an infrared radiometer from a space borne platform are measured by satellite sensors such as LANDSAT, AVHRR, MODIS and ASTER. Converting radiometric temperatures to kinetic temperature requires considerations about surface emissivity (λE), preferably from ground measurements. Remotely LST is subject to atmospheric effects which are primarily associated with the absorption of infrared radiation by atmospheric water vapor and which lead to errors of 3–5 K. A wide range of techniques have been developed to correct for atmospheric effects, including: single-channel methods; split-window techniques; multi-angle methods and combinations of split-window and multi-channel methods. Radiant and convective fluxes can be described: by considering the observed surface as a single component (single layer approaches); by separating soil and vegetation components with different degrees of canopy description in concordance with the number of vegetation layers (multilayer approaches). Net radiant energy depends on the incident solar radiation (R_g), incident atmospheric radiation over the thermal spectral domain (R_a), surface albedo (α_s), surface emissivity (ε_s) and surface temperature (T_s), according to the following equation:

$$R_n = (1 - \alpha_s)R_g + \varepsilon_s R_a - \varepsilon_s \sigma T_s^4 \tag{2}$$

For single layer models, R_n is related to the whole surface and in the case of multiple layer models, R_n is linked with both soil and vegetation layers. For single approaches, sensible heat flux H is estimated using the aerodynamic resistance between the surface and the reference height in the lower atmosphere (usually 2 m) above the surface. Aerodynamic resistance (r_a) is a function of wind speed, atmospheric stability and roughness lengths for momentum and heat. For multiple layer models, H is characterized taking into account the soil and canopy resistance, with the corresponding temperature:

$$H = \rho c_p \frac{(T_s - T_a)}{r_a} \tag{3}$$

Eq. (3) shows that the estimation of λE parameter can be made using the residual method, which induces that λE is linearly related to the difference between the surface temperature (T_s) and air temperature (T_a) at the time of T_s measurement if the second order dependence of r_a on this gradient is ignored.

$$\lambda E = R_n - G - \rho cp \frac{(T_s - T_a)}{r_a} \tag{4}$$

Equation (4) is usually used to estimate λE. At midday, it provides a good indicator regarding the plant water status for irrigation scheduling. For λE estimation over longer periods (daily, monthly, seasonal estimations), the use of ground-based ET from weather data is necessary to make temporal interpolation. Some studies have used the trend for the evaporative fraction (EF), such as the ratio of latent heat flux to available energy for convective fluxes, to be almost constant during the daytime. This allows estimating the daytime evaporation from one or two estimates only of EF at midday, for example at the satellite acquisition time (Courault et al., 2005).

$$EF = \frac{\lambda E}{R_n - G}, \quad ET_{24} = EF * R_{n24} \tag{5}$$

ET can be estimated from air vapor pressure (p_a) and a water vapor exchange coefficient (h_s) according to the following equation:

$$\lambda E = \rho c_p h_s (p_s^*(T_s) - e_a) \tag{6}$$

Usually this method is used in models simulating Soil-Vegetation-Atmosphere Transfers (SVAT). $p_s^*(T_s)$ represent the saturated vapor pressure at the surface temperature T_s and h_s is the exchange coefficient which depends on the aerodynamic exchange coefficient ($1/r_a$), soil surface and stomatal resistances of the different leaves in the canopy. Katerji & Perrier (1985) estimated a global canopy resistance (r_g) including both soil and canopy resistances (equation 6)

$$r_g = \frac{1}{\dfrac{1}{r_{veg} + r_w} + \dfrac{1}{r_0 + r_s}} \tag{7}$$

where: r_{veg} is the resistance due to the vegetation structure, r_w the resistance of the soil layer depending on the soil water content, r_0 the resistance due to the canopy structure and r_s the bulk stomatal resistance. To calculate this parameters it necessary to have information regarding the plant structure like LAI and fraction of vegetation cover (FC), the minimum stomatal resistance (r_{smin}). Many studies proposed various parameterizations of the stomatal resistance taking into account climatic conditions and soil moisture (Jacquemin & Noilhan, 1990). This proves that the ($T_s - T_a$) is related to ET term, and that Ts can be estimated using thermal infrared measurements (at regional or global scale using satellite data, and at local scale using ground measurements).

The second approach uses vegetation indices (VI) derived from canopy reflectance data to estimate basal crop coefficient (K_{cb}) that can be used to convert reference ET to actual crop ET, and requires local meteorological and soil data to maintain a water balance in the root zone of the crop. The VIs is related to land cover, crop density, biomass and other vegetation characteristics. VIs such as the Normalized Difference Vegetation Index (NDVI), the Soil Adjusted Vegetation Index (SAVI), the Enhanced Vegetation Index (EVI) and the Simple Ratio (SR), are measures of canopy greenness which may be related to physiological processes such as transpiration and photosynthesis. Among the relatively new satellite sensors it has to be mentioned the advantages of using MODIS/Aqua that offer improved spectral and radiometric resolution for deriving surface temperatures and vegetation indices, as well as increased frequency of evaporative fraction and evaporation estimates when compared with other sensors. The observed spatial variability in radiometric surface temperature is used with reflectance and/or vegetation index observations for evaporation estimation. For ET estimation from agricultural crops the most direct application is to substitute the VIs for crop coefficients (defined as the ratio between actual crop water use and reference crop evaporation for the given set of local meteorological conditions). Negative observing correlations between the NDVI and radiometric surface temperature could be linked to evaporative cooling, although for most landscapes variations in fractional vegetation cover, soil moisture availability and meteorological conditions will cause considerable scatter in those relationships. The methods associated with this approach generate spatially distributed values of K_{cb} that capture field-specific crop development and are used to adjust a reference ET (ET_o) estimated daily from local weather station data.

The third approach uses remotely sensed LST with Land Surface Models (LSMs) and Soil-Vegetation-Atmosphere (SVAT) models, developed to estimate heat and mass transfer at the land surface. LSMs contain physical descriptions of the transfer in the soil-vegetation-atmosphere continuum, and with proper initial and boundary conditions provide continuous simulations when driven by weather and radiation data. The energy-based LSMs are of particular interest because these approaches allow for a strong link to remote sensing applications. The use of the spatially distributed nature of remote sensing data as a calibration source has been limited, with the focus placed on data assimilation approaches to update model states, rather than inform the actual model structure. Data assimilation is the incorporation of observations into a numerical model(s) with the purpose of providing the model with the best estimate of the current state of a system. There are two types of data assimilation: (i) sequential assimilation which involves correcting state variables (e.g. temperature, soil moisture) in the model whenever remote sensing data are available; and (ii) variation assimilation when unknown model parameters are changed using data sets obtained over different time windows. Remotely sensed LSTs have been assimilated at point scales into various schemes for estimating land surface fluxes by comparing simulated and observed temperatures and adjusting a state variable (e.g. soil moisture) or model parameters in the land surface process model. Such use of remote sensing data has highlighted problems of using spatial remote sensing data with spatial resolutions of tens or hundreds of kilometers with point-scale SVAT models and has led to the search for "effective" land surface parameters. There exist no effective means of evaluating ET spatially distributed outputs of either remote sensing based approaches or LSMs at scales greater than a few kilometers, particularly over non-homogeneous surfaces. The inability to evaluate remote sensing based estimates in a distributed manner is a serious limitation in broader scale applications of such approaches. It must be noted here that ET evaluation of remote sensing based approaches with ground based data tends to favour those few clear sky days when fluxes are reproduced most agreeably, and on relatively flat locations.

In this case the radiation budget is given by the following equation (Kalma et al., 2008):

$$R_n = K\downarrow - K\uparrow + L\downarrow - L\uparrow \tag{8}$$

where K↓ is the down-welling shortwave radiation and it depends on atmospheric transmissivity, time of the day, day of the year and geographic coordination. K↑ represents the reflected shortwave radiation which depends on K↓ and surface albedo (a), L↓ is the down-welling long wave radiation and L↑ is the up-welling long wave radiation. L↓ depends on the atmospheric emissivity (which in turn is influenced by amounts of atmospheric water vapor, carbon dioxide and oxygen) and by air temperature. L↑ si influenced by land surface temperature and emissivity

3. Direct methods using difference between surface and air temperature

Mapping daily evapotranspiration over large areas considering the surface temperature measurements has been made using a simplified relationship which assumes that it is possible to directly relate the daily (λE_d) to the difference $(T_{rad} - T_a)_i$ between (near) mid-day observations (i) of surface temperature and near-surface air temperature (Ta) measured at midday as follows:

$$\lambda E_d = (R_n)_d - B(T_{rad} - T_a)_i^n \tag{9}$$

B is a statistical regression coefficient which depends on surface roughness. n depends on atmospheric stability. Equation 9 was derived from Heat Capacity Mapping Mission (HCMM) observations over fairly homogeneous irrigated and non-irrigated land surfaces, with areas between 50 and 200 km^2 (Seguin et al. 1982a, b). Some authors as Carlson et al. (1995a) proposed a simplified method based on Eq. 9 which uses the difference $(T_{rad} - T_a)$ at 50 m at the time of the satellite overpass. They showed that B coefficient and n are closely related to fractional cover f_c that can be obtained from the NDVI–T_{rad} plots. B values vary from 0.015 for bare soil to 0.065 for complete vegetation cover and n decreased from 1.0 for bare soil to 0.65 for full cover.

4. Surface energy balance models

Surface energy balance models (SEBAL) assume that the rate of exchange of a quantity (heat or mass) between two points is driven by a difference in potential (temperature or concentration) and controlled by a set of resistances which depend on the local atmospheric environment and the land surface and vegetation properties. In the review made by Overgaard et al. (2006) regarding the evolution of land surface energy balance models are described the following approaches: the combination approach by Penman (1948) which developed an equation to predict the rate of ET from open water, wet soil and well-watered grass based on easily measured meteorological variables such as radiation, air temperature, humidity, and wind speed; the Penman–Monteith "one-layer", "one-source" or "big leaf" models (Monteith 1965) which recognize the role of surface controls but do not distinguish between soil evaporation and transpiration; this approach estimates ET rate as a function of canopy and boundary layer resistances; "two-layer" or "two-source" model such as described by Shuttleworth and Wallace (1985) which includes a canopy layer in which heat and mass fluxes from the soil and from the vegetation are allowed to interact; multi-layer models which are essentially extensions of the two-layer approach.

4.1 The Penman–Monteith, "one-source" SEB models

The Penman–Monteith (PM) approach combines energy balance and mass transfer concepts (Penman, 1948) with stomatal and surface resistance (Monteith, 1981). Most "one source" SEB models compute λE by evaluating R_n, G and H and solve for λE as the residual term in the energy balance equation (see Eq. 10). The sensible heat flux (H) is given by:

$$H = \rho C_p \left[\frac{(T_{ad} - T_a)}{r_a}\right] \tag{10}$$

Where: ρ = air density (kg*m^{-3}); C_p = specific heat of air at constant pressure (J kg^{-1} K^{-1}); T_{ad} = aerodynamic surface temperature at canopy source height (K); T_a = near surface air temperature (K); r_a = aerodynamic resistance to sensible heat transfer between the canopy source height and the bulk air at a reference height above the canopy (s m^{-1}). The r_a term is usually calculated from local data on wind speed, surface roughness length and atmospheric stability conditions. According to Norman and Becker (1995), the aerodynamic surface temperature (T_{ad}) represent the temperature that along with the air temperature and a resistance calculated from the log-profile theory provides an estimate H. The key issue of PM approach is to estimate an accurately sensible heat flux. T_{ad} is obtained by extrapolating the logarithmic air temperature profile to the roughness length for heat transport (z_{oh}) or, more precisely, to (d + z_{oh}) where d = zero-plane displacement height. Usually, due to the fact that T_{ad} cannot be measured using remote sensing, it is replaced with T_{rad}. As it is demonstrated by Troufleau et al. (1997), for dense canopy T_{rad} and T_{ad} may differ with 1-2 K and much more for sparse canopy. Surface temperature (T_{rad}) is related to the kinetic temperature by the surface emissivity (ε) (Eq, 11) and it depends on view angle (θ) (Norman et. al, 2000). On the other hand T_{ad} and aerodynamic resistance are fairly difficult to obtain for non-homogenous land surfaces.

$$T_{rad} = \varepsilon^{1/4} * T_k \tag{11}$$

The aerodynamic resistance r_a can be calculated with the following equation:

$$r_a = \frac{1}{k^2} u \left[ln \frac{z-d}{z_{oh}} - \Psi_h \frac{z-d}{L} \right] \left[ln \frac{z-d}{z_{om}} - \Psi_m \frac{z-d}{L} \right] \tag{12}$$

where: k = 0.4 (von Karman's constant); u = wind speed at reference height z (m s^{-1}); d = zero-plane displacement height (m); z_{oh} and z_{om} = roughness lengths (m) for sensible heat and momentum flux, respectively; Ψ_h and Ψ_m = stability correction functions for sensible heat and momentum flux, respectively; L = Monin-Obukhov length L (m). The Ψ_h = 0 and Ψ_m = 0 if near surface atmospheric conditions are neutrally stable. Usually, the aerodynamic resistance is estimated from local data, even that area averaging of roughness lengths is highly non-linear (Boegh et al. 2002). Several studies, such as Cleugh at al. (2007) used these equations for evapotranspiration landscape monitoring. Their approach estimates E at 16-day intervals using 8-day composites of 1 km MODIS T_{rad} observations and was tested with 3 years of flux tower measurements and was obtained significant discrepancies between observed and simulated land surface fluxes, generated by the following factors: the estimation of H with Eqs. 9 and 10 is not constrained by the requirement for energy conservation; errors in z_{oh} determination; use of unrepresentative emissivities; using time-averages of instantaneous T_{rad}, T_a and R_n, the non-linearity of Eq. 9 may cause significant errors; standard MODIS data processing eliminates all cloud-contaminated pixels in the composite period. Bastiaanssen et al. (1998a) developed a calibration procedure using image data to account for the differences between T_{aero} and T_{rad}, which are important, mainly for incomplete vegetation covers. Other authors, such as Stewart et al. (1994) and Kustas et al. (2003a), made empirical adjustments to aerodynamic resistance, related to z_{oh} (eq. 13).

$$H = \rho C_p \left[\frac{T_{rad}(\Theta) - T_a}{r_a - r_{ex}} \right] \tag{13}$$

where: T_{rad} (θ) =radiometric surface temperature (K) at view angle θ derived from the satellite brightness temperature; r_{ex} = excess resistance (s m^{-1}) (reflects differences between momentum and sensible heat transfer. According to Stewart et al. (1994) r_{ex} is function of the ratio of roughness lengths for momentum z_{om} and for sensible heat z_{oh} and the friction velocity u* (m s^{-1}) (eq. 14):

$$r_{ex} = \frac{kB^{-1}}{ku^*} = ln\frac{z_{om}/z_{oh}}{ku^*} \tag{14}$$

where kB^{-1} = dimensionless ratio determined by local calibration. Eq. 14 assumes that the ratio z_{om}/z_{oh} may be treated as constant for uniform surfaces, although kB^{-1} has been found to be highly variable (Brutsaert 1999).

In the case of the one source Surface Energy Balance System (SEBS) (Su, 2002) the surface heat fluxes are estimated from satellite data and available meteorological data. There are three sets of input data in SEBS: the first set includes the following parameters: α, ε, T_{rad}, LAI, fractional vegetation coverage and the vegetation height (if the vegetation information is not explicitly available, SEBS can use as input data the Normalized Difference Vegetation Index (NDVI)); the second set includes T_a, u, actual vapour pressure (e_a) at a reference height as well as total air pressure; the third set of data consists of measured (or estimated) $K\downarrow$ and $L\downarrow$. For R_n, G, and the partitioning of (R_n - G) into H and λE, SEBS use different modules (Fig. 3): H is estimated using Monin–Obukhov similarity theory; in the case of u and vegetation parameters (height and LAI) is used the Massman (1997) model to to estimate the displacement height (d) and the roughness height for momentum (z_{om}); the equations proposed by Brutsaert (1982, 1999) are used when only the height of the vegetation is available. The SEBS was successfully tested for agricultural areas, grassland and forests, across various spatial scales. Several studies used flux tower method and data from Landsat, ASTER ad Modis sensors (Su et al. 2005, 2007, McCabe and Wood 2006).

The Fig. 4 shows the time series, determined during the Soil Moisture Atmosphere Coupling Experiment 2002 (SMACEX-02) (Kustas et al. 2005). These time series illustrates latent heat fluxes and sensible heat fluxes measured with in situ eddy-covariance equipment (closed) together with SEBS model (open) over a field site (corn) from Iowa. The gaps in the time series are caused either the missing ancillary data or absence of flux measurements. Many factors influence the single-source approach: there are uncertainties due to atmospheric and emissivity effects; because of the vegetation properties and of the angle view, the relationship between T_{ad} and T_a is not unique; this approach requires representative near-surface T_a and other meteorological data measured (or estimated) at the time of the satellite overpass at a location closely with the T_{rad} observation. This can generate errors in defining meteorological parameter for each satellite pixel from a sparse network of weather stations (at the time of satellite overpass), mainly for areas with high relative relief and slopes. Another important factor is that the accuracy of any of the estimates depends on the performance of the algorithm used for temperature retrieval.

The major advantages of SEBS are: uncertainty due to the surface temperature or meteorological variables can be limited taking into account the energy balance at the limiting cases; through the SEBS was formulated a new equation for the roughness height for heat transfer, using fixed values; a priori knowledge of the actual turbulent heat fluxes is not required. Another single-source energy balance models, developed based on the conception of SEBAL, are S-SEBI (Simplified-SEBI), METRIC (Mapping EvapoTranspiration at high Resolution with Internalized Calibration), etc. The main difference between such kinds of models is the difference in how they calculate the sensible heat, i.e. the way to define the dry (maximum sensible heat and minimum latent heat) and wet (maximum latent heat and minimum sensible heat) limits and how to interpolate between the defined upper and lower limits to calculate the sensible heat flux for a given set of boundary layer parameters of remotely sensed data (T_s, albedo, NDVI, LAI) and ground-based air temperature, wind speed, humidity. The assumptions in all these models are that there are few or no changes in atmospheric conditions (especially the surface available energy) in space and sufficient surface horizontal variations are required to ensure dry and wet limits existed in the study area.

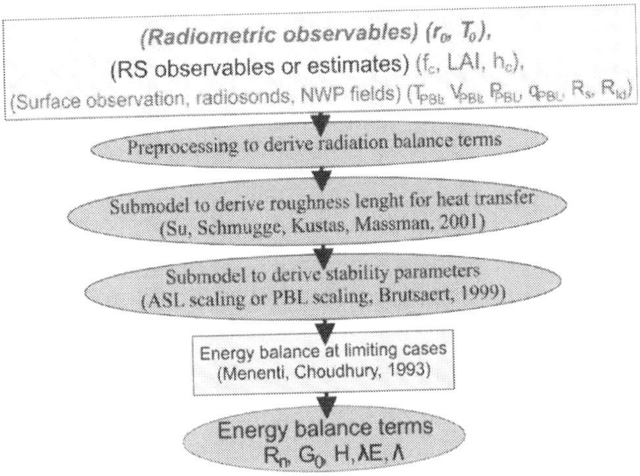

Fig. 3. Schematic representation of SEBS (after Su, 2008)

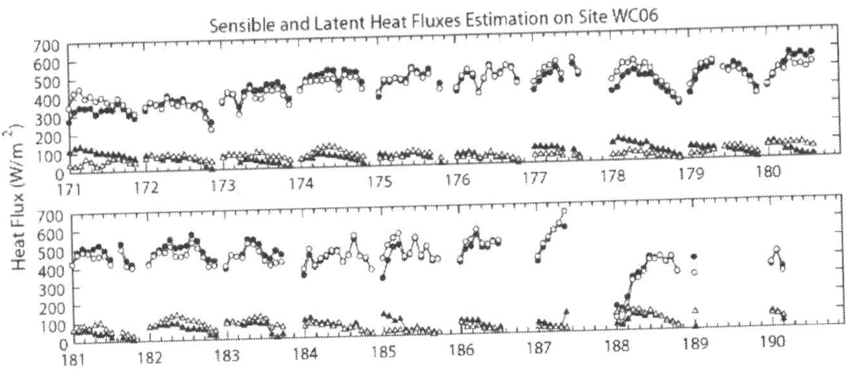

Fig. 4. Reproduction of surface flux development with a one-source model (SEBS) (after Kalma, 2008)

4.2 Two-source SEB models

The equations 10 and 13 make no difference between evaporation soil surface and transpiration from the vegetation and from this reason the resistances are not well defined.

To solve this problem two-source models have been developed for use with incomplete canopies (e.g. Lhomme et al. 1994; Norman et al. 1995; Jupp et al. 1998; Kustas and Norman 1999). These models consider the evaporation as the sum of evaporation from the soil surface and transpiration from vegetation. For example, Norman et. Al. (1995) developed a two-source model (TSM) based on single-time observations which eliminate the need for r_{ex} as used in equations 13 and 14. They reformulated the equation 10 as:

$$H = \rho C_p \frac{T_{rad}(\theta) - T_a}{r_r} \tag{15}$$

where: T_{rad} = directional radiometric surface temperature obtained at zenith view angle θ; r_r = radiometric-convective resistance (s m^{-1}). The radiometric convective resistance is calculated according to the following formula:

$$r_r = \frac{T_{rad}(\theta) - T_a}{\left[\frac{(T_c - T_a)}{r_a} + \left(\frac{(T_s - T_a)}{r_a + r_s}\right)\right]} \tag{16}$$

where: T_c = canopy temperature; T_s = soil surface temperature; R_s = soil resistance to heat transfer (s m-1). To estimate the T_c and T_s variables, Norman et al. used fractional vegetation cover (fc) which depends on sensor view angle (Eq. 17):

$$T_{rad}(\theta) \approx [f_c(\theta)T_c^4 + \{1 - f_c(\theta)\}T_s^4]^{\frac{1}{4}} \tag{17}$$

H variable is divided in vegetated canopy (H_c) and soil (H_s) influencing the temperature in the canopy air-space. Other revisions of TSM compared flux estimates from two TSM versions proved that thermal imagery was used to constrain T_{rad} and H and microwave remote sensing was employed to constrain near surface soil moisture. The estimations resulting from those two models were compared with flux tower observations. The results showed opposing biases for the two versions that it proves a combination between microwave and thermal remote sensing constraints on H and λE fluxes from soil and canopy. Compared to other types of remote sensing ET formulations, dual-source energy balance models have been shown to be robust for a wide range of landscape and hydro-meteorological conditions.

5. Spatial variability methods using vegetation indices

Visible, near-infrared and thermal satellite data has been used to develop a range of vegetation indices which have been related to land cover, crop density, biomass or other vegetation characteristics (McVicar and Jupp 1998). Several vegetation indices as the Normalized Difference Vegetation Index (NDVI), the Soil Adjusted Vegetation Index (SAVI), the Enhanced Vegetation Index (EVI) and the Simple Ratio (SR), are indicators of canopy greenness which can be related to physiological processes such as transpiration and photosynthesis (Glenn et al., 2007).

5.1 Vegetation indices, reflectance and surface temperature

The SEBAL approach used remotely sensed surface temperature, surface reflectivity and NDVI data. It has been developed for the regional scale and it requires few ground level observations from within the scene. K↓ and L↓ are computed using a constant atmospheric transmissivity, an appropriate atmospheric emissivity value and an empirical function of T_a, respectively. G is calculated as a fraction of R_n depending on T_{rad}, NDVI and α (Bastiaanssen 2000). The instantaneous values of sensible heat flux are calculated in three main steps. First step makes the difference between T_{ad} and T_{rad} and assumes that the relationship between T_{rad} and the near-surface temperature gradient ($\Delta T = T_{ad} - T_a$) is quasi-linear. Therefore wet and dry extremes can be identified from the image. These extremes fix the quasi-linear relationship relating ΔT to T_{rad}, allowing ΔT to be estimated for any T_{rad} across the image. In the second step, a scatter plot is obtained for all pixels in the entire image of broadband α values versus T_{rad}. Low temperature and low reflectance values correspond to pixels with large evaporation rates, while high surface temperatures and high reflectance values correspond to the areas with little or no evaporation rates. Scatter plots for large heterogeneous regions frequently show an ascending branch controlled by moisture availability and evaporation rate, and a radiation-controlled descending branch where evaporation rate is negligible. The ascending branch indicates that the temperatures increase with increasing α values as water availability is reduced and evaporation rate becomes more limited. For the descending branch the increasing of α induce a decreasing of surface temperature. If the radiation-controlled descending branch is well defined, r_a may be obtained from the (negative) slope of the reflectance–surface temperature relationship. The last step use the local surface roughness (z_{om}) based on the NDVI; is assumed that the z_{om}/z_{oh} ratio has a fix value and H can be calculated for every pixel with λE as the residual term in Eq. 1. The SEBAL models have been used widely with satellite data in the case of relatively flat landscapes with and without irrigation.

The Mapping EvapoTranspiration with high Resolution and Internalized Calibration (METRIC) models, derived from SEBAL are used for irrigated crops (Allen et al. 2007a, b). METRIC model derive ET from remotely sensed data (LANDSAT TM) in the visible, near-infrared and thermal infrared spectral regions along with ground-based wind speed and near surface dew point temperature. In this case extreme pixels are identified with the cool/wet extreme comparable to a reference crop, the evaporation rates being computed wit Penman-Monteith method. The ET from warm/dry pixel is calculated using soil water budget having local meteorological data as input parameters. METRIC model can be used to produce high quality and accurate maps of ET for areas smaller than a few hundred kilometers in scale and at high resolution (Fig. 5). In their study, Boegh et al. (1999) presented an energy balance method for estimating transpiration rates from sparse canopies based on net radiation absorbed by the vegetation and the sensible heat flux between the leaves and the air within the canopy. The net radiation absorbed by the vegetation is estimated using remote sensing and regular meteorological data by merging conventional method for estimation of the land surface net radiation with a ground-calibrated function of NDVI.

SEBAL and METRIC methods assume that the temperature difference between the land surface and the air (near-surface temperature difference) varies linearly with land surface temperature. Bastiaanssen et al. (1998) and Allen and al. (2007) derive this relationship based on two anchor pixels known as the hot and cold pixels, representing dry and bare agricultural fields and wet and well-vegetated fields, respectively. Both methods use the linear relationship between the near-surface temperature difference and the land surface temperature to estimate the sensible heat flux which varies as a function of the near-surface temperature difference, by assuming that the hot pixel experiences no latent heat, i.e., ET = 0.0, whereas the cold pixel achieves maximum ET.

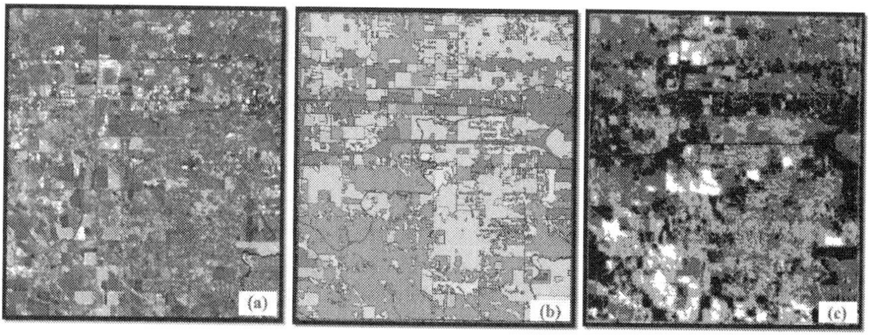

Fig. 5. (a) Landsat color infrared image of T3NR1E of the Boise Valley; (b) Land use/land cover polygons in T3NR1E of the Boise Valley; (c) ET image of T3NR1E the Boise Valley (after R.G. Allen et al., 2007)

The sensible heat flux is assessed like a linear function of the temperature difference between vegetation and mean canopy air stream. The surface temperature recorded by satellite comprises information from soil and from vegetation; therefore the vegetation temperature is estimated taking into account the linear relationship between NDVI and surface temperature. The difference between the surface temperature and the mean canopy air stream temperature is linearly related to the difference between surface temperature and the air temperature above the canopy with the slope coefficient which depend on the canopy structure. This relationship was used to evaluate the mean canopy air stream temperature. The method was used in the Sahel region for agricultural crops, natural vegetation, forest vegetation, with ground based, airborne and satellite remote sensing data and validated with sapflow and latent heat flux measurements. Agreement between remote sensing based estimates and ground based measurements of λE rates is estimated to be better than 30–40 W m^{-2}.

5.2 Reflectance and surface temperature

The Simplified Surface Energy Balance Index (S-SEBI) proposed by Roerink et al. (2000) estimate the instantaneous latent heat flux (λE_i) with (Kalma, 2008):

$$\lambda E_i = \Lambda_i(R_{ni} - G_i) \tag{18}$$

where: (R_{ni} - G_i) = available energy at the time of the satellite overpass; Λ_i = the evaporative fraction. The S-SEBI algorithm has two limitations: the atmospheric conditions have to be almost constant across the image and the image has to contain borh dry and wet areas. Λ_i was obtained from a scatter plot of observed surface temperature (T_{rad}) and Landsat TM derived broadband a values across the single scene. Λ_i is with:

$$\Lambda_i = \frac{T_H - T_{rad}}{T_H - T_{rad}} \tag{19}$$

where: T_{rad} = observed surface temperature for a given pixel; T_H = temperature for the upper boundary (dry radiation controlled conditions - all radiation is used for surface heating and α decreases with increasing surface temperature (T_H - where λE = 0 (W m^{-2})); $T_{\lambda E}$ = temperature at the lower boundary (evaporation controlled wet conditions - all energy is used for λE and α increases with an increase of surface temperature ($T_{\lambda E}$ -where H = 0 W m^{-2})). This method does not need any additional meteorological data.

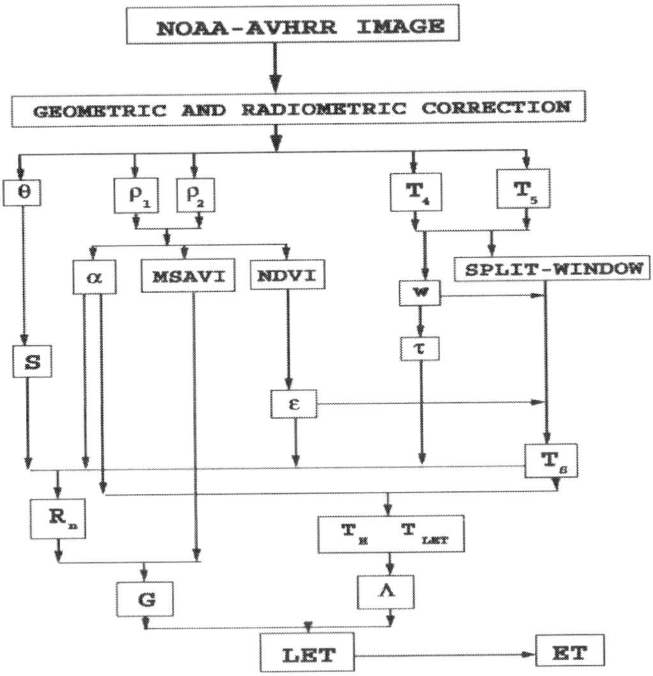

Fig. 6. Flowchart of the proposed methodology to obtain ET from NOAA-AVHRR data (after Sobrino et al., 2007)

Sobrino et. al (2007) use S-SEBI algorithm to estimate the daily evapotranspiration from NOAA-AVHRR images for the Iberian Penisnula. The Figure 6 present the flowchart used by Sobrino et al. (2007) to obtain ET from NOAA-AVHRR. Daily evapotranspiration (ET_d) is given by:

$$ET_d = \frac{\Lambda_i C_{di} R_{ni}}{L} \tag{20}$$

where: R_{nd} = daily net radiation; R_{ni} = instantaneous net radiation: L = 2.45 MJ kg^{-1} = latent heat vaporization; $C_{di}=R_{nd}/R_{ni}$. In this case the daily ground heat flux was considered close to 0. There are several studies which proposed methods for C_{di} calculation. For example Seguin and Itier (1983) proposed a constant value for C_{di} = (0.30±0.03). Wassenaar et al. (2002) showed that this ratio have a seasonal variation 0.05 in winter to 0.3 in summer, following a sine law. In the Sobrino et al. (2007) study, C_{di} was calculated using net radiation fluxes measured at the meteorological station of located on the East coast of the Iberian Peninsula (El Saler area). The ET estimation from high spectral and spatial resolution data (~5 m) was adapted to the low resolution data NOAA-AVHRR (1 km spatial resolution) based on the evaporative fraction concept proposed by Roerink et al. (2007). The main advantage of the Sobrino et al. (2007) methodology is that the method requires only satellite data to estimate ET.

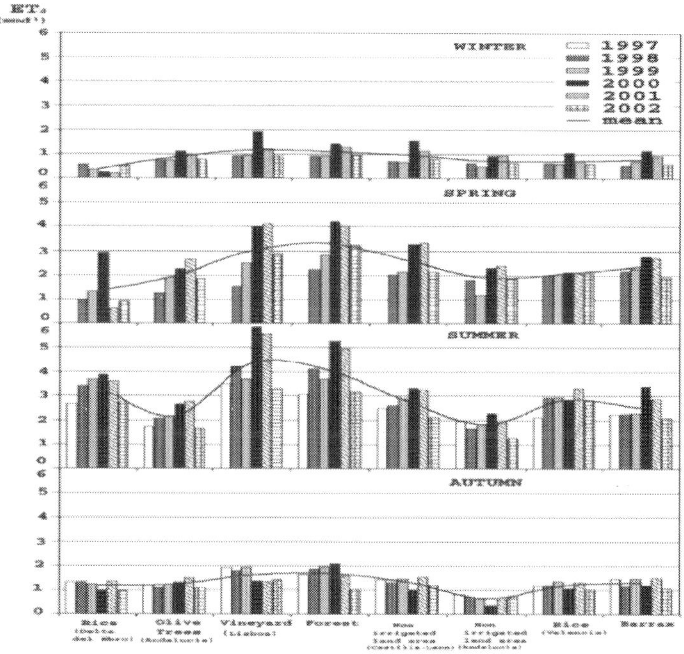

Fig. 7. Monthly evolution (from June 1997 to November 2002) of the daily evapotranspiration (ET_d) in the eight selected zones. There is represented also the temporal mean for the six years of analyzing (after Sobrino et al., 2007).

Its major disadvantage is represented by the requiring that satellite images must have extreme surface temperatures. The method was tested over agricultural area using high resolution values, with errors lower than 1.4 mm d^{-1}. As it can be observed from Fig. 7, regarding the monthly and seasonal evolution of ET the highest values (~6 mm d^{-1}) were obtained in the West of the Iberian Peninsula, which is the most vegetated area. Taking into account the impact of incoming solar energy the higher values of ET was obtained in spring and summer and the lower values in autumn and winter. Seasonal ET was obtained by averaging daily ET over the season. Figure 8 shows as an example the monthly ET maps obtained from the NOAA-AVHRR images acquired in 1999. Fig. 9 also indicates that the highest ET values were obtained in the summer and spring, in the north and west of Iberian Peninsula. To map land surface fluxes and surface cover and surface soil moisture, Gillies and Carlson (1995) combined two model, SVAT and ABL and run it for vegetative cover with the maximum known NDVI and for bare soil conditions with the minimum known NDVI in the scene for a range of soil moisture values until AVHRR observed (T_{rad}) and simulated (T_{ad}) surface temperatures corrected, at which stage the actual fractional vegetation cover (f_c) and surface soil moisture were estimated.

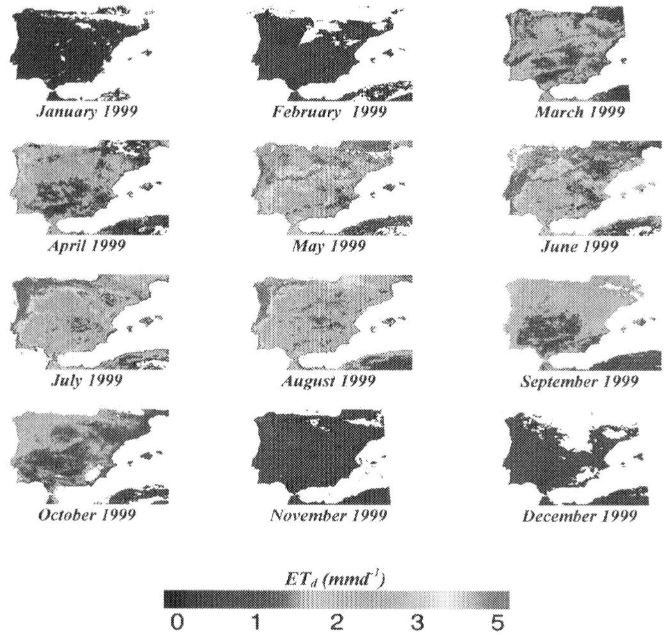

Fig. 8. Monthly mean for the daily evapotranspiration obtained from NOAA-AVHRR data over the Iberian Peninsula in 1999. Pixels in black color correspond to sea and cloud masks and red correspond to higher value of ET (after Sobrino et al., 2007).

5.3 Vegetation indices and surface temperature

Several studies shown the efficiency of "triangle method" (Carlson et al. (1995a, b); Gillies et al. 1997; Carlson 2007) to estimate soil moisture from the NDVI-T_{rad} relationship. The major advantages of the remotely sensed VI-T_s triangle method are that: the method allows an accurate estimation of regional ET with no auxiliary atmospheric or ground data besides the remotely sensed surface temperature and vegetation indices; is relatively insensitive to the correction of atmospheric effects. Its limitations are: determination of the dry and wet edges requires a certain degree of subjectivity; to make certain that the dry and wet limits exist in the VI-T_{rad} triangle space most of pixels over a flat area with a wide range of soil wetness and fractional vegetation cover are required. So, the boundaries of this triangle are limiting conditions for H and λE. Other studies suggest the dependence of T_{rad} variability on the remote sending data resolution, thus higher resolution data means that the variations of T_{rad} and NDVI is more related to the land cover type. Lower resolution data show the dependency of the NDVI and T_{rad} variations to agricultural practices and rainfall. Jiang and Islam (2001) proposed a triangle method based on the interpolation of the Priestley-Taylor method (Priestley and Taylor, 1972) using the triangular (T_{rad}, NDVI) spatial variation. The Priestley-Taylor expression for equilibrium evaporation from a wet surface under conditions of minimal advection (λE_{PT}) is given by:

$$\lambda E_{PT} = \alpha_{PT}(R_n - G)\frac{\Delta}{\Delta + \gamma} \tag{21}$$

where: Δ = slope of the saturated vapour pressure curve at the prevailing Ta ((Pa K^{-1}); γ = psychrometric constant (Pa K^{-1}); α_{PT} = Priestley-Taylor parameter defined as the ratio between actual E and equilibrium E. For wet land surface conditions, α_{PT} = 1.26. Its value is affected by global changes in air temperature, humidity, radiation and wind speed. Jiang and Islam (2001) replaced α_{PT} with parameter ϕ which varies for a wide range of r_a and r_c values. The warm edge of the (T_{rad}, NDVI) scatter plot represents pixels with the highest T_{rad} and minimum evaporation from the bare soil component, while E_a can vary function of the vegetation type. Linear interpolation between the sides of the triangular distribution of T_{rad} - NDVI allows to derive ϕ for each pixel using the spatial context of remotely sensed T_{rad} and NDVI. The ϕ values are related to surface wetness, r_s and T_{rad}. Therefore, the minimum value of ϕ is 0 for the driest bare soil pixel and the maximum value is 1.26 for a densely vegetated, well-watered pixel. Thus the actual ϕ value for each pixel in a specified NDVI interval is obtained from the observed $(T_{rad})_{obs}$ with the following:

$$\phi = \phi_{max}\frac{(T_{rad})_{max} - (T_{rad})_{obs}}{(T_{rad})_{max} - (T_{rad})_{min}} \tag{22}$$

where $(T_{rad})_{min}$ and $(T_{rad})_{max}$ are the lowest and highest surface temperatures for each NDVI class, corresponding to the highest and lowest evaporation rates, respectively. The evaporative fraction can be calculated with:

$$\Lambda = \phi\frac{\Delta}{\Delta + \Upsilon} \tag{23}$$

Based on the Jiang and Islam (2001) approach, Wang et al. (2006) obtained better results using the spatial variation (ΔT_{rad}, NDVI), where ΔT_{rad} represent the day–night difference in T_{rad}, obtained from MODIS data. However, to convert Λ into E, the method described above still requires estimation/ measurement of net radiation (R_n) and soil heat flux (G). In a later work, Jiang and Islam (2003) consider the fractional vegetative cover (f_c) as a more suitable generalized vegetation index calculated from the normalized NDVI with (Kalma et al. 2008):

$$f_c = \left(\frac{NDVI - NDVI_{min}}{NDVI_{max} - NDVI_{min}}\right)^2 \tag{24}$$

They assumed that the evaporative fraction $\Lambda = \lambda E/(R_n - G)$ is linearly related to $\Delta T = T_{rad} - T_a$, inside a certain class f_c. The reason for this assumption is that theΔT is more representative for sensible heat flux H. Thus the evaporative fraction can be estimated from f_c and ΔT, for a given set of ΔT_{max}, ΔT_e ($\Delta T_e = \Delta T_{max}$ for $f_c = 1$) and a stress factor (β). In their study, they used NOAA-AVHRR data and obtained better results using the aerodynamic resistance-energy balance method represented by Eq. 13, this equation including atmospheric stability corrections and using an iterative procedure to reach the most appropriate kB^{-1} value.

Serban et al. (2010) used the Priestly-Taylor equation modified by Jiang and Islam (2001) in their study to estimate the evapotranspiration using remote sensing data and Grid Computing. The most advantage of Priestly-Taylor equation is that the all terms can be calculated using remotely sensed data. Grid computation procedure has two major advantages: strong data processing capacity and the capability to use distributed computing resources to process the spatial data offered by a satellite image. According to Jiang and Islam (2001) the parameter α_{PT} parameter is obtained by two-step linear interpolation: in the first step is obtained upper and lower bounds of α_{PT} for each specific NDVI class (determined from the land use/land cover map); in the second step the parameter α_{PT} is ranged within each NDVI class between the lowest temperature pixel and the highest temperature pixel. According to land use/land cover map, for this paper, was considered four main land uses: vegetation, water, barren land and urban. Each NDVI value corresponds to a certain NDVI class. In this case the relationship between LST and NDVI is used. Thus, the parameter α_{PT} is calculated with:

$$\alpha_{PT} = \left(\frac{\Delta + \gamma}{\Delta}\right)\left(\frac{LST_i^{max} - LST}{LST_i^{max} - LST_i^{min}}\right)\left(\frac{NDVI_i^{max} - NDVI_i^{min}}{NDVI_i^{max}}\right) + \left(\frac{\Delta + \gamma}{\Delta}\right)\left(\frac{NDVI_i^{min}}{NDVI_i^{max}}\right) \quad (25)$$

where: LST = surface temperature for current pixel; LST_i^{max} and LST_i^{min} = maximum and minimum surface temperature within NDVI class which has the current pixel; $NDVI_i^{max}$ and $NDVI_i^{min}$ are the maximum and minimum NDVI within NDVI class which has the current pixel. They calculated the daily value of ET with the following (Fig. 9):

$$\lambda E_{daily} = \alpha_{PT}\frac{2DL(R_i - G_i)}{\pi \sin\left(\pi \frac{t}{DL}\right)} \quad (26)$$

where: DL = total day length (hours); t = time beginning at sunrise. To obtain the 24 hours totals, the daily ET values are multiplied by 1.1 for all days. LST was computed using Jimenez- Munoz and Sobrino's algorithm which requires a single ground data (the total atmospheric water vapor content – w) (Fig. 10):

$$LST = \gamma[LSE^{-1}(\psi_1 L_{sensor} + \psi_2) + \psi_3] + \delta \quad (27)$$

$$\gamma = \left[\frac{c^2 L_{sensor}}{T_{sensor}^2}\left(\frac{\lambda^4}{c_1}L_{sensor} + \lambda^{-1}\right)\right] \quad (28)$$

$$\delta = \gamma L_{sensor} + T_{sensor} \quad (29)$$

$$L_{sensor} = gain * DN + bias - spectral\ radiance \quad (30)$$

$$T_{sensor} = \frac{K_2}{ln\left(\frac{K_1}{L_{sensor}} + 1\right)} \quad (31)$$

where: LSE = land surface emissivity = 1.0094+0.047*ln(NDVI); λ = effective wavelength; DN = digital number of a pixel; T_{sesnor} = brightness temperature; $c^1 = 1.19104*10^8$ Wμm^4 m^{-2}sr^{-1}; c^2 = 14387.7μmK; ψ_i (i = 1, 2, 3) = atmospheric parameters, which depend on total atmospheric water vapor content (w). Besides satellite data, this study uses two ground meteorological data: the total atmospheric water vapor content - w, used in LST estimation algorithm, and the air temperature - T_{air}. To estimate evapotranspiration, Serban et al. (2010) used one subset of Landsat ETM+ (7th June 2000) for Dobrogea area corresponding to Constanta weather station, which was atmospherically corrected.

From the bands ETM+ 3 and 4 were analyzed the NDVI values, the band ETM+ 6 was processed to determine LST, and the other bands (ETM+ 1, 2, 5 and 7) were used to estimate the albedo values. The difference between the actual mean soil surface temperature at the time when satellite passed and the remote sensed mean land surface temperature (0.73°C) is considered acceptable. The evapotranspiration (Fig. 10) ranges between 0.33 and 5.24mm/day. According to Constanta weather station, the multi-annual average of the evapotranspiration in June is between 4.5 and 5.6 mm/day, so the estimation error is eligible.

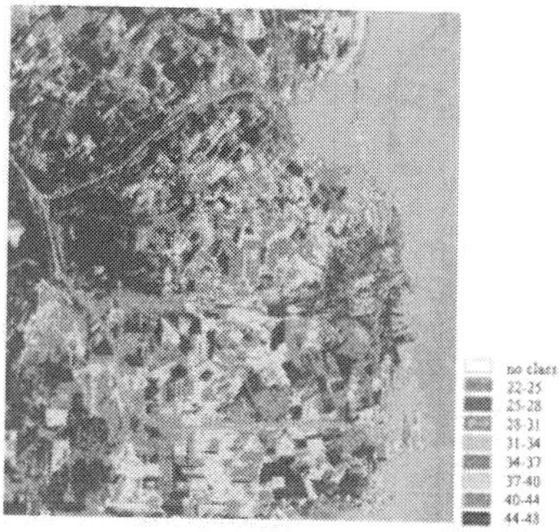

Fig. 9. LST Image - Dobrogea region, 2000 (After Serban et al., 2010)

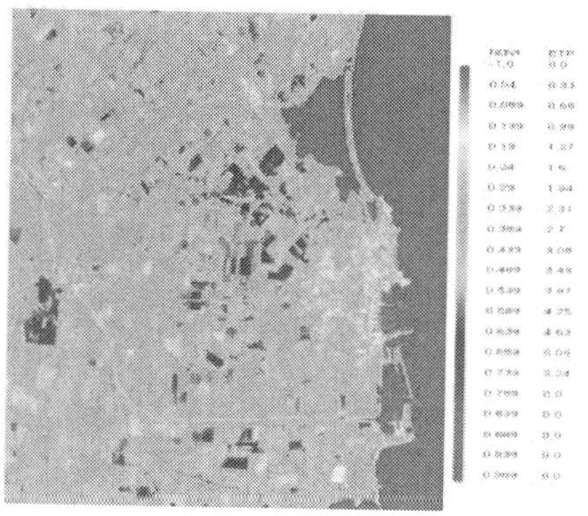

Fig. 10. ETP Image - Dobrogea region, 2000 (After Serban et al., 2010)

6. ET estimation using meteorological data

6.1 Crop evapotranspiration

At a crop level, ET may not occur uniformly because variations in crop germination, soil water availability, and other factors such as non-uniform water and nutrient applications and an uneven distribution of solar radiation within the canopy. Usually, the top leaves are more active in transpiration than the lower leaves because they receive more light. Also, the bottom leaves mature and age earlier and they may have lower transpiration rates than the greener and younger top leaves. Thus, weather parameters, crop characteristics, environmental and management aspects are the factors which influence the evaporation and transpiration processes. The main weather parameters influencing evapotranspiration are radiation, air temperature, humidity and wind speed. Several algorithms have been developed to estimate the evaporation rate from these parameters. The evaporation power of the atmosphere is expressed by the reference crop evapotranspiration (ET_o) which represents the evapotranspiration from a standardized vegetated surface (Allen et al., 1998). The reference surface is a hypothetical grass reference crop with specific characteristics. Because ET_o is affected by only climatic parameters, it is a climatic parameter and may be computed from weather data. Thus ET_o is the evaporating power of the atmosphere at a specific location and time of the year and does not take into account the crop characteristics and soil factors.

Crop water requirement is defined as the amount of water required to compensate the evapotranspiration loss from the cropped field. Even the values for crop evapotranspiration are identical with crop water requirement (CWR), crop evapotranspiration refers to the amount of water that is lost by evapotranspiration, while CWR refers to the amount of water that needs to be supplied. Thus, the irrigation water requirement represents the difference between the crop water requirement and effective precipitation and also includes additional water for leaching of salts and to compensate for non-uniformity of water application (Allen et al., 1998). Several empirical methods have been developed over the last five decades in order to estimate the evapotranspiration from different climatic variables. Testing the accuracy of the methods under a new set of conditions is laborious, time-consuming and costly, and yet evapotranspiration data are frequently needed at short notice for project planning or irrigation scheduling design. To meet this need, guidelines were developed and published in the FAO Irrigation and Drainage Paper No. 24 'Crop water requirements'. From different data availability, four methods are usually used to estimate the reference crop evapotranspiration (ET_o): the Blaney-Criddle, radiation, modified Penman and pan evaporation methods. From these four methods, the modified Penman-Monteith method offer the best results with minimum possible error in relation to a living grass reference crop. The radiation method can be used for areas where available climatic data include measured air temperature and sunshine, cloudiness or radiation, but not measured wind speed and air humidity. The Blaney-Criddle method is better to be applying for areas where available climatic data cover air temperature data only. The pan method gives acceptable estimates, depending on the location of the pan. Based on the original Penman- FAO proposed a standard parameterization of the Penman–Monteith method for estimating the evaporation from a -irrigated, homogenous, 0.12 m grass cover considered as a "reference crop" (Allen et al., 1998) (Fig. 11).

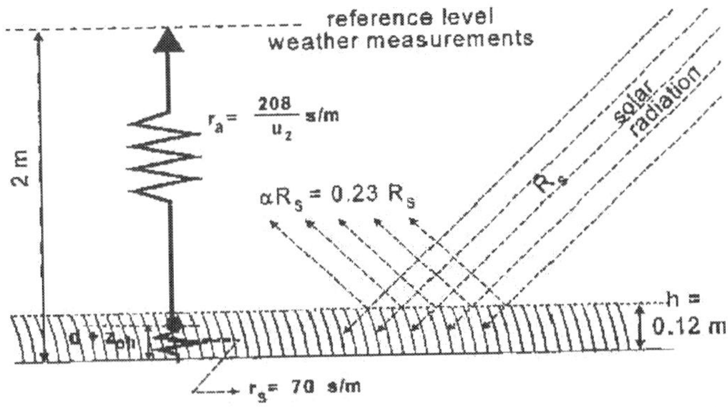

Fig. 11. Characteristics of the hypothetical reference crop (after Allen et al., 1998)

Monteith equation and the equations of the aerodynamic and surface resistance, the FAO Penman-Monteith method to estimate ET_o is the following:

$$ET_0 = \frac{0.408\Delta(R_n - G) + \gamma\left(\frac{900}{T+273}\right)u_2(e_s - e_a)}{\Delta + \gamma(1 + 0.34u_2)} \tag{32}$$

where: ET_0 = reference evapotranspiration [mm day^{-1}]; R_n = net radiation at the crop surface [MJ m^{-2} day^{-1}]; G = soil heat flux density [MJ m^{-2} day^{-1}]; T = mean daily air temperature at 2 m height [°C]; u_2 = wind speed at 2 m height [m s^{-1}]; e_s = saturation vapour pressure [kPa]; e_a = actual vapour pressure [kPa]]; e_s - e_a = saturation vapour pressure deficit [kPa]; Δ = slope vapour pressure curve [kPa $°C^{-1}$]; γ = psychrometric constant [kPa $°C^{-1}$]. The equation uses standard climatological records of solar radiation (sunshine), air temperature, humidity and wind speed. To obtain correct estimations of ET_0, the weather measurements should be made at 2 m (or converted to that height) above an extensive surface of green grass, shading the ground and not short of water. The psychrometric constant, γ, is calculated with:

$$\gamma = \frac{C_p P}{\varepsilon\lambda} = 0.665 * 10^{-3} \tag{33}$$

Where: P = atmospheric pressure [kPa]; λ = latent heat of vaporization, 2.45 [MJ kg^{-1}]; c_p = specific heat at constant pressure, 1.013 10-3 [MJ kg^{-1} $°C^{-1}$]; ε = ratio molecular weight of water vapour/dry air = 0.622. For standardization, T_{mean} for 24 hour is defined as the mean of the daily maximum (T_{max}) and minimum temperatures (T_{min}) rather than as the average of hourly temperature measurements.

$$T_{mean} = \frac{T_{max} - T_{min}}{2} \tag{34}$$

The temperature is given in degrees Celsius (°C), Fahrenheit (°F) or in Kelvin (K =C° + 273,16).

$$P = 101.3\left(\frac{293 - 0.0065z}{293}\right)^{526} \tag{35}$$

where: z = elevation above sea level [m].

6.2 CROPWAT model

CROPWAT is a decision support system developed by the Land and Water Development Division of FAO for planning and management of irrigation. The main functions of CROPWAT model are: to calculate the reference evapotranspiration, crop water requirements and crop irrigation requirements; to develop irrigation schedules under different management conditions and water supply schemes; to estimate the rainfed production and drought effects; to evaluate the efficiency of irrigation practices.

The input data of the model are the following climatic, crop and soil data: reference crop evapotranspiration: (ET_o) values measured or calculated using the FAO Penman-Montieth equation based on monthly climatic average data of the minimum and maximum air temperature (°C), relative humidity (%), sunshine duration (h) and wind speed (m/s); rainfall data: (daily/monthly data); monthly rainfall is divided for each month into a number of rainstorms; a cropping pattern: crop type, planting date, crop coefficient data files (including K_c values, stage days, root depth, depletion fraction, K_y values) and the area planted (0– 100% of the total area); a set of typical crop coefficient data files are provided in the program; soil type: total available soil moisture, maximum rain infiltration rate, maximum rooting depth, and initial soil moisture depletion (% of the total available moisture);scheduling criteria: several options can be selected regarding the calculation of the application timing and application depth.

The output parameters for each crop are crop reference crop evapotranspiration Et_0 (mm/period), crop K_c (average values of crop coefficient for each time step, effective rain (mm/period) (the amount of water that enters in the soil); water requirements (CWR) or ET_m (mm/period); irrigation requirements (IWR - mm/period); actual crop evapotranspiration (ET_c - mm); effective rain (mm/period) which represents the amount of water that enters into the soil; daily soil moisture deficit (mm); estimated yields reduction due to crop stress (when ET_c/ET_m falls below 100%).

The CROPWAT model can compute the actual evapotranspiration using the FAO Penman-Monteith equation or using directly the evapotranspiration measurements values. The crop water requirements (CWR) or maximum evapotranspiration (ET_m) (mm/period) are calculated as:

$$CWR = ET_0 * CropK_c \tag{36}$$

This means that the peak CWR in mm/day can be less than the peak Et_o value when less than 100% of the area is planted in the cropping pattern.

The average values of the crop coefficient (K_c) for each time step are estimated by linear interpolation between the K_c values for each crop development stage. The "Crop K_c" values are calculated as:

$$CropK_c = K_c * CropArea \tag{37}$$

where CropArea is the area covered by the crop. So, if the crop covers only 50% of the area, the "Crop Kc" values will be half of the Kc values in the crop coefficient data file.

The CROPWAT model operates in two modes: computing the actual evapotranspiration using climatic parameters and using directly the evapotranspiration measurements values.

Possibilities to use the satellite-based data as input into the CROPWAT model are limited, because this model was not developed to use satellite-derived information directly. But this information can be useful for the comparison/validation procedures of some model input/output data, as precipitation, sunshine duration and evapotranspiration. Satellite based data can be used by CROPWAT model in different ways: measured evapotranspiration may be replaced with estimations derived from satellite data; for comparison and validation procedures; satellite-derived evapotranspiration values may bring better accuracy for the specialization of the punctual computing values; satellite information may be used for the assessment of the some reference parameters of the actual evapotranspiration (e.g. Land surface temperature, vegetation indexes, etc.).

6.3 Using earth observation data and CROPWAT model to estimate the actual crop evapotranspiration

There is a strong dependence between evapotranspiration and surface temperature on the, thus thermal images meteorological satellites (METEOSAT, NOAA, MODIS, LANDSAT) adequate for mapping of regional evapotranspiration. Several works have been done to determine regional evapotranspiration from satellite data (Batra et al., 2006; Courault et al., 2005; Wood et al., 2003). The application of NOAA AVHRR data seems to be more successful because of the higher spatial and spectral resolution (Stancalie et al., 2010). Multichannel algorithms are routinely used for atmospheric correction of the AVHRR data.

Efforts are directed towards the estimation of surface temperatures by considering the effects of emissivity (Lagouarde and Brunet, 1991; Li and Becker, 1993). The method used for the estimation of the daily crop actual evapotranspiration, ET_{cj}, is based on the energy balance of the surface. The method uses the connection between evapotranspiration, net radiation and the difference between surface and air temperatures measured around 14:00 h (the time of the satellite passage), local time. The first version of the method used a simplified linear relationship as:

$$ET_{cj} - R_{nj} = A - B * (T_s - T_{amax}) \tag{38}$$

where R_{nj} is the daily net radiation; T_s and T_{amax} is the surface and air maximum temperature; A, B are coefficients which depend on the surface type and the daily mean wind speed. Coefficients A and B may be determined either analytically, on the basis of the relationships given by Lagouarde and Brunet (1991), or statistically. The coefficients A and B are stable in the case of mature crop vegetation cover and in clear sky conditions. The coefficient B vary considerably, function of the land vegetation cover percent. In case of soil with great thermal inertia, the heat flux changed by conduction at the soil-atmosphere interface can be neglected and the computing relationship for daily actual crop evapotranspiration can be expressed in a version 2 of the proposed method:

$$ET_{cj} = R_{nj} - B' * (T_s - T_{amax}) \tag{39}$$

$$B' = 0.0253 + \left[\frac{1.0016}{log2(2/zh)}\right] v \tag{40}$$

$$zh = [1 - exp(-LAI)] \left[exp\left(-\frac{LAI}{2}\right)\right] \tag{41}$$

where: v = daily average wind speed; zh = vegetation roughness and LAI the foliar index.
One possible use of satellite information is to replace the measured evapotranspiration by estimations made from satellite information. Because the estimations made from satellite information are available only for clear sky conditions, it was not possible to estimate the monthly average evapotranspiration, as input data in the CROPWAT model. For this reason, the satellite-derived data have been used for comparison/validation procedures of the CROPWAT model output data, like evapotranspiration. Fig. 12 presents the comparison between daily crop evapotranspiration values computed by the CROPWAT model and those computed through the energy balance method (Version 1), using remotely sensed data at the Alexandria and Craiova test-areas (situated in the south-western part of Romania), in the conditions of the year 2000 (Stancalie et al., 2010, 2010).
Analysis of model results concerning comparison of daily actual crop evapotranspiration calculated by using climatic data vs. satellite estimations based on the surface energetic balance (Version 1) showed that ET_c values from satellite information are in general higher than those simulated by the model, the differences being from +0.45 - 1.9 mm/day. Preliminary results highlighted a good correlation between the simulated values (CROPWAT) and those derived from the satellite data; with relative errors from +20% - 18% at Craiova site and from +13% -17% at Alexandria site (Stancalie et al., 2010).
Fig. 13 shows a comparison between ET_c simulated daily by the CROPWAT model over the whole maize-growing season and by the energy balance method (Version 2) respectively, using satellite data, at Alexandria and Craiova test-areas. The ET_c calculated by the model is very similar to the estimated one. The results obtained can constitute the premise of an ET_c data validation process, determined by the CROPWAT model (Stancalie et al., 2010).

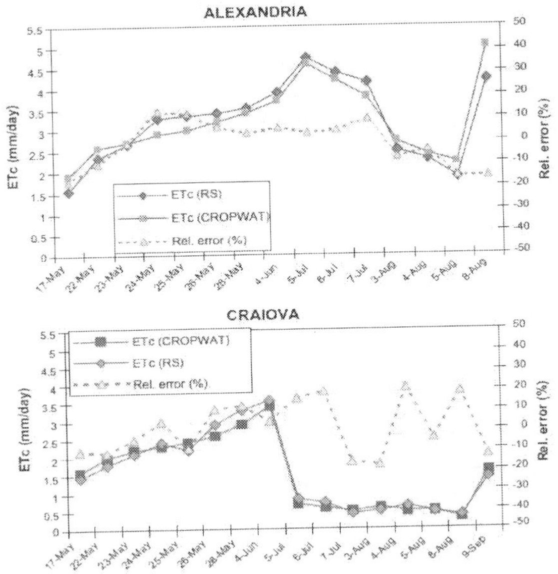

Fig. 12. Comparison between daily crop evapotranspiration values computed by the CROPWAT model and by the energy balance method (Version 1) using satellite data at the Alexandria and Craiova test-areas (after Stancalie et al., 2010)

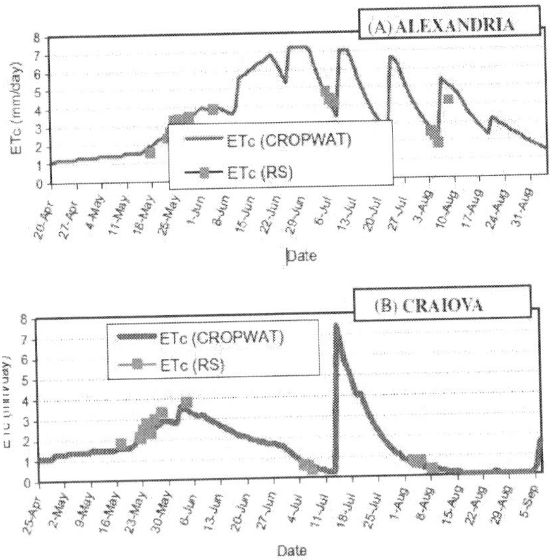

Fig. 13. Comparison between daily crop et values computed by the CROPWAT model and by the energy balance method (Version 2) using satellite data, at Alexandria (A) and Craiova (B) test-areas, for the maize vegetative development period in 2000 (Stancalie et al., 2010).

7. Conclusions

The use of the multispectral satellite data can improve the classical methods applied in determining the agrometeorological parameters, including evapotranspiration.
Estimating evapotranspiration using remote sensing methodologies have a significant role in irrigation management and crop water demand assessment, for plant growth, carbon and nutrient cycling and for production modeling in dry land agriculture and forestry. Also it can have an important role in catchment hydrology, and larger scale meteorology and climatology applications. In the last years, due to the exceptional developments of satellite technology, a wide range of remote sensing-based evapotranspiration (ET) methods/models have been developed and evaluated. The use of remote sensing data for ET estimation is mainly based on land surface temperature (LST) and reflectivity (using different spectral regions) due to satellite ability to spatially integrate over heterogeneous surfaces at a range of resolutions and to routinely generating areal products once long time-series data availability issues are overcome. The chapter reviews some main methods for estimating crop evapotranspiration based on remotely sensed data, and highlights uncertainties and limitations associated with those estimation methods. This paper is focused on Surface Energy Balance models (SEB), spatial variability methods using vegetation indices and ET estimation using meteorological data through CROPWAT model. The analysis and critical issues are supported by the dedicated literature and specific case-studies. This review provides information of temporal and spatial scaling issues associated with the use of optical and thermal remote sensing for estimating evapotranspiration. Improved temporal scaling procedures are required to extrapolate estimates to daily and longer time periods and gap-filling procedures are needed when temporal scaling is affected by intermittent satellite coverage. It is also noted that analysis of multi-resolution data from different satellite/sensor systems is able to assist the development of spatial scaling and aggregation approaches. Approaches differ in: (i) type and spatial extent of application (e.g. irrigation, dry-land agriculture); (ii) type of remote sensing data; and (iii) use of ancillary (micro-) meteorological and land cover data. The integration of remotely sensed data into methods/models of ET facilitates the estimation of water consumption across agricultural regions. There are important limitations for using remote sensing data in estimating evapotranspiration.
Usually evapotranspiration is computed using land surface temperature and air temperatures. All this methods are affected by errors induced by estimation or measurements of those temperatures. The accuracy of T_{rad} observations is influenced by atmospheric factors, surface emissivity or view angle. Emissivity information is useful in estimating of the radiative temperature of the land surface. Several direct methods (which atmospheric variables are coupled with radiative transfer models) or indirect algorithms (use only remote sensing data) to make atmospheric corrections in order to obtain the brightness temperature that represents the temperature of a black body that would have the same radiance as that observed by the radiometer. The uncertainties of surface temperature have a strong influence in determination of sensible heat flux H. The difference between surface and air temperatures depends on many factors, including vegetation type, fractional cover f_c and view angle. Another important limitation of various spatial variability methods is considered the fact according to the highest and lowest surface temperatures observed in the one scene are assumed to represent very dry and very wet pixels. Usually the available energy (R_n - G) is obtained from ground based point observations of R_n: R_n is estimated

based on observations of K↓, α, LAI, emissivity of land surface and atmosphere, and T_{rad}. Such kind of estimation generates errors in the calculation of long and short wave components. G can be estimated for example as function of NDVI. An alternative method would be to assume that soil heat flux is a constant fraction of net radiation flux, but this estimation doesn't take into account the diurnal variation. Many models for ET estimation need ground based meteorological data, mainly air temperature and wind speed. For that models which based on computing the difference between Tad and Ta, the time and location of air temperature (Ta) observations and their spatial representativeness are very important).

Incomplete vegetation cover generates also errors in evapotranspiration estimation. The two source models require parameterizations for the segmentation of the computed surface temperature between vegetation and soil, for the turbulent exchange of heat and mass between soil and atmosphere and between vegetation and atmosphere. Also, these models require some assumptions regarding solar transmittance, extinction coefficients and canopy emissivity in order to compute the variation of net radiation flux inside the canopy.

Another important limitation, regarding the spatial variability methods is that a large number of pixels are required over the area of interest with a wide range of soil wetness and fractional vegetation cover. The identification of vegetation limits for bare soil or full vegetation cover can be easily done using high resolution images which display a wide range in surface wetness conditions and land cover conditions

Remote sensing data is a useful tool that provides input data in land surface model (NDVI, LAI, f_c – fraction cover) and can be used to correct the state variables of the models.

The frequency of spatial resolution imagery is also very significant: satellites which provide high resolution data usually have lower temporal frequency while low spatial resolution images have higher temporal frequency. Some applications require different spatial and temporal coverage rates and need different "turn-around" times. If acquiring the satellite data and ET estimation method are more time consuming, the method are not very convenient for operational applications like determining water requirements for irrigated agriculture.

Another significant limitation for using remote sensing is the presence of clouds that generates intermittent coverage. Cloudy days are characterized by a diffuse light, whereas while direct light is dominant on clear days when most TIR data are acquired for use in modeling applications. Most SEB models have been developed for use in cloud-free conditions and do not makes difference between direct and diffuse radiation; they use only daytime data obtained for clear-sky conditions. For a continuously monitoring of water balance, the effects of an increased diffuse fraction should be taking into account, because the diffuse radiation is used by vegetation more efficiently than direct radiation. For water use efficiency, to ignore difference between direct and diffuse radiation can induce significant differences in ET estimations.

8. References

Allen RG, Pereira LS, Raes D, Smith M (1998), Crop evapotranspiration - guidelines for computing crop water requirements. FAO irrigation and drainage paper 56, Rome, Italy http://www.fao.org/docrep/X0490E/X0490E00.htm

Allen RG, Tasumi M, Trezza R (2007a) Satellite-based energy balance for mapping evapotranspiration with internalized calibration (METRIC): model. J Irrig Drain Eng 133(4):380–394. doi:10.1061/(ASCE) 0733-9437(2007)133(4):(380)

Allen RG, Tasumi M, Trezza R (2007b) Satellite-based energy balance for mapping evapotranspiration with internalized calibration (METRIC): applications. ASCE J Irrig Drain Eng 133(4):395–406

Bastiaanssen WGM, Menenti M, Feddes RA, Holtslag AAM (1998a) A remote sensing surface energy balance algorithm for land. I. Formulation. J Hydrol (Amst) 212/213:198–212. doi:10.1016/S0022-1694(98)00253-4

Bastiaanssen WGM (2000) SEBAL-based sensible and latent heat fluxes in the irrigated Gediz Basin, Turkey. J Hydrol (Amst) 229:87–100. doi:10.1016/S0022-1694(99)00202-4

Batra N, Islam S, Venturini V, Bisht G, Jiang L (2006) Estimation and comparison of evapotranspiration from MODIS and AVHRR sensors for clear sky days over the Southern Great Plains. Remote Sens Environ 103:1–15. doi:10.1016/j.rse.2006.02.019

Boegh E, Soegaard H, Hanan N, Kabat P, Lesch L (1999) A remote sensing study of the NDVI–Ts relationship and the transpiration from sparse vegetation in the Sahel based on high-resolution satellite data. Remote Sens Environ 69:224–240. Doi: 10.1016/S0034-4257(99)00025-5

Boegh E, Soegaard H, Thomsen A (2002) Evaluating evapotranspiration rates and surface conditions using Landsat TM to estimate atmospheric resistance and surface resistance. Remote Sens Environ 79:329–343. doi:10.1016/S0034-4257(01)00283-8

Brutsaert W (1999) Aspects of bulk atmospheric boundary layer similarity under free-convective conditions. Rev Geophys 37:439–451. Doi: 10.1029/1999RG900013.

Burba G., Hubart J.A., Pidwirny M. (2010), Evapotranspiration, Encyclopedia of Earth, Eds. Cutler J. Cleveland (Washington, D.C.: Environmental Information Coalition, National Council for Science and the Environment, August 3, 2010, http://www.eoearth.org/article/Evapotranspiration

Carlson TN, Capehart WJ, Gillies RR (1995a) A new look at the simplified method for remote sensing of daily evapotranspiration. Remote Sens Environ 54:161–167. Doi: 10.1016/0034-4257(95)00139-R

Carlson TN (2007) An overview of the "triangle method" for estimating surface evapotranspiration and soil moisture from satellite imagery. Sensors 7:1612–1629

Cleugh HA, Leuning R, Mu Q, Running SW (2007) Regional evaporation estimates from flux tower and MODIS satellite data. Remote Sens Environ 106:285–304. doi:10.1016/j.rse.2006.07.007

Courault D, Seguin B, Olioso A (2005) Review on estimation of evapotranspiration from remote sensing data: from empirical to numerical modelling approaches. Irrig Drain Syst 19:223–249. Doi: 10.1007/s10795-005-5186-0

Gillies RR, Carlson TN (1995) Thermal remote sensing of surface soil water content with partial vegetation cover for incorporation into climate models. J Appl Meteorol 34:745–756. doi :10.1175/1520-0450(1995)034\0745:TRSOSS[2.0.CO;2

Gillies RT, Carlson TN, Cui J, Kustas WP, Humes KS (1997) A verification of the "triangle" method for obtaining surface soil water content and energy fluxes from remote measurements of the Normalized Difference Vegetation Index (NDVI) and surface radiant temperatures. Int J Remote Sens 18(15):3145–3166. doi:10.1080/014311697217026

Glenn EP, Huete AR, Nagler PL, Hirschboeck KK, Brown P (2007) Integrating remote sensing and ground methods to estimate evapotranspiration. Crit Rev Plant Sci 26(3):139–168. doi:10.1080/07352680701402503

Hope AS, Petzold DE, Goward SN, Ragan RM (1986) Simulated relationships between spectral reflectance, thermal emissions, and evapotranspiration of a soybean canopy. Water Resour Bull 22:1011–1019

Huntington, T. 2006. Evidence for intensification of the global water cycle: review and synthesis. J Hydr. 319: 83–95.

Hutley, L., O'Grady, A., and Easmus, D. 2001. Monsoonal influences on evapotranspiration of savanna vegetation of northern Austalia. Oecologia 126: 434–443.

Huxman, T., Smith, M., Fay, P., Knapp, A., Shaw, M., Loik, M., Smith, S., Tissue, D., Zak, J., Weltzin, J., Pockman, W., Sala, O., Haddad, B., Harte, J., Koch, G., Schwinning, S., Small, E.,Williams, D. 2004. Convergence across biomes to a common rain-use efficiency. Nature 429: 651–654.

Huxman, T.,Wilcox, B., Breshears, D., Scott, R., Snyder, K., Small, E., Hultine, K., Pockman, W., and Jackson, R. 2005. Ecohydrological implications of woody plant encroachment. Ecology 86: 308–319.

Jacquemin, B.&Noilhan, J. 1990. Sensitivity study and validation of land surface parametrization using the Hapex-Mobilhy data set. Bound-Layer Meteorology 52: 93–134.

Jiang L, Islam S (2001) Estimation of surface evaporation map over southern Great Plains using remote sensing data. Water Resour Res 37:329–340. doi:10.1029/2000WR900255

Jupp DLB, Tian G, McVicar TR, Qin Y, Fuqin L (1998) Soil moisture and drought monitoring using remote sensing I: theoretical background and methods. CSIRO Earth Observation Centre, Canberra http://www.eoc.csiro.au/pubrep/scirpt/jstc1.pdf

Kalma, J. D., McVicar, T. R. and McCabe, M. F. (2008). "Estimating land surface evaporation: A review of methods using remotely sensed surface temperature data." Surveys in Geophysics 29(4-5): 421-469.

Katerji N., Perrier A.1985 - Détermination de la résistance globale d'un couvert végétal à la diffusion de vapeur d'eau et de ses différentes composantes. Approche théorique et vérification expérimentale sur une culture de luzerne. Agric. For Meteorol., 34, 2-3, 105-120.

Kustas WP, Norman JM (1996) Use of remote sensing for evapotranspiration monitoring over land surfaces. Hydrol Sci J 41(4):495–516

Kustas WP, Norman JM (1999) Evaluation of soil and vegetation heat flux predictions using a simple two source model with radiometric temperatures for partial canopy cover. Agric For Meteorol 94:13–29. doi:10.1016/S0168-1923(99)00005-2

Kustas WP, French AN, Hatfield JL, Jackson TJ, Moran MS, Rango A (2003a) Remote sensing research in hydrometeorology. Photogramm Eng Remote Sensing 69(6):613–646

Kustas WP, Hatfield JL, Prueger JH (2005) The Soil Moisture–Atmosphere Coupling Experiment (SMACEX): background, hydrometeorological conditions, and preliminary findings. J Hydrometeorol 6:825–839. doi:10.1175/JHM460.1

Lagouarde, J.P., Brunet, Y., 1991. A simple model for estimating the daily upward longwave surface radiations from NOAA-AVHRR data. International Journal of Remote Sensing 12, 1853-1864.

Lhomme JP, Monteny B, Amadou M (1994) Estimating sensible heat flux from radiometric temperature over sparse millet. Agric For Meteorol 44:197–216

Li, Z.L., Becker, F., 1993. Feasibility of land surface temperature and emissivity determination from AVHRR data. Remote Sensing of Environment 43, 67–85.

Li F, Kustas WP, Prueger JH, Neale CMU, Jackson TJ (2005) Utility of remote sensing based two-source energy balance model under low and high vegetation cover conditions. J Hydrometeorol 6(6):878–891. doi:10.1175/JHM464.1

Li Z.L., Tang R., Wan Z., Bi Y., Zhou C., Tang B., Yan G. and Zang X. (2009), A Review of Current Methodologies for Regional Evapotranspiration Estimation from Remotely Sensed Data, Sensors 2009, 9, 3801-3853; doi:10.3390/s90503801

McCabe MF, Wood EF (2006) Scale influences on the remote estimation of evapotranspiration using multiple satellite sensors. Remote Sens Environ 105(4):271–285. doi:10.1016/j.rse.2006.07.006

McVicar TR, Jupp DLB (1998) The current and potential operational uses of remote sensing to aid decisions on drought exceptional circumstances in Australia: a review. Agric Syst 57:399–468. doi:10.1016/S0308-521X(98)00026-2

Monteith JL (1965) Evaporation and the environment. In: Fogg GE (ed) The state and movement of water in living organisms, 19th symposium of the society for experimental biology. University Press, Cambridge, pp 205–234

Monteith, J.L., 1981. Evaporation and surface temperature. Quart. J. Roy. Meteorolog. Soc., 107, 1-27.

Norman JM, Kustas WP, Humes KS (1995) A two-source approach for estimating soil and vegetation energy fluxes from observations of directional radiometric surface temperature. Agric For Meteorol 77:263–293. doi:10.1016/0168-1923(95)02265-Y

Norman JM, Kustas WP, Prueger JH, Diak GR (2000) Surface flux estimation using radiometric temperature: a dual-temperature-difference method to minimize measurement errors. Water Resour Res 36:2263–2274. doi:10.1029/2000WR900033

Overgaard J, Rosbjerg D, Butts MB (2006) Land-surface modelling in hydrological perspective—a review. Biogeosciences 3:229–241

Penman HL (1948) Natural evaporation from open water, bare soil and grass. Proc R Soc Lond A Math Phys Sci 193:120–146. doi:10.1098/rspa.1948.0037

Roerink CJ, Su Z, Menenti M (2000) S SEBI: a simple remote sensing algorithm to estimate the surface energy balance. Phys Chem Earth, Part B Hydrol Oceans Atmos 25(2):147–157. doi:10.1016/S1464-1909(99)00128-8

Seguin B, Baelz S, Monget JM, Petit V (1982a) Utilisation de la thermographie IR pour l'estimation de l'e´vaporation re´gionale I Mise au point me´thodologique sur le site de la Crau. Agronomie 2(1):7–16. doi:10.1051/agro:19820102

Serban C., Maftei C., Barbulescu A. (2010), Assessment of Evapotranspiration Using Remote Sensing Data and Grid Computing and Apglication, WSEAS Transactions on computers, ISSN: 1109-2750, Issue 11, Volume 9, November 2010, pg.1245-1254

Shuttleworth WJ, Wallace JS (1985) Evaporation from sparse crops-an energy combination theory. Q J R Meteorol Soc 111:839–855. doi:10.1256/smsqj.46909

Smith, R. C. G. & Choudhury, B. J. (1991) Analysis of normalized difference and surface temperature observations over southeastern Australia. Int. J. Remote Sens. 12, 2021-2044.

Sobrino JA, Gomez M, Jimenez-Munoz JC, Olioso A (2007) Application of a simple algorithm to estimate daily evapotranspiration from NOAA-AVHRR images for the Iberian Peninsula. Remote Sens Environ 110:139–148. doi:10.1016/j.rse.2007.02.017

Stancalie GH, Marica A, Toulios L, (2010) Using earth observation data and CROPWAT model to estimate the actual crop evapotranspiration. Physics and Chemistry of the Earth, 35:25-30, doi:10.1016/j.pce.2010.03.013.

Stewart JB, Kustas WP, Humes KS, Nichols WD, Moran MS, De Bruin HAR (1994) Sensible heat flux–radiometric surface temperature relationships for eight semi-arid areas. J Appl Meteorol 33:1110–1117. doi :10.1175/1520-0450(1994)033\1110:SHFRST [2.0.CO;2

Su Z (2002) The Surface Energy Balance System (SEBS) for estimation of turbulent heat fluxes. Hydrol Earth Syst Sci 6(1):85–99 (HESS)

Su H, McCabe MF, Wood EF, Su Z, Prueger JH (2005) Modeling evapotranspiration during SMACEX: comparing two approaches for local- and regional-scale prediction. J Hydrometeorol 6(6):910–922. doi:10.1175/JHM466.1

Su Z (2008) The Surface Energy Balance System (SEBS) for estimation of turbulent heat fluxes and evapotranspiration, Dragon 2, Advanced Trainig Course in Land Remote Sensing, http://dragon2.esa.int/landtraining2008/pdf/D3L2b_SU_SEBS.pdf

Wang K, Li Z, Cribb M (2006) Estimation of evaporative fraction from a combination of day and night land surface temperature and NDVI: a new method to determine the Priestley–Taylor parameter. Remote Sens Environ 102:293–305. doi:10.1016/j.rse.2006.02.007

Wood, E.F., Hongbo, Su, McCabe, M., Su, B., 2003. Estimating evaporation from satellite remote sensing. In: Geoscience and Remote Sensing Symposium 2003. IGARSS Proceedings of the IEEE International, vol. 2, pp. 163–1165.

21

Operational Remote Sensing of ET and Challenges

Ayse Irmak[1], Richard G. Allen[2], Jeppe Kjaersgaard[2], Justin Huntington[3], Baburao Kamble[4], Ricardo Trezza[2] and Ian Ratcliffe[1]
[1]School of Natural Resources University of Nebraska–Lincoln, HARH, Lincoln NE
[2]University of Idaho, Kimberly, ID
[3]Desert Research Institute, Raggio Parkway, Reno, NV
[4]University of Nebraska-Lincoln, Lincoln, NE
USA

1. Introduction

Satellite imagery now provides a dependable basis for computational models that determine evapotranspiration (ET) by surface energy balance (EB). These models are now routinely applied as part of water and water resources management operations of state and federal agencies. They are also an integral component of research programs in land and climate processes. The very strong benefit of satellite-based models is the quantification of ET over large areas. This has enabled the estimation of ET from individual fields among populations of fields (Tasumi et al. 2005) and has greatly propelled field specific management of water systems and water rights as well as mitigation efforts under water scarcity. The more dependable and universal satellite-based models employ a surface energy balance (EB) where ET is computed as a residual of surface energy. This determination requires a thermal imager onboard the satellite. Thermal imagers are expensive to construct and more a required for future water resources work. Future moderate resolution satellites similar to Landsat need to be equipped with moderately high resolution thermal imagers to provide greater opportunity to estimate spatial distribution of actual ET in time. Integrated ET is enormously valuable for monitoring effects of water shortage, water transfer, irrigation performance, and even impacts of crop type and variety and irrigation type on ET. Allen (2010b) showed that the current 16-day overpass return time of a single Landsat satellite is often insufficient to produce annual ET products due to impacts of clouds. An analysis of a 25 year record of Landsat imagery in southern Idaho showed the likelihood of producing annual ET products for any given year to increase by a factor of NINE times (from 5% probability to 45% probability) when two Landsat systems were in operation rather than one (Allen 2010b).

Satellite-based ET products are now being used in water transfers, to enforce water regulations, to improve development and calibration of ground-water models, where ET is a needed input for estimating recharge, to manage streamflow for endangered species management, to estimate water consumption by invasive riparian and desert species, to estimate ground-water consumption from at-risk aquifers, for quantification of native

American water rights, to assess impacts of land-use change on wetland health, and to monitor changes in water consumption as agricultural land is transformed into residential uses (Bastiaanssen et al., 2005, Allen et al., 2005, Allen et al. 2007b).

The more widely used and operational remote sensing models tend to use a 'CIMEC' approach ("calibration using inverse modeling of extreme conditions") to calibrate around uncertainties and biases in satellite based energy balance components. Biases in EB components can be substantial, and include bias in atmospheric transmissivity, absolute surface temperature, estimated aerodynamic temperature, surface albedo, aerodynamic roughness, and air temperature fields. Current CIMEC models include SEBAL (Bastiaanssen et al. 1998a, 2005), METRIC (Allen et al., 2007a) and SEBI-SEBS (Su 2002) and the process frees these models from systematic bias in the surface temperature and surface reflectance retrievals. Other models, such as the TSEB model (Kustas and Norman 1996), use absolute temperature and assumed air temperature fields, and so can be more susceptible to biases in these fields, and often require multiple times per day imagery. Consequently, coarser resolution satellites must be used where downscaling using finer resolution reflectance information is required.

Creating 'maps' of ET that are useful in management and in quantifying and managing water resources requires the computation of ET over monthly and longer periods such as growing seasons or annual periods. Successful creation of an ET 'snapshot' on a satellite overpass day is only part of the required process. At least half the total effort in producing a quantitative ET product involves the interpolation (or extrapolation) of ET information between image dates. This interpolation involves treatment of clouded areas of images, accounting for evaporation from wetting events occurring prior to or following overpass dates, and applying a grid of daily reference ET with the relative ET computed for an image, or a direct Penman-Monteith type of calculation, over the image domain for periods between images to account for day to day variation in weather. The particular methodology for estimating these spatial variables substantially impacts the quality and accuracy of the final ET product.

2. Model overview

Satellite based models can be separated into the following classes, building on Kalma et al. (2008):

- Surface Energy Balance
    - Full energy balance for the satellite image: $\lambda E = R_n - G - H$
    - Water stress index based on surface temperature and vegetation amounts
    - Application of a continuous Land Surface Model (LSM) that is partly initialized and advanced, in time, using satellite imagery
- Statistical methods using differences between surface and air temperature
- Simplified correlations or relationships between surface temperature extremes in an image and endpoints of anticipated ET
- Vegetation-based relative ET that is multiplied by a weather-based reference ET

where λE is latent heat flux density, representing the energy 'consumed' by the evaporation of water, R_n is net radiation flux density, G is ground heat flux density and H is sensible heat flux density to the air.

Except for the LSM applications, none of the listed energy balance methods, in and of themselves, go beyond the creation of a 'snapshot' of ET for the specific satellite image date. Large periods of time exist between snapshots when evaporative demands and water availability (from wetting events) cause ET to vary widely, necessitating the coupling of hydrologically based surface process models to fill in the gaps. The surface process models employed in between satellite image dates can be as simple as a daily soil-surface evaporation model based on a crop coefficient approach (for example, the FAO-56 model of Allen et al. 1998) or can involve more complex plant-air-water models such as SWAT (Arnold et al. 1994), SWAP (van Dam 2000), HYDRUS (Šimůnek et al. 2008), Daisy (Abrahamsen and Hansen 2000) etc. that are run on hourly to daily timesteps.

2.1 Problems with use of absolute surface temperature

Error in surface temperature (T_s) retrievals from many satellite systems can range from 3 - 5 K (Kalma et al. 2008) due to uncertainty in atmospheric attenuation and sourcing, surface emissivity, view angle, and shadowing. Hook and Prata (2001) suggested that finely tuned T_s retrievals from modern satellites could be as accurate as 0.5 K. Because near surface temperature gradients used in energy balance models are often on the order of only 1 to 5 K, even this amount of error, coupled with large uncertainties in the air temperature fields, makes the use of models based on differences in absolute estimates of surface and air temperature unwieldy.

Cleugh et al. (2007) summarized challenges in using near surface temperature gradients (dT) based on absolute estimates of T_s and air temperature, T_{air}, attributing uncertainties and biases to error in T_s and T_{air}, uncertainties in surface emissivity, differences between radiometrically derived T_s and the aerodynamically equivalent T_s required as a sourcing endpoint to dT.

The most critical factor in the physically based remote sensing algorithms is the solution of the equation for sensible heat flux density:

$$H=\rho_a c_p \frac{T_{aero}-T_a}{r_{ah}} \tag{1}$$

where ρ_a is the density of air (kg m^{-3}), c_p is the specific heat of air (J kg^{-1} K^{-1}), r_{ah} is the aerodynamic resistance to heat transfer (s m^{-1}), T_{aero} is the surface aerodynamic temperature, and T_a is the air temperature either measured at standard screen height or the potential temperature in the mixed layer (K) (Brutsaert et al., 1993). The aerodynamic resistance to heat transfer is affected by wind speed, atmospheric stability, and surface roughness (Brutsaert, 1982). The simplicity of Eq. (1) is deceptive in that T_{aero} cannot be measured by remote sensing. Remote sensing techniques measure the radiometric surface temperature T_s which is not the same as the aerodynamic temperature. The two temperatures commonly differ by 1 to 5 °C, depending on canopy density and height, canopy dryness, wind speed, and sun angle (Kustas et al., 1994, Qualls and Brutsaert, 1996, Qualls and Hopson, 1998). Unfortunately, an uncertainty of 1 °C in T_{aero} - T_a can result in a 50 W m^{-2} uncertainty in H (Campbell and Norman, 1998) which is approximately equivalent to an evaporation rate of 1 mm day^{-1}. Although many investigators have attempted to solve this problem by adjusting r_{ah} or by using an additional resistance term, no generally applicable method has been developed.

Campbell and Norman (1998) concluded that a practical method for using satellite surface temperature measurements should have at least three qualities: (i) accommodate the difference between aerodynamic temperature and radiometric surface temperature, (ii) not require measurement of near-surface air temperature, and (iii) rely more on differences in surface temperature over time or space rather than absolute surface temperatures to minimize the influence of atmospheric corrections and uncertainties in surface emissivity.

2.2 CIMEC Models (SEBAL and METRIC)

The SEBAL and METRIC models employ a similar inverse calibration process that meets these three requirements with limited use of ground-based data (Bastiaanssen et al., 1998a,b, Allen et al., 2007a). These models overcome the problem of inferring T_{aero} from T_s and the need for near-surface air temperature measurements by directly estimating the temperature difference between two near surface air temperatures, T_1 and T_2, assigned to two arbitrary levels z_1 and z_2 without having to explicitly solve for absolute aerodynamic or air temperature at any given height. The establishment of the temperature difference is done via inversion of the function for H at two known evaporative conditions in the model using the CIMIC technique. The temperature difference for a dry or nearly dry condition, represented by a bare, dry soil surface is obtained via $H=R_n - G- \lambda E$ (Bastiaanssen et al., 1998a):

$$T_1 - T_2 = \Delta T_a = \frac{H\, r_{ah_{1-2}}}{\rho_a c_p} \tag{2}$$

where $r_{ah,1-2}$ is the aerodynamic resistance to heat transfer between two heights above the surface, z_1 and z_2. At the other extreme, for a wet surface, essentially all available energy R_n - G is used for evaporation λE. At that extreme, the classical SEBAL approach assumes that $H \approx 0$, in order to keep requirements for high quality ground data to a minimum, so that $\Delta T_a \approx 0$. Allen et al. (2001, 2007a) have used reference crop evapotranspiration, representing well-watered alfalfa, to represent λE for the cooler population of pixels in satellite images of irrigated fields in the METRIC approach, so as to better capture effects of regional advection of H and dry air, which can be substantial in irrigated desert. METRIC calculates $H = R_n - G - k_1 \lambda ET_r$ at these pixels, where ET_r is alfalfa reference ET computed at the image time using weather data from a local automated weather station, and ΔT_a from Eq. (2) , where $k_1 \sim 1.05$. In typical SEBAL and METRIC applications, z_1 and z_2 are taken as 0.1 and 2 m above the zero plane displacement height (d). z_1 is taken as 0.1 m above the zero plane to insure that T_1 is established at a height that is generally greater than $d + z_{oh}$ (z_{oh} is roughness length for heat transfer). Aerodynamic resistance, r_{ah}, is computed for between z_1 and z_2 and does not require the inclusion and thus estimation of z_{oh}, but only z_{om}, the roughness length for momentum transfer that is normally estimated from vegetation indices and land cover type. H is then calculated in the SEBAL and METRIC CIMEC-based models as:

$$H = \rho_a c_p \frac{\Delta T_a}{r_{ah_{1-2}}} \tag{3}$$

One can argue that the establishment of ΔT_a over a vertical distance that is elevated above $d + z_{oh}$ places the r_{ah} and established ΔT_a in a blended boundary layer that combines influences of sparse vegetation and exposed soil, thereby reducing the need for two source modeling and problems associated with differences between radiative temperature and aerodynamic temperature and problems associated with estimating z_{oh} and specific air temperature associated with the specific surface.

Evaporative cooling creates a landscape having high ΔT_a associated with high H and high radiometric temperature and low ΔT_a with low H and low radiometric temperature. For example, moist irrigated fields and riparian systems have much lower ΔT_a and much lower T_s than dry rangelands. Allen et al. (2007a) argued, and field measurements in Egypt and Niger (Bastiaanssen et al., 1998b), China (Wang et al., 1998), USA (Franks and Beven, 1997), and Kenya (Farah, 2001) have shown the relationship between T_s and ΔT_a to be highly linear between the two calibration points

$$\Delta T_a = c_1 T_s - c_2 \tag{4}$$

where c_1 and c_2 are empirical coefficients valid for one particular moment (the time and date of an image) and landscape. By using the minimum and maximum values for ΔT_a as calculated for the nearly wettest and driest (i.e., coldest and warmest) pixel(s), the extremes of H are used, in the CIMEC process to find coefficients c_1 and c_2. The empirical Eq. (4) meets the third quality stated by Campbell and Norman (1998) that one should rely on differences in radiometric surface temperature over space rather than absolute surface temperatures to minimize the influence of atmospheric corrections and uncertainties in surface emissivity.

Equation (3) has two unknowns: ΔT_a and the aerodynamic resistance to heat transfer $r_{ah,1\text{-}2}$ between the z_1 and z_2 heights, which is affected by wind speed, atmospheric stability, and surface roughness (Brutsaert, 1982). Several algorithms take one or more field measurements of wind speed and consider these as spatially constant over representative parts of the landscape (e.g. Hall et al., 1992; Kalma and Jupp, 1990; Rosema, 1990). This assumption is only valid for uniform homogeneous surfaces. For heterogeneous landscapes a unique wind speed near the ground surface is required for each pixel. One way to meet this requirement is to consider the wind speed spatially constant at a blending height about 200 m above ground level, where wind speed is presumed to not be substantially affected by local surface heterogeneities. The wind speed at blending height is predicted by upward extrapolation of near-surface wind speed measured at an automated weather station using a logarithmic wind profile. The wind speed at each pixel is obtained by a similar downward extrapolation using estimated surface momentum roughness z_{0m} determined for each pixel.

Allen et al. (2007a) have noted that the inverted value for ΔT_a is highly tied to the value used for wind speed in its CIMEC determination. Therefore, they cautioned against the use of a spatial wind speed field at some blending height across an image with a single ΔT_a function. The application of the image-specific ΔT_a function with a blending height wind speed in a distant part of the image that is, for example, double that of the wind used to determine coefficients c_1 and c_2 can estimate higher H than is possible based on energy availability. In those situations, the 'calibrated' ΔT_a would be about half as much to compensate for the larger wind speed. Therefore, if wind fields at the blending height (200 m) are used, then fields of ΔT_a calibrations are also needed, which is prohibitive. The single ΔT_a function of SEBAL and METRIC, coupled with a single wind speed at blending height, transcends these problems. Gowda et al., (2008) presented a summary of remote sensing based energy balance algorithms for mapping ET that complements that by Kalma et al. (2008).

Aerodynamic Transport. The value for $r_{ah,1,2}$ is calculated between the two heights z1 and z2 in SEBAL and METRIC. The value for $r_{ah,1,2}$ is strongly influenced by buoyancy within the boundary layer driven by the rate of sensible heat flux. Because both $r_{ah,1,2}$ and H are unknown at each pixel, an iterative solution is required. During the first iteration, $r_{ah,1,2}$ is computed assuming neutral stability:

$$r_{ah_{1-2}} = \frac{ln\left(\frac{z_2}{z_1}\right)}{u_* \ k} \quad (5)$$

where z_1 and z_2 are heights above the zero plane displacement of the vegetation where the endpoints of dT are defined, u_* is friction velocity (m s^{-1}), and k is von Karman's constant (0.41). Friction velocity u_* is computed during the first iteration using the logarithmic wind law for neutral atmospheric conditions:

$$u^* = \frac{k \ u_{200}}{ln\left(\frac{200}{z_{om}}\right)} \quad (6)$$

where u_{200} is the wind speed (m s^{-1}) at a blending height assumed to be 200 m, and z_{om} is the momentum roughness length (m). z_{om} is a measure of the form drag and skin friction for the layer of air that interacts with the surface. u_* is computed for each pixel inside the process model using a specific roughness length for each pixel, but with u_{200} assumed to be constant over all pixels of the image since it is defined as occurring at a "blending height" unaffected by surface features. Eq. (5) and (6) support the use of a temperature gradient defined between two heights that are both above the surface. This allows one to estimate $r_{ah,1-2}$ without having to estimate a second aerodynamic roughness for sensible heat transfer (z_{oh}), since height z_1 is defined to be at an elevation above z_{oh}. This is an advantage, because z_{oh} can be difficult to estimate for sparse vegetation.

The wind speed at an assumed blending height (200 m) above the weather station, u_{200}, is calculated as:

$$u_{200} = \frac{u_w ln\left(\frac{200}{z_{omw}}\right)}{ln\left(\frac{z_x}{z_{omw}}\right)} \quad (7)$$

where u_w is wind speed measured at a weather station at z_x height above the surface and z_{omw} is the roughness length for the weather station surface, similar to Allen and Wright (1997). All units for z are the same. The value for u_{200} is assumed constant for the satellite image. This assumption *is required* for the use of a constant relation between dT and T_s to be extended across the image (Allen 2007a).

The effects of mountainous terrain and elevation on wind speed are complicated and difficult to quantify (Oke, 1987). In METRIC, z_{om} or wind speed for image pixels in mountains are adjusted using a suite of algorithms to account for the following impacts (Allen and Trezza, 2011):

- Terrain roughness – the standard deviation of elevation within a 1.5 km radius is used to estimate an additive to zom to account for vortex and channeling impacts of turbulence
- Elevation effect on velocity – the relative elevation within a 1.5 km radius is used to estimate a relative increase in wind speed, based on slope.
- Reduction of wind speed on leeward slopes – when the general wind direction aloft can be estimated in mountainous terrain, then a reduction factor is made to wind speed on leeward slopes, using relative elevation and amount of slope as factors.

These algorithms have been developed for western Oregon and are being tested in Idaho, Nevada and Montana and are described in an article in preparation (Allen and Trezza, 2011). Allen and Trezza (2011) also refined the estimation of diffuse radiation on steep mountainous slopes.

Iterative solution for $r_{ah,1\text{-}2}$. During subsequent iterations for the solution for H, a corrected value for u_* is computed as:

$$u_* = \frac{u_{200}k}{ln\left(\dfrac{200}{z_{0m}}\right) - \Psi_{m(200m)}} \tag{8}$$

where $\psi_{m(200m)}$ is the stability correction for momentum transport at 200 meters. A corrected value for $r_{ah,1\text{-}2}$ is computed each iteration as:

$$r_{ah,1,2} = \frac{ln\left(\dfrac{z_2}{z_1}\right) - \Psi_{h(z_2)} + \Psi_{h(z_1)}}{u_* \times k} \tag{9}$$

where $\psi_{h(z2)}$ and $\psi_{h(z1)}$ are the stability corrections for heat transport at z_2 and z_1 heights (Paulson 1970 and Webb 1970) that are updated each iteration.

Stability Correction functions. The Monin-Obukhov length (*L*) defines the stability conditions of the atmosphere in the iterative process. *L* is the height at which forces of buoyancy (or stability) and mechanical mixing are equal, and is calculated as a function of heat and momentum fluxes:

$$L = -\frac{\rho_{air}\, c_p\, u_*^{\,3} T_s}{kgH} \tag{10}$$

where g is gravitational acceleration (= 9.807 m s^{-2}) and units for terms cancel to m for *L*. Values of the integrated stability corrections for momentum and heat transport (ψ_m and ψ_h) are computed using formulations by Paulson (1970) and Webb (1970), depending on the sign of *L*. When $L < 0$, the lower atmospheric boundary layer is unstable and when $L > 0$, the boundary layer is stable. For $L<0$:

$$\Psi_{m(200m)} = 2ln\left(\frac{1 + x_{(200m)}}{2}\right) + ln\left(\frac{1 + x_{(200m)}^{\;2}}{2}\right) - 2ARCTAN\left(x_{(200m)}\right) + 0.5\pi \tag{11}$$

$$\Psi_{h(2m)} = 2ln\left(\frac{1 + x_{(2m)}^{2}}{2}\right) \tag{12a}$$

$$\Psi_{h(0.1m)} = 2ln\left(\frac{1 + x_{(0.1m)}^{2}}{2}\right) \tag{12b}$$

where

$$x_{(200m)} = \left(1 - 16\frac{200}{L}\right)^{0.25} \tag{13a}$$

$$x_{(2m)} = \left(1 - 16\frac{2}{L}\right)^{0.25} \tag{13b}$$

$$x_{(0.1m)} = \left(1 - 16\frac{0.1}{L}\right)^{0.25} \tag{14}$$

Values for $x_{(200m)}$, $x_{(2m)}$, and $x_{(0.1m)}$ have no meaning when $L \geq 0$ and their values are set to 1.0. For $L > 0$ (stable conditions):

$$\Psi_{m(200m)} = -5\left(\frac{2}{L}\right) \tag{15}$$

$$\Psi_{h(2m)} = -5\left(\frac{2}{L}\right) \tag{16a}$$

$$\Psi_{h(0.1m)} = -5\left(\frac{0.1}{L}\right) \tag{16b}$$

When $L = 0$, the stability values are set to 0. Equation (15) uses a value of 2 m rather than 200 m for z because it is assumed that under stable conditions, the height of the stable, inertial boundary layer is on the order of only a few meters. Using a larger value than 2 m for z can cause numerical instability in the model. For neutral conditions, $L = 0$, $H = 0$ and ψ_m and $\psi_h = 0$.

2.2.1 The use of inverse modeling at extreme conditions during calibration (CIMEC)

In METRIC, the satellite-based energy balance is internally calibrated at two extreme conditions (dry and wet) using locally available weather data. The auto-calibration is done for each image using alfalfa-based reference ET (ET_r) computed from hourly weather data. Accuracy and dependability of the ET_r estimate has been established by lysimetric and other studies in which we have high confidence (ASCE-EWRI, 2005). The internal calibration of the sensible heat computation within SEBAL and METRIC and the use of the indexed temperature gradient eliminate the need for atmospheric correction of surface temperature (T_s) and reflectance (albedo) measurements using radiative transfer models (Tasumi et al., 2005b). The internal calibration also reduces impacts of biases in estimation of aerodynamic stability correction and surface roughness.

The calibration of the sensible heat process equations, and in essence the entire energy balance, to ET_r corrects the surface energy balance for lingering systematic computational biases associated with empirical functions used to estimate some components and uncertainties in other estimates as summarized by Allen et al. (2005), including:

- atmospheric correction
- albedo calculation
- net radiation calculation
- surface temperature from the satellite thermal band
- air temperature gradient function used in sensible heat flux calculation
- aerodynamic resistance including stability functions
- soil heat flux function
- wind speed field

This list of biases plagues essentially all surface energy balance computations that utilize satellite imagery as the primary spatial information resource. Most polar orbiting satellites orbit about 700 km above the earth's surface, yet the transport of vapor and sensible heat from land surfaces is strongly impacted by aerodynamic processes including wind speed, turbulence and buoyancy, all of which are essentially invisible to satellites. In addition, precise quantification of albedo, net radiation and soil heat flux is uncertain and potentially biased. Therefore, even though best efforts are made to estimate each of these parameters as accurately and as unbiased as possible, some biases do occur and calibration to ET_r helps to compensate for this by introducing a bias correction into the calculation of H. The end result is that biases inherent to R_n, G, and subcomponents of H are essentially cancelled by the subtraction of a bias-canceling estimate for H. The result is an ET map having values ranging between near zero and near ET_r, for images having a range of bare or nearly bare soil and full vegetation cover.

2.3 Calculation of evapotranspiration

ET at the instant of the satellite image is calculated for each pixel by dividing LE from $LE = R_n - G - H$ by latent heat of vaporization:

$$ET_{inst} = 3600 \frac{LE}{\lambda \rho_w} \tag{17}$$

where ET_{inst} is instantaneous ET (mm hr^{-1}), 3600 converts from seconds to hours, ρ_w is the density of water [~1000 kg m^{-3}], and λ is the latent heat of vaporization (J kg^{-1}) representing the heat absorbed when a kilogram of water evaporates and is computed as:

$$\lambda = \left[2.501 - 0.00236(T_s - 273.15)\right] \times 10^6 \tag{18}$$

The reference ET fraction (ET_rF) is calculated as the ratio of the computed instantaneous ET (ET_{inst}) from each pixel to the reference ET (ET_r) computed from weather data:

$$ET_rF = \frac{ET_{inst}}{ET_r} \tag{19}$$

where ET_r is the estimated instantaneous rate (interpolated from hourly data) (mm hr^{-1}) for the standardized 0.5 m tall alfalfa reference at the time of the image. Generally only one or two weather stations are required to estimate ET_r for a Landsat image that measures 180 km x 180 km, as discussed later. ET_rF is the same as the well-known crop coefficient, K_c, when used with an alfalfa reference basis, and is used to extrapolate ET from the image time to 24-hour or longer periods.

One should generally expect ET_rF values to range from 0 to about 1.0 (Wright, 1982; Jensen et al., 1990). At a completely dry pixel, $ET = 0$ and therefore $ET_rF = 0$. A pixel in a well established field of alfalfa or corn can occasionally have an ET slightly greater than ET_r and therefore $ET_rF > 1$, perhaps up to 1.1 if it has been recently wetted by irrigation or precipitation. However, ET_r generally represents an upper bound on ET for large expanses of well-watered vegetation. Negative values for ET_rF can occur in METRIC due to systematic errors caused by various assumptions made earlier in the energy balance process and due to random error components so that error should oscillate about $ET_rF = 0$ for completely dry pixels. In calculation of ET_rF in Equation (19), each pixel retains a unique value for ET_{inst} that is derived from a common value for ET_r derived from the representative weather station data.

24-Hour Evapotranspiration (ET_{24}). Daily values of ET (ET_{24}) are generally more useful than the instantaneous ET that is derived from the satellite image. In the METRIC process, ET_{24} is estimated by assuming that the instantaneous ET_rF computed at image time is the same as the average ET_rF over the 24-hour average. The consistency of ET_rF over a day has been demonstrated by various studies, including Romero (2004), Allen et al., (2007a) and Collazzi et al., (2006).

The assumption of constant ET_rF during a day appears to be generally valid for agricultural crops that have been developed to maximize photosynthesis and thus stomatal conductance. In addition, the advantage of the use of ET_rF is to account for the increase in 24-hour ET that can occur under advective conditions. The impacts of advection are represented well by the Penman-Monteith equation. However, the ET_rF may decrease during afternoon for some native vegetation under water short conditions where plants endeavor to conserve soil water through stomatal control. In addition, by definition, when the vegetation under study is the same as or similar to the vegetation for the surrounding region and experiences similar water inputs (natural rainfall, only), then (by definition) no advection can occur. This is because as much sensible heat energy is generated by the surface under study as is generated by the region. Therefore, the net advection of energy is nearly zero. Therefore, under these conditions, the estimation by ET_r that accounts for impacts of advection to a *wet* surface do not occur, and the use of ET_rF to estimate 24-hour ET may not be valid. Instead, the use of evaporative fraction, EF, that is used with SEBAL applications may be a better time-transfer approach for rainfed systems. Various schemes of using EF for rainfed portions of Landsat images and ET_rF for irrigated, riparian or wetland portions were explored by Kjaersgaard and Allen (2010). When used, the EF is calculated as:

$$EF = \frac{ET_{inst}}{R_n - G} \tag{20}$$

where ET_{inst} and R_n and G have the same units and represent the same period of time. Finally, the ET_{24} (mm/day) is computed for each image pixel in SEBAL as:

$$ET_{24} = (EF)\,(R_{n\_24}) \tag{21}$$

and in METRIC as:

$$ET_{24} = C_{rad}\left(ET_rF\right)\left(ET_{r\_24}\right) \tag{22}$$

where ET_rF (or EF) is assumed equal to the ET_rF (or EF) determined at the satellite overpass time, $ET_{r\text{-}24}$ is the cumulative 24-hour ET_r for the day of the image and C_{rad} is a correction term used in sloping terrain to correct for variation in 24-hr vs. instantaneous energy availability. C_{rad} is calculated for each image and pixel as:

$$C_{rad} = \frac{R_{so(inst)Horizontal}}{R_{so(inst)Pixel}} \cdot \frac{R_{so(24)Pixel}}{R_{so(24)Horizontal}} \tag{23}$$

where R_{so} is clear-sky solar radiation (W m^{-1}), the "*(inst)*" subscript denotes conditions at the satellite image time, "*(24)*" represents the 24-hour total, the "*Pixel*" subscript denotes slope and aspect conditions at a specific pixel, and the "*Horizontal*" subscript denotes values calculated for a horizontal surface representing the conditions impacting ET_r at the weather station. For applications to horizontal areas, C_{rad} = 1.0.

The 24 hour R_{so} for horizontal surfaces and for sloping pixels is calculated as:

$$R_{so(24)} = \int_0^{24} R_{so\_i} \tag{24}$$

where R_{so_i} is instantaneous clear sky solar radiation at time i of the day, calculated by an equation that accounts for effects of slope and aspect. In METRIC, $ET_{r\ 24}$ is calculated by summing hourly ET_r values over the day of the image.

After ET and ET_rF have been determined using the energy balance, and the application of the single dT function, then, when interpolating between satellite images, a full grid for ET_r is used for the extrapolation over time, to account for both spatial and temporal variation in ET_r. The ET_r grid is generally made on a 3 or 5 km base using as many quality-controlled weather stations located within and in the vicinity of the study area as available. Depending on data availability and the density of the weather stations various gridding methods including krieging, inverse-distance, and splining can be used.

Seasonal Evapotranspiration ($ET_{seasonal}$). Monthly and seasonal evapotranspiration "maps" are often desired for quantifying total water consumption from agriculture. These maps can be derived from a series of ET_rF images by interpolating ET_rF on a pixel by pixel basis between processed images and multiplying, on a daily basis, by the ET_r for each day. The interpolation of ET_rF between image dates is not unlike the construction of a seasonal K_c curve (Allen et al., 1998), where interpolation is done between discrete values for K_c.

The METRIC approach assumes that the ET for the entire area of interest changes in proportion to change in ET_r at the weather station. This is a generally valid assumption and is similar to the assumptions used in the conventional application of K_c x ET_r. This approach is effective in estimating ET for both clear and cloudy days in between the clear-sky satellite image dates. Tasumi et al., (2005a) showed that the ET_rF was consistent between clear and cloudy days using lysimeter measurements at Kimberly, Idaho. ET_r is computed at a specific weather station location and therefore may not represent the actual condition at each pixel. However, because ET_r is used only as an index of the relative change in weather, specific information at each pixel is retained through the ET_rF.

Cumulative *ET* for any period, for example, month, season or year is calculated as:

$$ET_{period} = \sum_{i=m}^{n} \left[\left(ET_r F_i \right) \left(ET_{r24i} \right) \right] \tag{25}$$

where ET_{period} is the cumulative *ET* for a period beginning on day m and ending on day n, ET_rF_i is the interpolated ET_rF for day i, and ET_{r24i} is the 24-hour ET_r for day i. Units for ET_{period} will be in mm when ET_{r24} is in mm d^{-1}. The interpolation between values for ET_rF is best made using a curvilinear interpolation function, for example a spline function, to better fit the typical curvilinearity of crop coefficients during a growing season (Wright, 1982). Generally one satellite image per month is sufficient to construct an accurate ET_rF curve for purposes of estimating seasonal *ET* (Allen et al., 2007a). During periods of rapid vegetation change, a more frequent image interval may be desirable. Examples of splining ET_rF to estimate daily and monthly ET are given in Allen et al. (2007a) and Singh et al. (2008).

If a specific pixel must be masked out of an image because of cloud cover, then a subsequent image date must be used during the interpolation and the estimated ET_rF or K_c curve will have reduced accuracy.

Average ET_rF over a period. An average ET_rF for the period can be calculated as:

$$ET_r F_{period} = \frac{\sum_{i=m}^{n} \left[\left(ET_r F_i \right) \left(ET_{r24i} \right) \right]}{\sum_{i=m}^{n} ET_{r24i}} \tag{26}$$

Moderately high resolution satellites such as Landsat provide the opportunity to view evapotranspiration on a field by field basis, which can be valuable for water rights management, irrigation scheduling, and discrimination of *ET* among crop types (Allen et al., 2007b). The downside of using high resolution imagery is less frequent image acquisition. In the case of Landsat, the return interval is 16 days. As a result, monthly ET estimates are based on only one or two satellite image snapshots per month. In the case of clouds, intervals of 48 days between images can occur. This can be rectified by combining multiple Landsats (5 with 7) or by using data fusion techniques, where a more frequent, but more coarse system like MODIS is used as a carrier of information during periods without quality Landsat images (Gao et al., 2006, Anderson et al., 2010).

2.4 Reflectance based ET methods

Reflectance based ET methods typically estimate relative fractions of reference ET (ET_rF, synonymous with the crop coefficient) based on some sort of vegetation index, for example, the normalized difference vegetation index, NDVI, and multiply the ET_rF by daily computed reference ET_r (Groeneveld et al., 2007). NDVI approaches don't directly or indirectly account for evaporation from soil, so they have difficulty in estimating evaporation associated with both irrigation and precipitation wetting events, unless operated with a daily evaporation process model. The VI-based methods are therefore largely blind to the treatment of both irrigation and precipitation events, except on an average basis. In contrast, thermally based models detect the presence of evaporation from soil, during the snapshot, at least, via evaporative cooling. VI-based methods also do not pick up on acute water stress caused by drought or lack of irrigation, which is often a primary reason for quantifying ET. These models can be run with a background daily evaporation process model, similar to the EB-based models, to estimate evaporation from precipitation between satellite overpass dates.

2.5 Challenges with snapshot models

The SEBAL, METRIC, and other EB models, that can be applied at the relatively high spatial resolution of Landsat and similar satellites, despite their different relative strengths and weaknesses, all suffer from the inability to capture evaporation signals from episodic precipitation and irrigation events occurring between overpass dates. In the case of irrigation events, which are typically unknown to the processer in terms of timing and location, the random nature of these events in time can be somewhat accommodated via the use of multiple overpass dates during the irrigation season (Allen et al. 2007a). In this manner, the ET retrieval for a specific field may be biased high when the overpass follows an irrigation event, but may be biased low when the overpass just precedes an irrigation event. Allen et al. (2007a) suggested that monthly overpass dates over a seven month growing season, for example, can largely compensate for the impact of irrigation wetting on individual fields, especially when it is total growing season ET that is of most interest. The variance of the error in ET estimate caused by unknown irrigation events should tend to decrease with the square root of the number of images processed during the irrigation season.

The impact by precipitation events is a larger problem in converting the 'snapshot' ET images from energy balance models or other methods into monthly and longer period ET. Precipitation timing and magnitudes tend to be less random in time and have much larger variance in depth per wetting event than with irrigation. Because of this, the use of snapshot ET models to construct monthly and seasonal ET maps is more likely to be biased high (if a number of images happen to be 'wet' following a recent precipitation event) or low (if images happen to be 'dry', with precipitation occurring between images). The latter may often be the case since the most desired images for processing are cloud free.

One important use of ET maps is in the estimation of ground water recharge (Allen et al., 2007b). Ground water recharge is often uncertain due to uncertainty in both precipitation and ET, and is usually computed using the difference between P and ET, with adjustment for runoff. It is therefore important to maintain congruency between ET and P data sets or 'maps'. Lack of congruency can cause very large error in estimated recharge, especially in the more arid regions.

3. Adjusting for background evaporation

Often a Landsat or other image is processed on a date where antecedent rainfall has caused the evaporation from bare soil to exceed that for the surrounding monthly period. Often, for input to water balance applications, it is desirable that the final ET image represent the average evaporation conditions for the month. In that case, one approach is to adjust the 'background' evaporation of the processed image to better reflect that for the month or other period that it is to ultimately represent. This period may be a time period that is half way between other adjacent images.

An example of a sequence of Landsat images processed using the METRIC surface energy balance model for the south-western portion of the Nebraska Panhandle (Kjaersgaard and Allen, 2010a) is shown in Figure 1 along with daily precipitation from the Scottsbluff High Plains Regional Climate Center (HPRCC) weather station. The August 13 image date was preceded by a wet period and followed by a very dry period, thus the evaporation from non-irrigated areas at the satellite image date is not representative for the month.

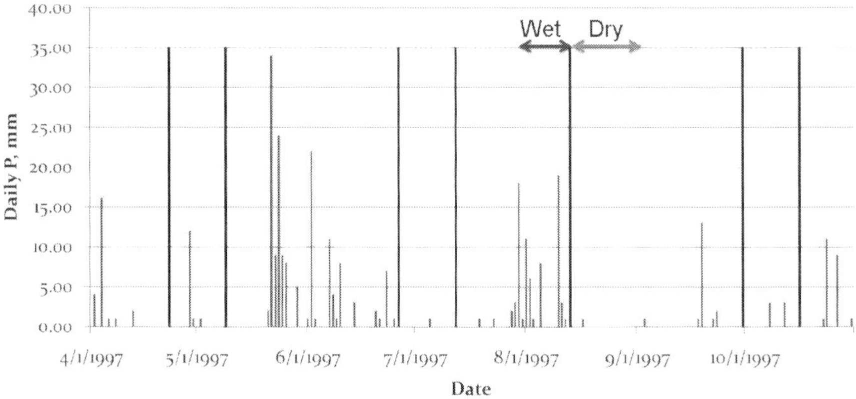

Fig. 1. Image dates of nearly cloud free Landsat 5 path 33 row 31 images from the Nebraskan Panhandle in 1997 (black vertical bars) and precipitation recorded at the Scottsbluff HPRCC weather station (red bars). After Kjaersgaard and Allen (2010).

In making the adjustment for background evaporation, the background evaporation on the overpass date is subtracted out of the image and the average background evaporation is substituted in. Full adjustment is made for areas of completely bare soil, represented by $NDVI = NDVI_{bare\ soil}$, with no adjustment to areas having full ground covered by vegetation, represented by $NDVI = NDVI_{full\ cover}$, and with linear adjustment in between.

The following methodology is taken from a white paper developed by the University of Idaho during 2008 and 2009 (Allen 2008, rev. 2010). The ET_rF of the Landsat image is first adjusted to a 'basal' condition, where the evaporation estimate is free of rainfall induced evaporation, but still may contain any irrigation induced evaporation:

$$(ET_rF_i)_b = ET_rF_i - \left(ET_rF_{background}\right)_i \left(\frac{NDVI_{full\,cover} - NDVI_i}{NDVI_{full\,cover} - NDVI_{bare\,soil}}\right) \tag{27}$$

where $(ET_rF_{background})_i$ is the background evaporation on the image date (i) for bare soil, computed using a gridded FAO-56 two-stage evaporation model of Allen et al. (1998) with modification to account for 'flash' evaporation from the soil skin (Allen 2010a) or some other soil evaporation model such as Hydrus or DAISY. The soil evaporation model is on a daily timestep using spatially distributed precipitation, reference ET, and soil properties. $(ET_rF_i)_b$ is the resulting 'basal' ET image for a particular image date, representing a condition having NDVI amount of vegetation and a relatively dry soil surface. This parameter represents the foundation for later adjustment to represent the longer period.

3.1 Adjustment for cases of riparian vegetation

For riparian vegetation and similar systems, where soil water stress is not likely to occur due to the frequent presence of shallow ground water, an adjusted ET_rF is computed for the image date that reflects background evaporation averaged over the surrounding period in proportion to the amount of ground cover represented by NDVI:

$$\left(ET_rF_i\right)_{adjusted} = \left(ET_rF_i\right)_b + \overline{\left(ET_rF_{background}\right)}\left(\frac{NDVI_{full\,cover} - NDVI_i}{NDVI_{full\,cover} - NDVI_{bare\,soil}}\right) \tag{28}$$

where $\overline{\left(ET_rF_{background}\right)}$ is the average evaporation from bare soil due to precipitation over the averaging period (e.g., one month), calculated as:

$$\overline{\left(ET_rF_{background}\right)} = \frac{\sum_1^n \left(ET_rF_{background}\right)_i}{n} \tag{29}$$

Equations 5 and 6 can be combined as:

$$\left(ET_rF_i\right)_{adjusted} = ET_rF_i + \left(\overline{\left(ET_rF_{background}\right)} - \left(ET_rF_{background}\right)_i\right)\left(\frac{NDVI_{full\,cover} - NDVI_i}{NDVI_{full\,cover} - NDVI_{bare\,soil}}\right) \tag{30}$$

with limits $NDVI_{bare\,soil} \le NDVI_i \le NDVI_{full\,cover}$.

The outcome of this adjustment is to preserve any significant evaporation stemming from irrigation or ground-water and any transpiration stemming from vegetation, with adjustment only for evaporation stemming from precipitation to account for differences between the image date and that of the surrounding time period. In other words, if the initial ET_rF_i, prior to adjustment is high due to evaporation from irrigation or from high ground-water condition, much of that evaporation would remain in the adjusted ET_rF_i estimate.

3.2 Adjustments for non-riparian vegetation

The following refinement to Eq. 30 is made for application to non-riparian vegetation, to account for those situations where, during long periods (i.e., months), soil moisture may have become limited enough that even transpiration of vegetation has been reduced due to moisture stress. If the Landsat image is processed during that period of moisture stress, then the ET_rF value for vegetated or partially vegetated areas will be lower than the potential (nonstressed) value. This can happen, for example, during early spring when winter wheat may go through stress prior to irrigation or a rainy period, or in desert and other dry systems.

This causes a problem in that the method of Eq. 8 attempts to 'preserve' the ET_rF of the vegetated portion of a pixel that was computed by METRIC on the image date. However, when a rain event occurs following the image date, not only will the ET_rF of exposed soil increase, but any stressed vegetation will equally 'recover' from moisture stress and the ET_rF of the vegetation fraction of the surface will increase. This situation may occur for rangeland and dryland agricultural systems. It is therefore assumed that the ET_rF of nonstressed vegetation will be at least as high as the ET_rF of bare soil over the same time period, since it should have equal access to shallow water. An exception would be if the vegetation were sufficiently stressed to not recover transpiration potential. However, this amount of stress should be evidenced by a reduced NDVI. A minimum limit is therefore placed, using the background ET_rF $\left(\overline{ET_rF_{background}}\right)$ for the period.

To derive a modified Eq. 8, it is useful to first isolate the 'transpiration' portion of the ET_rF. On the satellite image date, the bulk ET_rF computed by METRIC for a pixel, is decomposed to:

$$ET_rF_i = (1 - f_c)\left(ET_rF_{background}\right)_i + f_c\left(ET_rF_{transpiration}\right)_i \tag{31}$$

where $ET_rF_{transpiration}$ is the apparent transpiration from the fraction of ground covered by vegetation, f_c. The f_c is estimated as 1 – f_s, where f_s is the fraction of bare soil, and for consistency with equations 30, f_s is estimated as:

$$f_s = \left(\frac{NDVI_{full\,cover} - NDVI_i}{NDVI_{full\,cover} - NDVI_{bare\,soil}}\right) \tag{32a}$$

so that:

$$f_c = 1 - \left(\frac{NDVI_{full\,cover} - NDVI_i}{NDVI_{full\,cover} - NDVI_{bare\,soil}}\right) \tag{32b}$$

Eq. 31 is not used as is, since ET_rF_i comes from the energy balance-based ET image (i.e., from METRIC, etc.). However, one can rearrange Eq. 31 to solve for $ET_rF_{transpiration}$:

$$f_c\left(ET_rF_{transpiration}\right)_i = ET_rF_i - (1 - f_c)\left(ET_rF_{background}\right)_i \tag{33}$$

Now, if $ET_rF_{transpiration}$ is limited to the maximum of the $ET_rF_{transpiration}$ on the day of the image, or the $\left(\overline{ET_rF_{background}}\right)$ for the period, then:

$$\left(ET_rF_{transpiration}\right)_{adjusted} = \max\left[\left(ET_rF_{transpiration}\right)_i, \left(\overline{ET_rF_{background}}\right)\right] \tag{34}$$

Then the new ET_rF adjusted value becomes:

$$\left(ET_rF_i\right)_{adjusted} = (1 - f_c)\left(\overline{ET_rF_{background}}\right) + f_c\left(ET_rF_{transpiration}\right)_{adjusted}$$
$$or \tag{35}$$
$$\left(ET_rF_i\right)_{adjusted} = (1 - f_c)\left(\overline{ET_rF_{background}}\right) + f_c \max\left[\left(ET_rF_{transpiration}\right)_i, \left(\overline{ET_rF_{background}}\right)\right]$$

where $\left(\overline{ET_rF_{background}}\right)$ is the average evaporation from bare soil due to precipitation over the averaging period (e.g., one month) and $ET_rF_{transpiration}$ is the original transpiration computed from Eq. 33. Eq. 33 and 35 can be combined so that:

$$\begin{aligned}\left(ET_rF_i\right)_{adjusted} &= (1-f_c)\left(\overline{ET_rF_{background}}\right)+ \\ &+\max\left[\left(ET_rF_i-(1-f_c)\left(\overline{ET_rF_{background}}\right)_i\right)_i, f_c\left(\overline{ET_rF_{background}}\right)\right]\end{aligned} \tag{36}$$

Only areas with bare soil or partial vegetation cover are adjusted. Pixels having full vegetation cover, defined as when NDVI > 0.75, are not adjusted. An example of an image date where the adjustment increased the ET_rF for bare soil and partially vegetated areas is shown in Figure 2. Figure 3 shows an example of an image date where the ET_rF from bare soil and partial vegetation cover was decreased by the adjustment.

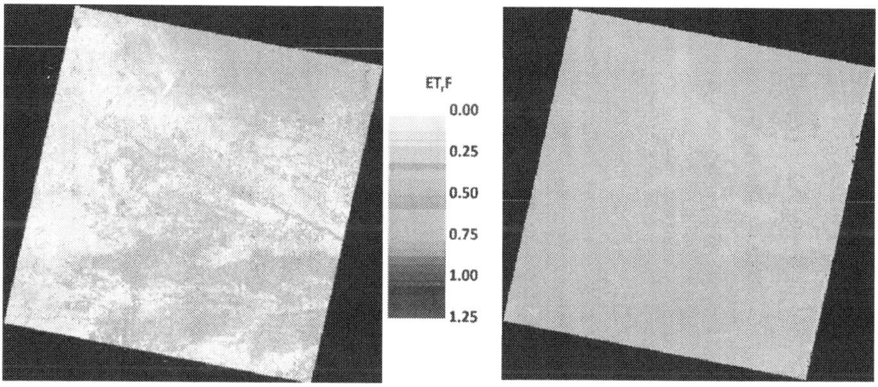

Fig. 2. ET_rF in western Nebraska from May 9 1997 before (left) and after (right) adjustment for background evaporation representing the time period (~month) represented by that image. After Kjaersgaard and Allen (2010).

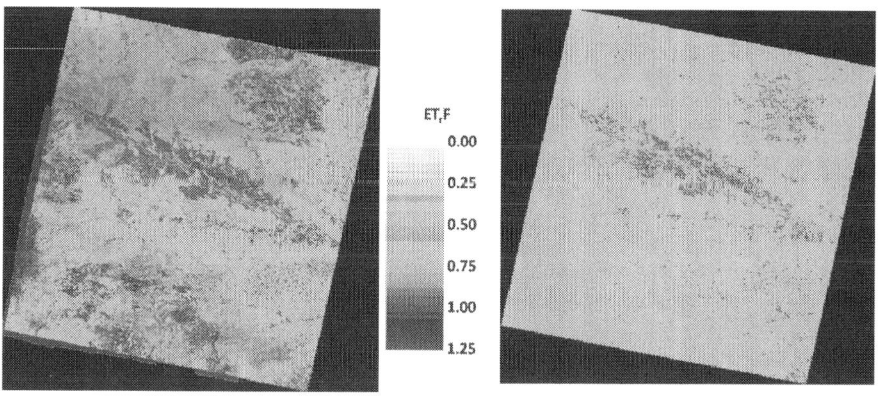

Fig. 3. ETrF in western Nebraska on August 13 1997 before (left) and after (right) adjustment to reflect soil evaporation occurring over the time period (~ 1 month) represented by that image. Note that irrigated fields with full vegetation cover having a substantial transpiration component were not affected by the adjustment. After Kjaersgaard and Allen (2010).

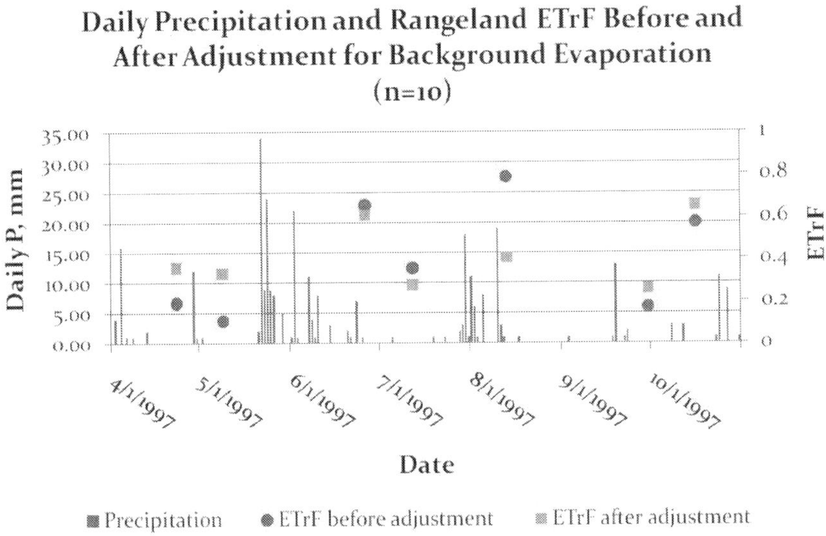

Fig. 4. Average ET_rF from ten rangeland locations in western Nebraska before and after adjustment. Also shown is the precipitation from the Scottsbluff HPRCC weather station (after Kjaersgaard and Allen 2010).

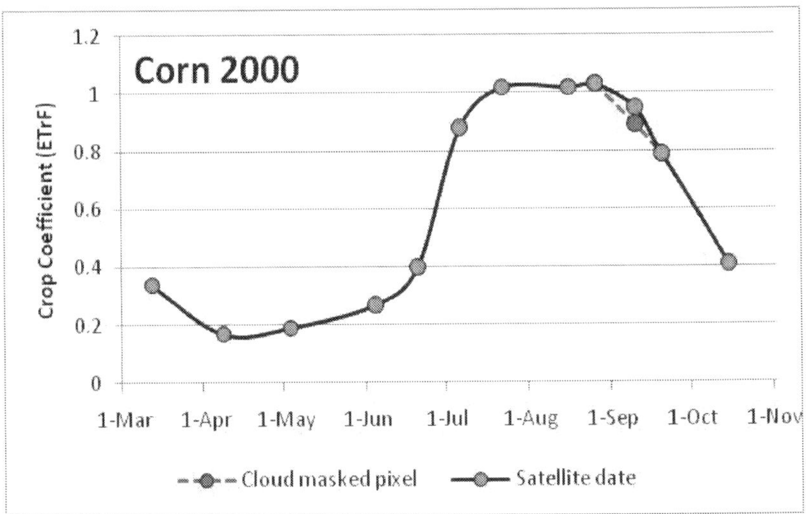

Fig. 5. Schematic representation of the linear cloud gap filling and the cubic spline used to interpolate between image dates for a corn crop. The green points represent image dates and the black line is the splined interpolation between points; the red point represents the value of ET_rF that is interpolated linearly from the two adjacent image dates had the field had cloud cover on September 10.

Average ET_rF on image dates before and after adjustment for background evaporation is shown in Figure 4 from ten rangeland locations in western Nebraska. For some image dates, such as early and late in the season, the adjusted ET_rF values are "wetter" than that represented by the original image. Similarly, for other images dates, such as in the middle of the growing season, the images were "drier". The adjustment for one image in August reduced the estimated ET for the month of August by nearly 50%, which is considerable.

It is noted, that the images no longer represent the ET from the satellite overpass dates after the adjustment for background evaporation. The images are merely an intermediate product that is used as the input into an interpolation procedure when producing ET estimates for monthly or longer time periods.

4. Dealing with clouded parts of images

Satellite images often have clouds in portions of the images. ET_rF cannot be directly estimated for these areas using surface energy balance because cloud temperature masks surface temperature and cloud albedo masks surface albedo. Generally ET_rF for clouded areas must be filled in before application of further integration processes so that those processes can be uniformly applied to an entire image. The alternative is to directly interpolate ET_rF between adjacent (in time) image dates or to run some type of daily ET process model that is based on gridded weather data.

In METRIC applications (Allen et al. 2007b), ET_rF for clouded areas of images is usually filled in prior to interpolating ET_rF for days between image dates (and multiplying by gridded ET_r for each day to obtain daily ET images). A linear interpolation, as shown in Figure 5, is used to fill in ET_rF for clouded portions of images rather than curvilinear interpolation that is used to interpolate ET_rF between nonclouded image portions because some periods between cloud-free pixel locations can be as long as several months. Often, the change in crop vegetation amount and thus ET_rF is uncertain during that period. Thus, the use of curvilinear interpolation can become speculative.

Image processing code can be created to conduct the 'filling' of cloud masked portions of images. The code used with METRIC accommodates up to eight image dates and corresponding ET_rF, with conditionals used to select the appropriate set of images to interpolate between, depending on the number of consecutive images that happen to be cloud masked for any specific location. Missing (clouded) ET_rF for end-member images (those at the start or end of the growing season) must be estimated by extrapolation of the nearest (in time) image having valid ET_rF, or alternatively, for end-member images, a 'synthetic' image can be created, based on daily soil water balance or other methods, to be used to substitute for cloud-masked areas. Often, the availability of images for early spring is limited due to clouds. In these cases, the ET_rF values in the synthetic image are based on a soil-water balance–weather data model, such as the FAO-56 evaporation model or Hydrus or DAISY, applied over the month of April, for example, to provide an improved estimate of ET_rF over the early season. The synthetic image(s) are strategically placed, date-wise, so that the cloud-filling process and the subsequent cubic spline process used to interpolate final ET_rF has end-points early enough in the year to provide ET_rF for all days of interest during the growing period.

Examples of cloud masking for a METRIC application in western Nebraska are shown in Figure 6. Black portions within each image are the areas masked for clouds. ET_rF for cloud masked areas was filled in for individual Landsat dates prior to splining ET_rF between images. The cloud mask gap filling and interpolation of ET between image dates entails interpolating the ET_rF for the missing area from the previous and following images that have ET_rF for that location.

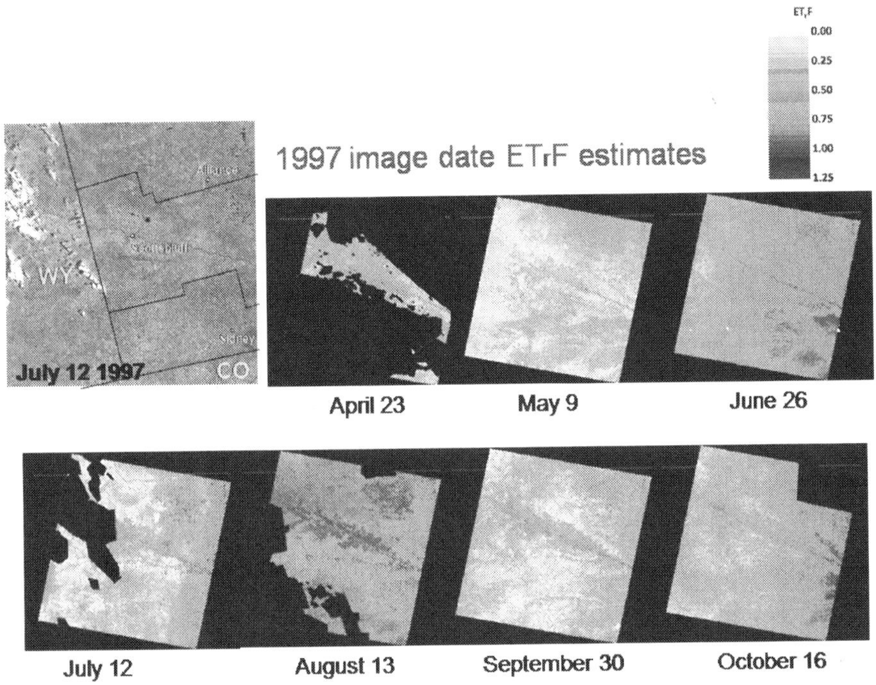

Fig. 6. Maps of cloud masked ET_rF from seven 1997 images dates. The geographical extent of the North Platte and South Platte Natural Resource Districts boundaries and principal cities is shown on the image in the top left corner (after Kjaersgaard and Allen 2010).

In current METRIC applications, gaps in the ET_rF maps occurring as a result of the cloud masking are filled in using linear time-weighted interpolation of ET_rF values from the previous image and the nearest following satellite image date having a valid ET_rF estimate, adjusted for vegetation development. The NDVI is used to indicate change in vegetation amount from one image date to the next. The principle is sketched in Figure 7, where a location in the two nearest images (i-1 and i+1) happen to be clouded. During the gap filling, the interpolated values for the clouded and cloud-shadowed areas are adjusted for differences in residual soil moisture between the image dates occurring as a result of heterogeneities in precipitation (such as by local summer showers) in inverse proportion to NDVI and by adding an interpolated 'basal' ET_rF from the previous and following satellite image dates. This procedure is needed to remove artifacts of this precipitation-derived evapotranspiration that are unique to specific image dates but that may not be representative of the image date that is to be represented by the ET_rF from the previous and the following images. A comparison between cloud gap filling without and with adjustment for background evaporation is shown in Figure 8. An additional example from Singh et al. (2008) is shown in Figure 9 for central Nebraska, where filled in areas that were clouded are difficult to detect due to the adjustment for background evaporation via a daily process model.

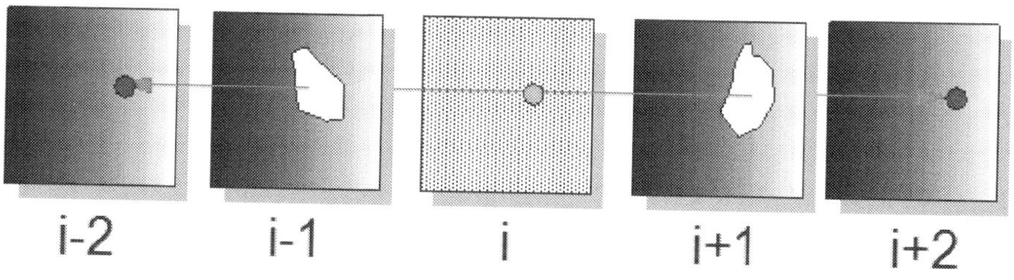

Fig. 7. Principle of cloud gap filling. "i" is the image having cloud masked areas to be filled; "i-1" and "i-2" are the two earlier images than image I; "i+1" and "i+2" are the two following images.

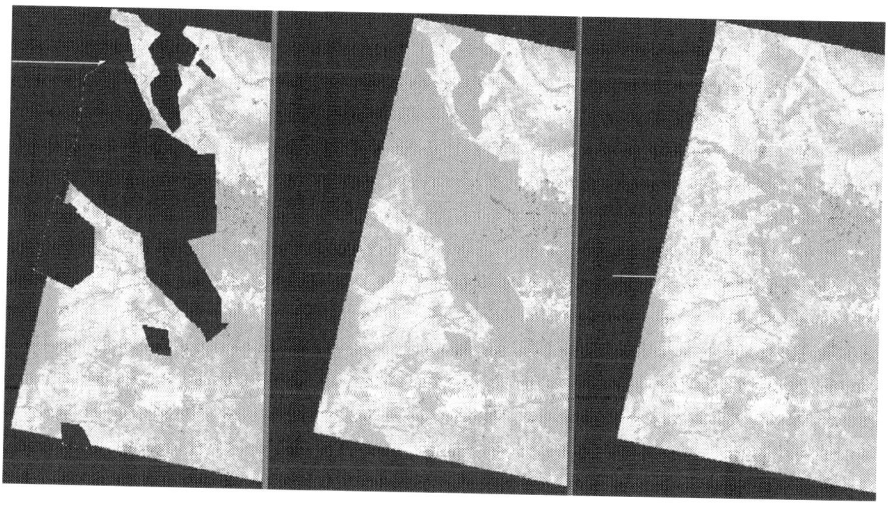

Fig. 8. Maps of ET_rF from Landsat 5, July 12 1997, in western Nebraska after cloud masking (left) (black indicate areas removed during cloud masking or background); and after cloud gap filling without (center) and with (right) adjustment for vegetation amount and background evaporation from antecedent rainfall. The August 13 image from which part of the ET_rF data was borrowed was quite wet from precipitation, and thus had high ET_rF for low-vegetated areas, and therefore created substantially overestimated ET_rF for July 12 in the filled areas (center). After Kjaersgaard and Allen (2010).

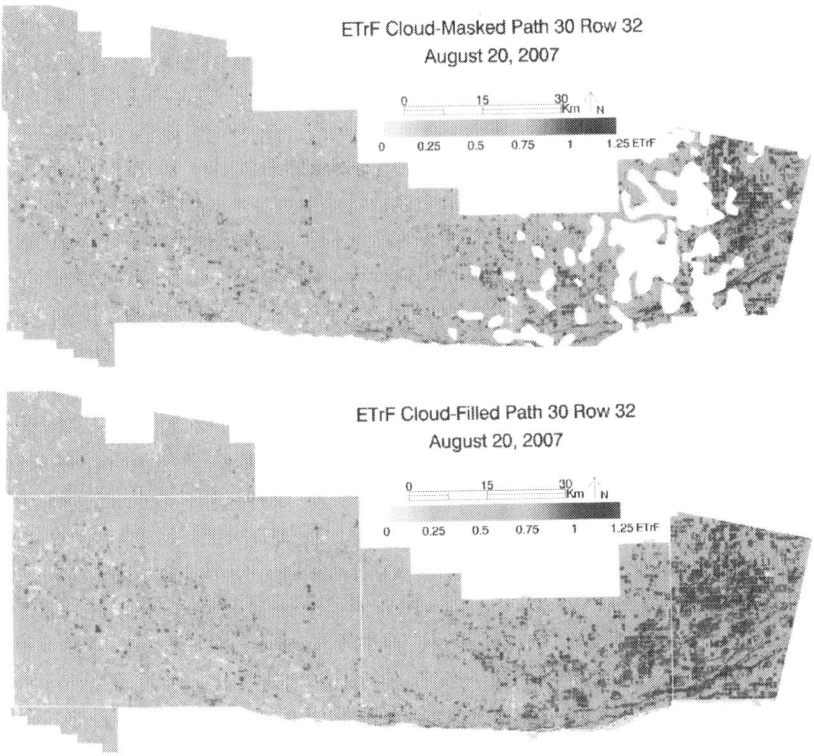

Fig. 9. ETrF product for August 20, 2007 over the Central Platte Natural Resources District, Nebraska, with clouded areas masked (top) and filled (bottom) using a procedure that adjusted for background evaporation from antecedent precipitation events (after Singh et al., 2008).

5. Other remaining challenges with operational models for spatial ET

In addition to challenges in producing daily time series of spatial ET, as described in the previous section, other challenges remaining with all models, snapshot and process models alike include the following. These were described by Allen et al., (2010) and include estimation of aerodynamic roughness at 30 m scale; aerodynamic roughness and wind speed variation in complex terrain and in tall, narrow vegetation systems such as riparian systems; and estimation of hemispherical reflectance from bi-direction reflectance in deep vegetation canopies from nadir-looking satellites such as Landsat. Other remaining challenges include estimation of soil heat and aerodynamic sensible heat fluxes in sparse desert systems and in playa and estimation of ET over 24-hour periods using one-time of day observation (for example ~1000 solar time for Landsat) based on energy balance, especially where substantial stomatal control exists (desert and forest). METRIC capitalizes on using weather-based reference ET to make this transfer over time, which has been shown to work well for irrigated crops, especially in advective environments (Allen et al. 2007a). However, the evaporative fraction, as used in early SEBAL (Bastiaanssen et al. 1998a) and other models may perform best for rainfed systems where, by definition, advection can not exist. Therefore, a mixture of ET_rF and EF may be optimal, based on land-use class.

6. Conclusions

Satellite-based models for determining evapotranspiration (ET) are now routinely applied as part of water and water resources management operations of state and federal agencies. The very strong benefit of satellite-based models is the quantification of ET over large areas. Strengths and weaknesses of common EB models often dictate their use. The more widely used and operational remote sensing models tend to use a 'CIMEC' approach ("calibration using inverse modeling of extreme conditions") to calibrate around uncertainties and biases in satellite based energy balance components. Creating 'maps' of ET that are useful in management and in quantifying and managing water resources requires the computation of ET over monthly and longer periods such as growing seasons or annual periods. This requires accounting for increases in ET from precipitation events in between images. An approach for estimating the impacts on ET from wetting events in between images has been described. This method is empirical and can be improved in the future with more complex, surface conductance types of process models, such as used in Land surface models (LSM's). Interpolation processes involve treatment of clouded areas of images, accounting for evaporation from wetting events occurring prior to or following overpass dates, and applying a grid of daily reference ET with the relative ET computed for an image, or a direct Penman-Monteith type of calculation. These approaches constitute a big step forward in computing seasonal ET over large areas with relatively high spatial (field-scale) definition, where impacts of intervening wetting events and cloud occurrence are addressed.

7. References

Abrahamsen, P. and S. Hansen . 2000. Daisy: an open soil-crop-atmosphere system model. *Environmental Modelling and Software*, 15(3):313-330.

Allen, R.G. 2008, rev. 2010. Procedures for adjusting METRIC-derived ETrF Images for Background Evaporation from Precipitation Events prior to Cloudfilling and Interpretation of ET between Image Dates. Internal memo., University of Idaho. 11 pages. Version 7, last revised April 2010.

Allen, R.G., 2010a. Modification to the FAO-56 Soil Surface Evaporation Algorithm to Account for Skin Evaporation during Small Precipitation Events. Memorandum prepared for the UI Remote Sensing Group, revised May 16, 2010, August 2010, 9 pages.

Allen, R.G. 2010b. Assessment of the probability of being able to produce Landsat resolution images of annual (or growing season) evapotranspiration in southern Idaho - and effect of the number of satellites. Memorandum prepared for the Landsat Science Team. 5 p.

Allen, R.G. and Wright, J.L. (1997). Translating Wind Measurements from Weather Stations to Agricultural Crops. *J. Hydrologic Engineering*, ASCE 2(1): 26-35.

Allen, R.G. and Trezza, R. (2011). New aerodynamic functions for application of METRIC in mountainous areas. Internal memo. University of Idaho, Kimberly, ID. 12 p.

Allen, R. G., Pereira, L., Raes, D., and Smith, M. 1998. *Crop Evapotranspiration*, Food and Agriculture Organization of the United Nations, Rome, It. ISBN 92-5-104219-5. 300 p.

Allen, R.G., W.B.M. Bastiaanssen, M. Tasumi, and A. Morse. 2001. Evapotranspiration on the watershed scale using the SEBAL model and LandSat Images. Paper Number 01-2224, ASAE, Annual International Meeting, Sacramento, California, July 30-August 1, 2001.

Allen, R.G., M.Tasumi, A.T. Morse, and R. Trezza. 2005. A Landsat-based Energy Balance and Evapotranspiration Model in Western US Water Rights Regulation and Planning. *J. Irrigation and Drainage Systems*. 19: 251-268.

Allen, R.G., M. Tasumi and R. Trezza. 2007a. Satellite-based energy balance for mapping evapotranspiration with internalized calibration (METRIC) - Model. ASCE *J. Irrigation and Drainage Engineering* 133(4):380-394.

Allen, R.G., M. Tasumi, A.T. Morse, R. Trezza, W. Kramber, I. Lorite and C.W. Robison. 2007b. Satellite-based energy balance for mapping evapotranspiration with internalized calibration (METRIC) - Applications. ASCE *J. Irrigation and Drainage Engineering* 133(4):395-406.

Allen,R.G., J. Kjaersgaard, R. Trezza, A. Oliveira, C. Robison, and I. Lorite-Torres. 2010. Refining components of a satellite based surface energy balance model to complex-land use systems. Proceedings of the Remote Sensing and Hydrology Symposium, Jackson Hole, Wyo, IAHS. Oct. 2010. 3 p.

Anderson, M.C., Kustas, W.P., Dulaney, W., Feng, G., Summer, D., 2010. Integration of Multi-Scale Thermal Satellite Imagery for Evaluation of Daily Evapotranspiration at the Sub-Field Scale. Remote Sensing and Hydrology Symposium 2010. Jackson Hole, Wyoming, USA. p. 62.

Arnold, J.G., J.R. Williams, R. Srinivasan, K.W. King and R.H. Griggs. 1994. SWAT--Soil and Water Assessment Tool--User Manual. Agricultural Research Service, Grassland, Soil and Water Research Lab, US Department of Agriculture.

ASCE - EWRI. (2005). *The ASCE Standardized reference evapotranspiration equation*. ASCE-EWRI Standardization of Reference Evapotranspiration Task Comm. Report, ASCE Bookstore, ISBN 078440805, Stock Number 40805, 216 pages.

Bastiaanssen, W.G.M., M. Menenti, R.A. Feddes, and A.A. M. Holtslag. 1998a. A remote sensing surface energy balance algortithm for land (SEBAL). Part 1: Formulation. *J. of Hydrology* 198-212.

Bastiaanssen, W.G.M., H. Pelgrum, J. Wang, Y. Ma, J.F. Moreno, G.J. Roerink, R.A. Roebeling, and T. van der Wal. 1998b. A remote sensing surface energy balance algortithm for land (SEBAL). Part 2: Validation. *J. of Hydrology* 212-213: 213-229.

Bastiaanssen , W.G.M., E.J.M. Noordman , H. Pelgrum, G. Davids, B.P. Thoreson and R.G. Allen. 2005. SEBAL model with remotely sensed data to improve water resources management under actual field conditions. *J. Irrig. Drain. Engrg*, ASCE 131(1): 85-93.

Brutsaert, W. 1982. *Evaporation into the atmosphere*. Reidel, Dordrecht, The Netherlands.

Brutsaert, W., A.Y. Hsu, and T.J. Schmugge. 1993. Parameterization of surface heat fluxes above a forest with satellite thermal sensing and boundary layer soundings. *J. Appl. Met.* 32: 909-917.

Campbell, G.S. and J.M. Norman. 1998. *An introduction to environmental biophysics.* Sec. Edition. Springer, New York.

Cleugh, H.A., R. Leuning, Q. Mu and S.W. Running. 2007. Regional evaporation estimates from flux tower and MODIS satellite data. *Remote Sens. Environ.* 106:285-304.

Colaizzi, P.D., S. R. Evett, T. A. Howell, J. A. Tolk. 2006. Comparison of Five Models to Scale Daily Evapotranspiration from One-Time-of-Day Measurements. Trans. ASABE. 49(5):1409-1417.

Farah, H.O. 2001. Estimation of regional evaporation under different weather conditions from satellite and meteorological data. A case study in the Naivasha Basin, Kenya. Doctoral Thesis Wageningen University and ITC.

Franks, S.W. and K.J. Beven. 1997. Estimation of evapotranspiration at the landscape scale: a fuzzy disaggregation approach. *Water Resour. Res.* 33:2929-2938.

Gao, F., Masek, J., Schwaller, M., Hall, F., 2006. On the Blending of the Landsat and Modis Surface Reflectance: Predicting Daily Landsat Surface Reflectance. IEEE Trans on Geosci. and Remote Sens. 44, 2207-2218.

Gowda, P.H., J.L. Chávez, P.D. Colaizzi, S.R. Evett, T.A. Howell, and J.A. Tolk. 2008. ET mapping for agricultural water management: present status and challenges. *Irrig. Sci.* 26:223-237

Groeneveld, D.P., W.J. baugh, J.S. Sanderson and D.J. Cooper. 2007. Annual groundwater evapotranspiration mapped from single satellite scenes. *J. Hydrol.* 344:146-156.

Hall, F.G., K.F. Huemmrich, S.J. Goetz, P.J. Sellers, and J.E. Nickeson. 1992. Satellite remote sensing of the surface energy balance: success, failures and unresolved issues in FIFE. *J. Geophys. Res.* 97(D17):19061-19090.

Hook, S. and A.J. Prata. 2001. Land surface termpature measured by ASTER-First results. *Geophys. Res. Abs.*, 26th Gen. Assemb. 3:71.

Jensen, M.E., Burman, R.D. and Allen, R.G. (eds.) (1990). *Evapotranspiration and Irrigation Water Requirements,* ASCE Manuals and Reports on Engineering Practice No. 70. ISBN 0-87262-763-2, 332 p. ASCE, Reston, VA.

Kalma, J.D. and D.L.B. Jupp. 1990. Estimating evaporation from pasture using infrared thermography: evaluation of a one-layer resistance model. *Agr. and Forest Met.* 51:223-246.

Kalma, J.D., T.R. McVicar, and M.F. McCabe. 2008. Estimating land surface evaporation: a review of methods using remotely sensed surface temperature data. *Surv. Geophys* 29:421-469.

Kjaersagaard, J. and R.G. Allen. 2010. Remote Sensing Technology to Produce Consumptive Water Use Maps for the Nebraska Panhandle. Final completion report submitted to the University of Nebraska. 60 pages.

Kustas, W. P., and J. M. Norman. 1996. Use of remote sensing for evapotranspiration monitoring over land surfaces. *Hydrol. Sci. J.* 41(4): 495-516.

Kustas, W. P., Moran, M. S., Humes, K. S., Stannard, D. I., Pinter, J., Hipps, L., and Goodrich, D. C. 1994. Surface energy balance estimates at local and regional scales using optical remote sensing from an aircraft platform and atmospheric data collected over semiarid rangelands. *Water Resources Research,* 30(5): 1241-1259.

Oke, T.R, (1987). *Boundary Layer Climates.* 2nd Ed., Methuen, London, 435 pp, ISBN 0-415-04319-0.

Paulson, C.A. (1970). The mathematical representation of wind speed and temperature profiles in the unstable atmospheric surface layer. *Appl. Meteorol.* 9:857-861.

Qualls, R., and Brutsaert, W. (1996). "Effect of vegetation density on the parameterization of scalar roughness to estimate spatially distributed sensible heat fluxes." *Water Resources Research*, 32(3): 645-652.

Qualls, R., and Hopson, T. (1998). "Combined use of vegetation density, friction velocity, and solar elevation to parameterize the scalar roughness for sensible heat." *J. Atmospheric Sciences,* 55: 1198-1208.

Romero, M.G. (2004) *Daily evapotranspiration estimation by means of evaporative fraction and reference ET fraction*. Ph.D. Diss., Utah State Univ., Logan, Utah.

Rosema, A. 1990. Comparison of meteosat-based rainfall and evapotranspiration mapping of Sahel region. *Rem. Sens. Env.* 46: 27-44.

Šimůnek, J., M.T. van Genuchten, and M. Šejna. 2008. Development and applications of the HYDRUS and STANMOD software packages and related codes. *Vadose Zone J.* 7:587-600.

Singh, R.K., A. Irmak, S. Irmak and D.L. Martin. 2008. Application of SEBAL Model for Mapping Evapotranspiration and Estimating Surface Energy Fluxes in South-Central Nebraska. J. Irrigation and Drainage Engineering 134(3):273-285.

Su, Z. 2002. The surface energy balance system (SEBS) for estimation of turbulent fluxes. *Hydrol. Earth Systems Sci.* 6(1): 85-99.

Tasumi, M., R. G. Allen, R. Trezza, J. L. Wright. 2005. Satellite-based energy balance to assess within-population variance of crop coefficient curves, *J. Irrig. and Drain. Engrg,* ASCE 131(1): 94-109.

Van Dam, J.C., 2000. Field-scale water flow and solute transport: SWAP model concepts, parameter estimation and case studies. Proefschrift Wageningen Universiteit.

Wang, J., W.G.M Bastiaanssen, Y. Ma, and H. Pelgrum. 1998. Aggregation of land surface parameters in the oasis-desert systems of Northwest China. *Hydr. Processes* 12:2133-2147.

Webb, E.K. (1970). Profile relationships: the log-linear range, and extension to strong stability. *Quart. J. Roy. Meteorol. Soc.* 96:67-90.

Wright, J.L. (1982) New Evapotranspiration Crop Coefficients. *J. of Irrig. and Drain. Div.* (ASCE), 108:57-74.

22

Adaptability of Woody Plants in Aridic Conditions

Viera Paganová and Zuzana Jureková
Slovak University of Agriculture in Nitra
Slovak Republic

1. Introduction

Ecological conditions and sources such as water, temperature, solar radiation, and carbon dioxide concentration are factors that limit plant growth, development, and reproduction. Deviations from the optimal values of these factors can cause stress. Plants are subjected to multiple abiotic and biotic stresses that adversely influence plant survival and growth by inducing physiological dysfunctions (Kozlowski & Pallardy, 2002). On the other hand, plants use different strategies for survival that are important for their distribution throughout various regions. Plants differ widely in their ability to adjust to a changing environment and the associated stress (Itail et al., 2002), including the ability to cope with drought (Kozlowski & Pallardy 1997).

Water deficiency is the most significant stress factor for plant growth and reproduction. Drought is mostly associated with the dieback of trees within various regions and throughout the world (Mc Dowel et al., 2008). However, physiological mechanisms of woody plant survival have not yet been described. According to Passioura (2002a), all mechanisms that support physiological functions of plants under conditions of limited water availability are mechanisms of stress resistance. These mechanisms have developed over a long period of time as part of plant adaptability. According to Jones (1993), there are three mechanisms for plant drought resistance. The first mechanism consists of avoiding water deficit and involves the limitation of transpiration and maximisation of root uptake. The second mechanism involves the tolerance to water deficit (Passioura, 2002b; Gielen et al., 2008), and the third mechanism optimises the utilisation of water (Jones 2004).

Plant water stress is the result of a disproportionate balance between the amount of received and released water through various interactions with plant growth, development, and biomass production. The interactions are modified by genetic properties of the specimen and by the character and degree of plant adaptation. The amount of water that a plant can receive depends on the water supply in the soil and on eco-physiological characteristics of plant roots. The transport of water enables for a potential water gradient between the atmosphere and soil, and depends on the hydraulic resistance of the root and stem vascular system. Another component of the water regime of plants - release through transpiration - is a function of the physiological availability and mobility of the water. Plant regulation of the stomata opening and transpiration depend on the pressure potential and other influencing factors. Maintenance of a positive pressure potential is therefore conditional for the survival of plants under drought. The water regime of plants is therefore an ensemble of the physical and physiological rules of the water transport within the soil-plant-atmosphere continuum.

2. Distribution of the wild pear *Pyrus pyraster* (L.) Burgsd. and service tree *Sorbus domestica* L. in Slovakia

The wild pear and service tree are members of the rare woody plants in Slovakia. The wild pear often grows in the scattered vegetation of the landscape, but also on the forest margins mainly in communities of oak stands. The service tree appears mostly in the rural landscape, and mainly in vineyards and fruit orchards. In many European countries, wild pear and service tree are often sought after by landscape designers and foresters, because both species have aesthetic influence in the landscape, a good growth rate, and provide valuable timber.
The vertical distribution of wild pear has been documented mainly at lower altitudes up to 400 m (Hofmann, 1993; Schmitt, 1998; Wilhelm, 1998). The highest location found was at an altitude of 754 m in bundesland Süd-Niedersachsen und Nordhessen (Schmitt, 1998).
In Slovakia, wild pear grows in the lowlands to sub-mountain areas, up to an approximate altitude of 950 m (Peniašteková, 1992), and in some cases up to an altitude of 1163 m (Blattný & Šťastný, 1959). A detailed study on the environmental conditions of stands where wild pear naturally occurs was conducted in 1994-1999 (Paganová, 2003). The basic data were obtained from 64 locations (Fig. 1).
Stands with wild pear were located mostly on grazing lands, meadows, and in the scattered woodlands. Wild pear populations were also found along a dry stream channel (locality 8) and in a thin forest (location 21). Wild pear often grows on the forest edge (locations 30, 34, 41, and 48), or on former grazing land that gradually changed to woodlands (location 42, 55, and 56).
The majority of stands with wild pear (80%) were found at altitudes up to 500 m. The lowest location in Slovakia was at an altitude of 100 m (12 Solnička), and the highest analysed stand was at an altitude of 800 m (19 Jezersko) (Paganová, 2003).

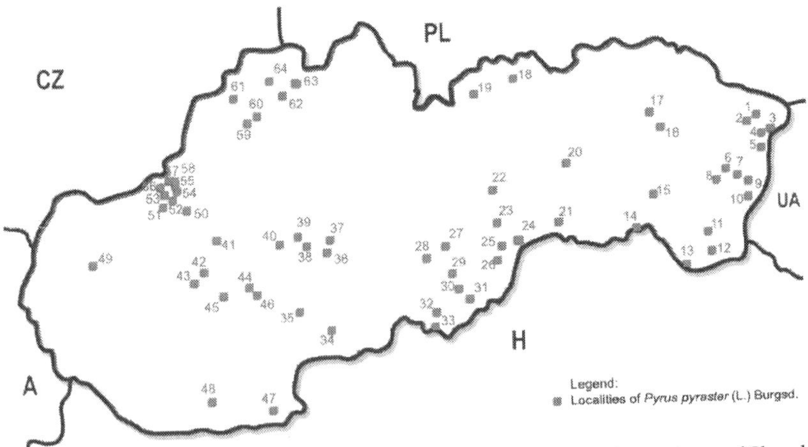

Fig. 1. Distribution of the wild pear *(Pyrus pyraster)* populations in the territory of Slovakia (Paganová, 2003).

The service tree is one of the rare autochthonous woody plants in the entire area of natural distribution. The area of natural distribution of the service tree reaches the northern part of Asia Minor and Africa as well as the northern border crosses of North Rhine-Westphalia, Lower Saxony, Saxony-Anhlat, and Thüringen, Bavaria. The northernmost occurrence is located in the Federal Republic of Germany at approximate latitude 51° of north width (Haeupeler & Schönfelder, 1988), and then continues to South Moravia and Slovakia, Hungary, Romania, and Crimea Mt.

According to Májovský (1992), the service tree has higher demands for light and high temperatures. In Slovakia, it is cultivated in the uplands on sunny south and southwest exposed stands. The vertical distribution of this woody plant occurs at an altitude of 109 m (Benčať, 1995) or 175 m (Michalko, 1961) up to an altitude of 610 m (Michalko, 1961; Benčať, 1995).

In 1996-2000, the environmental conditions of 24 locations of the service tree were analysed (Paganová, 2008). In Slovakia, the service tree appears in the southern regions in warmer stands at lower altitudes (Fig. 2). The distribution of the analysed stands containing the service tree confirmed its occurrence mainly at lower altitudes. The lowest stand with the service tree was found at an altitude of 200 m (location 23 - Vinné), and the highest stand was found at an altitude of 490 m (location 1 - Predpoloma) (Fig. 2).

The majority of the analysed stands containing the service tree (50%) were located in an open landscape near vineyards and fruit orchards (location 10, 11, 12, 13, 14, 15, 16, 17, 18, 19, 21, and 22). The service tree was frequently (46% of analysed stands) found in abandoned fruit orchards or on grazing lands as well (location 1, 2, 3, 4, 5, 6, 7, 19, and 20). Only a few plants were found in woodlands (location 23 and 24) and one stand of service tree (location 8) was located in an oak forest.

According to the analysis of the vertical distribution (Fig. 3), the service tree grows mainly on uplands in Slovakia. Approximately 66% of the analysed stands were found at an

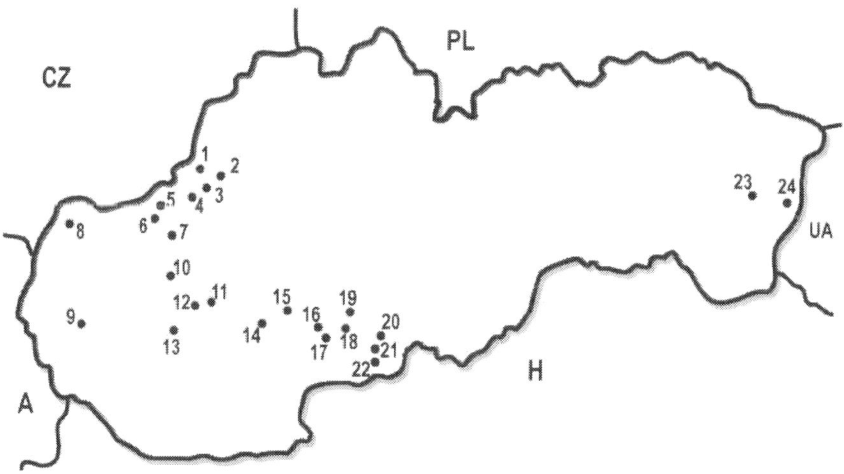

Fig. 2. Location of the stands with a higher number of service trees (*Sorbus domestica*) in Slovakia (Paganová, 2008).

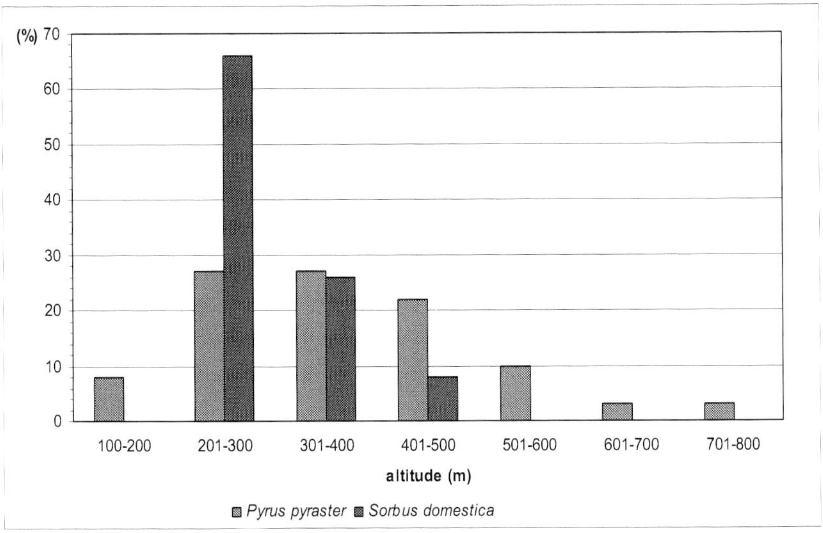

Fig. 3. Distribution of wild pear (*Pyrus pyraster*) and service tree (*Sorbus domestica*) in Slovakia according to stand altitude.

altitude of 201–300 m, and 26% of the stands were at an altitude of 301–400 m. One location was at altitudes of 450 m and 490 m. Compared to the wild pear (Fig. 3), the service tree is absent from the lowlands, and the occurrence of this woody plant at altitudes above 400 m is rare. According to Kárpáti (1960), the service tree is frequently found within communities of oak forests at lower altitudes and on fertile soils in communities of *Lithospermo-Quercetum, Melico (uniflorae)-Quercetum petraeae,* and others. Michalko (1961) confirmed the findings by Kárpáti, but according to his opinion, the service tree grows at higher altitudes only in extreme communities of *Corneto-Quercetum* (*pubescentis* and *petraeae*), and can even be found in the beech woodland *Corneto-Fagetum* and in relict pinewoods growing on limestone and dolomite parent rock.

Similar to the data mentioned above, very similar findings regarding the range of altitudinal distribution were found in Switzerland. In this country, the service tree was found within an altitude of 384 m in the Basel region and 675 m in the Schaffhausen region (Brütsch & Rotach, 1993). In the southeast section of the Wiener Wald in the area of Merkenstein, the service tree has been found up to an altitude of 550 m (Steiner, 1995). At the northern border of its natural distribution in Germany in the region of Sachsen-Anhalt, the service tree is distributed from 140 m to 310 m, and predominantly within an altitude of 161-240 m (Steffens, 2000). On the Plateau of Lorraine, the service tree appears in forest crops at an altitude of 200-400 m (Wilhelm, 1998).

In southern regions of the natural distribution, the service tree grows at higher altitudes than in Slovakia. For example, in Spain, it grows at altitudes up to 1400 m, in Greece up to 1350 m, in Turkey up to 1300 m, and in southern Bulgaria from 300 to 800 m (Kausch, 2000). In southern Italy (Mt. Vesuvius), the service tree grows from the banks of the sea up to an altitude of 800 m (Bignami, 2000).

3. Ecological characteristics of the stands with *Pyrus pyraster* (L.) Burgsd. and *Sorbus domestica* L.

The wild pear is considered to be a light-demanding woody plant, which prefers warm stands with a sufficient amount of sunlight (Namvar & Spethman 1986; Hofmann, 1993; Wagner, 1995; Kleinschmit & Svolba 1998; Schmitt, 1998; Rittershoffer, 1998; Roloff, 1998; Wilhelm, 1998). Hofmann (1993) previously created a diagram for the occurrence of 300 wild pear plants according to the stand exposure. The plants were predominantly in locations with south and southwest exposures. In support of these findings, Roloff (1998) also found that the most frequent occurrence of wild pear plants was on slopes with a south or west exposure.

The service tree is explicitly regarded as a light-demanding woody plant (Michalko, 1961; Májovský, 1992; Brütsch & Rotach, 1993; Pagan, 1996; Wilhelm, 1998). In Slovakia, 96% of stands with the service tree were found in the open landscape with solitary trees. Two stands were on the margin of woodlands with a few service trees in the crop, and only one location was an oak forest, with service trees found in the upper tree canopy or slightly above it. In all of these stands, the individual service trees grew under nearly full light without competition from other woody plants (Paganová, 2008). The service tree is intolerant to shading at an early age, and similar to the wild pear, will die quickly without a minimum light supply (Wilhelm, 1998).

Based on data obtained on the distribution of 507 wild pear plants in Slovakia, the majority of stands (80%) had a south, southeast, or southwest exposure. However, a limited number of stands (14%) containing wild pears were also found to have west, east, or northwest exposures, and four locations (6%) were on a plain stand (Fig. 4).

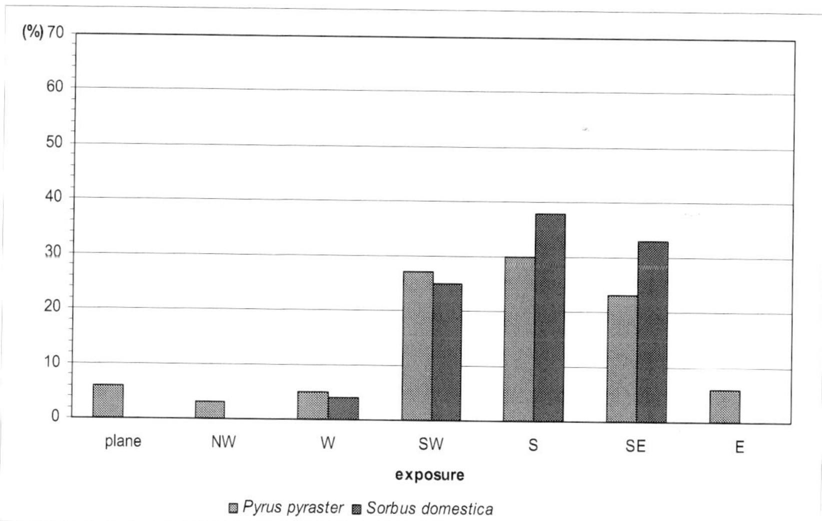

Fig. 4. The distribution of locations with wild pear (*Pyrus pyraster*) and service tree (*Sorbus domestica*) in Slovakia according to stand exposure.

In comparison to the wild pear, a majority of stands with service trees (38%) had a southern exposure, and many locations also had southeast (33%) and southwest (25%) exposures. One location (4%) had a western exposure. According to measurements by Geiger (1961), slopes with northern exposure obtain just half of the absolute total light emission as slopes with southern exposure. The prevalent distribution of service trees in southern-exposed stands supports the hypothesis regarding their high demand of light and warm climate. In Slovakia, none of the analysed locations had northern exposure. In Switzerland, 74% of the locations with service trees had southern exposure (Brütsch & Rotach, 1993).

The ecological-climatic amplitude of the wild pear locations in Slovakia is relatively wide. In stands with wild pear plants, the conditions range from plain and fold climates, to a mountain climate (Paganová, 2003). Stands were classified to climate-geographic types and subtypes according to Tarábek (1980) and Špánik et al. (1999).

Within the analysed scale of the wild pear stands in Slovakia, the average January temperatures range from -1.4°C to -5.8°C and the average July temperatures range from 13.5°C to 20.4°C. The annual sum of precipitation reaches values ranging from 570 mm to 900 mm. The majority of the pear locations (53%) fall within the climate-geographic type of mountain climate (Fig. 5), which is humid or very humid with rare temperature inversion. These stands were most frequently found in the warm and moderately warm subtypes of the mountain climate at an altitude of 250-550 m. The average January temperatures range from -1.4°C to -5.0°C, the average July temperatures range from 17.0°C to 20.4°C, and the annual sum of precipitation for these stands ranges from 580 mm to 790 mm.

Stands within the warm subtype of the fold climate are observed quite frequently. The fold climate is semi-humid to semi-arid with a remarkable inversion of temperatures. These locations are at altitudes of 210-450 m. The average January temperatures range from -2.0°C to -4.0°C, the average July temperatures range from 18.0°C to 19.2°C, and the annual sum of precipitation ranges from 628 mm to 765 mm in the respective locations.

The lowest number of wild pear locations was documented for stands in the warm or mostly warm subtypes of the plane climate, which is arid and semi-arid. Locations were registered at altitudes of 120-400 m. The average January temperatures in these stands range from -1.5°C to -3.3°C, and the average July temperatures range from 17.2°C to 20.1°C. The annual sum of precipitation is 570-700 mm.

The climate-geographic characteristics of stands with service trees are slightly different than stands containing wild pears (Fig. 5). The climate with the highest number of locations (42%) belongs to the mostly warm subtype of the plane climate, which is arid or semi-humid with a mild inversion of air temperatures. These stands are at an altitude of 200-300 m. The average annual temperature ranges from 8.3°C to 9.0°C, and the annual sum of precipitation is 610–650 mm.

A high number of stands (33%) were classified in the mountain climate with a mild inversion of air temperatures (the climate is rather humid). These stands are at an altitude of 220-490 m. The average annual temperature is 8.3°C and the annual sum of precipitation ranges from 650 to 620 mm.

The climate with the fewest locations of service trees was the fold climate (25%), which has a markedly high inversion of air temperatures (with an arid or even humid climate). The stand altitude ranges from 250 m to 380 m, the annual average temperature ranges between 8.1–8.5°C, and the annual sum of precipitation is 620–700 mm.

According to a recent climatic evaluation (Škvarenina et al., 2004), the annual average temperature of the majority of locations with service trees is above 8°C, with the exception

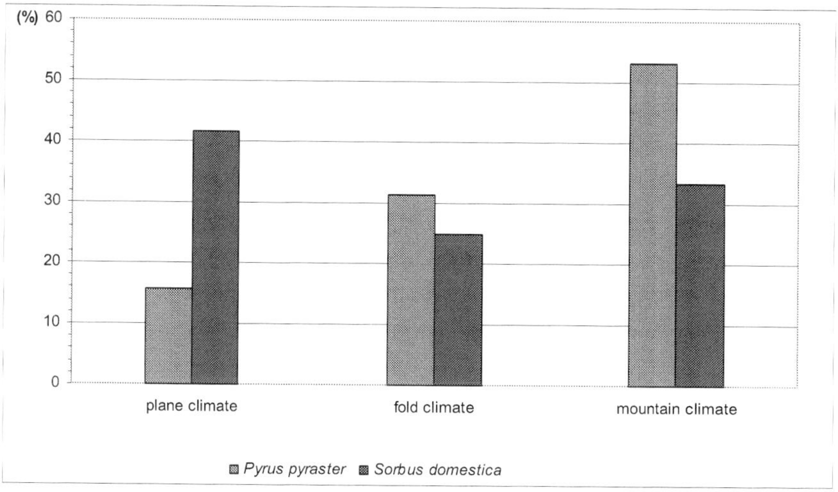

Fig. 5. Wild pear (*Pyrus pyraster*) and service tree (*Sorbus domestica*) stand classification in Slovakia according to climate-geographic types.

of two of the analysed locations (1 and 5), where the annual average temperature ranges from 7.5°C to 7.7°C. The average annual sum of precipitation for the majority of the stands is 610–700 mm, with the exception of the two mentioned locations, where this parameter reaches 790 mm and 750 mm, respectively. The potential evapotranspiration amount in the majority of the analysed stands was 600–750 mm during one year. Considering that the annual average sum of precipitation is 610–700 mm, it is possible that the service tree has to obtain enough moisture during the main growing season predominantly from water resources in soil. The deficit of rain during the summer occurs in the majority of the stands with this woody plant.

Warm and arid (southeast, south, southwest, and even west) stand exposures play an important role in the formation of the arid microclimate and mezzo-climate of the mentioned locations. According to climate classification in Slovakia (Špánik et al., 1999), the analysed locations with service trees were in warm and even moderately warm as well as semi-humid and even semi-arid climates. Compared to the wild pear, the service tree prefers stands at lower altitudes and is prevalent in warm and arid climates. The wild pear has wider ecological amplitude and grows at higher altitudes in stands with different water regimes and climate extremes.

Based on the brief pedology characteristics of our experimental plots, we can hypothesize that these soils are very well fertile (Chernozem, Fluvi-mollic soils, Cambisols, and Orthic Luvisols) or well supplied with nutrients (Luvisols, Pararendzinas, and Fluvisols). In addition, browned Rendzinas can be considered as relatively favourable soils.

According to the ecological scheme of Ellenberg (1978), the wild pear is a woody plant with broad ecological amplitude that grows in nearly all soil types, with the exception of extreme acidic soils. Rittershoffer (1998) found that mildly acidic or mildly alkaline soils were optimal for wild pear growth. According to information from Westfalen-Lippe (Germany), the wild pear prefers soils developed on limestone or on the rich nutrient parent rocks (80% of stands) (Schmitt, 1998). In the area of Süd-Niedersachsen und Nordhessen (Germany), the wild pear frequently grows in shallow rendzinas from mussel limestone or from lime sandstone, and very rarely appears on deeper brown forest soils. More than 92% of all natural stands were on rich basic rocks (Hofmann, 1993). In the forest on the Plateau Lorraine, the wild pear grows mainly on parent rock of the mussel limestone and on keuper sediments. There are deep terra fusca soils and shallow Rendzinas (Wilhelm, 1998). The colective data from different areas of the natural distribution indicate that suitable growth conditions for the wild pear are mainly basic and rich nutrient soils with occasional water deficits.

In Slovakia, the wild pear grows on fertile soils (Chernozem, Fluvi-mollic soils, Cambisols, and Orthic Luvisols), or soils well supplied with nutrients (Albic Luvisol, Pararendzina, and Fluvisol). In addition, in some stands it grows on soils that are rich in minerals but under conditions of unbalanced soil chemistry with little fertility (Paganová, 2003).

In general, Fluvi-mollic and Cambisol soils have a sufficient water supply. At lower altitudes, the water deficit appears mainly in Rendzinas. The water deficit in Luvisols is usually a result of a lower amount of precipitation and higher evaporation. Orthic Luvisols have a lower water supply, and therefore the possibility of their aridization is higher. In addition, a fluctuating water regime appears within the Planosols and Fluvisols (Šály, 1988).

According to the ecological scheme by Ellenberg (1978), the wild pear has optimal growth conditions on fresh basic soils (its potential optimum). Another more frequent existence optimum of this woody plant is near the xeric forest limit, where the wild pear grows in arid soils rich in bases as well as in moderately acidic soils. Some authors have placed the wild pear among xerophytic woody plants according to its lower demands on soil humidity (Bouček, 1954). When under competition with some woody plants, it grows on its synecological optimum in extreme arid stands - rocky hilltops, stands of the xeric forest limit close to steppe communities, and in the sparse xerophytic oak woodlands (Rittershoffer, 1998). However, another existence optimum of this woody plant occurs in the hydric forest boundary in stands of the hardwood floodplain forests, where wild pear growth is limited by inundation (Rittershoffer, 1998). Based on these findings, the wild pear is a flexible woody plant with tolerance to a large range of soil humidity.

In Slovakia, stands with the service tree have favourable physical characteristics, good saturation, and are very fertile (Orthic-Luvisols and Cambisols), or have soils that are well supplied with nutrients (Rendzinas). However, under conditions of unbalanced soil chemistry, there is little fertility, and the pH of this soil is moderately acidic, neutral, or moderately alkaline. Cambisols generally have a sufficient water supply, and Orthic-Luvisols have a lower water supply with the possibility of aridization. Water deficiency can appear in Rendzinas as a result of the water penetration, so the water supply in this soil is usually low (Šály, 1988).

According to Wilhelm (1998), the service tree grows on mussel limestone and on keuper sediments on the Plateau Lorraine. These soils are well or very well supplied with nutrients. On slopes based with mussel limestones are deep terra fusca soils, and in the upper parts of the slopes are shallow Rendzinas. On the keuper, there are abundant, deep Vertic Cambisols with water deficiency during summer.

In east Austria, the service tree grows in the oak forest communities and is considered to be a woody plant of the uplands with less demands on soil humidity, but with quite high demands on the nutrient content of the soils (Kirisits, 1992). In the Wiener Wald (Steiner, 1995), the service tree appears on limestone and dolomite parent rock with prevalence in semi-humid and arid Rendzinas. In addition, the service trees in Switzerland are found mainly in arid soil with less skeleton that is rich in bases (Landolt, 1977; Brütsch & Rotach, 1993), as determined from detailed studies of the service tree stands in Canton Genf, which refer to the medium deep and deep skeletal Cambisols and Luvisols with slower water penetration and possible water logging. In the Bassel region, the service tree also grows on Rendzinas or Lithosols, which are shallow and extreme skeletal soils that have a very low water capacity. In the Schaffhausen, approximately 92% of the service tree plants grow on limestones, and the rest of the stands grow on gravels of the high terrace that belong to Riss. In the deeper strata, there are limestones that are part of the morena and gravels. Various soils, even acidic soils, can appear randomly on small areas of the parent rocks . These data document quite a broad range of soil conditions for the stands containing service trees, and tolerance of the taxon to periodic or rare occurrences of water deficit in the soils is evident. On some stands within the area of its natural distribution, the service tree grows under conditions of a soil drought.

4. Potential adaptability of the analysed woody plants to progressive drought

Drought can be considered in meteorological, agricultural, hydrological, and socio-economic terms (Wilhite & Glantz, 1985). Meteorological drought reflects one of the primary causes of drought. It is usually defined as precipitation less than a long-term average (defined as normal) over a specific period of time. Agricultural drought is expressed in terms of the moisture availability at a particular time during the growing season for a particular crop. Hydrological drought is usually expressed as a deficiency in surface and subsurface suppliers, and refers to a period when stream flows are unable to supply the established users under a given water management system. Socio-economic definitions of drought relate to the supply and demand of specific goods. Importantly, humans can create a drought situation through land-use choices or an excess demand for water (Wilhite & Glantz, 1985).

According to Škvarenina et al. (2009a), drought is a temporary aberration that differs from aridity, which is restricted to low rainfall regions and is a permanent feature of the climate. The altitude and topography are significant climate-differentiating factors. In Slovakia, a considerably broken topography plays an important role in the variability of climate conditions. The increase in altitude causes changes in solar radiation as well as thermal and water balance of the land (Škvarenina et al., 2009a). Vertical differentiation of the climate conditions has a significant influence on species structure of the natural vegetation. The biogenocenoses can be classified into nine vegetation stages described by Zlatník (1976) based on altitude, exposure, and topography, which are named after woody plants that are dominant in the area.

Škvarenina et al. (2009a) analysed trends in the occurrence of dry and wet periods in altitudinal vegetation stages in Slovakia between 1951 and 2005. The authors considered relative evapotranspiration (E/E_0), which is defined as the rate of the actual evapotranspiration (E) to potential evapotranspiration (E_0), as an excellent measure of water sufficiency for vegetation. According to their findings, the smallest annual values of (E/E_0) were recorded in the Danube lowland (1st Oak vegetation stage) with relatively high totals of potential evapotranspiration (E_0) above 700 mm and with annual precipitation totals below (P) 550 mm. The lowest value of the relative evapotranspiration (approximately 60%) was recorded in the lowest areas of Slovakia with an altitude up to 200 m. Relative evapotranspiration reached higher values towards higher vegetation stages (above 90% in the 4th Beech vegetation stage at altitudes above 650 m).

In addition to relative evapotranspiration, the drought index (E_0/P) has also been used to describe the relationship between the energy and precipitation (P) inputs within particular vegetation stages. Warm forest-steppe stands in Slovakia with oak communities have drought index values (E_0/P) of approximately 1. The predominant areas of Slovak forests are stands with drought index values up to 0.3. Moreover, the vegetation stages with $E_0/P < 0.3$ are within the mountain climate (Škvarenina et al., 2009b).

In Slovakia, wild pear stands are distributed from lowlands up to an altitude of 800 m. Specimens also appear in 1st (oak) and 2nd (beech-oak) vegetation stages with a water deficit during the growing season. The stands in these vegetation stages are classified as a territory with a dry (arid) climate according to the relative evapotranspiration and drought index. On the other hand, the wild pear is also distributed in stands at higher altitudes in the 4th (beech) and 5th (fir-beech) vegetation stages, which have a higher humidity (higher relative evapotranspiration). This type of distribution shows that the wild pear is tolerant to different conditions of water sufficiency.

The service tree is predominantly distributed in the 1st (oak), 2nd (beech oak), and 3rd (oak-beech) vegetation stages in Slovakia, avoids lowland stands, and appears mainly on slope terrain of the forest steppe stands. This taxon often grows in conditions of warm oak communities with an arid climate. At higher altitudes, the service tree most likely avoids the consequences of a strong beech competition. In the Slovak lowlands, the absence of the service tree is most likely due to the higher underground water level and the intensive agricultural utilization of the land.

According to Škvarenina et al. (2009a), a markedly severe drought between 1951 and 2005 was only identified in the Danube Lowland (1st Oak vegetation stage) and in the Záhorská lowland (2nd Beech-oak vegetation stage) of Slovakia. Considering the natural distribution of the wild pear and tolerance to a wider range of water supply, this woody plant has the potential to adapt to the decreasing humidity of the Danube Lowland. The service tree has similar qualities and the potential to grow in arid conditions; however, this taxon is mainly found on the slopes of forest-steppe stands.

According to a drought analysis of the Slovak territory conducted on the climatic data obtained from 1960-1990, agricultural regions become more sensitive to conditions of climate change upon drought occurrence (Šiška & Takáč, 2009). The authors used two indices for spatial evaluation of drought conditions in Slovakia: the climatic index of drought and the evapotranspiration deficit. The climatic index of drought (K) was applied for the entire growing season (GS10 period) and $K_{GS10} = \Delta E$, where E_0 is the potential evapotranspiration during GS10 and R is the rainfall during GS10. The evapotranspiration deficit ΔE during the growing season was calculated as $\Delta E_{GS10} = E_0 - E$, where E_0 is the potential evapotranspiration during the main growing season (GS10) and E is the actual evapotranspiration during the main growing season.

Two very dry and hot regions were classified in Slovakia, the Danubian and east Slovakian lowlands, which represent maize production areas with a water deficit that exceeds 250 mm during the growing season. These evapotranspiration deficit values will most likely be present in river valleys up to altitudes of 300 m as well (Šiška & Takáč, 2009).

The findings described here support the hypothesis that a higher frequency of drought occurs in agroclimatic regions of the Slovak Republic. In the future, it is important to elaborate on several concepts of the stabilization of agricultural production against water deficit and soil aridity. With the exception of breeding programs that focus on developing new crop varieties that can tolerate the changed climatic conditions and development of integrated irrigation systems, there are also possibilities for landscape stabilization using non-forest woodlands. These types of woodlands should be established with woody plants that are tolerant to water deficit and that are adaptable to dynamic changes of water regimes. The taxa analysed here, including the wild pear and service tree, belong among the prospective woody plant species that are suitable for planting in regions potentially endangered by droughts.

The described research focused on an analysis of the physiological parameters of two woody plant species (wild pear and service tree) under conditions of a regulated water regime and water stress. The aims of the study were to verify the adaptive potential of both taxa to drought, and to obtain information on the mechanisms used by these woody plants under conditions of water deficit.

5. Interspecific differences of the selected physiological parameters of woody plants

Woody plants make different ecological adjustments to water deficit, and can modify their physiological functions and anatomical structures for adaptation. Adaptability is a rather complex quality, and the explicit function of a typical plant response to water deficit is very difficult to define. Therefore, we established experiments that regulated the water regime of juvenile (two-year old) wild pear and service tree plants under semi-controlled conditions.

The plants were planted in pots (content 2 L) with mixed peat substrate enriched with clay (content of clay 20 $kg.m^{-3}$; pH 5.5-6.0; fertilizer 1.0 $kg.m^{-3}$). The potted plants were placed under a polypropylene cover with 60% shading. The plants were regularly watered and maintained on 60% of the full substrate saturation for 28 days. In the phenological stage of shoot elongation (at the beginning of June), the plants of both taxa were divided in two variants according to a differentiated water regime. Variant "stress" was supplied with water at 40% of full substrate saturation and "control" at 60% of full substrate saturation. The model of the differentiated water regime was maintained for 126 days (to the end of September). Sampling was performed at 14 day periods for both conditions.

The size of the leaf area (A) and leaf water content (LWC) were measured, and a determination of fresh weight (F_W) and dry weight (D_W) was done gravimetrically. The size of leaf area (A) was calculated from leaf scans using ImageJ software (http://rsbweb.nih.gov/ij/). The LWC and specific leaf area (SLA) were calculated according to the methods described by Larcher (2003). For metabolic characteristics, the total chlorophyll and carotenoid content were determined according to the methods described by Šesták & Čatský (1966).

Data were analysed from three growing seasons in 2008-2010 for each taxon under two variations of water regimes (40% and 60% substrate saturation). The relationship between SLA and LWC of the plants under stress and control conditions as well as changes in the assimilatory pigments during water stress were also analysed. A statistical assessment of these parameters was conducted by regression analysis using the statistical software Statgraphics Centurion XV (StatPoint Technologies, USA). A $P < 0.05$ was consisted statistically significant.

5.1 The influence of water stress on the production of leaf dry mass

The different reactions of the analysed taxa (wild pear and service tree) to water stress were confirmed by the dry mass (DM) measurements taken under controlled and stress experimental conditions (Table 1). Under control conditions, the increment of leaf dry mass of the wild pear was 14.67 mg p^{-1} d^{-1} and the increment of leaf dry mass of the service tree was 18.37 mg p^{-1} d^{-1}. Under conditions of water deficit (stress), the increment of the leaf dry mass of wild pear plants was 12.78 mg p^{-1} d^{-1} and the increment of leaf dry mass for the service tree plants decreased to 3.04 mg p^{-1} d^{-1}. The impact of water stress on the wild pear was less significant, and this plant is probably more tolerant to drought. Importantly, the relationship to photosynthesis economy depends on the leaf structure. The wild pear is a typical taxon of sunny and arid stands, and contains heterobaric leaves. Parenchyma (or sclerenchyma) cells without chloroplasts accompany the vascular system, and similar to ribs, lead to the top (adaxial) or bottom (abaxial) epidermis (Essau, 1977; Fahn, 1990; Terashima, 1992). The tips (ribs) of the vascular bundles divide leaf mesophyll hermetically into compartments that are reciprocally isolated against gas exchange. In the compartments, the intercellular space is relatively small with low chlorophyll content. The compartments are similar to "open windows", which transmit visible light into the internal layers of the mesophyll (Liakoura et al., 2009). Heterobaric leaf structures are also significant because they allow for easier transport of water to the epidermis due to increased hydraulic conductivity. One predominant factor that limits plant transpiration is leaf area. The reduction of leaf area during water deficit is typical for plants from arid stands. Several authors (Reich et al., 2003, Wright et al., 2004; Niclas & Cobb, 2008) have confirmed the narrow relationship between leaf structure and function. Our comparison of the leaf area ratio to dry weight of the leaves (SLA) of the analysed species confirmed the interspecific differences (Table 1). Wild pear leaves with higher values of SLA were thinner than leaves of the service tree under control conditions. The leaf water content per unit of dry weight in pear leaves was higher than service tree leaves. In experiments with fast growing woody plants, Dijkstra (1989) confirmed the thinner leaves of these species as well as the presence of larger vacuoles in the cells, which accumulate a larger amount of water per unit of dry mass. In our experiments with wild pear, the values of SLA decreased after 70 days under both conditions (stress and control), and the pear leaves became xeromorphous. There were no significant differences in SLA values of the pear leaves after 70 days under the differentiated water regime or due to water stress (Table 1).

The different functional qualities of the leaves can be effected by 1) changes in the leaf structure, and 2) different compositions of the leaf, including sclerenchyma elements and organic compounds (lignins and phenols), which increase leaf dry mass as described by Mooney & Gulmon (1982) and Lin & Harnly (2008).

Interspecific differences in the reaction to water deficit were not confirmed in the analysed taxa of this study. However, at the beginning of the experiments and after 70 days of cultivation, the values of LWC of the wild pear and service tree plants were different, and these values did not change under conditions of water stress (Table 1).

Based on our analysis of the relationship between SLA and LWC, both of the analysed taxa maintained higher LWC with increasing values of the specific leaf area, regardless of the level of substrate saturation (Fig. 6, 7, 8, and 9). In addition, a significant linear correlation was observed between SLA and LWC under control and stress conditions without interspecific differences.

Physiological characteristics	Taxon							
	Pyrus pyraster				*Sorbus domestica*			
	control		stress		Control		stress	
	0 day	70 day	0 day	70 day	0 day	70 day	0 day	70 day
Size of the leaf area A (mm^2)	23 312	34 570	23 312	32 578	59 499	78 713	59 499	51 210
Specific leaf area SLA ($mm^2.mg^{-1}$)	19.13	15.14	19.13	15.80	16.57	16.71	16.57	16.06
Leaf dry weight DW_l (mg)	1 176	2 203	1 176	2 071	3 488	4 774	3 488	3 275
Leaf water content LWC (%)	66.3	57.0	66.3	57.6	45.4	52.2	45.4	51.5
Chlorophyll content ($mg.mm^{-2}$)	515.7 $.10^{-6}$	679.2 $.10^{-6}$	515.7 $.10^{-6}$	779.7 $.10^{-6}$	333.9 $.10^{-6}$	470.5 $.10^{-6}$	333.9 $.10^{-6}$	452.0 $.10^{-6}$
Carotenoid content ($mg.mm^{-2}$)	110.2 $.10^{-6}$	138.4 $.10^{-6}$	110.2 $.10^{-6}$	147.4 $.10^{-6}$	76.3 $.10^{-6}$	105.4 $.10^{-6}$	76.3 $.10^{-6}$	101.6 $.10^{-6}$

Table 1. Physiological characteristics of leaves taken from 2-year old potted plants of wild pear (*Pyrus pyraster*) and service tree (*Sorbus domestica*) grown in conditions of differentiated water regime - control (60% of the full substrate saturation) and stress (40% of the full substrate saturation) conditions.

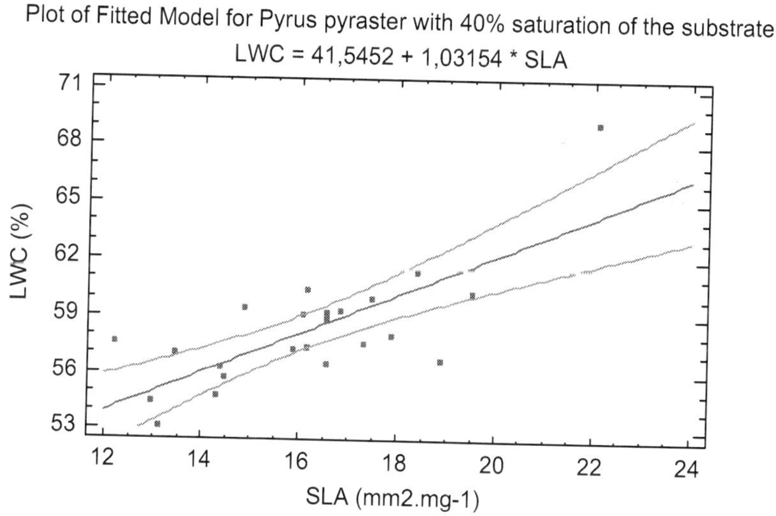

Fig. 6. Positive linear correlation between SLA ($mm^2.mg^{-1}$) and LWC (%) of wild pear (*Pyrus pyraster*) leaves under conditions of water stress. Correlation coefficient (r) = 0.760432, p value = 0.0000.

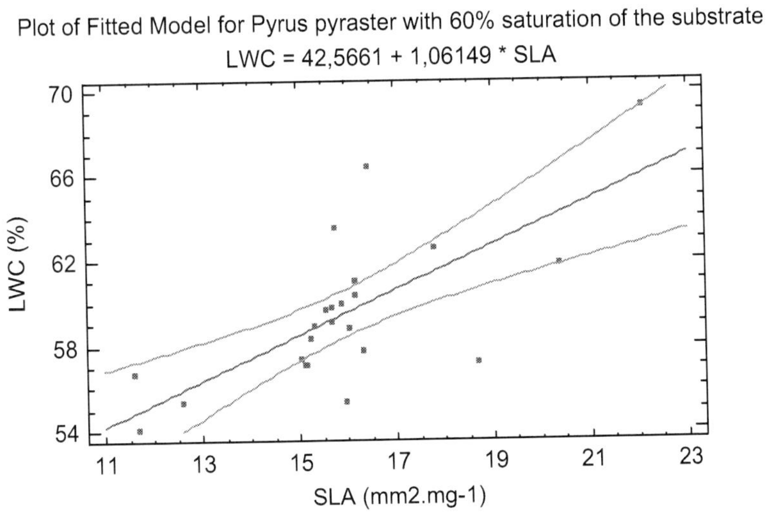

Fig. 7. Positive linear correlation between SLA ($mm^2.mg^{-1}$) and LWC (%) of wild pear (*Pyrus pyraster*) leaves under control conditions. Correlation coefficient (r) = 0.704177, p value = 0.0002.

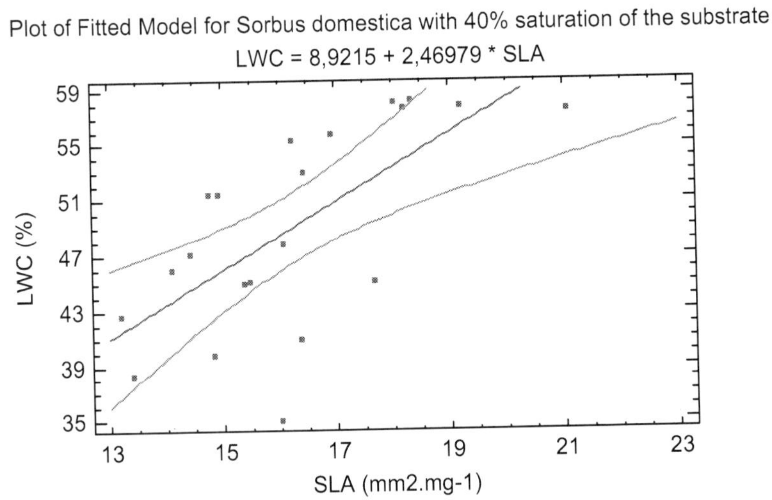

Fig. 8. Positive linear correlation between SLA($mm^2.mg^{-1}$) and LWC (%) parameters of service tree (*Sorbus domestica*) leaves under water stress. Correlation coefficient (r) = 0.669898, p value = 0.0009.

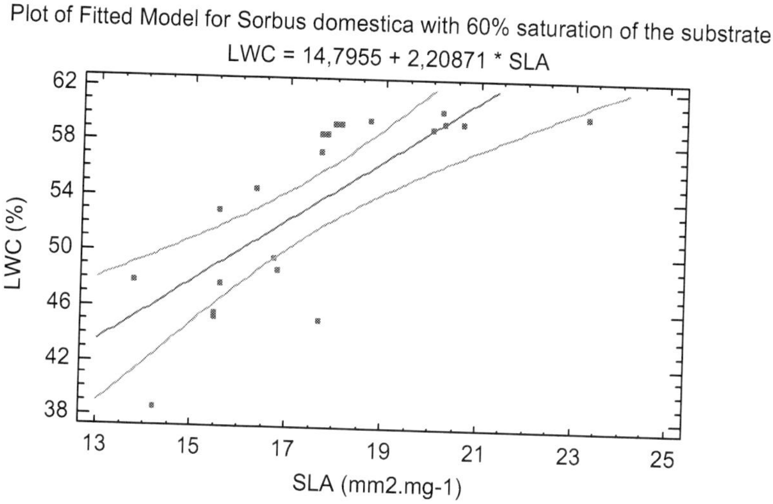

Fig. 9. Positive linear correlation between SLA ($mm^2.mg^{-1}$) and LWC (%) parameters of service tree (*Sorbus domestica*) leaves under control conditions. Correlation coefficient (r) = 0.76925, p value = 0.0000.

5.2 Changes in the assimilatory pigment content in leaves under conditions of water stress

The content of assimilatory pigments is an important factor that has a significant influence on thermal characteristics of the leaves. Leaves with lower chlorophyll content have higher reflexion, and the leaf surface temperature can have relatively lower values than the temperature of leaves with a higher content of assimilatory pigments. In addition, leaves with a higher content of carotenoids should have a relatively higher resistance against water stress. On the other hand, the ability of a plant to maintain a higher content of assimilatory pigments during stress can be very important for the functional activity of the leaves. Our analysis confirmed a different content profile of assimilatory pigments (chlorophyll a and chlorophyll b), β-carotene, and neoxantine in the leaves of the wild pear and service tree. There was a significant positive linear correlation between carotenoid and chlorophyll content in the leaves of both analysed taxa, regardless of the level of water saturation of the substrate (Table 2). This relationship is illustrated in Figure 10 for the wild pear plants at 40% substrate saturation. The results of the regression analysis for the wild pear under the control condition as well as for the service tree under both conditions are shown in Table 2. The SLA values of the service tree leaves did not change significantly under the differentiated water regime or under conditions of water stress (Table 1). The values of SLA in wild pear leaves decreased during the differentiated water regime under both conditions (control and stress). The decrease of SLA was most likely influenced by the specific quality of the taxon, which produces so called "summer leaves" during twig elongation. Two-year old plants of the service tree created leaves on terminal shoots only, and the values of SLA were not significantly changed in both variants of the water regime (control and stress) within the analysed period of time. During summer, the chlorophyll content in leaves of the wild pear increased under control and water stress conditions. The chlorophyll content in

taxon/substrate saturation	wild pear/40	wild pear/60	service tree/40	service tree/60
correlation coefficient r	0.973681	0.964724	0.982228	0.974045
p value	0.0002	0.0004	0.0005	0.0001

Table 2. Results of a simple regression between total chlorophyll content and carotenoid content in leaves of the analysed taxa wild pear (*Pyrus pyraster*) and service tree (*Sorbus domestica*) under two conditions of substrate saturation. Legend: 40 – conditions of water stress (40% substrate saturation); 60 – control conditions (60 % of substrate saturation).

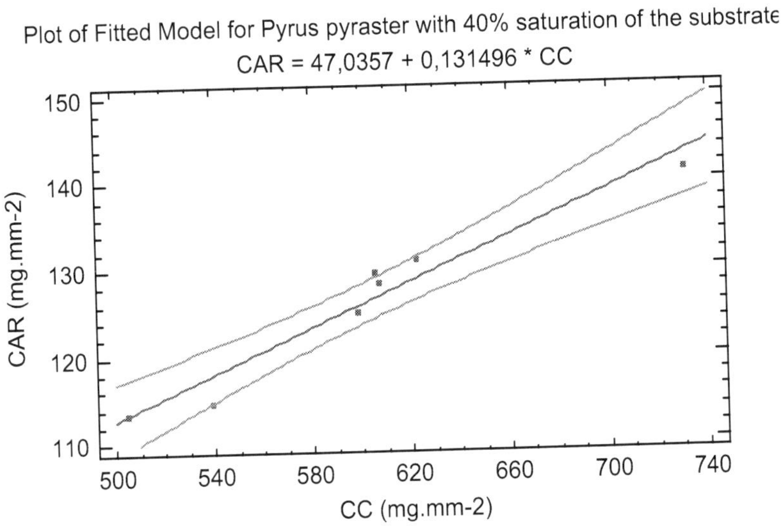

Fig. 10. Positive linear regression between total chlorophyll content (CC) and carotenoid content (CAR) in leaves of wild pear (*Pyrus pyraster*) plants growing under conditions of water stress. The correlation is quite close, with a correlation coefficient (r) = 0.973681 and statistically significant p value = 0.0002.

service tree leaves also increased; however, under water stress conditions, the chlorophyll content was lower than in the leaves of the control plants.

We confirmed a statistically significant relationship between SLA values and chlorophyll content in the leaves of the service tree under conditions of water stress, and this relationship was described by a polynomial curve of the second order (Figure 11). These data showed that the service tree maintained a balanced content of chlorophyll in leaves with a lower specific leaf area. In the stress variant, the chlorophyll concentration in service tree leaves varied between 340-470 mg.mm^{-2} within a 95% confidence level.

The relationship between SLA and chlorophyll content in the leaves of the wild pear under water stress conditions was also described as a polynomial function of the second order (Figure 12). However, this relationship was not significant. The leaf chlorophyll concentration ranged between 490-610 mg.mm^{-2} in the wild pear plants under conditions of lower substrate saturation (water stress).

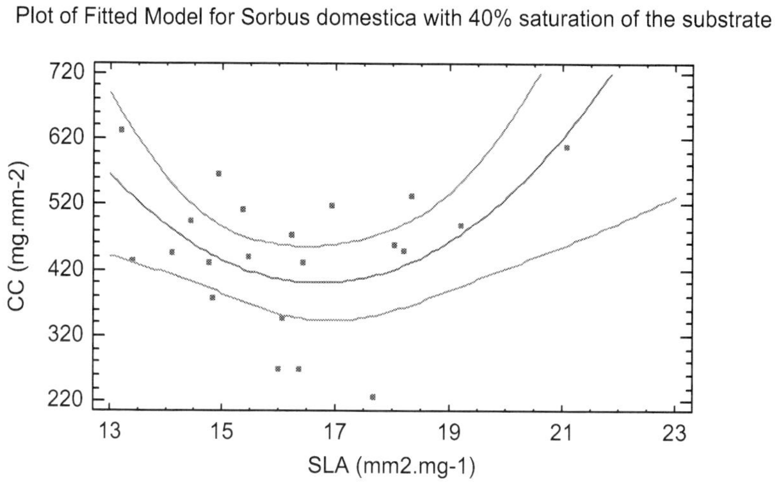

Fig. 11. Polynomial regression of the second order between specific leaf area (SLA) and chlorophyll content (CC) in the leaves of service tree (*Sorbus domestica*) plants grown under conditions of water stress. $R^2 = 30.92\%$; p = 0.0358.

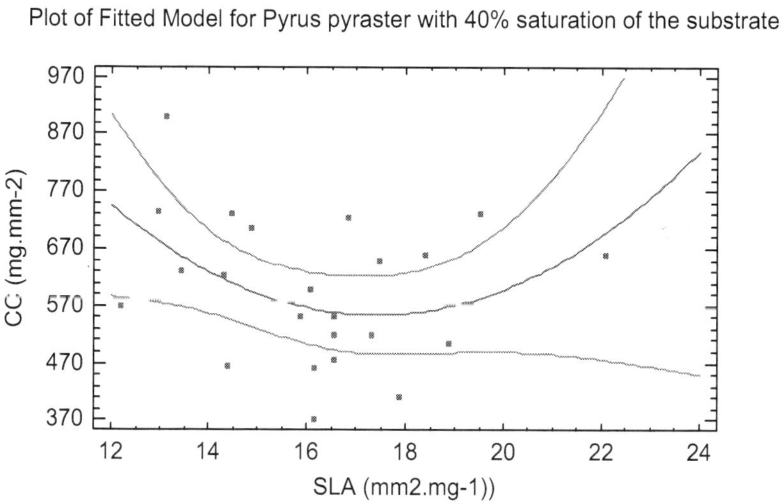

Fig. 12. Polynomial regression of second order between specific leaf area (SLA) and chlorophyll content (CC) in the leaves of wild pear (*Pyrus pyraster*) plants growing under conditions of water stress. $R^2 = 18.3086\%$; p = 0.1324

According to the results obtained from experiments with the differentiated water regime, we found a non-significant influence of low substrate saturation on the metabolic processes related to chlorophyll production in both of the analysed woody plant species.

6. Conclusion

With regard to progressive aridization, the research of resistant autochthonous woody plants that survive in extreme drought conditions is considerable. We have studied two taxa that naturally grow in the cultural landscape of Slovakia - the wild pear and service tree. Both species are light-demanding woody plants and occur in similar stands. Compared to the wild pear, the service tree prefers stands at lower altitudes, and is prevalent in warm and arid climates. The wild pear has wider ecological amplitude, and also grows at higher altitudes in stands with a different water regime and climate extremes.

Two-year old plants of the studied taxa were used in experiments with a regulated water regime. The plant material was grown from seeds collected directly from original stands in Slovakia, and the plants were maintained under semi-controlled conditions with 60% and 40% substrate saturation. Under these conditions, we analysed the following parameters: leaf dry mass, size of leaf area, leaf water content, specific leaf area, and the complex of assimilatory pigments.

Assessment of the analysed parameters confirmed interspecific differences in the physiological reactions of the woody plants under regulated conditions of a water regime. Each of the studied taxa utilized unique drought tolerance strategies. Under a differentiated water regime, the wild pear produced and increased leaf dry mass regardless of the level of substrate saturation (water regime). Based on these findings, the wild pear uses this mechanism to resist drought conditions.

Interspecific differences between the wild pear and service tree were confirmed by measuring the specific leaf area (SLA) and leaf water content (LWC). Compared to the service tree, the wild pear had higher SLA values when provided with a sufficient water supply. The SLA values of both taxa had a positive linear correlation with the leaf water content (LWC). Under water stress conditions, the wild pear reduced SLA, which was influenced not only by water deficit, but also by different morphogensis of the assimilation apparatus. During the experiment with the regulated water regime, the service tree had lower values of SLA than the wild pear and maintained them without significant changes, even under conditions of water stress.

A statistically significant relationship was confirmed between SLA values and chlorophyll concentration in the leaves of the service tree under conditions of water stress. This relationship was described as a polynomial curve of the second order. The relationship between SLA and chlorophyll concentration in the leaves of the wild pear under water stress conditions was also described as a polynomial function of the second order; however, this relationship was not significant. The low level of substrate saturation did not significantly influence metabolic processes related to chlorophyll production in both of the analysed taxa.

The water regime of the analysed woody plants is the decisive factor that affects their distribution and survival in conditions of progressive aridization. Considering the natural distribution of these woody plants and their tolerance to a wide range of water supply, the wild pear exhibits good adaptability to decreasing humidity. The service tree has similar qualities and the potential to adapt to arid conditions; however, it is generally found on slopes of forest-steppe stands.

In the future, studies will focus on strategies of water utilization used by xerotermic woody plants under conditions of aridization. The photosynthetic activity and transpiration of woody plants will also be analysed under conditions of limited water supply. The research will focus on the photosynthesis, transpiration, stomatal resistance, structural leaf elements, and root system of woody plants.

7. Acknowledgment

The research was supported by research grant project VEGA 1/0426/09 „Plant adaptability and vitality as criteria of their utilization in urban environment and in the landscape" from the Slovak Grant Agency for Science.

8. References

Blattný, T &, Šťastný, T. (1959). The natural distribution of forest woody plants in Slovakia. Bratislava (in Slovak), Slovenské vydavateľstvo pôdohospodárskej literatúry, 402 pp.

Benčať, F. (1995). Distribution and originality of *Sorbus domestica* L. in Slovakia (in Slovak). Zb. ref." Výsledky botanických záhrad a arborét pri záchrane domácej dendroflóry a II. Dendrologické dni".Vyd. TU vo Zvolene, Zvolen, p. 136–149.

Bignami, C. (2000). Der Speierling in Süditalien: Erforschung der Hänge des Vesuvs (Kampanien). Bovenden. Corminaria, *14*: 7–10.

Bouček, B. (1954). Pear (in Czech). In: Lesnická práce, roč. 33, č.2, p. 57-62.

Brütsch, U. (1993). Der Speierling, ein Vielfältiger Baum mit Zukunft. Wald und Holz, 13:8 - 11.

Brütsch,U. & Rotach, P. (1993). Der Speierling (*Sorbus domestica* L.) in der Schweiz: Verbreitung, Ökologie, Standortsansprüche, Konkurrenzkraft und waldbauliche Eignung. Schweiz. Z. Forswes., *144*, 12: 967–991.

Dijkstra, P. (1989). Cause and effect of differences in SLA. In: Lambers, H., Cambridge, M.,L. Konings, I.L. Pons, eds, Causes and Consequencess of variation in Growth rate and productivity of Higher Plants, SPB Academic Publishing, TheHague, This volume, pp.125-140.

Ellenberg, H. (1978). Vegetation Mitteleuropas mit den Alpen in okologischer Sicht. Phytologie IV/2. - Stuttgart: Ulmer

Essau, K. (1977). Anatomy of seed plants, 2nd ed. John Wiley, new York, New York, USA., 576 pp. ISBN 9780471245209

Fahn, A. (1990). Plant anatomy, Pergamon Press, Oxford,UK, , 600 pp., ISBN 978-0080374918

Geiger, R. (1961). Das Klima der bodennahen Luftschicht. Ed. 4. Vieweg and Sohn, Braunschweig, 646 pp.

Gielen, B., Brognolas, F., Jian, G., Johnson, J.D., Karnosky, D.F., Polle, A., Scarascia-Mugnozza, G., Schroeder, W.R., Ceulemans, R. (2008). Poplars and willows in the world. International Poplar Commission Thematic Papers, Workong Paper IPC/9-7, FAO, Rome, Italy

Haeupeler, H. & Schönfelder, P. (1988). Atlas der Farn-und Blütenpflanzen der Bundesrepublik Deutschland. Eugen Ulmer Verlag, Stuttgart, 95 pp.

Hofmann, H. (1993). Zur Verbreitung und Ökologie der Wildbirne (*Pyrus communis* L.) in Süd Niedersachsen und Nordhessen sowie ihrer Abgrenzung von verwilderten

Kulturbirnen (*Pyrus domestica* Med.). In Mitteilungen der Deutschen Dendrologischen Gesellschaft. *81*, p. 27-69.

Itai, Ch., Nejidat, A., Bar-Zvi, D. (2002). Stress Physiology in the New Millenium. In: Plant Physiology in the New Millenium, Belgrade, p.31-39 ISBN 86-7384-011-2,

Jones, H.G. (1993). Drought tolerance and water-use efficiency. In: J.A.C. Smith & Griffiths, eds. Water deficit: Plant Responses From Cell To Community, pp.193-201. Oxford, UK, Bios Scientific Publishers. 345 pp.

Jones, H.G. (2004). What is water use efficiency ? In: M.A. Bacon& J.A. Roberts eds. Water use Efficiency in Plant Biology. Blackwell Publishing, CRC Press, Oxford, UK, pp.27-41

Kárpáti, Z. (1960). Die Sorbus-Arten Ungarns und der angrenzenden Gebeite. - Feddes Repertorium 62:71-334

Kausch, W. (1992). Der Speierling. Goltze-Druck & Co. GmbH., Göttingen, 224 pp.

Kausch, W. (2000) Der Speierling. Eigenverlag , Bovenden, 184 pp.

Kirisits, T. (1992). Der Speierling - Die seltensteeinheimische Baumart. Österreichische Forstzeitung, *10*: 55–56.

Kozlowski, T.T. & Pallardy, S.G. (1997). Physiology of Woody Plants. Ed.2. Academic Press, San Diego, CA.

Kozlowski, T.T. & Pallardy, S.G. (2002). Acclimation and Adaptive responses of Woody plants to Environmental Stresses. In: The Botanical Review 68(2):207-334.

Kleinschmit, J. & Svolba, J. (1998). Auslese von Wildbirne (*Pyrus pyraster*) und Rückführung in den Wald. In: Kleinschmit, J., Soppa, B. & Fellenberg,U. (eds) 1998: Die Wildbirne, *Pyrus pyraster* (L.) Burgsd. Tagung zum Baum des Jahres am 17. und 18.3. 1998 in Göttingen. Frankfurt am Main, J.D.Sauerländers, p.83-96

Landolt, E. (1977) Okologisce Zeigerwerte zur Scheier Flora. Verrff. Geobot. Inst. ETH, Stifung Riibel, Zürich 64:1-208

Larcher, W. (2003). Physiological Plant Ecology. Berlin : Springer Verlag, Berlin Heidelberg, 2003. 488 p. ISBN 3-540-43516-6

Liakoura, V., Fotelli, M. N., Rennenberg, H., Karabourniotis, G. 2009: Should structure-function relations be considered separately for homobaric vs. Heterobaric leaves?. American Journal of Botany 96(3) : 612-619,2009

Lin, L.,Z. & Harnly, J.N. (2008). Phenolic compounds and chromatographic profiles of pear (Pyrus). J.Agric. Food Chem.2008,56,9094-9101.

Májovský, J. (1992). *Sorbus* L. emend. Crantz. In Bertová L. (ed.), Flora of Slovakia, IV/3. (in Slovak) Veda, Bratislava, p. 405–408.

Mc Dowell, N., Pockman, W., Allen,D., Breshears, D., Cobb,N., Kolb, T., Plaut, J., Sperry, J.,West, A., Williams, D. & Yepez, E.A. (2008). Mechanisms of plant survival and mortality during drought: why do some plants survive while others succumb to drought? New Phytol.178: 719-739.

Michalko, J. (1961). Originality of the service tree (*Sorbus domestica* /L./) in oak communities of the Carpathian Mountains (in Slovak). Biológia, Bratislava, *16*, 4: 241–248.

Mooney, H.A., Gulmon, S.L. (1982). Constrains on leaf structure and function in reference to herbivory. Bioscience,32,198-206.

Namvar, K. & Spethmann,V. (1986). Die Wild- oder Holzbirne (*Pyrus pyraster*). In Allgemeine Forstzeitschrift, 21, p. 520-522.

Niclas,K. & Cobb, E.D. (2008). Evidence for "diminishing returns" from the scaling of stem diameter and specific leaf area. American Journal of Botany 95: 549-557.

Pagan, J., (1996). Forestry Dendrology (in Slovak).Vysokoškolské skriptá. Zvolen, Vyd. TU vo Zvolene, 378 pp. ISBN 80-228-0534-3.

Paganová, V. (2003). Wild pear *Pyrus pyraster* (L.) Burgsd. requirements on environmental conditions. In: Ecology (Bratislava) Vol. 22, č. 3 (2003), s. 225-241.

Paganová, V. (2008). Ecology and distribution of service tree *Sorbus domestica* (L.) in Slovakia. In: Ekológia, Bratislava, 2008, 27(2):152-168.

Passioura, J., B. (2002b). Soil conditions and plant growth. Plant, Cell and Environment 25:311-318

Passioura, J., B. (2002a). Environmental biology and crop improvement. Funct.Plant Biol.,29:537-546

Peniašteková, M, (1992). *Pyrus* L. Hruška. In Bertová L. (ed.): Flóra Slovenska, IV/3, Bratislava, Veda, p. 381-388.

Reich, P. B., Wright, I. J., Cavender-Bares, J., Craine, J., Oleksyn, M., Westoby, M. & Walters, M. B.(2003). The evolution of plant functional variation: Traits, spectra and strategies. International Journal of Plant Sciences 164:S143-S164.

Rittershoffer, B., (1998). Förderung seltener Baumarten im Wald. Auf den Spuren der Wildbirne. InAllgemeine Fosrtzeitschrift /Der Wald, 16, p. 860-862.

Roloff, A., (1998). Der Baum des Jahres 1998: die Wildbirne (*Pyrus communis* L. sp. *pyraster* Gams.) In Kleinschmit,J., Soppa, B., Fellenberg,U. (eds): Die Wildbirne, *Pyrus pyraster* (L.) Burgsd. Tagung zum Baum des Jahres am 17.und 18.3. 1998 in Göttingen. Frankfurt am Main, J.D.Sauerländers, p. 9-15.

Schmitt, H. P. (1998). Wildbirnen-Vorkommen in Westfalen-Lippe. In Kleinschmit, J., Soppa, B., Fellenberg, U. (eds), 1998: Die Wildbirne, *Pyrus pyraster* (L.) Burgsd. Tagung zum Baum des Jahres am 17. und 18.3. 1998 in Göttingen. Frankfurt am Main, J.D.Sauerländers, p.57-59.

Steffens, R., (2000). Der Speierling in Sachsen-Anhalt – Verbreitung, Ökologie und genetische Variation. Bovenden. Corminaria, *14*: 14–17.

Steiner, M., (1995). Speierlingskartierung im südöstlichen Wienerwald. Österreichische Forstzeitung, *6*: 50–51.

Šály, R., (1988). Pedology and microbiology (in Slovak). Vyd. VŠLD Zvolen, 378 pp.

Šesták, J. & Čatský J., (1966). Metody studia fotosyntetické produkce rostlin, Akademia, ČSAV Praha, 393 s.

Šiška, B. & Takáč, J. (2009). Drought naalyses of agricultural regions as influenced by climatic conditions in the slovak Republic. In: Időjárás. Quarterly Journal of the Hungarian Meteorological Service. Vol. 113, No. 1-2, January-June 2009, pp. 135-143.

Škvarenina, J., Križová, E. & Tomlain, J. (2004). Impact of the climate change on the water balance of altitudinal vegetation stages in Slovakia. Ekológia 23 (Suppl. 2/2004):13-19

Škvarenina, J., Tomlain, J., Hrvoľ, J. & Škvareninová, J. (2009a). Ocurrence of dry and wet periods in Altitudinal vegetation stages of West Carpatians in Slovakia: Time-series analysis 1951-2005. In: Strelcova, K; Matyas, C; Kleidon, A; Lapin, M; Matejka, F; Blazenec, M; Skvarenina, J; Holecy, (eds), Bioclimatology and natural hazards, Springer Science+Business Media B.V. 2009, p. 97-106. ISBN: 978-1-4020-8875-9